Spatial Modeling and Assessment of Urban Form

Biswajeet Pradhan
Editor

Spatial Modeling and Assessment of Urban Form

Analysis of Urban Growth: From Sprawl to Compact Using Geospatial Data

Springer

Editor
Biswajeet Pradhan
Department of Civil Engineering
University Putra Malaysia (UPM)
Serdang
Malaysia

ISBN 978-3-319-54216-4 ISBN 978-3-319-54217-1 (eBook)
DOI 10.1007/978-3-319-54217-1

Library of Congress Control Number: 2017935397

Printed on acid-free paper

This Springer imprint is published by Springer Nature
The registered company is Springer International Publishing AG
The registered company address is: Gewerbestrasse 11, 6330 Cham, Switzerland

To the memory of my loving dad Purna Chandra Pradhan

Preface

Although the total amount of urban areas covers insignificant percentage of the Earth's land surface, but still the growth of these areas is the main reason of various natural environmental related problems. Currently the influence of urban areas on Earth's resources consumption, environmental pollutions and climate changes is clearly observable. The continuous growth of manmade developments has increased these problems and produced several other negative effects on natural environments. Rapid growth of population and rural–urban migration due to higher quality of life especially in developing countries contribute to the horizontal and sprawl developments. Urban sprawl due to low density, large rural development, spatially segregated land uses and widespread commercial strip development does not provide a good quality of urban neighborhood. In addition, urban sprawl and unorganized horizontal city expansion because of high carbon emission, traffic congestion, agricultural and forest destruction, higher infrastructural provision costs, various public health problems and several other environmental, economic and social issues are not characterized as an acceptable and sustainable urban form. Hence, in recent decades there is a growing awareness about urban sprawl development and its negative consequences.

Due to these negative impacts, attaining urban sustainability is one of the most primary goals for planners and decision makers in urban-related applications. In general, sustainable development concerns about the consumption of natural resources in such a way that does not jeopardize the ability of future generations to use the same resources. With respective to urban perspectives, sustainable urban development concerns about the minimum inputs of energy and resources and minimum outputs of air pollution, water pollution, and wastes from an urban system. Hence, urban sustainability can also be defined as improving quality of life of human being within the availability of Earth's limited resources. Urban sustainability takes into account of three main aspects, namely; social, economic and environmental issues. Each of these aspects deals with separate issues of an urban system such as: security, livability and social equity; improve productivity, personal and public finances; pollution levels, the amount of reserve habitat and resource consumption respectively. Sustainable urban development can be achieved through an efficient land use growth and management by implementing proper planning and urban design. These tasks can be done by adopting various strategies and planning to minimize the energy consumption, protect biological diversity, reduce pollution, improve social interaction and develop more green landscapes. Therefore, the contribution of shape and form of the cities has become one of the main focal points to conduct these tasks.

Among various aspects of sustainable urban development, environmental protection especially agricultural and forest conservations are dominated in tropical regions. Particularly, small cities and towns with high potential of growth due to proximity to big metropolitan cities need to be controlled to avoid large horizontal urban expansion. Thus, it is important to propose various alternative development scenarios based on objectives of urban sustainability to avoid negative consequences of these urban sprawl developments. Compact city, transit oriented development (TOD) and smart growth are some examples of such development scenarios. Among these examples, compact city is widely accepted as one of the most promising solution for urban development pattern to achieve ultimate goals of urban sustainability. Compact urban development with high built-up density, land use diversity and intensified neighborhood aims to protect natural environment, reduce land consumption, decrease car dependency, support public transportation facilities, increase walking and cycling behavior and etc. These characteristics are seen to have contributed to the sustainable urban development in the form of social, economic and environmental aspects. In addition to environmental perspectives, compact city has several aspects to improve quality of life as social advantages.

Compact land use pattern is a relatively new terminology in the field of *Urban Planning* which, assumes that the new development should be built around the existing built-up areas in higher density, intensity and land use diversity and therefore promoting city compactness characteristics. The revitalization and redevelopment of existing brownfields and abandoned lands within the city borders is one of the most feasible and cost effective strategies in increasing city compactness. Evaluation of existing compactness and simulation of compact urban forms are the main step towards the implementation of compact development initiative to achieve ultimate goals of urban sustainability. Land use change modeling based on city compactness, or in a proper terminology, compact land use pattern modeling not only should consider various complexities of a conventional land use change processing, but also full fill different perspective of compact urban development concept and eventually sustainable urban development. In compact land use development human scale factors and quality of life has higher priority rather than other aspects, which made these kinds of development modeling more sophisticated.

Generally speaking, compact urban development is a complex and long-term project that requires a flexible law system and supportive government. Unfortunately, improper understanding and agreement about the definitions comprising several concepts and indicators make urban sustainability and compact city is an extremely difficult achievable task. Consequently, these complexities have influenced on each phase of sustainable and compact development processes such as modeling, implementation and measurement. For instance, to develop a compact city, the initial step is to assess and evaluate the various aspects of existing compactness in order to realize the current situation before any decision-making takes place. In this regard, there is no standard and consistent evaluation methodology exists in the literature. Moreover, city compactness has been assessed mainly based on data availability, local zoning manner and objective of the research itself. For instance, measuring urban density and land use diversity are usually based on census tracts, which vary in size and resolution. Therefore, the assessments are not comprehensive and reliable enough because the results can be different by various zoning manner, cell size, and type of input data. In addition, in large-scale regions such as country basis, urban compactness is generally measured based on the cellular concept and the concentration of the built-up cells in a specific area. Whereas city compactness apart from the urban built-up density (which is an implication of physical compactness) consists of various other aspects related to functional compactness which reveals valuable and useful information about the existing condition of cities. Moreover, evaluation of city compactness can be done through applying common statistical techniques to measure various entities such as mixed land use development indicating the land use richness of a local neighborhood. However, the distribution pattern which depends on the adjacency and relationship among various land use categories can only be evaluated using spatial and mapping based approaches.

In addition to form and shape of the cities, an understanding of spatial distribution of land use changes and the resulting impacts of this process on urban environment is one of the most important tasks. The lack of a clear understanding of this process leads to a level of uncertainty due to inclusive of several unknown and complicated parameters. Land use change phenomenon is a result of complex interaction of various environmental, physical, political, cultural, and other factors. Monitoring of these changes could reveal the flow of conversion from natural environment (forest lands) to agricultural fields and finally to built-up areas. Thus, the simulation and prediction of these changes provide insightful information and allow for more systematic analysis of the relationship between forms and process in several environmental and urban planning applications.

In this regard, evaluation of previous growth and extraction of development trend as historical components of land use change modeling is an essential task. This process is fundamental in order to simulate and predict the future growth and changes of various land use categories. However, the lack of proper understanding about urban systems, its related issues and several involved factors and stakeholders make modeling and prediction process a difficult task. Specifically, land use change arises from complex interaction of various factors and mainly dependent on spatial location, scale, and current state of land use. The existing modeling and prediction techniques cannot be solely applied for this complex phenomenon. A reliable and comprehensive modeling approach which can be created from integration of several modeling techniques should be proposed in order to tackle related issues and variables. In addition, the proposed hybrid models should be developed based on the core principles of land use change modeling. Similarly, the processing scale of the modeling is an important issue. In a large processing scale (low spatial resolution), the models can evaluate land use changes at a regional scale, thereby facilitating the definition of appropriate environmental policies. However, land use modeling at these resolutions is incapable of identifying subtle land use changes which is observable and effective in local neighborhood bases. Therefore, it is very important to propose a hybrid model at fine spatial resolution to deal with complexity of land use modeling and prediction.

First this book describes about the fundamental concept about urban growth and expansion, historical growth models, forms of urban growth, and its negative consequences on natural and green environments. Furthermore, the concept of sustainable development with an emphasis on urban sustainability and its relationship with two common urban forms (sprawl and compact development) will be discussed in detail. Generally, assessment and evaluation of various aspects of current pattern of urban areas is important. Hence, the current book has gone through a comprehensive urban form assessment in two physical and functional aspects. Especially, compactness assessment is discussed regarding urban density, land use diversity, and urban intensity evaluations. In this phase, two new terminologies, i.e., Degree of Compactness (DoC) and Trend of Compactness (ToC) which reveal the compactness growth pattern, will be proposed and explained. In addition to urban form and pattern evaluation, it is important to analyze the historical trend; and to model and predict the future trend of urban growth in a finer scale land use changes. This phase is presented with description about the effective factors through various applied techniques related to urban growth and land use changes, mainly based on two scenarios: "business as usual" and "compact land use pattern."

This book is organized into 14 chapters. The first three chapters and Chap. 8 present a theoretical information and introduction to urban growth and expansion, sustainable urban development, forms of urban growth, and common techniques applicable in land use change modeling. Rest of the chapters present the application of these theoretical concepts on specific case studies with detail explanation about input data, study areas, methodological processes, and results and discussion.

Chapter 1 provides introduction to urban growth and expansions with retrospective view on urbanization process and driving factors of urban expansion. Additionally, more descriptions are given on the forms of urban growth and expansion, historical modeling theories of urban growth such as Von Thünen theory, concentric zone theory, and central place theory, and urban growth and natural environment deterioration.

Chapter 2 provides general information about sustainable development with an emphasis on urban sustainability with respect to three main aspects i.e. environmental, economic and social sustainability. Next, as a case study, Malaysian perspectives of urban growth and sustainability is discussed, specifically related to Kuala Lumpur as capital city and Putrajaya as a newly developed city based on sustainable development paradigms.

Chapter 3 presents two main forms of urban growth: sprawl and compact development. First, the origin and various positive and negative aspects of sprawl development are explained, and then compact development is discussed in detail as an alternative solution to avoid the negative social, environmental, and economic consequences. Next, compact development is evaluated with respect to various aspects of sustainable urban development. Finally, a brief discussion is presented about Malaysian perspectives of compact urban development.

Chapter 4 presents an assessment and analysis approach of the spatiotemporal patterns of urban expansions in the Tripoli metropolitan area (Libya) based on the urban sprawl assessment concept. Urban expansion and sprawl are assessed and investigated as a pattern and process using Urban Expansion Intensity Index (UEII), population and urban expansion proportions, landscape metrics, entropy model, and degree of freedom model. Tripoli metropolis, which has not been studied before, was chosen as the area to discover its urban sprawl patterns, and assess well-established urban modeling techniques in a North African city. Next, the results of urban sprawl assessment are presented and discussed in detail with respect to the study area.

In contrast to Chap. 4, Chap. 5 presents the methodological process of city compactness assessment of Kajang city (Malaysia) based on main compact city paradigms (urban density, intensity, and land use diversity) for four temporal land use maps of this city (2004, 2008, 2012, and 2015). Kajang is a city located in the eastern part of Selangor province in the southwestern region of Peninsular Malaysia. City compactness assessment is performed as an initial step of compact city modeling based on physical and functional assessment by proposing two new terminologies; degree of compactness (DoC) which illustrates the level of compactness of the smallest pixels or cells of the study area, and trend of compactness (ToC) which shows the trend of the growth and loss of compactness of the study area. These two measurements are implemented and evaluated to reveal the growth pattern of compactness of the Kajang city. This assessment provides baseline information and guidelines for analysis of compact land use pattern.

Chapter 6 presents the methodological approaches dealing with the relationship between city compactness and residential land use growth. Residential land use is selected due to more significant growth of this land use type than other urban land use categories. This growth causes the destruction of large amount of green and natural environment, especially in sprawl urban expansion. Thus, a proper analysis of the reciprocal relationship between residential growth and compact development is necessary to predict and propose different future alternative scenarios. In this process, first, the city compactness of the study area is assessed with respect to residential land use changes. Second, the growth of residential areas is predicted by using two common land use change modeling approaches and the future residential maps are evaluated with respect to city compactness maps. In this manner, the performances of the selected models are also evaluated for land use change modeling applications in terms of model accuracy, complexity, and functional relationships between dependent and independent variables.

Chapter 7 presents a change detection process to discover the spatiotemporal analysis of urban land use change patterns and highlight the trend of historical development of Kajang city (Malaysia) during 2004–2015. Land use change assessments provide a clear understanding of the built-up growth through various land uses and land cover categories. These assessments reveal the rates, amount, and directions of the growth. Thus, significant growth and/or loss of a specific land use type can be highlighted precisely. Cross-tabulation analysis is applied to each pair of available land use maps of the study area (2004, 2008, 2012, and 2015) to implement this analysis for Kajang City.

Chapter 8 identifies and explains several common land use change modeling techniques in order to provide baseline knowledge for the methodological approaches applied in Chaps. 9–11. Various statistical-based approaches, agent-based models, rule-based models, artificial neural networks, cellular automata model, and decision tree models are explained and discussed in detail. In addition, validation of urban modeling techniques is also explained. Urban growth and land use changes are the main reasons for environmental, social, and economic issues, such as hydrological problems, destruction of forests and agricultural fields, natural and wildlife disturbance, and global warming. Thus, a proper understanding of the reason, degree, direction, and consequences of urban growth and expansion is essential for most urban application projects which are discussed with examples.

Chapter 9 presents a methodological process for land use change modeling for the Tripoli metropolis as case study. In this chapter, the simulation process of urban growth in the Tripoli metropolis is presented and explained to understand its pattern and the role of each urban driving force behind the urbanization process. In the simulation process, the frequency ratio (FR) model is first applied based on the real urban expansion rather than entire urbanized area to present the role of classes within each urban factor and reflect actual urban expansion tendency. Second, the evidential belief functions (Dempster–Shafer) model (EBF) is applied to provide further information by generating four maps representing belief, disbelief, uncertainty, and plausibility of predicted future urban growth. Third, the logistic regression (LR) model is applied to assess the overall effect of each urban driving factor, and subsequently combined with a simple growth ratio equation to present probable future scenarios. Fourth, the classic CA–Markov chain (MC) model was used to predict explicit future urban land use in Tripoli in 2020 and 2025. Finally, a novel hybrid model of CHAID–CA–Markov is proposed based on the advantages and shortcomings of the aforementioned models, and employed to model, explain, and predict explicit urban growth in 2020 and 2025. Several multi-temporal space-borne remote sensing data are used to conduct spatial analysis, modeling, and predictions for urban expansion such as Landsat image 1984, Landsat image 1996, Spot 5 image 2002, Spot 5 image 2010, road networks, population data, digital contour map, and topographic map.

Chapter 10 presents a methodological process of compact land use change modeling to simulate and predict future spatiotemporal urban growth in compact form. These processes are conducted to identify and assess the various aspects of land use change modeling, especially regarding statistical (factor analysis) and cellular-based concepts. A hybrid land use modeling approach based on applied modeling techniques is also developed to create a comprehensive projection of the future development pattern in two scenarios. The first scenario (business-as-usual scenario) is based on several urban-related factors and interaction among various land use categories through a historical trend of land use change and growth. Next, the results are integrated into the CA model to facilitate the application of contiguity filters and project future land use maps based on the neighborhood concept. In the second scenario (compact land use scenario), the proposed land use modeling approach and evaluation of degree of compactness (DoC) and trend of compactness (ToC) are considered in proposing and implementing a compact land use scenario using the city intensification process. The proposed model considers the advantages and disadvantages of the existing models and analyzes the interactions of urban factors as well as their interaction among various land use categories. The analyses and modeling approaches used in this study can be employed to guide the identification and measurements of the changes and growth likely to happen in urban areas. The output maps and results can likewise be helpful for town planning in order to design compact and eventually sustainable urban areas.

Chapter 11 proposes a brownfields land use change modeling process according to a compact city paradigm in a larger scale perspectives rather than local aspects. The proposed model is a statistical-based weights-of-evidence (WoE) approach in the GIS environment. The growth of three main land use types in Kajang, Malaysia was predicted using several compact development parameters and other urban and physical site characteristics. This process are aggregated with an existing brownfields map in order to project future land use types according to planning strategies, as well as compact development characteristics. It is concluded that the combination of land use change modeling techniques and compact urban development theory in GIS environment can provide a strong tool for brownfields redevelopment planning and strategies.

Chapter 12 presents a methodological process for extracting the land use/land cover of Karbala City in Iraq using high-resolution satellite images based on rule-based algorithm of the object-oriented classification method. Change detection analysis is implemented on the growth of built-up areas to evaluate the previous trends of land use change pattern. Furthermore, future urban growth and expansion of the study area are projected using the integrated cellular automata and Markov chain technique. Finally, a novel approach for building extraction and counting was presented using the eCognition rule-based method. The methodological process is validated using ground truth points and standard confusion matrix. These analyses indicate the logical and accepted performance of the methods. The projected and produced maps can help identify the spatial growth pattern of urban settlement. Such identification can be used to create adequate future planning for the proper provision of social and infrastructural facilities for the local residences.

Chapter 13 discusses the applications of geographical information system (GIS) and remote sensing (RS) in urban-related fields, especially urban development and planning perspectives. This chapter explains the fundamental concept about GIS and RS and their necessities and relationship with urban-related issues. The basic concept of GIS is explained regarding its main components, input data, capabilities, basic analysis tools, mapping, and visualization abilities. Remote sensing also is explained regarding its advantages with respect to in situ data collection, resolutions, various sensors, and its capabilities for urban problems. Specifically, application of radar imagery in building extraction is presented and explained with proper examples and references. Next, site suitability process as one of the main application of GIS in urban planning and design is discussed and presented in detail with a special focus on multicriteria decision-making (MCDM) and analytical hierarchical process (AHP) techniques in this field. Finally, brief information about GIS application in urban planning and development regarding Malaysian perspective from past to present is explained.

Chapter 14 presents a methodological process attempted to describe and quantify the spatial pattern of urban expansion of the selected study area using several landscape metrics. Four satellite images of the study area from the years 1984, 1996, 2002, and 2010 are used to conduct the analysis of urban sprawl patterns in the Tripoli metropolitan area. The applied spatial landscape metrics provided good insight into urban sprawl from different perspectives and presented a reliable urban sprawl investigation tool. The findings of this study are useful in directing prospective urban plans and urbanization policies in Tripoli.

Chapter 15 presents an interesting application on relationship between urbanization and urban heat island (UHI) effect. The UHI phenomenon affects the environment, regional climate, and socioeconomic development. In this study, Enhanced Thematic Mapper Plus (ETM+) and Landsat Thematic Mapper (TM) images acquired in 2002 and 2009, respectively, are used to evaluate changes in land surface temperature (LST) over different land cover (LC) types during those years in Putrajaya, a planned city in the south of Kuala Lumpur, Malaysia. Urban thermal characteristics were further analyzed by investigating the relationships between LST and two indices, namely, normalized difference vegetation index (NDVI) and normalized difference built-up index (NDBI). Results suggest an inverse relationship between NDVI and LST and a strong direct correlation between NDBI and LST in Putrajaya city. Therefore, detecting the amount of changes in the significant areas, such as vegetated and urban areas, is essential for future urban strategies related to decreasing LST.

In general, this book discusses about the application of geospatial data, geographic information system (GIS) and remote sensing (RS) technologies in analysis and modeling of urban growth process and its pattern, with a specific focus on sprawl and compact development. This book confirms that the proposed advanced modeling approaches, geospatial data and GIS are very practical for identifying urban growth, land use change patterns and their general trends in future. The analyses and modeling approaches presented in this book can be employed to guide in identifying and measuring the changes and growth likely to happen in urban areas. This book also can serve as a guiding text book for postgraduate students and researchers who are interested in urban growth modeling. Though a lot of work on urban growth and assessment has been published as individual papers in various scientific journals; however, there is also a disconnection between the urban growth modeling and compact city assessment using remote sensing data. This book can provide an easy path from theory to practical algorithms with many case studies. In addition, this book can be helpful for town planning and local development agencies in order to design urban areas in a compact form and eventually sustainable manner.

I could not have produced this book without the efforts of many people who I would like to thank here. Foremost among them are my own research team members at Department of Civil Engineering, Universiti Putra Malaysia and authors of each chapter who worked closely with me for meeting the deadlines in developing the scope of each chapter. These individuals are Saleh Abdullahi, Abubakr A.A. Al-sharif, Hossein Mojaddadi, Amer D. Salman Aal-shamkhi, and Marziyeh Zahabi. Thanks to all my coauthors of individual chapters of this book.

The publication of this book would not have been possible without an excellent cooperation from my colleagues at Springer Verlag, Germany. Special thanks to Dr. Nabil Khelifi for motivating and encouraging me to write this book. At Springer Verlag, the efforts of Reyhaneh Majidi are appreciable.

Finally, I would like to thank my family, wife Sheila, my two adorable kids, Krish Pradhan and Darsh Pradhan, for their wonderful support and patience in allowing me to spare time to complete this book.

Serdang, Malaysia
April 2017

Biswajeet Pradhan

Contents

Part I
Introduction

Introduction to Urban Growth and Expansion

1

Saleh Abdullahi, Biswajeet Pradhan and Abubakr A.A. Al-sharif

1.1 Retrospective View on Urbanization Process

A city or an urban area is a permanent concentrated human settlement that is managed or governed by a local or regional administrative body (Goodal 1987; Kuper 2013). This centralization pattern provides great opportunity for interaction among citizens, businesses, and activities. Urban environments are symbols of civilization and modernization because of the existence of several complex systems of infrastructures, services, and facilities. The accessibility of these community facilities is one of the factors in the growth of urban areas (Banister 2012; Yamu et al. 2015). In fact, the concentration of these facilities and services distinguishes cities from rural areas and villages.

From the global environmental point of view, urbanization is the conversion of natural spaces to build up areas for residential, commercial, and industrial land uses (Xie et al. 2005). Mubareka et al. (2011) defined urbanization as the growth of land demands to build up areas over a period. Urbanization is an obvious human behavior to obtain better quality of life, livability, security, and so on. The urbanization level of the urban environments can be measured by the complexity, extent, and capacity of these factors. Meanwhile, urbanization increases economic growth and industrialization (Fig. 1.1).

Historically, concentrated settlements with central management systems have been created by the agglomeration of industrial areas in central parts and attracted populations to live around these areas and along road networks (Yeh et al. 2001). In addition, higher quality of life, livability, safety, security, and protecting poor people from rising land costs and speculation were the main aim of urban planners in the first decades of the nineteenth century (Banister 2012). After the Second World War, this kind of concentrated urban shape was replaced by dispersed and decentralized development in suburban areas. This decentralization affected the job opportunities, distribution of facilities and services, and residential development from a clustering pattern to the suburban and city edges (Garreau 1991). Hence, the primitive human settlements changed from mono-centric to polycentric and dispersed urban patterns. Currently, this kind of urban pattern is known as sprawl and/or leapfrog development, in which the built-up areas that belong to each urban land use are separated by open spaces, such as natural and abandoned fields. Ottensmann (1977) defined urban sprawl as the spread of new spontaneous urban developments on isolated zones, which are separated from other areas by unused land. In fact, this transformation is the consequence of several factors, but the main factors are transportation and technological development and advancement (Archer et al. 1993).

The majority of Earth's population lived in rural areas for a long period before industrialization. Although urban environments have existed for thousands of years with historical kinds of planning and development, a very small portion of the population were interested to live in cities (Elkin et al. 1991). Nevertheless, technological and industrial revolution changed the urban shape and built environments that attracted rural populations to live and work in urban areas instead of living in villages and working on agricultural fields (Arbury 2005).

Newman (1992) summarized three different periods of urban environments development.

(1) Traditional, small, and less dense cities, with walking and cycling transport. Urban structures pattern and distribution, such as housing, social activities, and businesses, were tightly intermixed. In this stage, the overall size, amount of required area for each activity, and growth of urban areas were insignificant and limited within an appropriate walking distance (not more than 5 km), which can be observed in the most of the European, North American, Australian, and New

B. Pradhan (ed.), *Spatial Modeling and Assessment of Urban Form*,
DOI 10.1007/978-3-319-54217-1_1

Fig. 1.1 Kuala Lumpur City (Malaysia); Google Earth and aerial images

Zealand cities. In this period, a clear physical distinction between urban areas and villages existed (Herndon 2011).

(2) The technological development of mass public transportation in the later part of the nineteenth century, which was the initial stage of urban outward expansions. The limitation of growth caused by the slow transportation forms was reduced and cities could expand depending on the extent of the train and tram railways (up to 20–30 km). A large number of cities were shaped during this period, especially in Europe, North America, and Australia.

(3) The technological development of automobile, which began before the Second World War. A significant growth was observed after the war, when private vehicle and bus became the main transportation mode. Thus, the cities grew in all directions, especially along the road networks (more than 50 km). The prevalence of private cars and buses increased low-density residential neighborhoods because the people were not limited to live in high-density and congested city centers or near their working places.

Figures 1.2 and 1.3 depict one of the good example of the forms and extents of city expansions, especially based on transportation modes.

The expansion of London city is a good example to show these three periods. In the first years of eighteenth century, before the development of public transportation, 87% of the population of this city was living in the inner parts (957,000

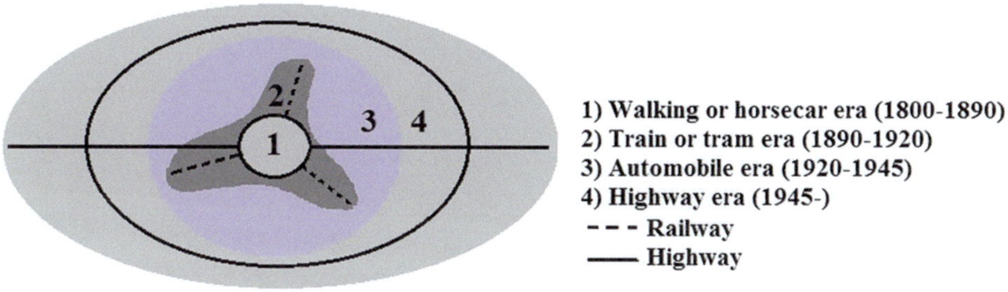

Fig. 1.2 The schematic illustration of urban growth based on transportation system (Muller 2004)

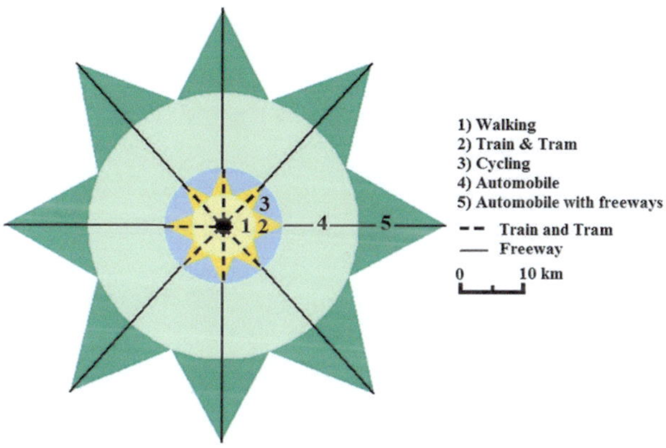

Fig. 1.3 The schematic illustration of urban expansion distance based on transportation system (Hugill 2002)

out of 1,100,000). In the first year of the nineteenth century, after the technological development of passenger trains, trams, and buses, 70% of the population of this city was living in the inner parts (4.5 million out of 6.5 million). Finally, in 2001 and 2011, 45 and 40% of the population were living in inner parts, respectively. Car ownership only started to become predominant in the 1970s. Hence, city expansion is significantly increased after the technological development of automobiles.

In addition to these periods, the shaping of urban areas and expansion can be assessed based on five main factors: industrialization, urbanization, advances in transportation, zoning ordinances, and growth of middle class affluent (Herndon 2011). Industrialization converted the agricultural-based business communities to manufacturing- and industrial-based communities. Technological advancement in machinery for various industries significantly increased the productivity and manufacturing profits. This encouraged the development of large industrial buildings in rural areas and farm fields; hence, the expansion of the city borders. The mass rural–urban migration process was triggered by the conversion of agricultural lands to industrial buildings that brought job opportunities. Rural–urban migration increased the urban population dramatically and expanded the cities horizontally and vertically to accommodate the newly migrated population. Thus, more natural and green environments were converted to urban land uses, especially for residential buildings. This rapid urbanization process created several problems in the urban areas, such as pollution, noise, congestion, and improper infrastructures and utilities. These intolerable problems in the urban areas encouraged wealthy and affluent residences to move away from the high-density parts of the cities to the outskirts for better quality of life in low-density neighborhoods. This aim was achieved, and the human limitations regarding placement and movement reduced significantly. This further horizontal expansion of urban areas reduced the distinction between urban and rural environments.

The environmental and social problems caused by the rapid urbanization forced the local planning authorities of urban areas (especially in the United States in the early nineteenth century and other European countries) to propose an alternative solution instead of moving from the central parts to the outskirts of the cities. They proposed to implement zoning ordinances in urban areas and separate various activities and land use categories. Their aim was to separate residential with non-residential (particularly heavy industrial land use with high pollution and noise) land uses to protect the public's health and for the sanitation and security of residential neighborhoods. The zoning ordinance in urban areas was first introduced in New York City in 1916 with the aim of bringing light and air back to the city and homes and providing assurance of what can only be developed next door (Gillham 2002). This zonal development and the increase of the average incomes of the middle class population because of industrialization caused the expansion of horizontal residential development with low density, large lot size detached homes in the suburban areas. These new suburban developments further segregated the urban growth and expansion with low population density (Bruegmann 2006). Consequently, these five factors (industrialization, urbanization, advances in transportation, zoning ordinances, and the growth of the middle class affluent), which historically formed the urban areas, influence the current characteristics of American cities (Herndon 2011), which are private vehicle dependent, separated and single land use

Fig. 1.4 General view of urban areas during industrial revolution

development, low population density in most neighborhoods, and blurred distinction between cities and countries (Fig. 1.4).

1.2 Driving Factors of Urban Expansion

In addition to the historical factors mentioned, several other factors that affect the shaping of urban growth and expansions regardless of periods exist. Cities are among the most complex structures created by human beings. Dynamism and continuous growth are the main characteristics of urban areas. Evaluating the driving factors behind the expansions is necessary to describe past urban patterns and predict future patterns (Abdullahi and Pradhan 2015). Urban growth and developments are consequences of many driving forces that control social, economic, and environmental variables (Liu et al. 2003; Verburg et al. 2006; Chen et al. 2014). From a practical point of view, many urbanization factors have been identified for urban spatial growth modeling and decision-making. In this section, a brief explanation will be presented. In the next chapters, more details on causative factors of urban growth with proper case studies will be discussed and evaluated using Geographical Information System aided statistical assessments.

- Environmental factors related to the environmental characteristics of a location can be a stimulation and constraint for urban growth and expansion. These factors include natural barriers, slope, risk of natural hazard, land cover categories, and so on.
- Local-scale neighborhood factors that are based on Tobler's first law of geography (1979); "Everything is related to each other, but near things are more related than distance things" (Sui 2004). The urban land use pattern usually has a strong influence on the urban dynamics of land use change and growth.
- Spatial characteristics of the city, such as size of the urban center, accessibility of the urban center, traffic flow and congestion, transportation system, and distribution of community facilities. For instance, new links in the road network may contribute greatly to the urban dynamics as an attraction for urban land use.
- Urban local and regional planning policies, such as zoning ordinance and resource allocation. The occupancy of the city by land use spaces is planned over time through land use zoning plans.
- Factors related to individual preferences, level of economic development, and socioeconomic and political issues. These driving factors are very complicated to model and understand. They are related to human decision-making processes, which are qualitative, time dependent, and can be transient in most cases. Consequently, they are difficult to define and calibrate as stochastic factors in urban dynamic modeling.

1.3 Forms of Urban Growth and Expansion

The growth and expansion of urban areas from the earliest stages until now are the results of several internal and external factors, such as industrialization revolution, transportation modes and extent, physical and geographical properties of site, environmental characteristics, and planning process. The investigation and evaluation of urban expansion forms based on these causative factors using various quantitative classification techniques has greatly contributed to the comparison of different urban areas (Huang et al. 2007).

In general, several forms of urban expansion exist, such as compact or sprawling, clustered or dispersed, leapfrog or continuous, self-organizing or spontaneous, and organic or planned (Bhatta 2010). Similarly, Clarke and Gaydos (1998) classified urban expansion into spread, organic, spontaneous, diffusive, and road-influenced forms. However, defining clear boundaries between these growth patterns, which certainly have some overlaps, is difficult (Yang et al. 2003). Compact and sprawl developments are the most general forms of urban growth. Other forms are normally defined and characterized based on these two forms.

Unlike compact development, which is characterized by centralization and high-density built-up area, sprawl development is mainly a low-density, scattered, and decentralized urban form (Burton 2000, 2002; Abdullahi et al. 2015). Urban density consists of various aspects, such as population density, building density, residential density, and road density (Abdullahi et al. 2015). Since building types in sprawl development are mainly single-story building with widely spaced; hence, most of the urban density aspects remains low especially building density. Urban sprawl develops the urban land use along the boundaries of existing cities, which require the extensions of essential urban infrastructure, such as sewers, roads, water, and power (Gillham 2002).

The three periods listed by Newman (1992) and other mentioned factors that affect urban growth all play different roles in shaping the cities from historical concentrated settlements to various horizontal expansion forms. Newman (1992) believed that the first stage in separating living from working places was made possible by the mass public transportation of trains and trams, which was especially for the middle class or high-income level. These fast transportation modes made it possible to escape from the high-density, congested, and polluted central parts to the low-density and green suburban areas with better neighborhoods. Furthermore, the advancement of electric streetcars in American and European cities at the end of the nineteenth century increased the suburban expansion. Public transportation using trains caused a more dispersed development growth because of longer distances between the train stations. Meanwhile, electric streetcars made contiguous forms of growth because of the shorter distances among the stations (Arbury 2005). Hence, the urban expansion form based on transportation systems and extents is usually known as the ribbon pattern (Fig. 1.5). This kind of urban form, which is considered as sprawl development, follows the main transportation routes outward from the urban centers. The widths of these expansion corridors were defined by the walking distances on both sides of the routes. Ribbon pattern was the main basis of urban growth in several older American, Australian, and New Zealand cities; hence, it is considered as the first stage of urban expansion from concentrated forms toward decentralized and sprawl development (Arbury 2005). Although this kind of urban pattern extended the traditional walking-based cities significantly, it is greatly different from the automobile-based expansion form in terms of density, physical appearance, and residential development pattern.

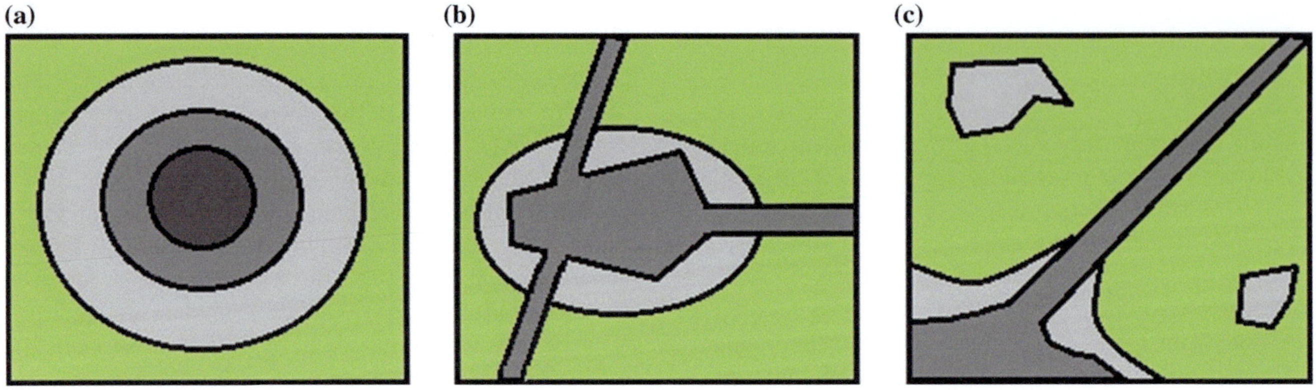

Fig. 1.5 **a** Low density of sprawl (radial sprawl), **b** ribbon sprawl, and **c** leapfrog development sprawl (Gillham 2002)

Similarly, road-influenced growth is a linear development that is influenced by new roads or corridors and generally surrounded by rural areas up to some distances from existing developed areas (Wilson et al. 2003). The advancement and popularization of car dependency in the early twentieth century was one of the main factors in the growth of urban areas of decentralized and dispersed forms. Over time, areas along the roads are converted to urban use as land values increase and infrastructural facilities are extended perpendicularly from the major roads (Gillham 2002). Similar to road-influenced urban form, commercial strip development, which is characterized by major roads, consists of various facilities, such as restaurants, shopping centers, and fuel stations. Commercial strip development normally has low density and surrounded by large parking spaces.

Meanwhile, the clustered type of urban form similar to historical concentrated human settlement is a neither a linear nor isolated urban growth and is typically a large, compact, and dense development (Wilson et al. 2003). In contrast, dispersed development is a decentralized kind of growth with low population density, widely separated buildings, rural developments, and no proper activity center (Schneider et al. 2008). Leapfrog pattern, which is a kind of dispersed and sprawl development, occurs beyond the urban fringes and creates isolated built-up areas; therefore, it is often a mixture of urban with non-urban uses (Schneider et al. 2008). This discontinuous growth can be considered as the most extreme urban sprawl, with its greater need for infrastructure and transportation, and inefficient use of lands (Gillham 2002). Strong private car dependency and high land consumption are some of the other characteristics of these kinds of horizontal expansions because of their low-density development.

Self-organizing growth is normally controlled by high-density and large-scale urban development (Bhatta et al. 2010). In contrast, spontaneous growth is normally controlled by small-scale dispersed development and occurs at the boundary of an existing settlement (Clarke et al. 1998). Unlike organic growth, planned urban growth is more likely manmade and controlled and developed by a pre-designed planning process. In general, most cities and towns seem to be a mixture of both (organic and planned) developments and usually contain elements of planned growth against a backcloth of natural growth (Batty et al. 1994; Batty 2008).

Adolphson (2010) explained the form of urban expansion mainly based on economic point of view. Centripetal and centrifugal forces are two factors that affect the growth of urban areas (Krugman 1996). The agglomeration of the same kinds of businesses in specific spatial locations or single business type developments (economies of localization-centrifugal forces) causes a segregated urban pattern, whereas the combination of various kinds of businesses or mixed business types developments (economies of urbanization-centripetal forces) causes an integrated urban pattern (McCann 2001). The ratio between these two forces (centripetal/centrifugal) has a direct relation to the size of the central area of the city and its capacity to attract more urban activities. Furthermore, any internal local changes, such as increase or decrease in the cost of transportation, will affect this ratio and consequently have an effect on the dispersion or clustering of urban growth and expansion. Thus, Adolphson (2010) summarized urban forms in nine main categories, which are mainly based on specific social–economical properties, as shown in Table 1.1. The first three are simple and the other six forms have structures that are more complex.

- Compact urban structure. The structure represents high density and is usually combined with mixed functionality.
- Dispersed urban structure. The structure is an apparent central business district (CBD) that is characterized by maximum urban density, maximum rents, maximum trip ends, and segregated land use (dominated by low-density residential suburbs) located in concentric zones around the CBD.

Table 1.1 Various urban forms expansion presented by Adolphson (2010)

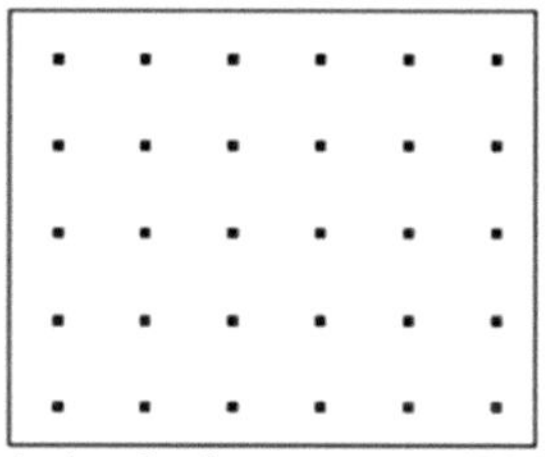 1. Areal urban structure (Wright 1935)	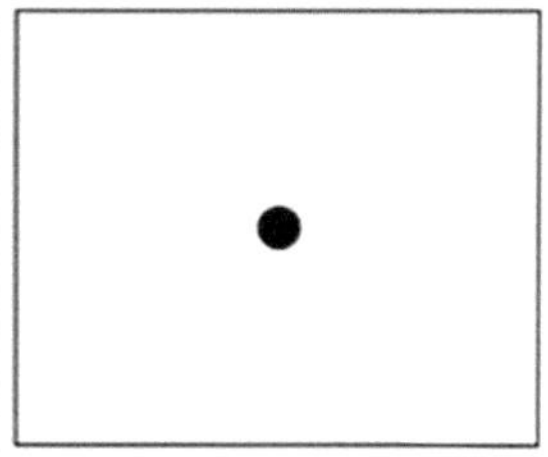 2. Point urban structure	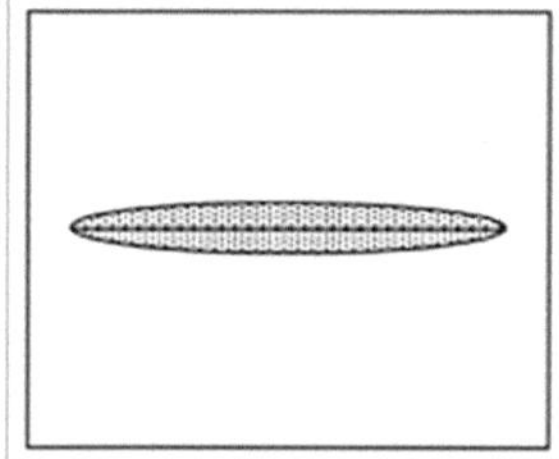 3. Linear urban structure (Salas 2009)
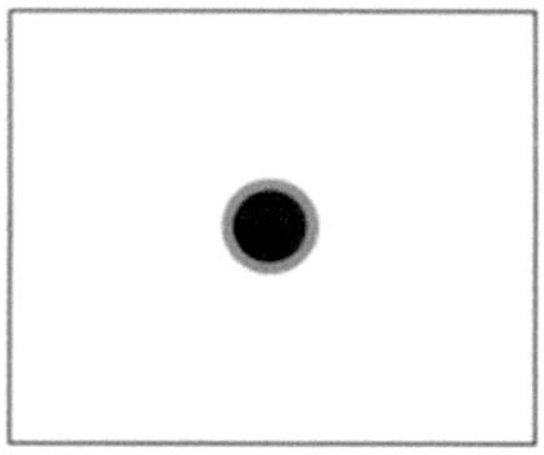 4. Compact urban structure	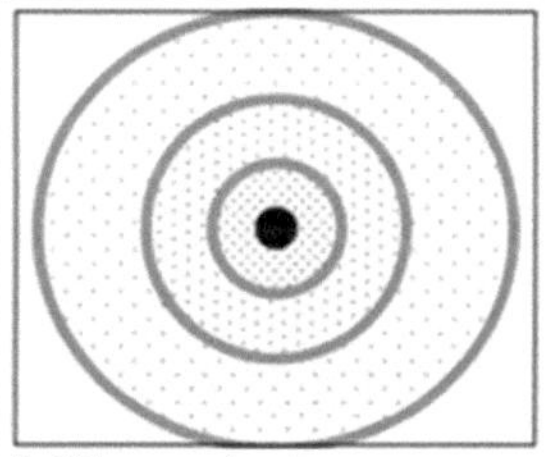 5. Dispersed urban structure	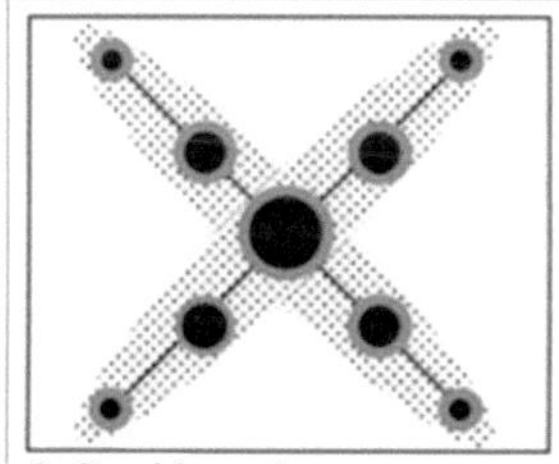 6. Corridor urban structure
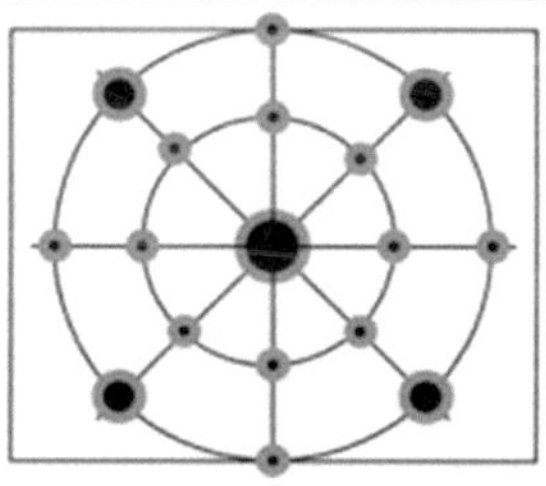 7. Multinucleated urban structure	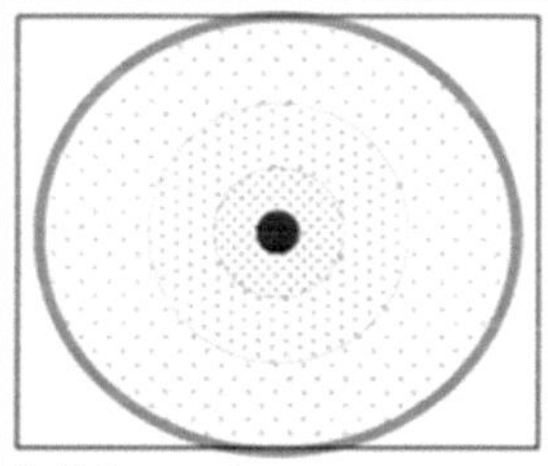 8. Fringe urban structure	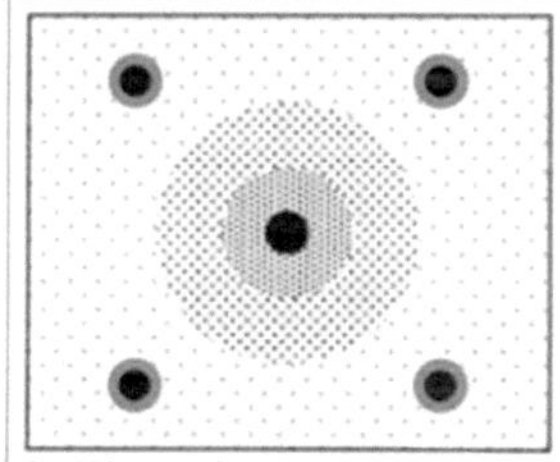 9. Ultra-urban structure

- Corridor urban structure. The intense land uses in this structure are extended out from the CBD along major transportation routes.
- Multinucleated urban structure. It is constituted by a number of sub-centers with local maxima according to floor space, population and employment density, rents, and trip ends (Anderson et al. 1996).
- Fringe urban structure. The structure develops when urban growth occurs at the urban–rural border.
- Ultra-urban structure. The structure is described as an urban structure that is "beyond something in space and time." It appears when communication technology eliminates the influence of space and time and hence a metropolis-based region emerges (Newton 1997; Wegener et al. 2004).

1.4 Modeling Theories of Urban Growth: A Preview

It is clear that a proper understanding of the reason, amount, direction, and consequences of urban growth and expansion is essential. Hence, some kind of models and simulation techniques are required to deal with these issues by expressing and explaining the growth process. The utilization of models in scientific research represents the natural behaviors and reactions in the real world (Liu 2008). Models are essential for understanding the dynamic behaviors, evaluating causative factors, analyzing the consequences, and supporting the planning and decision-making (Wang 2012). However, the behaviors of phenomena in the real world are very complex and multidimensional. Some simplifications and predefined assumptions are required to understand and investigate these processes. The proposed models should be comprehensive and applicable enough to support urban growth and create a better and clearer view of the function of this process. The models can be used as powerful tools to increase our mental capabilities regarding urban expansion and make more informed decisions (Costanza et al. 1998).

Historically, many models were applied to urban growth applications after the quantitative revolution in geographical science from the 1950s until the 1960s (Wrigley et al. 1981). The evolution of remote sensing (RS), geographic information science (GIS), and digital computing technology supported a means of working with very complex mathematical urban models. Recent models have provided artificial laboratories to conduct numerous analyses and research applications of the urban system to explain the processes and behaviors of urban expansion patterns and define the structure of metropolitan areas. Main urban modeling theories are discussed in the following sections.

1.4.1 Von Thünen Theory

Most of the models before the 1950s were based on spatial economic theory. A simple land use and urban expansion modeling theory was developed by Johann Heinrich Von Thünen in the eighteenth century (Wang 2012). The theory explains how market processes affect and control the spatial distribution of land use in a theoretical geographical context, as shown in Fig. 1.6 (Parker et al. 2003).

A simple agricultural land use context is used to explain this theory. The limiting assumptions of the model theory are as follows:

- The city is located centrally in isolation and no external influence exists.
- The city is surrounded by unoccupied lands.
- The land is wholly flat and there are no mountains or rivers to interrupt the terrain.
- The climate and soil quality are consistent.
- Farmers transport their own goods across the land directly to the city center using oxcarts, that is, the effect of roads is neglected.
- Farmers act to increase their incomes.

Von Thünen assumed that farming land use will be segregated into a spatially hierarchical configuration and intensive farming and dairying will take place near the city. In other words, fruits, vegetables, milk, and other dairy goods must be brought quickly to markets. Wheat can be situated further from the markets and the city center, while ranching can be located in the fringe areas surrounding the city. Beyond the ranch lands lay the unoccupied lands that are very far from the city center. Von Thünen's modeling theory is simple, but it is still an important theory in terms of geography. Moreover, the theory excellently illustrates the balance between transportation and land costs. The theory assumes that the price of land with its proximity to the city (Sinclair 1967; Candau 2002).

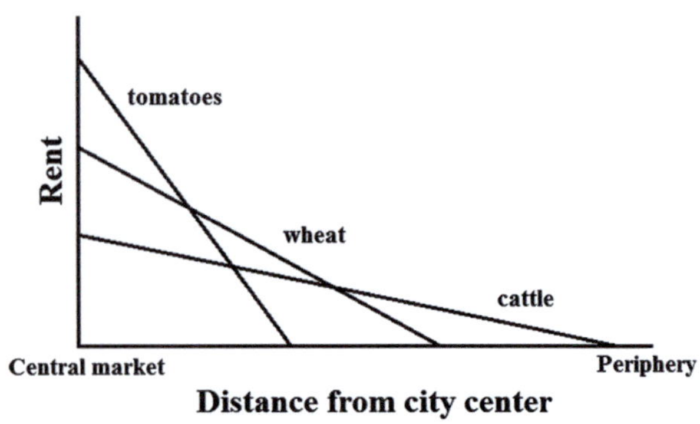

Fig. 1.6 The Von Thünen spatial organization of agricultural land use (Sinclair 1967)

1.4.2 Concentric Zone Theory

The concentric zone theory was proposed by Burgess in 1926 (King 1985). In this theory, the city is represented as a series of concentric land use circular zones, centered on the CBD. However, concentric zone theory offers a descriptive urban system formation rather than an analytical urban dynamics like in the von Thünen theory. Burgess proposed that the city grows by growing the circular concentric land use zones outward the CBD (Fig. 1.7).

The first zone represents the CBD, which lies at the city center. The second zone is composed of multi-use transitioning land use mixed with some migrant ghetto residences and manufacturing areas. The third zone is considered the working and residential class neighborhoods with few amenities. The fourth zone consists of better homes with more spaces than the third zone and is occupied by the middle class commuters. Finally, the fifth zone is dominated by higher quality housing and very good amenities. However, geographic space is generalized and restricted in the concentric zone theory. Topographic and transportation influences are also not considered, and the mono-centric urban expansion is insufficient to represent real urban land use patterns (Blumenfeld 1949; King 1985; Parker et al. 2003).

1.4.3 Central Place Theory

Central place theory was devised by geographer Walter Christaller in 1933 when he noticed that towns of a certain size were roughly equidistant. This theory attempts to

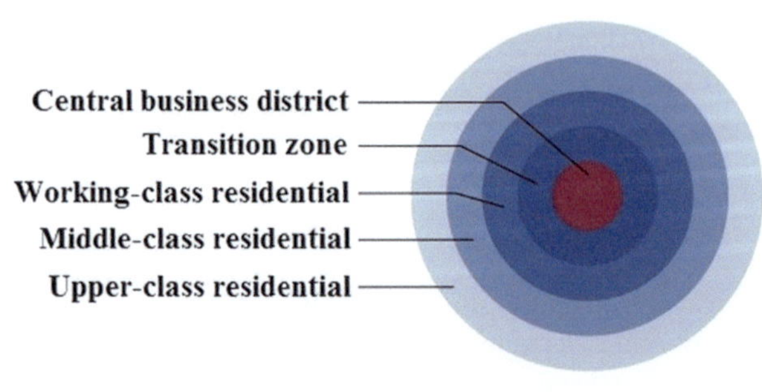

Fig. 1.7 Concentric zone theory (Candau 2002)

explain the size, number, spatial distribution, and hierarchical arrangement of cities. It is also concerned about the arrangements of cities in terms of providing retail and wholesale administrative functions and services to citizens. Christaller defined and examined the functions of each urban settlement structure and the size of the neighborhood land and found that modeling each urban settlement location pattern through geometric shapes, such as hexagons and triangles, is possible (White 1977; King 1985).

Central place theory defines the central place as an urban settlement with a number of smaller settlements each at an equivalent distance from it. The smaller settlements use the services and shops in the central place. The central place provides more goods and services than the smaller urban settlements (towns). This structure is based on simple rules:

- The lesser the number of settlements, the larger the domain of influence of its services.
- The larger the number of settlements, the fewer the services offered.

Figure 1.8 shows that the city is the main settlement. It has the largest number of services and a large hinterland. Hence, such cities rarely occur on urban landscapes. The towns, which have fewer services, are more abundant and have significantly lesser neighborhoods. This urban pattern continues in a hierarchical manner to include smaller urban settlements of villages. Each type of urban settlement places itself in relation to the subsequently greater urban settlement equidistant from urban settlements of a similar extent.

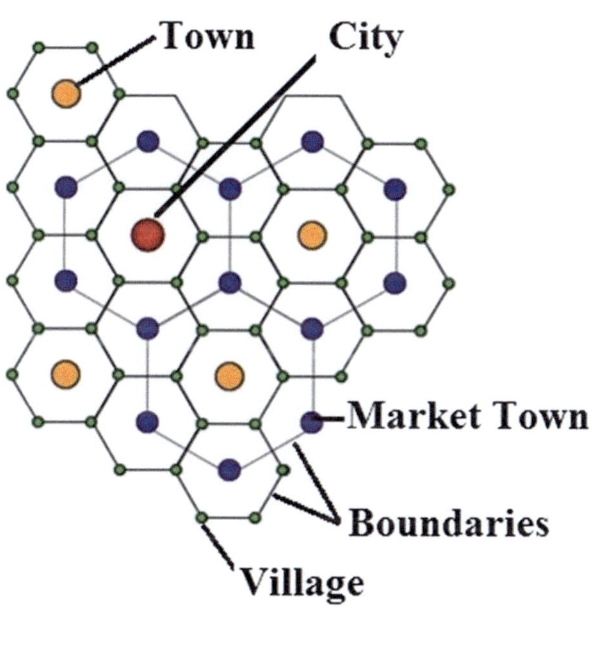

Fig. 1.8 Christaller's central place model (King 1985)

Therefore, a hexagonal pattern of urban settlements will disperse throughout the landscape (Preston 1991; Parr 2002).

1.4.4 Sector Theory

Sector theory is based on the idea that functional land use regions will expand in wedge-shaped zones outward from the CBD and in the concept that high-rental areas are initiated in wedge forms and spread out in radial sectors along lines from the CBD to the urban fringe. However, sector theory attempts to clarify the trends for several socioeconomic clusters to segregate them in terms of their housing locations. The theory also proposes that high-quality residential areas tend to grow outward from the urban center along the highways over time (Torrens 2000).

The sector model (Fig. 1.9) considers direction and distance as factors that form residential allocation. The model also recognizes that CBD is not the only main point of urban activity (Kivell 2002).

1.4.5 Multiple Nuclei Theory

Multiple nuclei theory is based on the simple fact that most big cities and many towns have various nubs that serve as centers of agglomerative growth instead of a simple CBD. This theory considers the urban spatial system as an urban area that includes several functional and industrial centers (Fig. 1.10). However, some of these centers are earlier settlements and others emerged from the urbanization process and external economies. Multiple nuclei theory surpassed the ideas that explain the spatial distribution of urban system activities by recognizing the significant influences on urban factors, such as accessibility, historical trends, and topography. Notably, the theory draws closer to clarifying and explaining why different urban spatial patterns occur in recognizing the polycentric structure of metropolises (Simmons 1965; Torrens 2000).

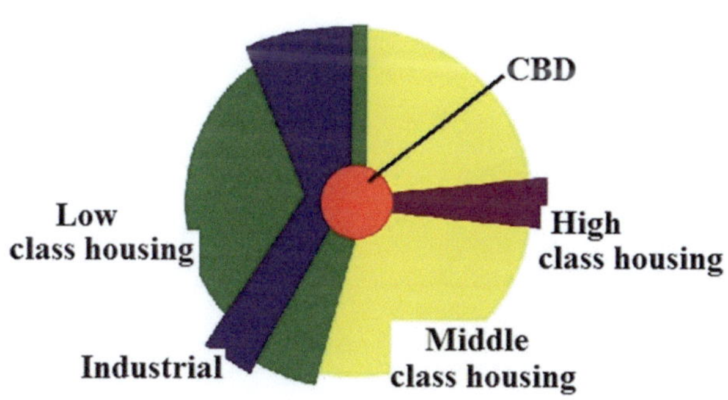

Fig. 1.9 Sector theory of urban expansion (Candau 2002)

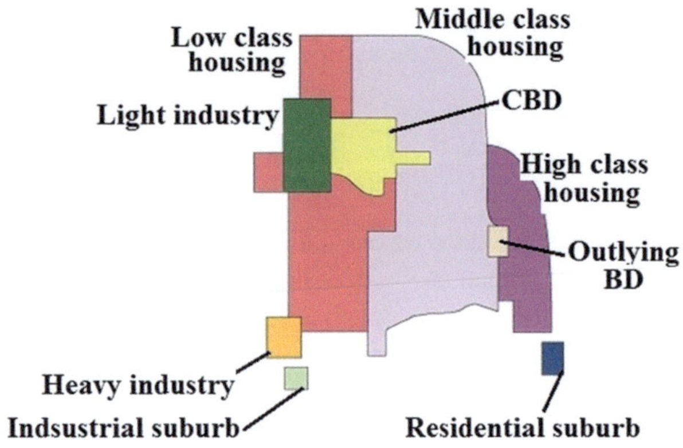

Fig. 1.10 Multiple nuclei model or urban expansion (Candau 2002)

1.4.6 Bid–Rent Theory

Bid–rent theory is based on von Thünen urban modeling theory and considers several urban factors like transportation. Given that transportation costs increase with the distance from the markets, rents generally tend to decrease correspondingly. Nevertheless, various forms of urban land uses (service, retail, housing, or industrial) generate different bid–rent curves (Fig. 1.11).

An urban land user prefers locations near the city center, but an urban developer will tend to accept locations that are further from the city center if rentals are lower. In essence, bid–rent theory is a study of housing compared to the amount of required land, and variations in the incomes used on land, transportation costs, and all services and goods. The price of land will decrease along with the increase in distance from the city center and the patterns of housing stock will emerge if the quality and quantity of services and goods are held constant. The amount of land that can be bought will increase as distance from the city center increases, but transportation costs will increase along with the distance from the city center. Based on these simple principles, the wealthy will select the services of lower density housing at the edge of the city and will pay a higher price for traveling over the distance. In comparison, the poor will remain in higher density residential areas near the city center. Bid–rent and von Thünen modeling theories reflect several aspects of the dynamicity of urban morphology (Torrens 2000).

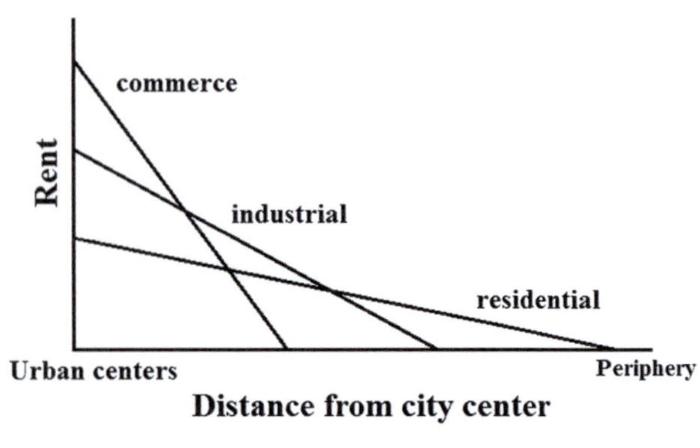

Fig. 1.11 Alonso model of housing stock based on bid–rent (Candau 2002)

1.5 Urban Growth and Natural Environment Deterioration

Population growth and rural–urban migration because of several advantages of urban areas for living and working purposes caused the continuous growth of cities (Xie et al. 2005; Nauman et al. 2015). Technological changes and industrial developments, especially from the late 1700s onwards, were one of the main reasons of massive migration to the urban areas (Arbury 2005). According to United Nation's Population Division report, about 38% of the Earth's population are in urban areas and this amount is expected to rise to 61% by 2025 (UNPD 2012; Nauman et al. 2015). This implies that about 5 billion people out of a total world population of 8 billion will be living in urban areas. In addition, 40 large cities will be added every five years so that there will be 639 metropolises with more than one million residents by 2025. Seventy-six percent of these metropolises will be in developing countries (UNPD 2012). Hence, the growth of urban environments in developing countries and tropical regions should be the main concern of urban scientists and researchers in this field (Fig. 1.12).

The growth of urban areas consists of different aspects or dimensions, such as population growth, physical growth, and economic growth. However, a strong and direct relation among these growth aspects exists. Economic and industrial developments attract huge amounts of population from rural areas, especially in developing countries, because of dissatisfaction with life in rural areas. Population growth significantly increases the physical expansion of urban areas. Nevertheless, the main concern of these growths from the environmental point of view is the physical expansion of built-up areas (urbanization processes of industry, commerce, and residence) through rural areas, which destroys a great extent of the natural environment (Kumar et al. 2007; Bhatta 2010; Abu Hammad et al. 2012). Xie et al. (2005) also defined urbanization as the conversion of natural environment to build up areas. Although the total amount of urban areas covers a very insignificant percentage of the Earth's land surface, the growth of these areas is still the main reason of various natural environment problems. Currently, the influence of urban areas on Earth's resources consumption, environmental pollutions, climate changes, and so on is clearly observable. The continuous growth of these manmade developments has magnified these problems and produced several other negative effects on the natural

Fig. 1.12 Metropolises in Europe and North Asia

environments. Hence, in the recent decades an awareness of the uncontrolled urban development and its negative consequences has grown (Hennig et al. 2015; Sisodia et al. 2016). Uncontrolled and horizontal development is considered as urban sprawl (Jiang et al. 2007). Urban sprawl has triggered various environmental and resource problems because of the fact that urban growth normally converts agricultural and forest lands to build up areas and increase wasteful use of land resources (Yeh et al. 2001; Blaikie et al. 2015). Wilson et al. (2003) stated from an overall perspective that unorganized urban development requires more pavements, hence increases air and water pollution; destroys forest lands, farmland, woodlots, and open space; disrupts ecosystems and fragment habitats; and increases fossil fuel consumption and emission of greenhouse gases.

Land use changes caused by urban growth have disturbed the biogeochemical life cycle to a great extent, increased water and food consumption, and consequently resulted in associated sewage and pollution problems (Xie et al. 2005). Unorganized urban expansion can affect agricultural production because of the conversion of farm lands into human settlements; affect the ecosystem because of the increase in emissions from transportation; have an effect on urban center dynamic behavior because of the further segregation or integration of urban structures; cause social welfare disadvantages because of the undersupply of collective consumption goods (Adolphson 2010).

Carbon dioxide emission is one of the main air pollutants from urban areas and human activities. About 61% of the Earth's population is expected to be living in urban environments by 2025 (UNPD 2012). About 70% of the overall carbon dioxide emissions are from these human settlements. Thus, dealing with this problem in a global and national scale is important to achieve more sustainable urban areas. Specifically, fast growing countries need to be more serious in monitoring and controlling the growth of and propose and implement low carbon societies (Fujita et al. 2009). As mentioned before, one of the main environmental effects is the growth of the physical aspects of urban areas. Residential buildings have more significant growth than other land use categories and building types. In this regard, different building construction materials have substantial effect on the rate of carbon emission. For instance, timber housing construction emits less carbon dioxide than the concrete buildings because of high carbon emission during cement production. Hence, considering the construction materials used for the development is important during the planning and implementation of low carbon societies, in addition to land use management and control of growth and expansion.

Numerous studies have been conducted regarding the effects of urban growth and expansion on surface temperature and global climate changes (Vörösmarty et al. 2000; Weng et al. 2004; Jenerette et al. 2007; Kahn 2009; Lemonsu et al. 2015). Heat waves are good examples of these negative effects in post-industrial countries (Kaveckis et al. 2014). Heat wave refers to a period of uncommon hot weather. Urban heat island (UHI) is a common and specialized terminology for this issue, which refers to the intensified heat produced by a relatively warmer local climate over the urban areas (Kaveckis et al. 2014). UHI can have a negative effect on the comfort and health of people, especially for those with heart diseases, the children, and the elderly (Svensson et al. 2002).

Flash flood caused by surface runoff is another environmental problem of urban expansion because of the conversion of vegetation land cover to impervious urban land use, such as buildings and paved surfaces. Hence, more frequent and severe floods can occur because of less infiltration of urban surface land uses, especially those developed in watersheds (Douglas 1983).

Therefore, the most common unorganized urban expansion, which is known as sprawl development, does not provide a good quality of urban neighborhood because of low density, large rural development, spatially segregated land uses, and widespread commercial strip development (Burchell et al. 2000). In addition, urban sprawl and unorganized horizontal city expansion are not considered acceptable and sustainable urban forms because of high carbon emission, traffic congestion, agricultural and forest destruction, higher infrastructural provision costs, various public health problems, and so on (Carruthers 2002; Gu et al. 2013; Litman 2015; Post et al. 2015). Hence, devoting considerable attempts on studying the proper balance between rural and urban areas for the preservation of natural environments while fulfilling the existing population initial needs (Xie et al. 2005) to attain sustained growth over a longer period of time is important.

References

Abdullahi, S., & Pradhan, B. (2015). Sustainable brownfields land use change modeling using GIS-Based weights-of-evidence approach. *Applied Spatial Analysis and Policy*, 1–18.

Abdullahi, S., Pradhan, B., & Jebur, M. N. (2015a). GIS-based sustainable city compactness assessment using integration of MCDM, Bayes theorem and RADAR technology. *Geocarto International, 30*(4), 365–387.

Abdullahi, S., Pradhan, B., Mansor, S., & Shariff, A. R. M. (2015b). GIS-based modeling for the spatial measurement and evaluation of mixed land use development for a compact city. *GIScience & Remote Sensing, 52*(1), 18–39.

Abu Hammad, A., & Tumeizi, A. (2012). Land degradation: socioeconomic and environmental causes and consequences in the eastern Mediterranean. *Land Degradation and Development, 23*(3), 216–226.

Adolphson, M. (2010). Kernel densities and mixed functionality in a multicentred urban region. *Environment and Planning. B, Planning & Design, 37*(3), 550.

Anderson, W. P., Kanaroglou, P. S., & Miller, E. J. (1996). Urban form, energy and the environment: A review of issues, evidence and policy. *Urban Studies, 33*(1), 7–35.

Arbury, J. (2005). *From urban sprawl to compact city: an analysis of urban growth management in Auckland.* Geography and Environmental Science, University of Auckland.

Archer, W. R., & Smith, M. T. (1993). Why do suburban offices cluster? *Geographical Analysis, 25*(1), 53–64.

Banister, D. (2012). Assessing the reality—Transport and land use planning to achieve sustainability. *Journal of Transport and Land Use, 5*(3), 1–14.

Batty, M. (2008). The size, scale, and shape of cities. *Science, 319* (5864), 769–771.

Batty, M., & Longley, P. A. (1994). *Fractal cities: A geometry of form and function*. Academic Press.

Bhatta, B. (2010). *Analysis of urban growth and sprawl from remote sensing data*. Springer Science & Business Media.

Bhatta, B., Saraswati, S., & Bandyopadhyay, D. (2010). Urban sprawl measurement from remote sensing data. *Applied Geography, 30*(4), 731–740.

Blaikie, P., & Brookfield, H. (2015). *Land degradation and society*. Routledge.

Blumenfeld, H. (1949). On the concentric-circle theory of urban growth. *Land Economics, 25*(2), 209–212.

Bruegmann, R. (2006). *Sprawl: A compact history*. University of Chicago press.

Burchell, R. W., Listokin, D., & Galley, C. C. (2000). Smart growth: More than a ghost of urban policy past, less than a bold new horizon. *Housing Policy Debate, 11*(4), 821–879.

Burton, E. (2000). The compact city: Just or just compact? A preliminary analysis. *Urban Studies, 37*(11), 1969–2006.

Burton, E. (2002). Measuring urban compactness in UK towns and cities. *Environment and Planning B, 29*(2), 219–250.

Candau, J. T. (2002). *Temporal calibration sensitivity of the SLEUTH urban growth model* (Master's thesis). University of California, Santa Barbara, CA.

Carruthers, J. I. (2002). The impacts of state growth management programmes: A comparative analysis. *Urban Studies, 39*(11), 1959–1982.

Chen, J., Chang, K.-T., Karacsonyi, D., & Zhang, X. (2014). Comparing urban land expansion and its driving factors in Shenzhen and Dongguan, China. *Habitat International, 43,* 61–71.

Clarke, K. C., & Gaydos, L. J. (1998). Loose-coupling a cellular automaton model and GIS: Long-term urban growth prediction for San Francisco and Washington/Baltimore. *International Journal of Geographical Information Science, 12*(7), 699–714.

Costanza, R., & Ruth, M. (1998). Using dynamic modeling to scope environmental problems and build consensus. *Environmental Management, 22*(2), 183–195.

Douglas, I. (1983). *The urban environment*. London: Edward Arnold Publisher.

Elkin, T., McLaren, D., & Hillman, M. (1991). *Reviving the City: Towards sustainable urban development*. Friends of the Earth London.

Fujita, Y., Matsumoto, H., & Siong, H. C. (2009). Assessment of CO_2 emissions and resource sustainability for housing construction in Malaysia. *International Journal of Low-Carbon Technologies, 4*(1), 16–26.

Garreau, J. (1991). *Edge city: Life on the frontier.* New York.

Gillham, O. (2002). *The limitless city: A primer on the urban sprawl debate*. Island Press.

Goodal, B. (1987). *The Penguin dictionary of human geography*. London.

Gu, Z., Sun, Q., & Wennersten, R. (2013). Impact of urban residences on energy consumption and carbon emissions: An investigation in Nanjing, China. *Sustainable Cities and Society, 7,* 52–61.

Hennig, E. I., Schwick, C., Soukup, T., Orlitová, E., Kienast, F., & Jaeger, J. A. (2015). Multi-scale analysis of urban sprawl in Europe: Towards a European de-sprawling strategy. *Land Use Policy, 49,* 483–498.

Herndon, J. D. (2011). *Mixed-use development in theory and practice: Learning from Atlanta's mixed experiences*. Unpublished report. Available online at https://smartech.gatech.edu/bitstream/handle/1853/40790/JoshuaHerndon_Mixed-Use%20Development%20in%20Theory%20and%20Practice.pdf. Accessed on September 20, 2015.

Huang, J., Lu, X., & Sellers, J. M. (2007). A global comparative analysis of urban form: Applying spatial metrics and remote sensing. *Landscape and Urban Planning, 82*(4), 184–197.

Hugill, P. J. (2002). The geographical imagination in America, 1880–1950 (review). *Journal of Interdisciplinary History, 33*(1), 141–142.

Jenerette, G. D., Harlan, S. L., Brazel, A., Jones, N., Larsen, L., & Stefanov, W. L. (2007). Regional relationships between surface temperature, vegetation, and human settlement in a rapidly urbanizing ecosystem. *Landscape Ecology, 22*(3), 353–365.

Jiang, F., Liu, S., Yuan, H., & Zhang, Q. (2007). Measuring urban sprawl in Beijing with geo-spatial indices. *Journal of Geographical Sciences, 17*(4), 469–478.

Kahn, M. E. (2009). Urban growth and climate change. *Annual Review of Resources and Economics, 1*(1), 333–350.

Kaveckis, G., & Bechtel, B. (2014). *Land use based urban vulnerability to climate change assessment.* Paper presented at the Environmental Engineering. Proceedings of the International Conference on Environmental Engineering. ICEE.

King, L. J. (1985). Central place theory. *Regional Research Institute*. Beverly Hills: Sage Publications.

Kivell P. (2002). *Land and the city: Patterns and processes of urban change*. Routledge.

Krugman, P. (1996). *The self-organizing economy*. New Jersey: Blackwell Publishers.

Kumar, J. A. V., Pathan, S., & Bhanderi, R. (2007). Spatio-temporal analysis for monitoring urban growth—A case study of Indore city. *Journal of the Indian Society of Remote Sensing, 35*(1), 11–20.

Kuper, A. (2013). *The social science encyclopedia*. Routledge.

Lemonsu, A., Viguié, V., Daniel, M., & Masson, V. (2015). Vulnerability to heat waves: Impact of urban expansion scenarios on urban heat island and heat stress in Paris (France). *Urban Climate, 14,* 586–605.

Litman, T. (2015). Analysis of public policies that unintentionally encourage and subsidize urban sprawl.

Liu Y. (2008). *Modelling urban development with geographical information systems and cellular automata*. CRC Press.

Liu, Y., & Phinn, S. R. (2003). Modelling urban development with cellular automata incorporating fuzzy-set approaches. *Computers, Environment and Urban Systems, 27*(6), 637–658.

McCann, P. (2001). Urban and regional economics. *OUP Catalogue*.

Mubareka, S., Koomen, E., Estreguil, C., & Lavalle, C. (2011). Development of a composite index of urban compactness for land use modelling applications. *Landscape and Urban Planning, 103* (3), 303–317.

Muller, P. O. (2004). Transportation and urban form-stages in the spatial evolution of the American metropolis. In S. Hanson (Ed.), *The geography of urban transportation* (2nd Edn., p. 29). New York: Guilford.

Nauman, E., VanLandingham, M., Anglewicz, P., Patthavanit, U., & Punpuing, S. (2015). Rural-to-urban migration and changes in health among young adults in Thailand. *Demography, 52*(1), 233–257.

Newman, P. (1992). The compact city: An Australian perspective. *Built Environment, 18*(4), 285–300.

Newton, P. (1997). Re-shaping cities for a more sustainable future. *Research Monograph, 6*. Canberra: Australian Academy of Technological Sciences and Engineering.

Ottensmann, J. R. (1977). Urban sprawl, land values and the density of development. *Land Economics, 53*(4), 389–400.

Parker, D. C., Manson, S. M., Janssen, M. A., Hoffmann, M. J., & Deadman, P. (2003). Multi-agent systems for the simulation of land-use and land-cover change: A review. *Annals of the Association of American Geographers, 93*(2), 314–337.

Parr, J. (2002). The location of economic activity: Central place theory and the wider urban system.

Post, C., Ritter, B., Akturk, E., Breedlove, A., Buchanan, R., Che, C., … Mikhailova, E., 2015. Analysis of factors contributing to abandoned residential developments using remote sensing and geographic information systems (GIS). *Urban Ecosystems, 18*(3), 701–713.

Preston, R. E. (1991). Central place theory and the Canadian urban system. *Canadian Cities in Transition,* 148–177.

Salas, J. F. (2009). La ciudad lineal del centenario: Los cien años de la utopía lineal. *Revista de Urbanismo, 20.*

Schneider, A., & Woodcock, C. E. (2008). Compact, dispersed, fragmented, extensive? A comparison of urban growth in twenty-five global cities using remotely sensed data, pattern metrics and census information. *Urban Studies, 45*(3), 659–692.

Simmons, J. W. (1965). Geographica: Descriptive Models of Urban Land Use. *The Canadian Geographer/Le Géographe Canadien, 9* (3), 170–174.

Sinclair, R. (1967). Von Thünen and urban sprawl. *Annals of the Association of American Geographers, 57*(1), 72–87.

Sisodia, P. S., Tiwari, V., & Dahiya, A. K. (2016). Prediction of urban sprawl using remote sensing, GIS and multilayer perceptron for the City Jaipur. *Intelligent Systems Technologies and Applications* (pp. 403–410). Springer.

Sui, D. Z. (2004). Tobler's first law of geography: A big idea for a small world? *Annals of the Association of American Geographers, 94*(2), 269–277.

Svensson, M. K., & Eliasson, I. (2002). Diurnal air temperatures in built-up areas in relation to urban planning. *Landscape and Urban Planning, 61*(1), 37–54.

Tobler, W. (1979). Cellular geography. *Philosophy in Geography* (pp. 379–386): Springer.

Torrens, P. M. (2000). *How cellular models of urban systems work* (1. Theory).

UNPD, U. N. P. D. (2012). World urbanisation prospects: The 1999 revision, data tables and highlights. (ESA/P/WP.161; 2000). New York: United Nations Population Division.

Verburg, P. H., Kok, K., Pontius, R. G., Jr. & Veldkamp, A. (2006). Modeling land-use and land-cover change. *Land-use and Land-cover Change* (pp. 117–135): Springer.

Vörösmarty, C. J., Green, P., Salisbury, J., & Lammers, R. B. (2000) Global water resources: Vulnerability from climate change and population growth. *Science, 289*(5477), 284–288.

Wang, F. (2012). *A cellular automata model to simulate land-use changes at fine spatial resolution* (Doctor of Philosophy). University of Calgary, Calgary (20365).

Wegener, M., & Fürst, F. (2004). Land-use transport interaction: State of the art. *Available at SSRN 1434678*.

Weng, Q., Lu, D., & Schubring, J. (2004). Estimation of land surface temperature–vegetation abundance relationship for urban heat island studies. *Remote Sensing of Environment, 89*(4), 467–483.

White, R. W. (1977). Dynamic central place theory: Results of a simulation approach. *Geographical Analysis, 9*(3), 226–243.

Wilson, E. H., Hurd, J. D., Civco, D. L., Prisloe, M. P., & Arnold, C. (2003). Development of a geospatial model to quantify, describe and map urban growth. *Remote Sensing of Environment, 86*(3), 275–285.

Wright, F. L. (1935). Broadacre City: A new community plan. *Architectural Record, 77*(4), 243–254.

Wrigley, N., & Bennett, R. J. (1981). *Quantitative geography: A British view*. Routledge.

Xie, C., Huang, B., Claramunt, C., & Chandramouli, C. (2005). *Spatial logistic regression and GIS to model rural-urban land conversion.* Paper presented at the Proceedings of PROCESSUS Second International Colloquium on the Behavioural Foundations of Integrated Land-use and Transportation Models: Frameworks, Models and Applications.

Yamu, C., & Frankhauser, P. (2015). *Spatial accessibility to amenities, natural areas and urban green spaces: Using a multiscale, multifractal simulation model for managing urban sprawl* (p. b130171p). Environment and Planning B: Planning and Design.

Yang, X., & Lo, C. (2003). Modelling urban growth and landscape changes in the Atlanta metropolitan area. *International Journal of Geographical Information Science, 17*(5), 463–488.

Yeh, A., & Li, X. (2001). A constrained CA model for the simulation and planning of sustainable urban forms by using GIS. *Environment and Planning B: Planning and Design, 28*(5), 733–753.

Sustainable Urban Development

2

Saleh Abdullahi and Biswajeet Pradhan

2.1 Introduction

The idea of sustainable development emerged in the late nineteenth century in the observation of the various critical environmental problems caused by the continuous growth of urban areas, especially in rural and natural environments. Meadows et al. (1972) were the first authors to use the sustainable terminology in their research regarding human development pattern. They stated that a tragic destruction of the global environment will occur in the 2000s if the current trend of human development and resource consumption continues. Hence, a fundamental solution is required to control the growth trend and define specific guidelines to address the ecological and economic aspect of the environment. In addition to this first spark of sustainable development concept, the 1972 UN conference on Human Environment and the 1973 oil crisis resulted in the strong agreement among scientists and stockholders that the existing development trend should not continue forever and proper preservation and care of the natural and environmental resources is required for future generations (Arbury 2005). After the publication of Our Common Future by the Brundtland Commission in 1987 (WCED 1987), the concept of sustainable development has become as an important objective to make a better quality of life in the economic, social, and environmental perspectives (Fig. 2.1). Based on this concept, the concern on the future of the world's environment and its resources became an established fact of life, and this was accompanied by expressions of good intention from governments worldwide (Burton et al. 2003).

In a general perspective, sustainable development can be defined as the utilization of natural resources for current human activities without jeopardizing the ability of future generations to use the same resources (WCED 1987). Considering how well we balance socioeconomic, environmental, and land use growth objectives is important when making decisions today. This statement recognizes the importance of ensuring that the needs of the world's current population are satisfied, with consideration for the needs of the future generation. In addition to the three main aspects of sustainable development, three basic principles should be considered related to inter-generational equity, social justice, and transfrontier responsibility (Haughton and Hunter 2004). These principles are significantly important, especially in the case of the urban perspective of sustainability. The inter-generational equity principle implies the main definition of sustainable development regarding the future generation's right to use the same resources. Social justice principle is more concerned with the social aspects of urban areas related to poverty, which should be tackled in the current generation because it is one of the main reasons of environmental destructions. This aim could be achieved through the proper distribution of resources and facilities, more comprehensive environmental conservation projects and guidelines, and social equity policies. The transfrontier responsibility principle deals with the social responsibility toward the natural environment at a global scale, and not limited to national borders (Haughton and Hunter 2004).

In general, an associated or linked action is believed capable of securing the ability of future generations to fulfill their own responsibilities (WCED 1987). This process should involve issues related to natural environment and resources, stable economy, the maintenance of quality of life, and the protection of the planning and development strategies of the society. The responsibilities of actors to achieve sustainable development in different scales, from the local to national levels, were specified by the United Nations Conference on Environment and Development in 1992 in Rio de Janeiro, when they were made part of Agenda 21 (Bleicher and Gross 2010). Thus, various attempts were implemented in the 1990s to specify the objective of sustainable development and determine indicators to control the growth and development projects. Moreover, several seminars and workshops were held to increase the general understanding of the sustainable development concept and its objectives and indicators. In addition, experts and scientists propose a sustainability benchmark rule for the comparison of enterprises related to sustainability, such as

B. Pradhan (ed.), *Spatial Modeling and Assessment of Urban Form*,
DOI 10.1007/978-3-319-54217-1_2

Fig. 2.1 Three general aspects of sustainable development (WCED 1987)

the one created by Van den Brick and Van der Woerd (2004).

The dependence of sustainable development on space, time, scale, and the actors involved should be realized (Bleicher and Gross 2010). Franz and Nathanail (2005) stated that "Sustainability is neither static in time nor does it imply a fixed spatial perspective. It cannot be seen as a destination but rather as a never ending journey." Therefore, sustainability for any project and action related to social, environmental, and economic should be involve with its specified characteristics in terms of its spatial, temporal, and thematic contexts (Olsson 2009).

2.2 Urban Sustainability

In urban perspectives, sustainable urban development is concerned with the minimum inputs and outputs from an urban system. Sustainability has become a key planning objective in urban growth and development since the sustainable development declaration by the International Union for the Conservation of Nature and Natural Resources (IUCN), the United Nations Environment Program, and the World Wildlife Fund in 1980, and the announcement regarding sustainable cities in the Toronto Declaration on World Cities and Their Environment in 1990 (Lin and Yang 2006). A sustainable city is defined as a city designed with environmental consideration, with minimum consumption of energy, water, and food and minimum output of waste and air, water, and soil pollutions. Urban sustainability can also be defined as the improvement of the quality of life of human beings within the capacity of Earth's limited resources. Hence, urban sustainability is the ability of cities to reduce the environmental effect of urban activities while improving social equity and livability in urban areas (Newman and Kenworthy 1999; Chiu 2008; van Wee and Handy 2016). Moreover, urban sustainability attempts to deal with the question of how societies develop and run their urban systems to ensure the preconditions of development for future generations. Thus, discovering how inhabitants can be educated and participate in developing a quality environment is important (Nurul 2015).

In the recent decade, more attention has been given to urban sustainability because these human-made settlements are the source of air, water, and land pollutions and the main consumers of natural land, food, and energy. Elkin et al. (1991) believed that urban areas have never been sustainable because of the linkage between urban growth process and desertification and significant disturbance in cyclical ecological systems caused by the extra production and extraction of food and materials. Although no strong agreement for this concept exists because of the lack of unique definition and scope, the concept is very important for human societies because of the continuous destruction of natural and environmental resources. The conversion of valuable agricultural and forest lands into build up areas has become the main issue for urban sustainability, especially in developing and tropical countries. For instance, countries with rapid population growth cannot risk losing large amounts of agricultural fields to feed their population. However, although rapid growth in developing countries seems to cause more unsustainability, resource consumption in large and developed cities is much more than that in developing cities (Arbury 2005). Thus, sustainability implementation in developed cities should be prioritized as well.

In this regard, the lack of land use change and urban growth management cause unnecessary destruction of huge amounts of natural environment (Abdullahi and Pradhan 2015). Hence, one of the main objectives of sustainable urban development is to avoid this issue based on four operational factors (Li and Yeh 2000):

- Avoid conversion of valuable natural environment at the initial stage of urban development.
- Perform proper land requirements analysis based on the land resource capacity.
- Implement land priority analysis to avoid destruction of valuable agricultural fields.
- Develop compact development with higher urban density and less land consumption.

The implementation of these factors and adoption of various other strategies and plans to minimize energy consumption, protect biological diversity, reduce pollution, improve social interaction, and so on are essential tasks to achieve urban sustainability (Kropp and Lein 2013). To achieve these objectives, urban sustainability is evaluated and implemented within three main perspectives, namely, environmental, social, and economic. Each of these aspects deals with separate issues of an urban system, such as security, livability, and social equity; productivity, personal, and public finances; and pollution levels, amount of reserve habitat, and resource consumption, respectively (Lin and Yang 2006) (Fig. 2.2).

2.2.1 Environmental Sustainability

Environmental sustainability is a principal concept in this theory and can be evaluated by measuring the pollution levels, amount of reserve habitat, and resource consumption (Lin and Yang 2006; Mellino and Ulgiati 2015). This concept places higher priority on a variety of plant and animal species, pollution reduction, and efficient utilization of resources. From the natural resources perspective, environmental effects depend on how these resources are prepared for utilization, how the produced energies are transmitted to consumers, the amount of wastes and pollutions created from consumption, and the effect of these processes on the natural resources. Many agree that the environmental problems of urban areas are caused by urban sprawl and spatial segregation (Chiu 2012). Kopfmüller et al. (2001) summarized a list of some environmental sustainability goals as follows:

- Sustainable utilization of renewable resources. The rate of utilization of these sources should not be more than their regeneration rate and should not endanger the capacity and dynamics of the corresponding ecosystem.
- Sustainable utilization of nonrenewable resources. These resources should be protected seriously.
- Avoiding the utilization of natural environment as a sink of urban waste and pollutions.
- Avoiding any kinds of disasters and negative effects to humans and natural environments.

Emission of carbon dioxide is one of the major pollutions that mainly arise from urban areas and human activities.

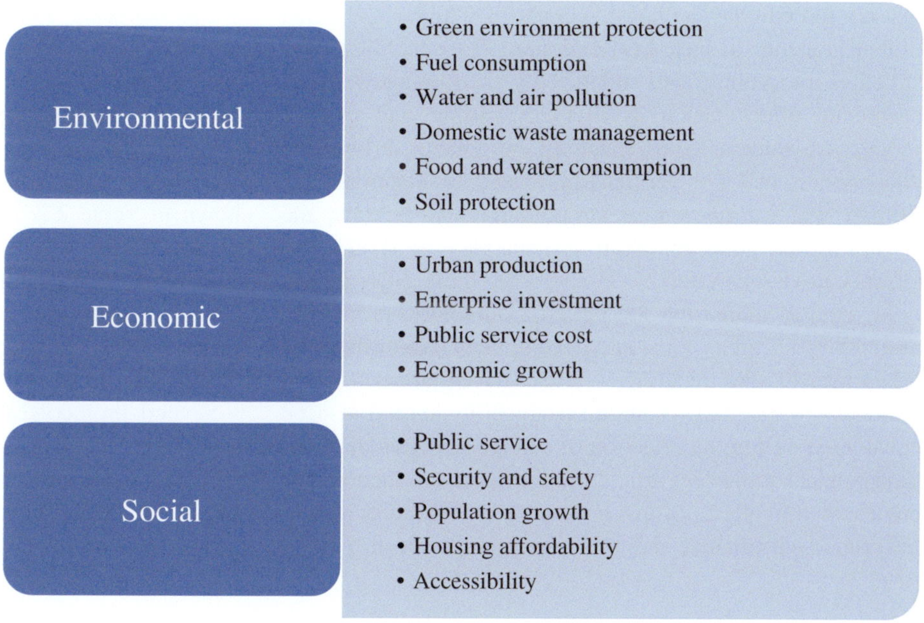

Fig. 2.2 General aspects of urban sustainability (Lin and Yang 2006)

These man-made environments are responsible for 70% of the overall carbon emission mainly because of fossil fuel consumption and land use changes (Ho and Kean 2007). The rapid growth of CO_2 emissions from the urban environment has become one of the main concerns of urban scientists and planners. Several research has shown that population and economic growths are the major causative factors of this emission in the recent decades. Shi (2001) estimated that 1% increase in population contributes to 1.28% increase in CO_2 emissions. In addition, economic growth causes increases in income level, industrial development, construction development, and so on. All these consequences lead to increase in carbon emissions in a variety of ways. For instance, income growth and advancement in production of affordable automobiles significantly increase private car ownership and reduce the usage of public transportation, which eventually increase CO_2 emissions in urban areas. Thus, addressing this issue and planning for low-carbon societies, especially for developing countries with rapid urban growth and expansion, are essential. The low-carbon society terminology was first used in 2003 when developed countries aimed to reduce CO_2 emissions to sustain the world's climate. The project to create a comprehensive view and definition for low-carbon society has been started by the Japan–England collaboration (Ho and Kean 2007). It involves the collaboration of several researches to review greenhouse gas (GHG) emission studies, analyze approaches to achieve a low-carbon society, and share knowledge and information among countries.

In addition to population and economic factors, different construction materials for urban development emit different amounts of CO_2. For instance, timber housing emits less CO_2 than reinforced-concrete housing because of the utilization of calcium carbonate as a raw material for cement production (Fujita et al. 2009). Gerilla et al. (2007) estimated that reinforced-concrete housing emits 23% higher CO_2 than timber housing. In fact, CO_2 emissions from building construction are mainly also supplied by other negative environmental effects, such as fuel combustion and cement production process. Thus, changing the policies in building construction that are particularly related to materials utilization will clearly reduce air pollution and achieve the environmental goals of urban sustainability. Several other efforts can also be effective in achieving these goals in this field, such as increasing the building durability, promoting more compact urban form and development, controlling the urban growth, and avoiding suburban development.

Meanwhile, forest resource sustainability should also be considered during the selection of construction materials for development projects. Particularly in timber production, the environmental effects of this task on forest resources should be properly estimated and minimized. Fujita et al. (2009) assessed the effect of building (particularly residential buildings) construction using timber on forest resources through the following procedures:

(1) The entire floor area of a newly constructed area was computed by considering the number of housing units and floor area of each housing unit.
(2) The timber consumption was assessed by considering the floor area of each housing unit and timber required for each floor area.
(3) The amount of forest area required for the construction of a residential building was evaluated based on forest productivity.

Different types of buildings based on the usage (residential, commercial, industrial), number and size of the building, interior and exterior designs, and other parameters consume different amounts of timber materials. Fujita et al. (2009) concluded that consuming forest resources in a sustainable way in a region of rapid population growth is possible because of the higher requirement of such resources, which exceed productivity. In addition, higher building durability reduces the negative effect on forest resources.

However, controlling the rapid population growth and urban expansion is more effective in reducing environmental effects, such as air pollution and forest destruction, than changing construction materials from concrete to timber.

In a wider perspective, a low-carbon city can be achieved by promoting low-carbon emission policies, such as urban growth and expansion control, fuel or automobile consumption regulations, and emission limitations. However, implementing and developing sustainable urban forms, such as compact city, eco-city, transit-oriented development (TOD), is also a good alternative to reduce carbon emission in a global scale. Every country emits different amounts of carbon dioxide (Table 2.1). Developed countries emit more than half of the total emissions. Meanwhile, the rapid population and economic growth of Asian countries has also led to the increase in their carbon emissions. Furthermore, the evaluation of the world average emissions per capita, as shown in Table 2.2, shows that the per capita emission of developed regions is more than that of the world average and developing countries (Ho and Kean 2007). Thus, these significant differences in the carbon emissions of countries caused urgency in addressing and proposing a proper solution for the global warming and climate change problems.

Several protocols and agreements have been signed by most of the countries as a solution to reduce environmental pollution, especially those related to global warming and climate changes, such as the Kyoto Protocol, Stockholm Convention on Persistent Organic Pollutants, Montreal

Table 2.1 Total CO_2 emissions by region, 1990–2003

Region	1990 (million metric tons)	2000 (million metric tons)	2003 (million metric tons)	% Change p.a. 1990–2003 (%)
World	21,283.38	23,832.70	25,575.99	1.6
Asia (excluding middle east)	5014.89	7272.53	8477.90	5.3
Central America and Caribbean	379.32	467.09	500.58	2.5
Europe	–	6002.02	6277.17	1.5
Middle East and North Africa	926.96	1474.34	1645.98	6.0
North America	5274.41	6232.06	6257.98	1.4
South America	537.47	757.03	740.45	2.9
Developed countries	–	14,623.79	15,043.57	1.0
Developing countries	5839.34	8475.59	9810.41	5.2
High-income countries	10,452.47	12,123.43	12,420.82	1.4
Middle-income countries	–	9204.17	12,420.82	1.1
Low-income countries	912.89	1494.26	1631.11	6.1

Table 2.2 CO_2 emissions per capita by region, 1990–2003

Region	1990 (metric tons per capita)	2000 (metric tons per capita)	2003 (metric tons per capita)	% Change 1990–2003 (%)
World	4.0	3.9	4.1	0.2
Asia (excluding middle east)	1.7	2.1	2.4	3.2
Central America and Caribbean	2.7	2.8	2.9	0.6
Europe	10.1	8.1	8.5	−1.2
Middle East and North Africa	3.0	3.9	4.1	2.8
North America	18.6	19.8	19.3	0.3
South America	1.8	2.2	2.0	0.9
Developed countries	12.0	11.0	11.1	−0.6
Developing countries	1.5	1.9	2.1	3.1
High-income countries	11.8	12.8	12.8	0.7
Middle-income countries	0.6	0.7	0.8	2.6
Low-income countries	3.3	3.2	3.5	0.5

Protocol on substances that deplete the ozone layer, Basel Convention on the trans boundary movement of hazardous waste and their disposal, Rotterdam Convention on prior consent procedure for hazardous chemical and pesticides in international trade, and the Cartagena Protocol on bio-safety (Ho and Kean 2007). Environmental implementation policies are mainly accomplished using several quality guidelines and rules, such as measurements of air pollutants to control air pollutions. Despite the fact that some actions, such as forest preservation, reduction and control of private car transportation, and other indirect reduction of carbon dioxide emissions, reduce these air pollutants, concentrating directly on carbon dioxide emissions and thinking toward achieving a low-carbon society seriously are more important.

2.2.2 Economical Sustainability

The US Environmental Protection Agency defined sustainable development as "the effort to reconcile the competing demands of regional development, namely, community integrity, economic development, and environmental protection" (EPA 2006). Although all three concepts of sustainability seem equally important, the diagram to describe these three concepts, which was presented by Krueger et al. (2012), shows the economic aspect as the central point of sustainable development, surrounded by the environment and social aspects (Fig. 2.3).

The economical sustainability objective is to improve the productivity of personal and public finances (Lin and Yang 2006; Bhattacharya et al. 2015). According to the literature (Kopfmüller et al. 2001; Bleicher and Gross 2010), several sustainability goals in the economic aspect are applicable, such as the following:

- Autonomous subsistence based on income from own work. All members of a society must be given the right to be able to secure their own livelihood (including bringing up their children and providing for old age) by means of a freely chosen occupation.
- Reducing high-income and wealth inequities. The variance and range between high and low incomes should be reduced.
- Sustainable development of man-made, human, and knowledge capitals. Capital goods, human capital, and knowledge capital are to be developed such that economic performance can be maintained or improved.

Therefore, an economic framework to implement these goals is required, should be applied and proposed by governments, and should be supported by local residents.

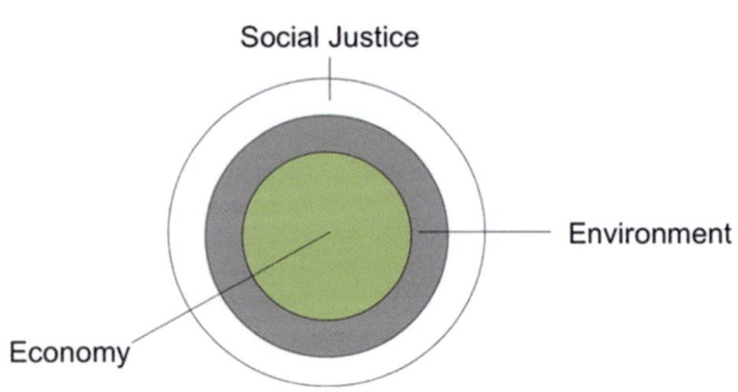

Fig. 2.3 Urban sustainability formulation (Krueger and Buckingham 2012)

In the Western context, the concept of sustainable development is mainly related to the need for adjustment of economic models to maintain a balance between economic growth and social requirements while protecting local ecologies and reducing the negative effect of growth on the global environment (Subeh and Al-Rawashdeh 2012). In contrast, other environmental and social sustainability objectives place second in terms of priority in developing countries because of several problems, such as economic growth, water scarcity, food security, and health. In the recent decades, most of the developing countries, especially in Eastern-Asia, have encountered significant growth and changes in economic and social aspects. Globalization, economic growth, and reconstitution have increased the burden and negative effects on the urbanization process. Thus, most of these countries started supporting and promoting sustainability principles to reduce and control these effects on the urbanization process. They attempted to control and manage the economic growth to maintain a balance among the three main aspects (environmental, social, and economic) of sustainable development (Subeh and Al-Rawashdeh 2012). Meanwhile, Grossman and Krueger (1995) stated that economic growth will ultimately benefit from natural environment preservation and conservation. In fact, sustainable economic growth is the main necessity for environmental and social improvement because the economic policies and activities of an urban area have significant effect on urban environmental and social conditions.

2.2.3 Social Sustainability

Social sustainability, as one the main aspects of urban sustainability, generally refers to how local residences behave within the physical environment of the city. In addition, social sustainability is concerned on security, livability, and social equity (Lin and Yang 2006). This aspect of sustainability can be identified by the access to community facilities (such as educational and health facilities), access to green and open spaces, job availability and accessibility, availability of public transportation, access to a proper walking and cycling environment, level of domestic living spaces, security condition, levels of social segregation, and availability of affordable housing (Burton 2000). In addition, social sustainability can be achieved through the following (Kopfmüller et al. 2001; Bleicher and Gross 2010):

- Protection of human health from hazards and risks from man-made environmental pollution;
- Securing the satisfaction of basic requirements (housing, nutrition, clothing, medical care, and so on) of all members of society;

- Equal access of people to all information, education, occupation, and social, political, and economic positions;
- Just distribution of for natural resources; and
- Participation in societal decision-making processes.

Thus, community safety, social equity, and a general acceptable level of quality of life are the main aspects of social sustainability. A sustainable society attempts to provide these properties to increase life satisfaction for the current and future generations. On the one hand, social equity focuses on narrower aspects, such as the accessibility and availability of community facilities and services and affordable housing, which are related to the characteristics of urban form and pattern. On the other hand, the broader aspects of social sustainability deal with social interaction, participation, and satisfaction from the living environment (Barton and Tsourou 2000). Quality of life is about good links between the living conditions with respect to working and community facilities. These links promote social interaction and a sense of community within the urban environment (Bramley and Kirk 2005). Security and safety, another aspect of social sustainability, are related to the extent of daily living activities of the residence without any fears, such as fear of being attacked, fear of being run-over, and fear of falling (Butterworth 2000). These negative feelings limit the eagerness to participate and interact with others in the community. Mixed land use development, proper and safe pedestrian, adequate street lighting, and well -maintained footways are some characteristics of a safe neighborhood.

With regard to the physical environment and development pattern of urban areas, land use diversity and pedestrian-friendly streets are important to increase social interaction and provide a sense of belonging to the community (Barton 2000). These properties have been applied in recent sustainable urban forms, such as new urbanism and compact city, where local residences have better opportunities of social interaction because of proximity and accessibility (Abdullahi et al. 2015b; Nurul 2015). Numerous studies have proven that urban form and pattern have significant effects on urban sustainability, especially in social and environmental sustainability. Urban form is related to the size, shape, and intensity of human settlements and the spatial distribution of various land use categories. Various aspects of urban forms, such as density, shape, degree of dispersion or concentration, and level of infrastructure for public transport, all have influence on social sustainability (Bramley and Kirk 2005). Social sustainability is directly linked to the behavior of local residents with respect to the characteristics of their neighborhoods. It also concerns the social relationship between society and natural environment over a long period of time (Barton and Tsourou 2000). Built environment and urban form play crucial roles in the urban health, well-being, social interaction, and participation of the residents (Littig and Grießler 2005). Porta (2001) illustrated the relationship between urban form and sustainability, with focus on the social aspects (Fig. 2.4). Social interaction emerges from street life in urban area, which indicates how urban form affects the living behavior of local residents in terms of the utilization of public spaces and their contribution to various social interactions (Nurul 2015).

Although urban areas and the number of cities are continuously increasing, that is, 40 large cities every 5 years (UNPD 2012), urban populations are not evenly distributed nor are cities at the same level of development. Therefore, following the sustainable development declaration, sustainability has become a key goal in urban planning. In a general environmental view, urbanization refers to the conversion of natural land cover to artificial man-made settlement. Hence, understanding this trend is important to evaluate the effects of urbanization at global and regional levels (Xie et al. 2005).

The descriptions of the different aspects of sustainable development depict sustainable development as a multidimensional concept that includes various perspectives (Figueira et al. 2005). Particularly, sustainable land development is a complex issue, which involves negotiations and compromises of various stakeholders (Li and Liu 2008). Barbier (1987) stated that sustainable development implies the simultaneous maximization of biological, economic, and social system goals. Hence, although fulfilling several objectives at the same time is impossible, adopting multi-disciplinary approaches, which can consider various

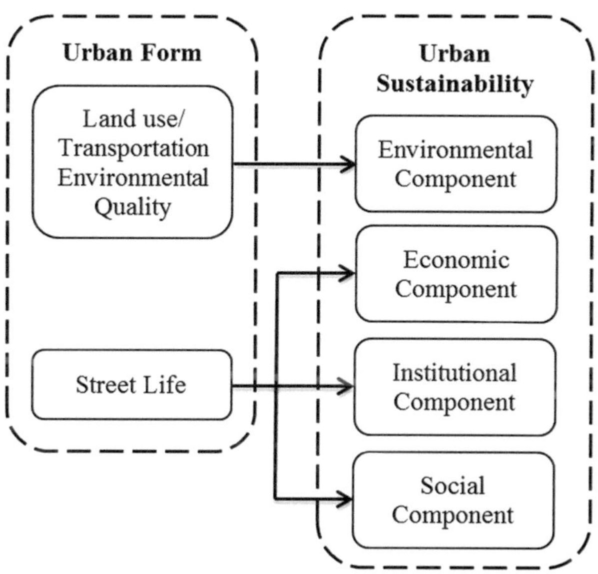

Fig. 2.4 The relationship between urban form and urban sustainability (Porta 2001)

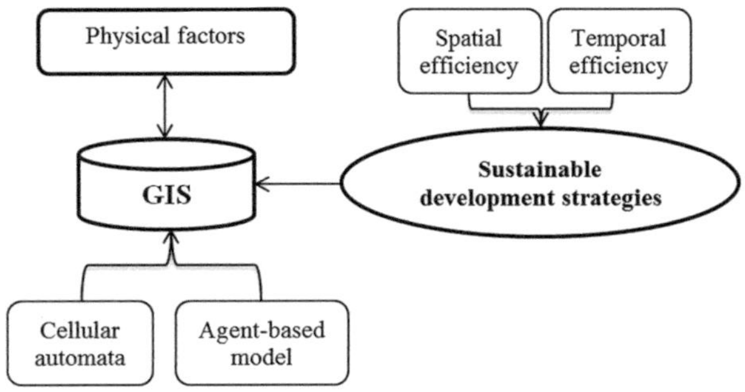

Fig. 2.5 Multi-disciplinary approaches for residential development based on sustainable development strategies (Li and Liu 2008)

perspectives of such a complex concept, is more rational. In addition, sustainable urban development and planning requires the analysis of extensive geospatial data to explore, design, modify, illustrate, and evaluate the proposed alternative scenarios (Henton and Studwell 2000). For instance, Li and Liu (2008) embedded sustainable development strategies with two other land use change modeling approaches to simulate planning options related to residential development (Fig. 2.5). Cellular automata (CA) and agent-based model were the two techniques used as spatial exploratory tools for generating alternative development patterns within a geographical information system (GIS). Sustainable development strategies were applied to regulate the model behaviors. The outputs will be compatible with environmental protection goals by controlling the proposed models and approaches based on sustainable urban development parameters. In most countries, the government is responsible for the sustainable use of land resources and determines the proper distribution of land requirements to various planning periods.

The idea of sustainable development has been extensively critiqued, especially the principles applied to urban areas. In fact, the idea is so general and thus impossible to contradict (Arbury 2005). Naess (2001) stated that "…a manifold range of strategies and projects are promoted with the claim that they are derived from the very concept of sustainable development. It has become practically impossible not to be a supporter of a sustainable development, so there is a clear danger that the concept will be watered out." In addition, how and when a proper sustainability of a specific project will be achieved is not clear because of the wide extent of the sustainable development concept and complexity of its principle.

Thus, a successful sustainable development can be accomplished through changes in the lifestyle of individual citizens and large-scale developments should be planned to be more environmental, economic, and socially sustainable while appealing to consumers as attractive places to live in. Among the various urban development forms, compact development provide more sustainable environment with respect to urban sprawl development because of its characteristics (Burton 2000; Burton et al. 2003; Arbury 2005; Abdullahi et al. 2015a).

2.3 Urban Growth and Urban Sustainability: Malaysian Perspectives

The rapid urbanization process in Malaysia has increased the concern on urban sustainable strategies considering quantitative emission, carbon footprint measurements, preservation of natural environments, and so on. After her independence in 1957, Malaysia rapidly grew with vast residential and township developments in the 1970s and 1980s (Ho et al. 2013). Consequently, huge amounts of natural spaces, especially agricultural lands at the peripheral of the cities, were converted to build up areas to accommodate the new urban populations. Furthermore, regional development authorities (RDAs) were established to implement urbanization strategies for less-developed states and expand the urban areas significantly. Several suburban towns, such as Bandar Tun Razak and Bandar Penawar, were developed to serve newly developed frontier regions. In the early 1990s, the concept of sustainable urban development was included in the national development plan of Malaysia; unfortunately, this move remained piecemeal and was only expressed quantitatively (Ho et al. 2013). More functional and applicable strategies are required to achieve a real sustainable development environment.

Fortunately, National Physical Plan 2025 and National Urbanization Plan 2006 (JPBD 2006) present strategic spatial policies on urban physical growth and land conservation (Ho et al. 2013). These plans aim to create more livable and sustainable Malaysian cities. These national planning frame works have eight objectives (JPBD 2010);

- Shaping the national spatial framework,
- Improvement of national economic competitiveness,
- Modernization of agricultural sector,
- Strengthening of tourism development,
- Management of human settlement,
- Conservation of wildlife and natural resources,
- Integration of all national transportation networks, and
- Installation of appropriate infrastructure.

The Malaysian government has always supported sustainable development objectives regarding environmental, economic, and social sustainability in all of their development projects (MGTC 2010). The aim of Malaysia to reach the developed nation status is included in her "vision 2020." The main objective of this vision, which was established in 1991 (during the Sixth Malaysian Plan), is for Malaysia to be a self-dependent industrialized nation by the year 2020 (PNMB 2010). This vision does not only consider the economic perspective, but also considers education, technological development, social satisfaction, and political goals. Urban planning strategies in the mid-1990s emphasized on mega projects and major township developments, such as Putrajaya and Cyberjaya, to create suitable economic and technological developments. All these developments were aimed to achieve urban sustainability without targeting some important issues, such as reduction in carbon emission (Ho et al. 2013). Thus, Malaysia included another objective in vision 2020, that is, the reduction of emission intensity of GDP by 40% from the 2005 level. This objective requires the support of government agencies, especially regional and local planning authorities. Table 2.3 illustrates the evolution of sustainable development and urban population from 1960 to 2020.

Malaysia started to implement this objective by creating roadmap projects and designing two new cities (Putrajaya and Cyberjaya) under the Green Technology and Water of the Ministry of Energy. These two cities were developed based on the urban sustainability principle (Green technology) as an example for other development projects. Green technology should be a motivation for further improvement in the national economy, reduction in environmental destruction and GHG emissions, and increase in the usage of renewable energies, and to promote overall sustainable urban development.

Other Malaysian strategy to achieve urban sustainability is to improve the environmental condition through the reduction of air and water pollution, solid waste management, and increase energy efficiency and promote renewable energies and environmental friendly technologies (EPU 2006). One way to achieve most of these sustainable development objectives and strategies, especially in the case of urban environments, is to develop the urban areas in a more compact manner. Recently, experts in The World Bank have also called on Malaysia to develop compact cities as part of its development efforts (MEM 2011). However, compact urban development is not an automatic task. It requires planning the urban environment to be of high density with proper distribution of facilities and a good public transportation system. Hence, various compact development plans are required to adopt the recommendations of the World Bank.

Historically, urban planning in Malaysia was started in 1929 by Charles Reade, who was responsible for improving the development of Kuala Lumpur (Samat 2006). Previously, "blue print" was the main approach in the preparation and monitoring of urban growth and development (Selamat et al. 2012). This approach was able to investigate the development problems, create development plans, and evaluate previous trends of growth and changes (Samat 2006). However, this approach had several drawbacks, such as difficulties in monitoring of uncontrolled urban growth, complexity, and time-consuming process (Yusoff et al. 2010), hence the emergence of new technologies, such as Geographical Information System (GIS), remote sensing, and several statistical and cellular bases, that addressed these problems using spatial and attribute data processing and analysis.

Currently, the Town and Country Planning Department (DTCP) is responsible for urban planning development and monitoring based on three levels of organization (Selamat et al. 2012):

Table 2.3 Evolution of sustainable development and urban population from 1960 to 2020

Vision	Colonial period: British colonial office	Post-independence: old economy policy	New economic policy (OPP1)	Vision 2020		
				National development policy (OPP2)	National vision policy (OPP3)	New economic model
Era		Natural resource and agricultural	Industrial	Information and communication technology + globalization		
Human settlement	– Traditional villages – Traditional towns and colonial towns	– New villages and estates – FELDA settlements – 1st Satellite town: Petaling Jaya	– RDAs settlements – Villages in urban area – 1st Satellite town: Petaling Jaya – Emerging new township	– Megaproject of multimedia super corridor, Cyberjaya, Putrajaya – The new mark on sustainable townships, housing estate, homes		
Urban–rural population	– Urban population increased from 27 to 72% – Rural population reduced from 73 to 28%					

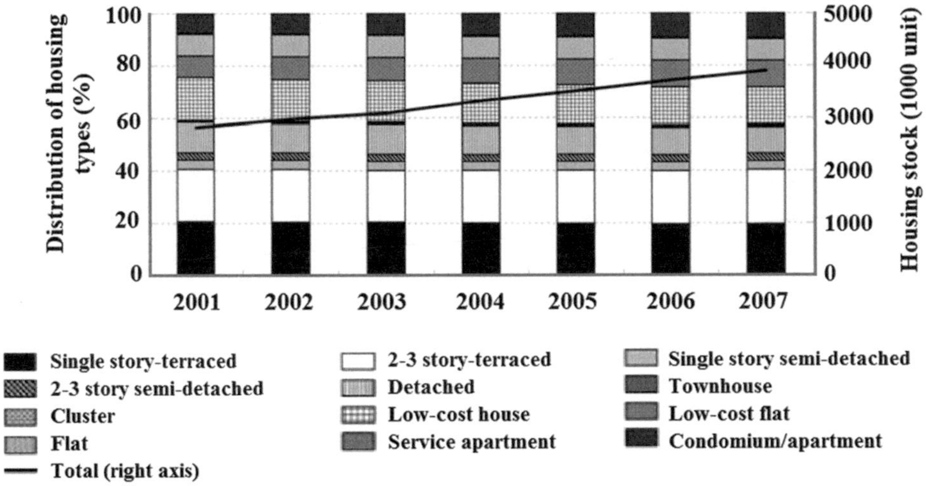

Fig. 2.6 Housing stock in Malaysia categorized by housing type (Fujita et al. 2009)

(a) the federal role is to advise the Federal Government on the issue of land development;
(b) the state as a state adviser on land planning; and
(c) the local level governs the use of land and buildings.

Providing a sustainable and livable environment is the main concern of the Malaysian government to ensure that the people's quality of life is protected. Thus, the planning and development of neighborhoods in such a manner is attempted to have potential to serve the community with proper social interactions and participations. The lack of these interactions would lead to some urban problems, such as threat to the safety and sense of security of residents. Unfortunately, Malaysian cities are facing a decline in quality of living in terms of safety (Nurul 2015). Proper social interaction increases an individual's well-being and reduces feeling of fear in the neighborhood (JPBD 2006). Nevertheless, governments recognize that urban form and pattern have significant influences on social sustainability in terms of accessibility, social interactions, quality of life, and satisfaction. Thus, as documented in the 10th Malaysian Plan, the government is committed to improve the overall quality of life. The Malaysian government has emphasized the need to ensure that urban areas are moving progressively toward building a vibrant and attractive living environment (Nurul 2015). This objective involves the improvement of the features of public transportation facilities, such as accessibility, security, and convenience. It also involves the important role of physical pattern and living environment characteristics in ensuring that the high quality of life of residences is protected. A study in Malaysia indicated that housing environment satisfaction is an important indicator of housing quality and condition, which affects the quality of life of individuals. It determines the way they respond to their residential neighborhood and environment (Salleh 2008).

In addition to the evaluation and analysis of housing affordability as a social aspect of urban sustainability, addressing the environmental aspect of residential growth and construction is essential. Figure 2.6 shows that Malaysia had a growth of 38% in unit numbers of housing stocks from 2001 to 2007 (Fujita et al. 2009).

In addition, the trend of utilizing concrete as a housing construction material has grown significantly, which has led to the rapid increase in CO_2 emissions. As mentioned in the environmental aspect of urban sustainability, different construction materials emit different amounts of CO_2. For instance, timber housing emits less CO_2 than reinforced-concrete housing because of the utilization of calcium carbonate as a raw material for cement production (Fujita et al. 2009). In this regard, the consumption of timber (by consideration of sustainable usage and under forest conservation guidelines) in Malaysia, which is a country of rich forest resource, for construction materials instead of concrete can reduce CO_2 emissions. Fujita et al. (2009) proposed an assessment approach for CO_2 emissions for housing construction based on material usage in Malaysia. This approach, as shown in Fig. 2.7, was implemented using life cycle analysis (LCA) data, residential interior designs, and estimated costs of types of structure materials for common residential buildings in Malaysia, such as terrace, semi-detached, low-cost, and detached houses. The results indicated that the CO_2 emissions of timber housing are about 30% lower than that of concrete housing (Fig. 2.8). In addition, housing of low building density, such as terrace housing, emits lower CO_2 because of lower structural weights and consequently lower timber construction material is required. Thus, changing the policy in building construction particularly related to material utilization will clearly reduce carbon emissions and help achieve environmental urban sustainable development.

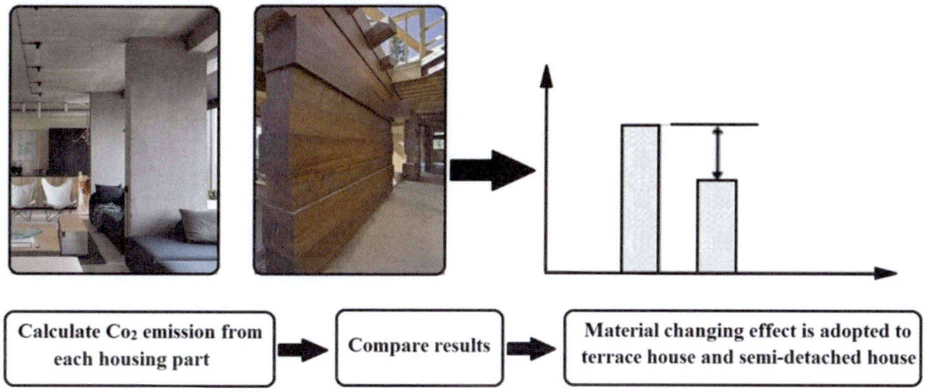

Fig. 2.7 The concept of methodological flowchart for approach using changed-material (concrete and timber wall in house construction) effect (Fujita et al. 2009)

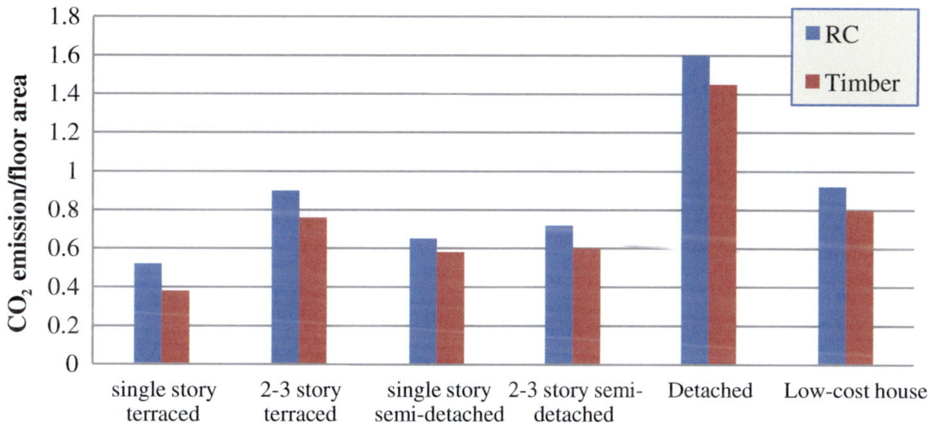

Fig. 2.8 Carbon dioxide emission for RC and timber construction materials (kt-CO_2/m^2)

Malaysia also signed the Kyoto Protocol of the United Nations Framework Convention on Climate Change on March 12, 1999, which was further ratified on September 4, 2002. However, this agreement does not ensure that all the countries involved will reduce their emissions significantly (Ho and Kean 2007). Table 2.4 shows that Malaysia, as a newly developed country, has a higher amount of emission than the world average, which is about 3.8 metric tons per capita.

The Malaysian government has been constantly insisting on environmental friendly development projects. The government contribution to these environmental conservation perspectives was enhanced, especially after the 8th Malaysian plan (2001–2005) (EPU 2001). Thus, Malaysia ranks 38th among the 146 countries in the world, with an environmental sustainability index (ESI) of 54, because of endeavors in various aspects to achieve sustainable development (Ho and Kean 2007). Malaysia and most other countries believe that economic growth should be achieved with the consideration of environmental issues. The Malaysian government has concentrated on promoting environmental quality in various aspects of air, water quality, and solid waste management and the usage of cleaner energies and technologies (EPU 2006). Particularly, the government focuses on global warming and climate change and implements various actions to decrease CO_2 emissions and promote energy efficiency. The promotion of energy efficiency and increase in consumption of renewable energy are also mentioned in the 9th Malaysian Plan (EPU 2006) as environment-friendly actions. The aim of the plan was to lead the development of the country based on sustainable development strategies to facilitate and manage natural environment resources. A fundamental action to achieve this objective was to set up a new Ministry of Natural Resources and Environment to organize and manage 10 environmental and natural resources agencies.

One of the main actions in terms of energy efficiency and sustainable energy consumption is to reduce petroleum products and replace the current fuels with renewable energies. In addition, the government aims to ensure a secure, reasonable cost, and effective supply of energy by focusing on various energy sectors to promote competitiveness and reliability of the economy. The highest amount of energy in Malaysia, like in most of the other countries, is mainly consumed by transportation followed by industrial, commercial, and residential purposes (Table 2.5). In the 9th Malaysia Plan, the amount of national average energy consumption was projected to increase by about 2217.9 PJ. In

Table 2.4 The list of the countries and the amount of CO_2 emissions in 2002

Country	CO_2 emissions (tons/capita)	Country	CO_2 emissions (tons/capita)
United State	19.9	Thailand	3.5
Saudi Arabia	18.1	Gabon	2.8
Australia	18.0	Egypt	2.2
Canada	14.2	China	2.2
Czech Republic	11.6	Brazil	1.9
Norway	11.2	Uruguay	1.7
Russia	9.9	Indonesia	1.5
UK	9.8	India	1.1
Germany	9.6	Philippines	1.0
Japan	9.5	Guatemala	0.9
South Africa	7.5	Pakistan	0.9
Ukraine	7.0	Yemen	0.6
Malaysia	**6.2**	Togo	0.5
France	6.1	Nigeria	0.4
Sweden	5.3	Bangladesh	0.3
Iran	4.9	Ethiopia	0.1
Mexico	4.5	Mozambique	0.1
Argentina	3.9	Uganda	0.1
Turkey	3.5	Mali	0.1

Source UNEP/GRID-Arendal (2007)

Table 2.5 Energy consumption of various sectors in Malaysia, 2000–2010 (EPU 2006)

Sources	Peta Joules (PJ)			Percentage of the total			
	2000	2005	2010	2000	2005	2010	Growth rate (% p.a.)
Industrial (include manufacturing, mining, and construction)	477.6	630.7	859.9	38.4	38.6	38.8	6.4
Transportation	505.5	661.3	911.7	40.6	40.5	41.1	6.6
Housing/commerce	162.0	213	284.9	13.0	13.1	12.8	6.0
Non-energy (include natural gas, bitumen, asphalt, industrial feedstock, and grease)	94.2	118.7	144.7	7.6	7.3	6.5	4.0
Agriculture/forestry	4.4	8.0	16.7	0.4	0.5	0.8	15.9
Sum	1243.7	1631.7	2217.9	100.0	100.0	100.0	6.3

addition, the amount of per capita of energy consumption was projected to grow from year 2000 to 2010 (EPU 2006). Several essential controls and tasks have been implemented by the government, such as Small Renewal Energy Power Program (SREP) and Malaysia Building Integrated Photovoltaic Technology Application Project (MBIPV), to address the sustainable energy consumption through the use of renewable energy resources (Ho and Kean 2007).

Malaysia and other countries should continue their efforts to reduce carbon dioxide emissions because these emissions and all kinds of pollutions will inevitably increase because economic and population growth, as shown in Fig. 2.9.

In continuing the efforts by promoting sustainable development policies, the 10th Malaysia plan (2011–2015) emphasized on environmental protection and conservation through the National Green technology Policy 2009 and National Climate Change Policy 2009 (Ho et al. 2013). The National Green technology Policy focuses on promoting the utilization of green technologies and the establishment of Green Technology Financing Scheme (GTFS). Meanwhile,

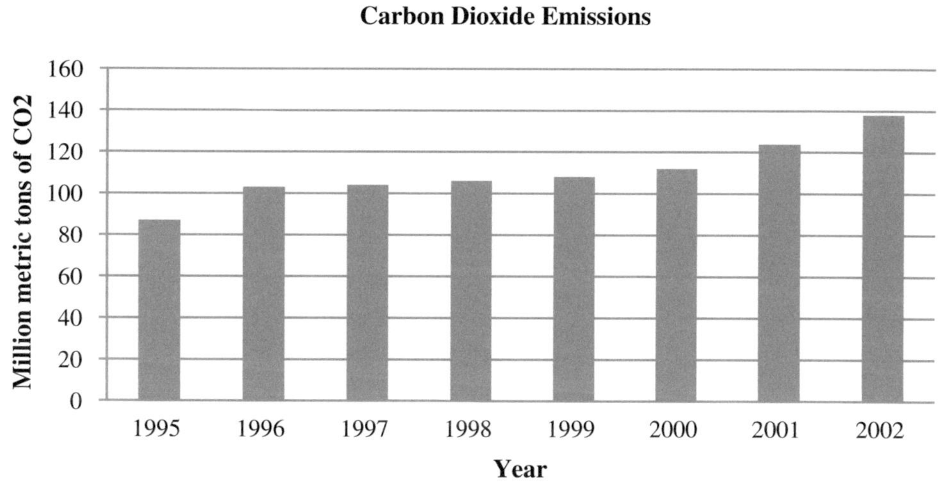

Fig. 2.9 Carbon dioxide emission in Malaysia, 1990–2002. *Source* Energy Information Administration (2007)

the National Climate Change Policy deals with the planning and implementation of the low-carbon economy principle (EPU 2010). Some of the actions to achieve this principle are as follows (Ho et al. 2013):

- Creating incentives for investments in renewable energy,
- Promoting energy efficiency to encourage productive use of energy,
- Improving solid waste management,
- Conserving forests, and
- Reducing emissions to improve air quality.

2.3.1 Kuala Lumpur Sustainable Development Planning

For example, Kuala Lumpur, the capital city of Malaysia, has a vision to be a world class city that promotes various aspects of urban sustainability, economic justice, a just and functional government, distribution of community facilities, and acceptable quality of life. The Kuala Lumpur Structure Plan 2020 (KLSP 2020) highlighted that the aim and objective of Kuala Lumpur involve implementing a sustainable city to ensure the planning and development of this city will maintain a balance among the physical, economic, social, and environmental aspects.

Some of the main policies and strategies to achieve sustainable development that is holistic embrace the universal principles of Islam Hadhari that are listed below:

- Faith and piety toward God;
- A fair and trustworthy government;
- Free and liberated people;
- A rigorous pursuit and mastery of knowledge;
- Balance and comprehensive economic development;
- Acceptable quality of life for local residents;
- Protection of the rights of minority groups and women;
- Cultural and moral integrity;
- Preservation of the natural and green environment; and
- Strong military powers.

Therefore, the foundation for the world class Kuala Lumpur is based on the commitment toward a holistic planning and development, and this city committed itself to sustainability as its main planning objective. In this regard, the National Physical Plan (NPP), a national level plan for up to year 2020, specifies the policies of physical development and preservation environment within Peninsular Malaysia. The main goal of the NPP is to create an efficient, equitable, and sustainable national spatial framework to guide the overall development of the country toward achieving a competitive developed nation status by the year 2020. The objectives of this plan are

- To rationalize national spatial planning for economic efficiency and global competitiveness;
- To optimize utilization of land and natural resources for sustainable development;
- To promote balanced regional development for national unity; and
- To secure spatial and environmental quality and diversity for a high quality of life (Fig. 2.10).

The Draft KL City Plan 2020 emphasizes on livability and quality of life for the people of Malaysia. It will secure the protection of residential neighborhoods to decrease the disparity problems of the urban poor. In addition, the city plan emphasizes on public transportation and quality of services to improve and increase investments and development. The environmental aspects of sustainable urban development are considered substantial paradigms for

Fig. 2.10 Aerial image of Kuala Lumpur city

growth and development in the proposed Draft. Environmental protection zones were determined and proper rules related to the conservation of these sites were proposed. The main objectives of the 9th Malaysian plan are

- To move the economy up the value chain,
- To raise the capacity for knowledge and innovation and nurture the "first class mentality,"
- To address persistent socioeconomic inequalities constructively and productively,
- To improve the standard and sustainability of quality of life,
- To strengthen the institutional and implementation capacity.

In addition, the National Urbanization Policy (NUP) is a fundamental framework for the Draft KL City Plan 2020. This policy calls for the creation of visionary cities, which promote livable communities and sustainable urban development. In terms of regional sustainability (which is an accepted agenda for KL and other neighboring planning organizations), development management based on a consultative approach is the main urban development activity. This involves issues relevant to road networks, public transportation networks, and solid waste management. These common agenda and related efforts emphasize on improving spatial and environmental quality and diversity. These efforts and the cooperation of KL and surrounding municipalities should be implemented progressively to solve the problems of regional sustainability. Some of these efforts are listed as follows:

- Promoting and preserving urban and green landscapes,
- Improving road and rail networks,
- Conserving forest environments,
- Controlling housing and development expansion,
- Managing solid waste and drainage issues,
- Monitoring commercial and industrial activities, and
- Controlling land use growth and changes of the city.

Therefore, the plan of Kuala Lumpur City for the future is generally concentrated on urban sustainability based on the main aspects related to environmental conservation, social equity, and stable economic growth. The consideration of these aspects provides guidelines for the management of land use growth and development and monitoring of environmental and economic activities.

2.3.2 Putrajaya City as a "City in a Garden"

Putrajaya is a city designed to serve the federal administrative center of Malaysia and located 25 km away from Kuala Lumpur (KL). This city is designed as an example of the future Malaysian sustainable urban development. The master plan of this city was established in October 1995, and the government units relocated from KL to Putrajaya in 1999 (Kang 2012). The construction project of this city was accomplished in 2012. However, the concept of changing the Federal Government's administrative function from Kuala Lumpur was proposed about 20 years ago. In the early 1970s, the capital city of Malaysia was under a struggle because of traffic congestion, weak infrastructure and utilities, and the numerous occurrences of flash floods along with the rapid growth in population and urban economy. Thus, the aim was to develop a "City in a Garden" and an "Intelligent City," which are intended to cater to the current and future projected population (Ariffini 2003) (Fig. 2.11).

This city is a great model of planned administrative city that illustrates the struggle of many former colonies to forge a distinct national identity that reflects both the values and aspirations of the new nation and differentiates itself from its colonial past (Moser 2010).

While Putrajaya City was developed as an act of promoting sustainable urban development principles, significant effort has not been exerted to reduce energy consumption and carbon dioxide emissions, except in the case of natural spaces surrounding the lake (Kang 2012). Malaysia has a daytime temperature of more than 30 °C and high level of humidity. Nevertheless, most of the buildings in the city are constructed and covered by steel and glass, which are aesthetically nice, but allows sunlight to enter the buildings, necessitating strong air conditioning to reduce the temperature inside (Moser 2010). Although higher density is recognized as one of the main characteristics of urban sustainability (to encourage walking and cycling), this city now has a low density with long distances (for walking mode) between daily destinations. The lack of sunshade on pedestrian roads discourages traveling by walking and cycling (particularly because of sunshine and heavy rain), which forces local residents to commute using their private vehicles (Moser 2010). Qureshi and Ho (2011) estimated that the CO^2 emission of Putrajaya will decrease by 2.4% if walking and cycling become the main transportation mode.

Another problem of this city is traffic and the lack of car parking places. The delay in the development of the rail transit system for public transportation increased the private car dependency, which accounts for more than 80% of the commuting population (Nor et al. 2006). Presently, the public transport of this city consists of three bus companies for inter-city public transport, which transfer commuters from KL to other city centers in the neighboring areas, and one other private bus service (Kang 2012). In addition to these buses, train systems also link KL, Putrajaya, and the international airport of Malaysia. However, the current transportation system does not fulfill the travel demand of the city (Nor et al. 2006). A huge volume of commuters to the central part of the city is expected from various parts of this city and other surrounding regions because most of the government offices, community facilities, and services are located in this area. However, not enough buses exist to transport local residences from residential neighborhoods to these city points of interest (Kang 2012) (Fig. 2.12).

Kang (2012) estimated the annual income of bus transportation systems in Putrajaya to be less than 0.6 million USD, causing them to suffer a loss of 5.80 million USD each year. Thus, improving public transportation systems and imposing penalties on commuting by private car to promote

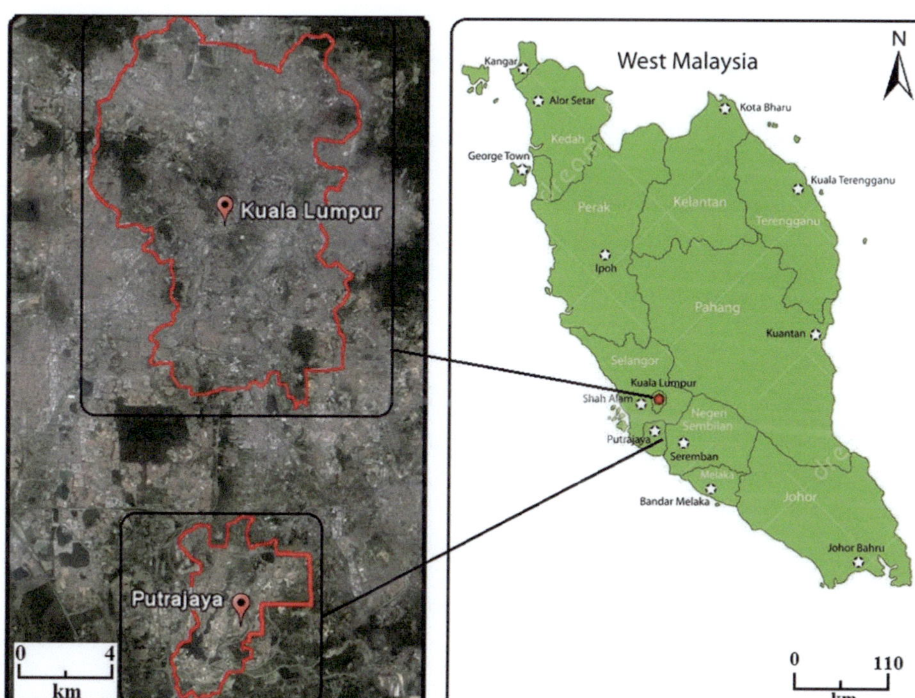

Fig. 2.11 The location of Putrajaya city

Fig. 2.12 Formalized structure of Putrajaya city

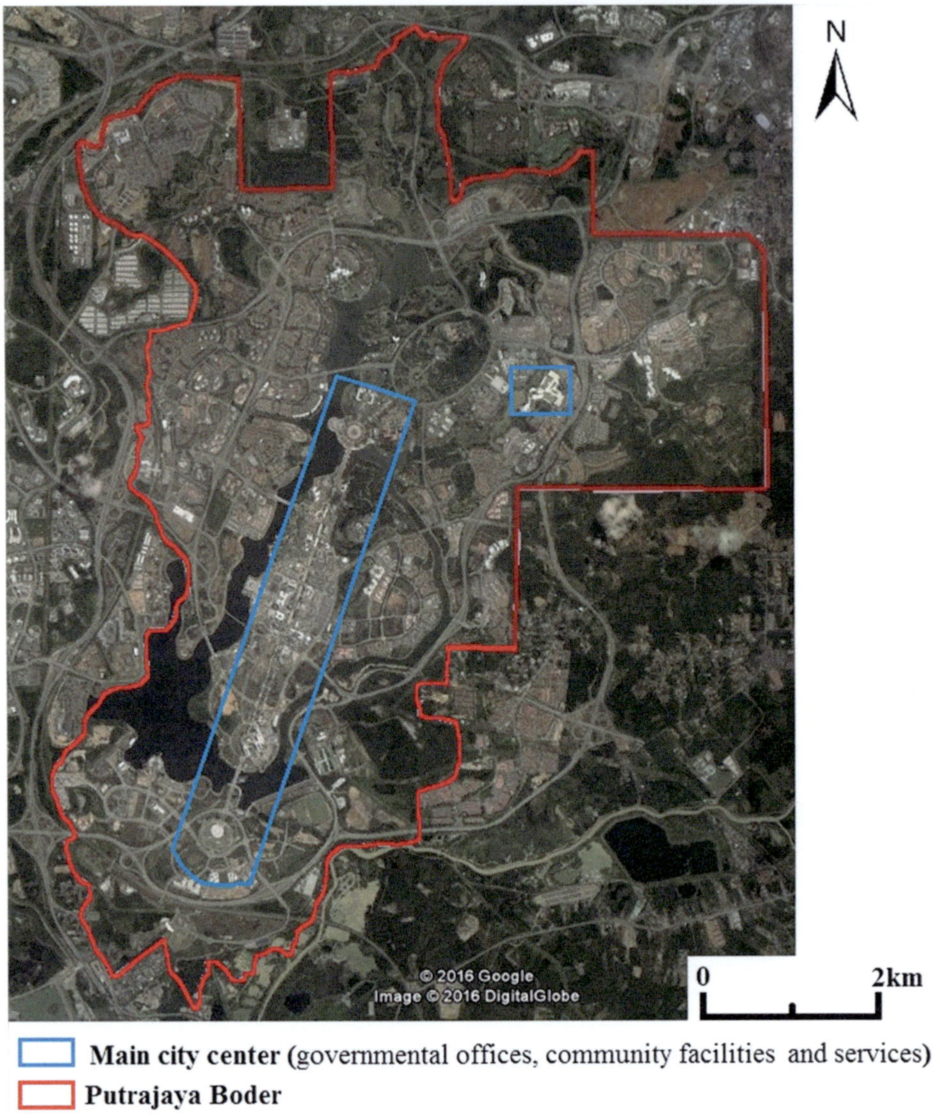

public transportation to 70% of the city transport, as set in the new plan for 2025. In spite of some deficiencies, the development effort on an administrative sustainable city with various environmental friendly policies and planning is an admirable project and should also be implemented in other high-density capital cities.

References

Abdullahi, S., & Pradhan, B. (2015). Sustainable brownfields land use change modeling using GIS-Based weights-of-evidence approach. *Applied Spatial Analysis and Policy*, 1–18.

Abdullahi, S., Pradhan, B., & Jebur, M. N. (2015a). GIS-based sustainable city compactness assessment using integration of MCDM, Bayes theorem and RADAR technology. *Geocarto International, 30*(4), 365–387.

Abdullahi, S., Pradhan, B., Mansor, S., & Shariff, A. R. M. (2015b). GIS-based modeling for the spatial measurement and evaluation of mixed land use development for a compact city. *GI Science and Remote Sensing, 52*(1), 18–39.

Arbury, J. (2005). *From urban sprawl to compact city: An analysis of urban growth management in Auckland.* Geography and Environmental Science–University of Auckland.

Ariffini, S. B. J. (2003). Putrajaya, Malaysia. *Australian Planner, 40*(3), 40–42.

Barbier, E. B. (1987). The concept of sustainable economic development. *Environmental Conservation, 14*(02), 101–110.

Barton, H. (2000). *Sustainable communities: The potential for eco-neighbourhoods*. Earthscan.

Barton, H., & Tsourou, C. (2000). *Health urban planning: A WHO guide to planning for people*. London/Copenhagen: World Health Organization Regional Office for Europe by Spon.

Bhattacharya, A., Oppenheim, J., & Stern, N. (2015). Driving sustainable development through better infrastructure: Key elements of a transformation program. *Global economy and development working paper*, 91.

Bleicher, A., & Gross, M. (2010). Sustainability assessment and the revitalization of contaminated sites: Operationalizing sustainable development for local problems. *International Journal of Sustainable Development and World Ecology, 17*(1), 57–66.

Bramley, G., & Kirk, K. (2005). Does planning make a difference to urban form? Recent evidence from Central Scotland. *Environment and Planning A, 37*(2), 355–378.

Burton, E. (2000). The compact city: Just or just compact? A preliminary analysis. *Urban Studies, 37*(11), 1969–2006.

Burton, E., Jenks, M., & Williams, K. (2003). *The compact city: A sustainable urban form?* Routledge.

Butterworth, I. (2000). The relationship between the built environment and wellbeing: A literature review. *Prepared for the Victorian Health Promotion Foundation.*

Chiu, R. L. (2008). Shanghai's rapid urbanization: How sustainable? *Built Environment, 34*(4), 532–546.

Chiu, R. L. (2012). Urban sustainability and the urban forms of China's leading mega cities: Beijing, Shanghai and Guangzhou. *Urban Policy and Research, 30*(4), 359–383.

Elkin, T., McLaren, D., & Hillman, M. (1991). *Reviving the city: Towards sustainable urban development*. Friends of the Earth London.

EPA, U. (2006). This is smart growth. Office of smart growth. US environmental protection agency. URL: http://www.smartgrowth.org/pdf/this_is_smart_growth.pdf. Accessed October 1, 2006.

EPU. (2001). *Eighth Malaysia plan 2001–2005*. Putrajaya: Economic Planning Unit.

EPU. (2006). *Ninth Malaysian plan 2006–2010*. Putrajaya: Economic Planning Unit.

EPU. (2010). *Tenth Malaysian plan 2011–2015*. Putrajaya: Economic Planning Unit.

Figueira, J., Greco, S., & Ehrgott, M. (2005). *Multiple criteria decision analysis: State of the art surveys* (Vol. 78). Springer Science and Business Media.

Franz, M., & Nathanail, C. P. (2005). *A sustainability assessment framework for brownfield regeneration* (na).

Fujita, Y., Matsumoto, H., & Siong, H. C. (2009). Assessment of CO_2 emissions and resource sustainability for housing construction in Malaysia. *International Journal of Low-Carbon Technologies, 4*(1), 16–26.

Gerilla, G., Teknomo, K., & Hokao, K. (2007). An environmental assessment of wood and steel reinforced concrete housing construction. *Building and Environment, 42*(7), 2778–2784.

Grossman, G., & Krueger, A. (1995). Economic environment and the economic growth. *Quarterly Journal of Economics, 110*(2), 353–377.

Haughton, G., & Hunter, C. (2004). *Sustainable cities*. Routledge.

Henton, D., & Studwell, K. (2000). *Informed regional choices: How California's regional organizations are applying planning and decision tools*. California Center for Regional Leadership.

Ho, C. S., & Kean, F. W. (2007). Planning for low carbon cities-the case of Iskandar development region, Malaysia. In *Toward establishing sustainable planning and governance II*. Seoul, Korea: Sustainable Urban Development Institute.

Ho, C. S., Matsuoka, Y., Simson, J., & Gomi, K. (2013). Low carbon urban development strategy in Malaysia-the case of Iskandar Malaysia development corridor. *Habitat International, 37,* 43–51.

JPBD. (2006). *National urbanization plan*. Kuala Lumpur: Federal Department of Town and Country Planning (JPBD).

JPBD. (2010). *National urbanization plan*. Kuala Lumpur: Federal Department of Town and Country Planning (JPBD).

Kang, J. (2012). *A Study on the future sustainability of Sejong, South Korea's multifunctional administrative city, focusing on implementation of transit oriented development* (p. 93). Sejong City, South Korea: Master, Uppsala University.

Kopfmüller, J., Brandl, V., Jörissen, J., Paetau, M., Banse, G., Coenen, R., et al. (2001). *Nachhaltige Entwicklung integrativ betrachtet: Konstitutive Elemente, Regeln, Indikatoren*: Ed. Sigma Berlin.

Kropp, W. W., & Lein, J. K. (2013). Scenario analysis for urban sustainability assessment: A spatial multicriteria decision-analysis approach. *Environmental Practice, 15*(02), 133–146.

Krueger, R., & Buckingham, S. (2012). Towards a 'consensual' urban politics? Creative planning, urban sustainability and regional development. *International Journal of Urban and Regional Research, 36*(3), 486–503.

Li, X., & Liu, X. (2008). Embedding sustainable development strategies in agent-based models for use as a planning tool. *International Journal of Geographical Information Science, 22*(1), 21–45.

Li, X., & Yeh, A. G.-O. (2000). Modelling sustainable urban development by the integration of constrained cellular automata and GIS. *International Journal of Geographical Information Science, 14*(2), 131–152.

Lin, J., & Yang, A. (2006). Does the compact-city paradigm foster sustainability? An empirical study in Taiwan. *Environment and Planning B: Planning and Design, 33*(3), 365.

Littig, B., & Grießler, E. (2005). Social sustainability: A catchword between political pragmatism and social theory. *International Journal of Sustainable Development, 8*(1–2), 65–79.

Meadows, D. H., Meadows, D. L., Randers, J., & Behrens, W. W. (1972). The limits to growth. *New York, 102*.

Mellino, S., & Ulgiati, S. (2015). Monitoring regional land use and land cover changes in support of an environmentally sound resource management. In *Sustainable development, knowledge society and smart future manufacturing technologies* (pp. 309–321). Springer.

MEM. (2011). *Smart cities*. Kuala Lumpur: Malaysian Economic Monitor.

MGTC. (2010). *National energy balance table 2009*. Malaysia (KeTTHA), Putrajaya: Ministry of Energy, Green Technology and Water.

Moser, S. (2010). Putrajaya: Malaysia's new federal administrative capital. *Cities, 27*(4), 285–297.

Naess, P. (2001). Urban planning and sustainable development. *European Planning Studies, 9*(4), 503–524.

Newman, P., & Kenworthy, J. (1999). *Sustainability and cities: Overcoming automobile dependence*. Island Press.

Nor, M., Ghani, N., Rahim, A., & Abdullah, A. Z. (2006). Predicting the impact of demand-and supply-side measures on bus ridership in Putrajaya, Malaysia. *Journal of Public Transportation, 9*(5), 4.

Nurul, W. M. R. W. (2015). Compact urban form for sociability in urban neighbourhoods. *International Journal of Social Science and Humanity, 5*(10), 822.

Olsson, J. (2009). Sustainable development from below: Institutionalising a global idea-complex. *Local Environment, 14*(2), 127–138.

PNMB. (2010). *New economic model for Malaysia*. Putrajaya: National Economic Advisory Council, Pencetak Nasional Malaysia Bhd. (PNMB).

Porta, S. (2001). Formal indicators: Quantifying the contribution of form to urban (social) sustainability. *Australia: Walking the 21st century*, pp. 67–79.

Qureshi, S., & Ho, C. (2011). Towards Putrajaya green city 2025 implementing neighbourhood walkability in Putrajaya. Retrieved April 5, 2012.

Salleh, A. G. (2008). Neighbourhood factors in private low-cost housing in Malaysia. *Habitat International, 32*(4), 485–493.

Samat, N. (2006). Applications of geographic information systems in urban land use planning in Malaysia.

Selamat, M. H., Selamat, A., Othman, M. S., Shamsuddin, N., & Zukepli, N. (2012). A review on geographical information system (GIS) in town planning: Malaysia experience. *Geoinformatica: An International Journal (GIIJ), 2*(2), 27–38.

Shi, A. (2001). *Population growth and global carbon dioxide emissions.* Paper presented at the IUSSP Conference in Brazil/session-s09.

Subeh, M. A., & Al-Rawashdeh, T. (2012). Urban sustainability, globalization and expansion organization in middle east. *Journal of Economics and Sustainable Development, 3*(14).

UNPD, U. N. P. D. (2012). *World urbanisation prospects: The 1999 revision, data tables and highlights (ESA/P/WP.161; 2000)*. New York: United Nations Population Division.

Van Den Brink, T. W., & van Der Woerd, F. (2004). Industry specific sustainability benchmarks: An ECSF pilot bridging corporate sustainability with social responsible investments. *Journal of Business Ethics, 55*(2), 187–203.

Van Wee, B., & Handy, S. (2016). Key research themes on urban space, scale, and sustainable urban mobility. *International Journal of Sustainable Transportation, 10*(1), 18–24.

WCED. (1987). *Our common future. The world commission on environment and development*. Oxford: Oxford University Press.

Xie, C., Huang, B., Claramunt, C., & Chandramouli, C. (2005). *Spatial logistic regression and GIS to model rural-urban land conversion.* Paper presented at the proceedings of PROCESSUS second international colloquium on the behavioural foundations of integrated land-use and transportation models: Frameworks, models and applications.

Yusoff, Z. M., Maassoumeh, B., Omar, D., Latif, Z. A., & Samad, A. M. (2010). *Applicability of geographical information system in assessing the accessibility and mobility of urban lower-income family living.* Paper presented at the signal processing and its applications (CSPA), 2010 6th international colloquium on.

Sprawl Versus Compact Development

3

Saleh Abdullahi, Biswajeet Pradhan and Abubakr A.A. Al-sharif

3.1 Introduction

Among all urban growth patterns, compact and sprawl developments are the most common patterns (Fig. 3.1). However, continuing debates on the advantages and disadvantages of these two urban forms exist. Compact city is a high-density built-up area with proximity among various land use types (Schwarz 2010), whereas urban sprawl is an inefficient urbanization with low density and higher car dependency, thus increasing air and water pollutions and ecological disturbance (Torrens and Alberti 2000). This chapter deals with these two forms of urban growth. First, the origin and various positive and negative aspects of sprawl development will be explained, and then compact development will be discussed in detail as an alternative solution to avoid the negative social, environmental, and economic consequences. Finally, compact development will be evaluated with respect to various aspects of sustainable urban development.

3.2 Sprawl Urban Growth

Historically, concentrated settlements were created by the agglomeration of industrial buildings surrounded by residential areas. After the Second World War, this traditional concentrated urban shape was replaced by dispersed and decentralized development in suburban areas because of several factors. This decentralization affected the job opportunities, distribution of facilities and services, and residential development (Garreau 1991). Hence, primitive human settlements were changed from monocentric to polycentric and dispersed urban patterns. Although Lewis Mumford (1961) believed that "the suburb becomes visible almost as early as the city itself," the current suburban development is significantly different from the limited one that existed in the nineteenth century. Currently, this kind of urban pattern is known as sprawl development, in which the built-up areas belong to each urban land use and are separated by open spaces, such as natural and abandoned fields.

Many factors affect the growth of urban areas, especially in low-density and horizontal expansion forms. Industrial revolution, technological development of transportation, either mass public transportation by trains and trams or automobile (private vehicles and buses), the zoning of the land use activities and categories, and the growth of middle class affluent were the main factors that affected the growth of urban environments. Gillham (2002) stated that the current suburban development is the consequence of industrial revolution and the advancement of transportation modes and communication in the nineteenth century. In addition to these advancements, various planning and development policies proposed by local authorities contributed to the growth of suburban development (Duany et al. 2001). For instance, the Federal Housing Administration and Veterans Administration loan program in the US provided mortgages for more than 11 million new residential units (Arbury 2005). Such planning policies encourage the development of new residential areas instead of the redevelopment or renovation of existing residential buildings. Furthermore, the implementation of the zoning ordinance in urban areas caused the growth of single land use development and separation of various activities and land use categories. The spatial separation of residential and working places increased car dependency and made automobile the first transportation mode. Thus, the combination of all these factors increased the growth of sprawl development and caused a blurred distinction between urban and rural areas.

Urbanization process is the complex phenomenon of transforming rural areas into urban lands, resulting in various effects on environmental structures (Weng 2007). Rapid urbanization leads to dispersed urban development surrounding existing urban areas causing the urban sprawl phenomenon (Adolphson 2010). Exponential growth in

B. Pradhan (ed.), *Spatial Modeling and Assessment of Urban Form*,
DOI 10.1007/978-3-319-54217-1_3

Fig. 3.1 **a** General view of compact urban development and **b** sprawl urban development

urban population requires more facilities, houses, and so on, thus forcing the expansion at the boundary of existing urban areas to the rural environment. Hence, urban sprawl is supported by piecemeal extensions of essential urban infrastructures, such as sewers, roads, water, and power (Gillham 2002). Historically, the escape of the middle class affluent of the population after industrialization period from the high-density central parts, with congestion, pollution, and so on, to the low-density suburban areas with better neighborhoods was the main reason of suburban expansion. Currently, economic issues force the low-income population to acquire and construct undeveloped and cheap lands for more affordable residential buildings, especially in developing countries. Minimum regulation for construction, maximum tax incentives, low commuting costs, and the role of individual choices are some other factors that promote fringe development (Deal and Schunk 2004).

The concept of sprawl development suffers from complexity and difficulties in definition (Angel et al. 2007; Bhatta 2010). Bhatta (2010) defined urban sprawl as a situation where urban growth adversely affects an urban environment that is neither an appropriate rural area for agricultural purposes nor a suitable urban condition. Irwin and Bockstael (2007) stated that sprawl is an expression that is often used loosely to explain various concepts linked both to pattern and process and to both causes and consequences. They decomposed this expression into eight dimensions, which consists of socioeconomic factors, such as accessibility, land use diversity, and concentration of community services. In addition to these, density, continuity, clustering, centrality, and nuclearity are also included in the dimensions of sprawl development (Galster et al. 2006; Frenkel and Ashkenazi 2008).

The term "urban sprawl" can be used both as a verb (process) and as a noun (condition), but it still requires a clearer definition despite many researchers claiming to "know it when they see it" (Galster et al. 2006). However, a general agreement exists that urban sprawl is the combined effect of growing affluence, changing life style, and increase in private mobility (Dieleman and Wegener 2004). Thus, a concern toward the interaction between transportation and urban form is important to understand sprawl development (Fig. 3.2).

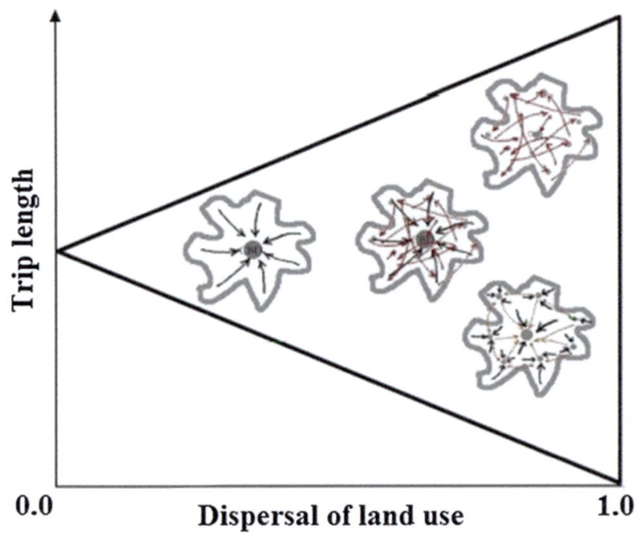

Fig. 3.2 The relationship between trip length, dispersal and urban form (Banister 2012)

After 1900, daily travel distance rose to 1 km/person/day, based on the analysis of Arnuld Grubler in France, and this value increased to 10 km/person/day in 1960 and 50 km/person/day in 2000 (Banister 2012). This analysis and other global assessments (Table 3.1) show travel distance increased significantly after 1950 because of the continuous growth in population and wealth, and globalization of the world economy (Schäfer 2009). Table 3.1 shows that car dependency is higher in industrialized countries than in

Table 3.1 Growth in global per capita travel distance (Schäfer 2009)

	1950	2005	2050	1950–2005	2005–2050
	PKT/cap	PKT/cap	PKT/cap	% Change	% Change
Industrialized economies	4530	18,400	42,200 (29,500)	+306.2	+129.3 (+60.3)
Reforming economies	947	5620	15,000 (16,300)	+493.4	+166.9 (+190.0)
Developing economies	388	3660	6800 (14,600)	+843.3	+85.8 (+298.9)
World	1420	6020	11,400 (16,400)	+323.9	+89.4 (+172.4)

Unit Passenger kilometers traveled per person per year by all modes including air. Projections for 2050 are based on economic growth rates of the MIT EPPA model reference run and those of the IPCC SRES-B1 scenario (in brackets)
Industrialized economies: North America, Pacific OECD, and Western Europe
Reforming economies: Eastern Europe and former Soviet Union

others, with some convergence in developing and reforming economies (Banister 2012).

The complexity of urban sprawl is also caused by the micro and macro perspectives of this phenomenon. At the micro level, changes in geography, climate, local public policy, and so on may all influence the expansion patterns of cities. At the macro level, urban sprawl reflects interregional migration, population growth, changes in transport systems of commuting, and increasing income among others. Unfortunately, no clear and distinguishable classification of sprawl development with significant regional and temporal variations exists (Cheng and Masser 2003). Consequently, sufficient data with detailed information on the micro spatial causes of sprawl are required to study and analyze urban sprawl accurately.

Urban sprawl development has several characteristics, as given by Gillham (2002), such as leapfrog or scattered, commercial strip, low density, and large expanses of single land use developments. Leapfrog or scattered development increases the growth of isolated built-up areas along the city borders. Similar to satellite towns, this kind of urban pattern is the most land resources-consuming form, with the highest car dependency and requirements of transportation network and other utilities. Commercial strip development is characterized by huge major roads lined with fast food restaurants, gas stations, shopping centers, drive-through banks, office complexes, and many other structures (Gillham 2002). This kind of urban pattern has low density and high car dependency, with long and low box configurations of retails surrounded by parking spaces. The development characterized by widespread, single story buildings with separate parking spaces and roadways in low density is the most obvious format of sprawl urban development. Urban density can be measured through a variety of aspects, such as population, building, residential, road, and density. Generally, a population density lower than 25 persons per hectare is assumed low, such as most of the North American and Australian cities. Meanwhile, most of the European cities with 50 persons per hectare and Asian cities often with more than 100 persons per hectare are assumed as high-density urban developments (Elkin et al. 1991; Arbury 2005). Finally, developing urban areas of single land use pattern with separation of different land use categories and urban activities is another important characteristic of sprawl development. Single land use development absolutely increases daily car dependency because of the separation of living, working, and recreational facilities locations.

Although urban sprawl development has several environmental, social, and economic effects on human life and natural environment because of its pattern and characteristics, it has some positive aspects as well. In the quality of life perspective, sprawl development provides single-family homes on large parcels with high movement freedom and green environment out of the city center with high density, traffic congestion, and high crime and poverty rates (Burchell and Mukherji 2003). A large house in a green and low-density neighborhood and multiple car ownership indicate affluence and wealth in most the cultures. Urban sprawl is the world of freedom of land ownership and wealth accumulation. In the economic profitability perspective, construction in sprawl development is a good option instead of the redevelopment of existing brownfield sites within city centers. Site clearance is the most important step in new development, which is more costly in brownfield redevelopment than in rural development. In addition, although brownfield redevelopment seems to have the advantages of infrastructure and utility accessibility, infrastructures are usually provided by local authorities. Hence, the absence of lack of infrastructures would not be a big issue in the case of rural development (Carruthers and Ulfarsson 2008). Suburban and urban fringes are more attractive for living purposes than inner cities because of such advantages. Gordon and

Richardson (1997) advocated sprawl development because of its low density and it is less crowded than the compact city center areas. Burchell and Mukherji (2003) believe that market, policy, and personal choices support sprawl development because of the relative abundance of resources and few care about the needs of the society.

The negative effects of sprawl development on human life and natural environment should not be forgotten because of these few personal benefits. Urban sprawl has become a major problem in rapidly growing and developing countries (Grant 2006; Dadi et al. 2016). Several developed countries have also experienced or are experiencing sprawl development. The negative impacts were highlighted further since the emergence of the sustainable urban development concept (Yeh and Li 2001). Therefore, this topic has gained considerable amount of attention among public and urban researchers because of various unsustainability characteristics. Public concern about this topic and its effects increased significantly after the 1990s (Bengston et al. 2005).

One of the main critical issues regarding sprawl development is the improper land resources management and energy consumption in rural developments. Particularly, one of the popular topics in the literature is the quantification of the infringement of built-up areas onto the open spaces and agricultural fields to evaluate the amount of encroachment of various development scenarios (Jaeger et al. 2010; Hayek et al. 2011). The advantage of having large parcel houses with private yards and parking spaces in a low-density neighborhood absolutely requires big lands, which means the conversion of natural environment to urban land. For instance, 24% of the counties in USA are affected by sprawl development and 55.8% of the projected houses for the period of 2000–2025 will be developed in suburban areas (Burchell and Mukherji 2003). Thus, in the recent decades, a wide expansion of sprawl development has also been observed in developed countries. These expansions destroy large areas of agricultural fields; the American Farmland Trust states that about 400,000 acres of agricultural areas converted to urban land are characterized by sprawl development (Gillham 2002). These growths through rural environments also destroy the natural habitat of many species and cause their endangerment. In addition, sprawl development entails higher car dependency because of the dispersed pattern of urban structures (buildings), and consequently, more land areas are consumed for road networks, parking spaces, and other infrastructures related to transportation (Duany et al. 2001).

The dispersion of urban structures and separation of human activities in sprawl development in addition to higher land consumption increase fuel consumption, traffic congestion, and commuting time. This also leads to the increased concern on air quality and the associated costs from human and environmental health issues (Deal and Schunk 2004). Specifically, concern on the global warming issue causes more attention to be focused on the air pollution and carbon emission from the automobile dependency of the sprawl development pattern. Although the carbon emission of current automobiles have been significantly reduced compared with older models (made before 1970) because of technological advancement, the spatial dispersion of urban patterns (which increase vehicle miles traveled—VMT) has still caused the emission of huge amounts of carbon into the air in the recent decades (Arbury 2005). Thus, considering that sprawl urban development increases VMT substantially is very important. Southworth (2001) identified three main factors of this growth in the past 25 years: (1) population growth, which mainly increases car ownership levels, (2) reduction in fuel prices, technological advancement of automobiles in terms of fuel consumption and improvement of roads and highway networks, and (3) changes in the land use pattern and distribution, which are affected by sprawl urban development characteristics: low density, single land use, and dispersed developments. The comfort of traveling by private vehicle rather than public transportation and higher priority given to living neighborhood condition rather than the proximity to the working places are other reasons of VMT growth in the recent years. The air pollution brought about by this rapid growth is significant; The US Environmental Protection Agency indicated that motor vehicles emitted over 50 million tons of carbon monoxide into the air, over seven million tons of nitrous oxides, over five million tons of volatile organic compounds, 320 tons of sulfur dioxide, and almost 15 million tons of road dust into the nation's air in 1997 (Nozzi 2003). In addition, Nozzi (2003) released some statistical information related to the negative health problems caused by these pollutions, such as 50–70 million respiratory-related restricted activity days, over 850 million headaches caused by carbon monoxide, 20,000–46,000 cases of chronic respiratory illnesses, 530 cases of cancer, and over 40,000 premature deaths. Unfortunately, pedestrians and cyclists are in danger more than the vehicle drivers themselves (Arbury 2005). In the central parts of the city, situations are even worse because of the concentration of activities, building arrangements, and higher traffic congestion. The construction and expansion of road networks to control the traffic congestion not only increase land consumption but also have no positive effect on the pollution levels, as experienced in USA (Nozzi 2003). Instead, the reduction of the number and length of automobile trips, which are mainly related to urban pattern and form, will significantly reduce air pollutions.

In economic perspectives, the economic segregation caused by a large monolithic development is most evident at the suburban development (Deal and Schunk 2004). The migration of a specific income population and moving of resources to the fringe cause a donut effect, social instability, and decline in the central parts of the cities. In addition, sprawl development increases the provision costs and inefficiency of infrastructures and utilities because of dispersed settlements. The provision cost of water and sewerage system is significantly high in horizontal and dispersed development (Arbury 2005). For instance, the American government is planning to spend about $190 billion for the provision of these two infrastructural systems in 2000–2025 for single-family detached residential neighborhoods (Burchell and Mukherji 2003). Obviously, the preparation of these infrastructural systems for compact residential buildings requires less amount of investment in both construction and maintenance. Burchell and Mukherji (2003) believe that more than 100 million gallons of water and sewer demand per day could be saved through a more compact development, without depriving residents of any fundamental facilities. Furthermore, the remarkable amount of road construction projects is another negative economic effect of urban sprawl development in case of infrastructural provision costs because of the dispersion of urban structures. The US government has planned to expend about 900 billion US dollars for road network expansion for sprawl development pattern from 2000 to 2025 (Burchell and Mukherji 2003).

In addition to infrastructural and utility provisions and maintenance costs, sprawl development reduces the efficiency of community facilities and services. The potential of these facilities to serve the community in more compact neighborhoods, especially in city centers, is higher because of their accessibility to local residents. Moreover, the distribution and allocation of such facilities in built-up areas are easier and cheaper, particularly in the brownfield redevelopment process. Burchell and Mukherji (2003) estimated that the American government spends about 143.2 billion US dollars annually for the provision of community facilities and services for sprawl urban pattern, in which 99.4 billion US dollars are compensated through the revenues from developments. Thus, 43.8 billion US dollars of financial resources are spent annually in sprawl development, whereas this could be reduced to 4.2 billion US dollars in compact urban pattern. These values are mainly the direct expenses in dispersed urban pattern. Several indirect financial costs are consequences of horizontal urban development, such as air and water pollution, health issues, and traffic congestion. Nozzi (2003) estimated that in 1991, the air pollution from motor vehicles caused up to $531 billion worth of health damage, $5 billion of crop damage, $44 billion of visibility damage, and $365 million of building damage.

Dispersed and horizontal urban pattern also has financial effects on private and personal businesses. It requires more funds for advertisement, delivery, maintenance, communication, and so on because of the distribution of their customers in suburban areas along the city boundary. Site suitability for the headquarters or branches of a particular business in a widely expanded city is more complicated and risky rather than that in the city center in a compact urban form. Meanwhile, sprawl development creates a segregated community because the central areas are occupied by those who are not able to migrate to urban fringes, which increases crime and poverty like in many American cities (Deal and Schunk 2004). Consequently, private and public investments decrease significantly, which leaves the city centers with abandoned infrastructures and utilities. Thus, the most unsustainable economic aspect of sprawl urban development is the migration from the inner cities to and investment on new infrastructures and utilities in suburban areas. However, individuals and the society also benefit from sprawl development. For instance, developers and personal purchasers in some cases prefer low-density development patterns because they are cheaper (Deal and Schunk 2004).

Sprawl urban development has several negative effects on social matters, which directly violate the social aspects of urban sustainability, such as health issues, segregated community, and social inequity (Gillham 2002; Nozzi 2003; Deal and Schunk 2004; Song and Knaap 2004; Abdullahi and Pradhan 2015). Some of these effects were already mentioned in previous explanations of the consequences of environmental or economical disadvantages of urban sprawl development. Such development pattern is highly dependent on private cars, but because of the cost of this transportation, poor people are limited to public transportation; forces the poor people to live in undesirable environments; increases the cases of mental illness because of fear of high traffic volumes; and segregates the community because of the daily travels of most residents to the suburban area (Hillman 1996; Arbury 2005). One of the worst effects of horizontal urban growth is the creation of gated or walled communities, which isolate and separate the living and social activities of local residents even in one neighborhood. A gated community consists of single- or two-story building residential streets with various facilities, such as park and kindergarten, which are closed by walls or fences (Fig. 3.3). Le Goix (2005) explained this type of urban community as "…a physical and obvious expression of the post-industrial societal changes (fragmentation, individualism, loss of communities) as part of a commoditization of urban public space and as a penetration of ideologies of fear and security supported by economic and political factors."

Although it is mainly based on the local planning authority's regulation and policy and cultural issues, gated communities have been growing because of several advantages, such as security, safety, available parking spaces, and low density. Nevertheless, the main concern about this type of community planning is the loss of public spaces and creation of a self-perpetuating segregationist pattern, where children may grow up with less or no sense of empathy for

Fig. 3.3 Several gated communities in one main local street. Photo of Malaysia taken by author

those living outside the gate and perceive them with suspicion and contempt (Duany et al. 2001; Arbury 2005). These types of community segregations reduce concern and responsibility for others beyond the subdivision walls.

However, gated communities of urban sprawl pattern are not the only factor or cause of segregated urban areas. Urban environments may be separated according to different types of residential areas with specific qualities, characteristics, and landscapes. People prefer to live in a neighborhood where the levels of the residents are similar. Thus, construction companies promote these separations based on building quality and neighborhoods to increase their profit from these classifications. In addition, local residents prefer to maintain the value of the area (avoid construction of lower quality housing in their neighborhoods) to prevent any negative effect on the values of their properties (Duany et al. 2001). This subdivision of urban areas into several homogenous neighborhood clusters limit the living location selection and force people to live in specific areas based on the level of their income and their status in society. Moreover, the lifestyle, social activities, such as school types for children, entertainment, and recreation, level of health care and insurance, and so on are also determined by the same level and position. In contrast, in concentrated and compact urban forms, facilities are properly distribution and the wider range of residential areas without walls and gates increase social equity by providing opportunity for interaction among the residents with different backgrounds and levels in the same environment. The wide range of residential types without classification pattern increases the affordability of houses in cities and consequently reduces the limitation of neighborhood selection and concentration of crime and poverty in a specific area.

In the medical and health perspectives, urban sprawl development creates several negative issues because of its physical characteristics. One of the main effects is higher car dependency, especially private vehicles, which significantly increases death while traveling. Nozzi (2003) reported that, "The number of people who die on US highways every year is the equivalent to a fully loaded Boeing 747 aircraft crashing every three days, killing everyone aboard. In 2000, almost 6.5 million motor vehicle crashes killed 41,821 people and injured more than three million." In addition to vehicular accidents, the car dependency of urban sprawl significantly reduces walking and cycling, thus increasing several physical illnesses, such as hypertension, heart diseases, and type-two diabetes (Kelly-Schwartz et al. 2004; Badland and Schofield 2005). In fact, a strong but quite complex relationship exists among the type of urban pattern, physical activities, and effect on the health of local residents (Kelly-Schwartz et al. 2004). Hence, considering the effect of urban forms properties, such as accessibility, density, dispersion, and centralization, on the lifestyle and activities of the residents is important.

Density, as one of the main aspects of urban forms, is influenced not only by the suburban development caused by sprawl growth, but also by the inner city land use intensities and population, residential, and employment densities (Clark 2013). From the energy consumption perspective, lower built-up density causes more radiant heat energy to surface heat island formation than higher density development (Stone and Rodgers 2001). Recently, the energy efficiency of urban sprawl has received much attention in the literature (Mindali et al. 2004; Norman et al. 2006; Milan and Creutzig 2016). Ewing et al. (2003) summarized the negative consequences of urban sprawl in three points: (1) de-investment in urban core areas and decline of the city center, (2) private vehicle dependency and thus increase in number of VMT, road congestion, and air pollution, and (3) the loss of open spaces and scenic areas in and close to the urban regions.

Unfortunately, the complexity of issues and disciplines prevent a public understanding of suburban development and its negative effects. New concepts, tools, technology, and methods are required to increase our understanding of dynamic urban growth and land use changes. Fortunately, the availability of sophisticated computational models, such as cellular automata, statistical approach, factor analysis techniques, and professional geographical and mapping system like GIS, help us to deal with these complex phenomena (Bhatta 2010; Bhatta et al. 2010).

Urban sprawl is not simply characterized by low density development. Walking and cycling behaviors, land use diversity, and a reduced car dependency are arguably more important determinants of sprawl development pattern. The contradictions of sprawl urban development and urban sustainability principles are significantly visible in the explanation of Calthorpe (1993) of urban areas as "…the city and suburb are now locked in a mutually negating evolution towards loss of community, human scale, and nature. In practical terms, these patterns of growth have created on one side congestion, pollution, and isolation and on the other urban disinvestment and economic hardship." Suburban development is no longer compatible with today's life because of the difference in household structures, living behaviors, workplace environments, and increasing environmental concerns. For instance, the growing number of working mothers requires short commutes and accessible neighborhoods because they cannot take their children to work every day (Arbury 2005). Calthorpe (1993) believed that traffic congestion and unaffordable housing are the two main reasons suburban development is no longer compatible with today's life. Thus, a proper development plan based on traditional walkable communities with high accessibility and minimum car dependency is required. In addition to these properties, a mixed and concentrated living manner from different levels and cultures reduces social segregation and polarization. Traditional neighborhoods that are more compact are likely to lead to better quality of life than those dominated by urban sprawl (Arbury 2005).

As listed above, urban sprawl development has several conflicts with urban sustainability in environmental, economic, and social perspectives. Thus, alternative scenarios regarding planning, controlling, and developing patterns all around the world need to be proposed and implemented, especially in developed countries, where local residents in North America consume 16 times more energy than those in Africa and over 8 times more than the residents in Asia and South America on average (Burton et al. 2003). Developed countries are able to propose more solution to achieve urban sustainability because of this difference (White 1994). However, cities are clearly the best location to apply and implement sustainable development rules because of their concentrated populations who are the main consumers of natural resources and major sources of environmental problems and pollution (Burton et al. 2003). Naess and Sandberg (1996) summarized several elements of urban development and spatial planning required to achieve sustainable urban development as follows:

- Significant devaluation in energy consumption and emission based on ecological and distributional parameters of urban sustainability at the global level;
- Reduction in land conversion and destruction of natural environment, ecosystem, and soil resources of food production;
- Reduction in the utilization of environmentally harmful construction materials;
- Replacement of open-ended flows, where natural resources are transformed into waste, with closed loops that rely on local resources to a greater extent;
- Providing a calm, healthy, and green environment for residents to experience and become emotionally related to nature.

A substantial revision in the urban system and process is necessary to achieve the goals of urban sustainability. Most of these goals are the exact drawbacks of sprawl urban development. Hence, most of the urban scientists and researchers in the 1990s focused on how to plan the urban form in a more sustainable manner. From this debate, compact cities are recognized as one of the best alternative scenarios to achieve urban sustainability. The Agenda 21 and Habitat Agenda of the UN both proposed a compact and concentrated urban pattern, natural environment preservation, reduction of car dependency, reduction of waste and pollution, creation of livable and community-oriented human environments, development of affordable residential areas, improvement of social equity, and development of a restorative local economy as the solution to achieve sustainable urban development (Wheeler 2000).

3.3 Compact Urban Development

Generally, sustainable urban development can be achieved through an efficient land use growth and management that implement proper planning and urban design. These tasks can be accomplished by adopting various strategies and planning to minimize the energy consumption, protect biological diversity, reduce pollution, improve social interaction, and develop more green landscapes (Kropp and Lein 2013). Therefore, the contribution of the shape and form of the cities has become one of the focus points to conduct these tasks. Many scholars and urban scientists believe that urban forms can be significantly linked to urban sustainability although it is not simple and straightforward. Therefore, much attention has been focused on the relationship between urban form and sustainability in the recent decades, that is, the implications of the shape and form of cities on sustainable development. From this debate, a strong agreement that compact cities are one of the most sustainable urban forms exists because of their various urban sustainability characteristics, such as less car dependency, public transportation promotion, rural development containment, and natural environment preservation (Livingstone and Authority 2003). These characteristics have contributed to the objectives of sustainable urban development in terms of social, economic, and environmental concerns. The popularization of sustainable development has contributed to the promotion of compact cities by enhancement of its ecological and environmental justifications.

In the 1970s, a round-shaped urban form with a population of about 250,000 persons in a radius of less than 3 km was considered as the most efficient urban form because of its high degree of compactness (Zagorskas et al. 2007). In the 1980s, the argument on urban compaction versus dispersion was the "order of the day" (Breheny 1996). In the 1990s, the European Union was the main arena for the debate on the advantages and disadvantages of compact city. For instance, Gordon and Richardson (1997) reiterate their warnings against compact urban form as a planning goal. Ewing (1997) believed that policy intervention should be involved in urban growth because sprawl development has several problems. Anderson et al. (1996) investigated the evidences on the relationship between urban forms and energy consumption for mobility. Bourne (1996) propounds compact city from the aspect of re-urbanization, while Breheny (1996) attempted to weigh the validity of the arguments presented by the "centrists" and the "de-centrists". Some scholars believed that high centralization and density should be the main objectives for environmental conservation as a goal of urban sustainability. Similarly, Adolphson (2010) mentioned that the compact city paradigm characterized by high density and land use diversity in this decade was promoted to achieve urban sustainability. Compact city preserves natural and rural environments, reduces private vehicle transportation, promotes public transportation, promotes walking and cycling, improves accessibility to community facilities, and increases urban vitality (Burton 2002). However, opponents accused compact city of suppressing human freedom and life style and creating problems, such as traffic congestion and air pollution. Newman and Kenworthy (1999) stated that most of the scholars agree that fuel consumption for traveling is reduced because of the proximity of various land use types in a compact city. Although the relationship is complicated, compact development can help reduce energy consumption and resource depletion.

The concept of compact city is related to the shape and pattern of urban features, such as spatial distribution, land use categories, and spatial pattern of road networks. In addition, it is related to activities and behaviors of the local residents of an urban region. In general, compact development can be defined as high-density urban settlements that promote central area revitalization, mixed land use development, rural development containment, public transportation facilities, and concentrated developments around transportation stations (Burton et al. 2003). This type of urban pattern has several advantages:

- Less car dependency, thus lower emissions,
- Reduced energy consumption,
- Better public transport services,
- Increased walking and cycling habits, thus healthy community,
- Increased overall accessibility,
- Reuse of infrastructure and previously developed land,
- Regeneration of existing urban areas and urban vitality,
- Higher quality of life,
- Preservation of green spaces,
- Creation of a proper environment for enhanced business and trading activities, and so on.

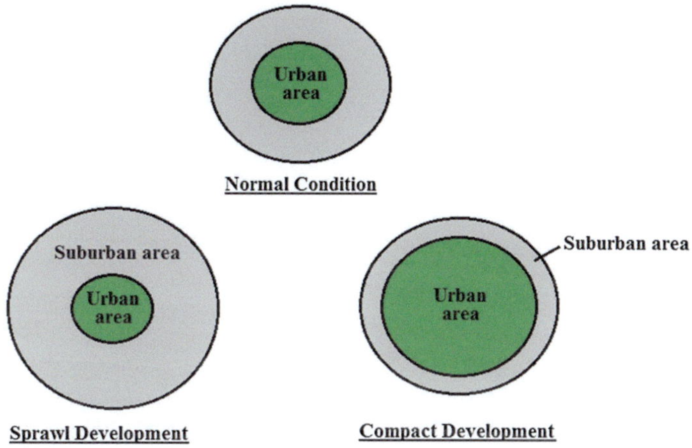

Fig. 3.4 General perspective of suburban areas in compact and sprawl development

The European Commission's Green Paper (CEC 2004, 2005) strongly advocates compact development, assuming that it makes urban areas more environmentally sustainable and improves quality of life. This type of urban development is being extensively promoted in European cities as a solution to the problem of sustainability (Livingstone and Authority 2003). Compact city has also received high attention in most of the Asian cities because of the significantly limited land resources and infrastructures and sensitivity to natural environment (Lin and Yang 2006). Doi (2005) stated that one of the most promising ways to achieve sustainability in towns and cities is by implementing a compact city with high density and intensity of urban forms and mixed land use development. Unlike that of sprawl development, the spatial containment strategies of a compact city have been viewed as a potential solution to the undesirable social and environmental effects, particularly when compact city is integrated with a suitable planning process (Neuman 2005). Concentrated development obviously encourages the redevelopment of existing brownfields and abandoned lands and provides opportunities to reuse existing infrastructures (Abdullahi and Pradhan 2015). Inner development reduces land consumption, protects rural environments, and revitalizes urban centers. Several examples of European cities show that the inside city can be a target of development (Zagorskas et al. 2007). In the land consumption aspect, more land areas are consumed in sprawl development for road networks and other infrastructures related to automobile because of the dispersed pattern of urban structures. This compares with about ten percent more of compact cities being devoted to the automobile, which would lead to a far smaller loss of productive land (Duany et al. 2001) (Fig. 3.4).

Compact city should be supported by multimodal transportation facilities, including a system oriented toward public transit, road network, cycling, and pedestrian. The compact city idea is concerned about the proximity of urban activities to ensure better access to services and community facilities via public transport, walking, and cycling, and more efficient utility and infrastructural provision (Doi 2005). Proper transport land use strategies encourage the use of alternatives to private vehicles. Empirical studies tend to confirm the transport and health benefits of densification, infill development, land use diversity, and job–housing balance (TRB 2005). Frumkin et al. (2004) and Stone et al. (2007) also stated that compact city is one way to reduce commuting and stimulate physical activities, thus reducing emissions and greenhouse gases and improving public health (Fig. 3.5).

Some of the main aims of compact city are to enhance the individual quality of life and achieve sustainable urban forms. Neighborhood characteristics, such as accessibility to community services, opportunities for recreational facilities, environmental quality, and mitigation of natural hazards are essential components of a good quality of life (Doi 2005). Compact city is expected to support these opportunities through more livable communities, increasing accessibility to various land use categories, and revitalizing old urban areas compared with sprawled or dispersed development.

Compact city does not necessarily imply a monocentric urban pattern. The focus of the sustainability debate has shifted toward a polycentric urban region because urban areas are integrated within regional settings; the perspective of "multicentred forms of compaction" (Burton et al. 2003).

Figure 3.6 shows that three main indicators are commonly used to describe and measure compact city: urban density, land use diversity, and urban intensity (Burton 2000; Lin and Yang 2006; Abdullahi et al. 2015). Each of these indicators can be further divided into and measured by some variables. The next sections will define and discuss these three indicators of a compact city.

Fig. 3.5 Schematic view of transportation advantages of compact city, **a** compact city and **b** non-compact city

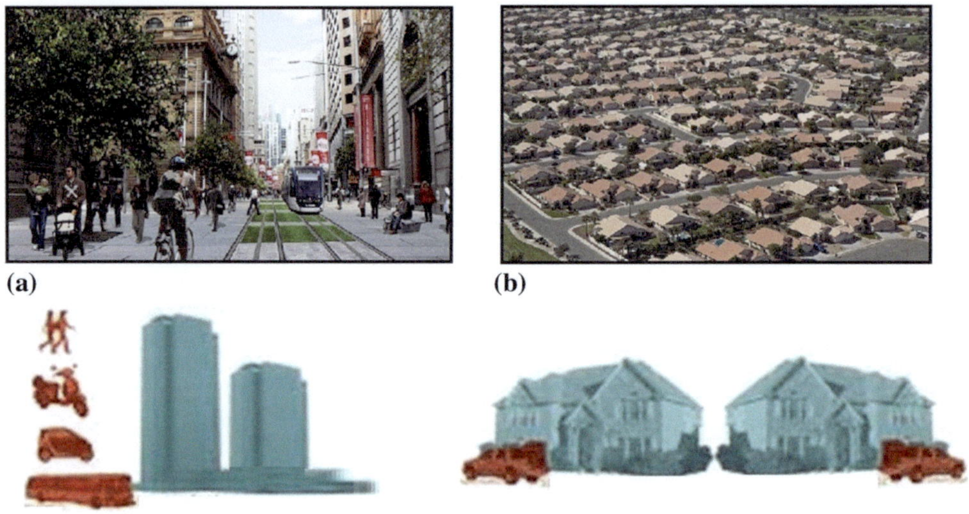

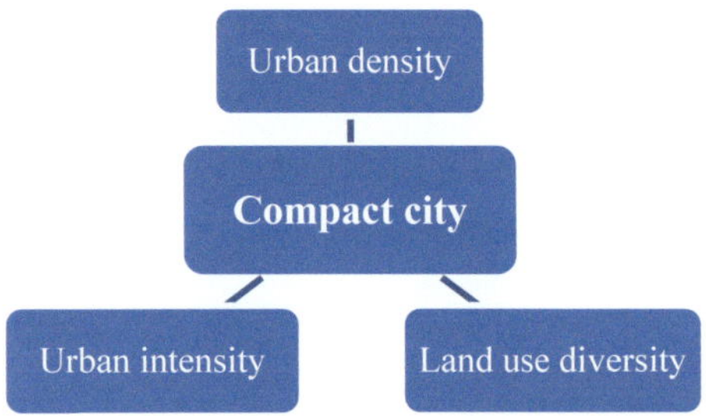

Fig. 3.6 Three main concepts of Compact city paradigm

3.3.1 Urban Density

Density has been considered as one of the most and important aspects of urban development. Historically, most of the urban forms, such as The Garden City (1900 in UK), The Neighborhood Unit (1920s in US), and Suburbia (from the 1920s onward in the USA), were developed by adopting a low-density format (Banister 2012). However, much higher density development was seen in the Urban Modernism era (new urban developments in Europe). In fact, it was widely proven that high density development is an effective characteristic to achieve urban sustainability (Carruthers 2002; Arifwidodo 2012). For instance, higher density development improves the government efficiency in financing developments and lowering infrastructural costs (Carruthers and Ulfarsson 2008); higher residential density may reduce car travel because of higher accessibility (Burton 2002; Macfarlane et al. 2015); medium or higher residential density increases public transportation efficiency and thresholds to support concentrations of economic activities, services, and facilities (CEC 1992); higher building densities reduce traffic jam and this can be the main solution to provide cities with environmental and quality of life benefits (Banister 2012); and higher population density promotes and supports public transportation, increases vitality and viable community facilities and services, such as recycling and local power regeneration (Force and Rogers 1999). Williams (1999) stated that higher density settlement improves social sustainability because of the more efficient use of community facilities and services, higher accessibility, and increases vitality, vibrancy, cultural activities, and social interaction. In the United States, the essential components of a walkable city are high-density development, which is the "new urbanist" antidote to car dependency and sprawl development (Calthorpe 1993). Banister (2012) conducted a comprehensive study on the relation between urban density and transportation efficiency to achieve urban sustainability. Adolphson (2010) also tested and evaluated the relation between urban density and land use diversity from a polycentric perspective. He used a kernel convolution to calculate the internal distribution of densities.

Although density is mainly considered as population density, building density is also one the most common interpretations of city compactness assessment in terms of physical compactness. Higher building density saves significant land, decreases energy consumption, and promotes affordable housing (DETR 1998). Moreover, higher energy efficiency can be achieved through compact housing forms, such as terraced housing and low rise blocks of flats (DETR 1998; Newton et al. 2000). However, the concept of residential density and housing form is a complex topic that requires the consideration of various issues, such as building design, public spaces, public fronts, private backs, crime rate, and safety (Burton 2002). In fact, density measurement is the most common interpretation of compact urban development (Fig. 3.7).

Although, higher density is one of the main concerns and characteristics of a compact city, the capacity and potential of the corresponding city to serve the high-density population, such as providing efficient public transportation, should

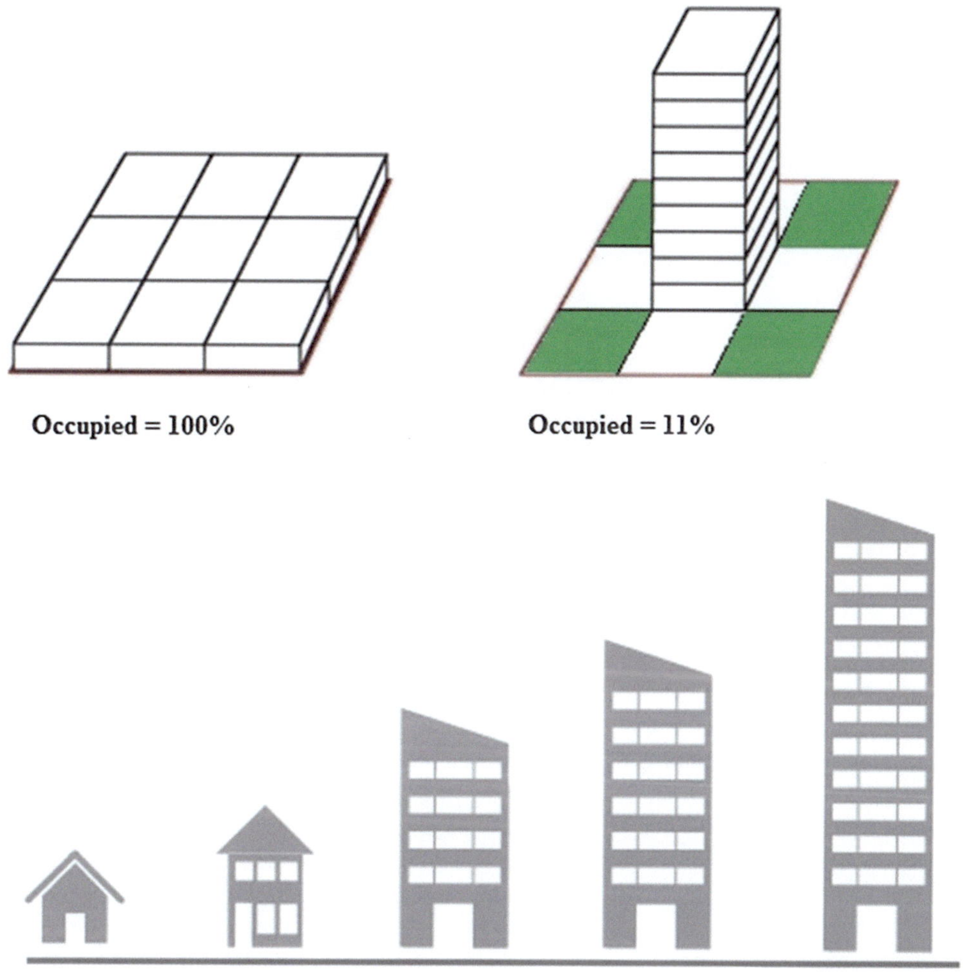

Fig. 3.7 High and low residential density

be studied carefully during planning and development. Compact city and its main aims toward urban sustainability cannot be achieved by simply increasing population density, building proximity, and number of residential units. Meanwhile, the measurement of density is another challenge, which the urban scientists and urban planners have not reached technical agreement on. Variation in density assessments, such as net, gross, overall, and zonal densities, causes a complex analysis and evaluation of the urban application process. For instance, several measures are assessed in England, such as "dwellings per hectare," "habitable rooms per hectare," or "bed spaces per hectare," to identify an optimum urban density value (Arbury 2005). Thus, considering the various aspects of urban density compatible with the characteristics of the study area is important.

3.3.2 Land Use Diversity

Land use diversity is another main component of compact cities. In a mixed land use area, different land use categories are mixed vertically in the same building or horizontally in the same neighborhood (Lin and Yang 2006; Abdullahi et al. 2015; Tian et al. 2015). Various land use categories, such as different commercial, residential, light industrial, and community facilities (school, hospital, recreational facilities, library, and so on) can be included within a mixed land use environment (Fig. 3.8).

Since 1990s, urban studies have led to the advocacy of urban areas that are spatially compact with higher land use diversity (Zagorskas et al. 2007). Historically, mixing various land uses has been the main feature of urban environments since the first human civilization. The traveling distances of local residences from the center living areas were limited to a small radius because walking was the main transportation mode. Thus, various daily and weekly destinations, such as facilities and services, commercial buildings, agricultural fields, and industrial areas, were all located in small spaces, which caused limitations on urban expansion and availability of site for required land uses. In this situation, most of the residential buildings were used as working places depending on the type of profession and trade, such as solicitorship and dressmaking. In fact, most of

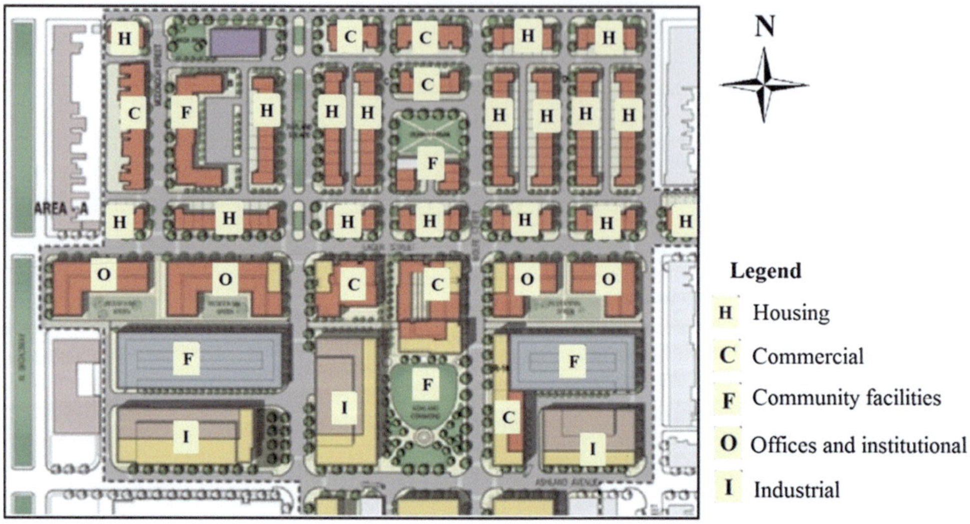

Fig. 3.8 A schematic example of horizontal mixed land use development

the urban areas in the world have the same characteristics, which are walking as the main transportation mode, mixing of land uses either in the same building or in a small neighborhood, and high population density in urban and low in suburban areas, which cause a clear distinction between urban and rural areas (Morris 2013). However, the industrialization period caused substantial changes in the development pattern that separated various land use categories, especially industrial areas from residential neighborhoods. This process was stimulated by other factors and periods, such as urbanization, advances in transportation, zoning ordinances, and the rise of an affluent middle class (Herndon 2011). However, in the recent decades, scientists and urban planners have realized the negative effects of this separated land use development (related to environmental, social, and economic aspects) and thus started to return to mixed land use development pattern, especially based on compact city paradigms. This modern idea of land use diversity is different with respect to the historical mixed development pattern, especially in relation to the Euclidian zonal manner. Although all land use categories in historical cities were together in a single zone or district, the current land use diversity is mainly single tiles within a mosaic of mostly single-use zoning classifications (Herndon 2011). Moreover, the historical land use diversity of urban developments has been growing gradually over long periods without predefined planning and design. In contrast, the current development is based on the master plan of the city, which was designed based on the urban capacity and local demands in a shorter period (Schwanke 2003). Thus, considering these differences and evaluating their consequences are important in planning and developing a mixed land use development.

Although the concept of mixed land use development seems to be simple and straightforward (combination of several land use categories), planning and implementing a proper mixed development zone is a very complex and challenging project. These challenges are related to the level of integration or degree of compatibility of land use types, vertical or horizontal land use diversity, and size and capacity of the site (Herndon 2011). Moreover, although a general agreement on the concept of mixed land use development exists, several scientists still argue that no clear and acceptable definition of this kind of development pattern exists, such as Rowley (1996), Grant (2002), Hoppenbrouwer and Louw (2005) and Rabianski et al. (2009). In this regard, in 1990s, Rowley (1996) proposed a conceptual model for land use diversity based on the aspect of the internal texture of settlements in 1990s. Figure 3.9 shows that this model mainly considers horizontal land use diversity or that within contiguous buildings. Rowley's model assumed that the physical pattern of mixed land use area is related to urban texture, setting, and site.

On the other hand, Hoppenbrouwer and Louw (2005) proposed a topology for the mixed development pattern based on spatial perspective that is managed by function, dimension, scale, and urban texture. Figure 3.10 shows that the function element is related to individual land use types, such as residential, commercial, and community facilities. The dimension consists of shared premise, horizontal, vertical, and time dimensions. The scale starts from the building, zone and continues to the city and region levels. Finally, urban texture includes grain, density, and the interweaving of functions.

With all these complexity and variables involved, land use diversity has been strongly suggested by several scientists and scholars to achieve sustainable urban development. Musakwa and Van Niekerk (2013) and Song and Rodríguez (2005) discussed the direct advantages of mixed land use development for urban sustainability. Numerous studies have proven that efficient transportation facilities and healthy

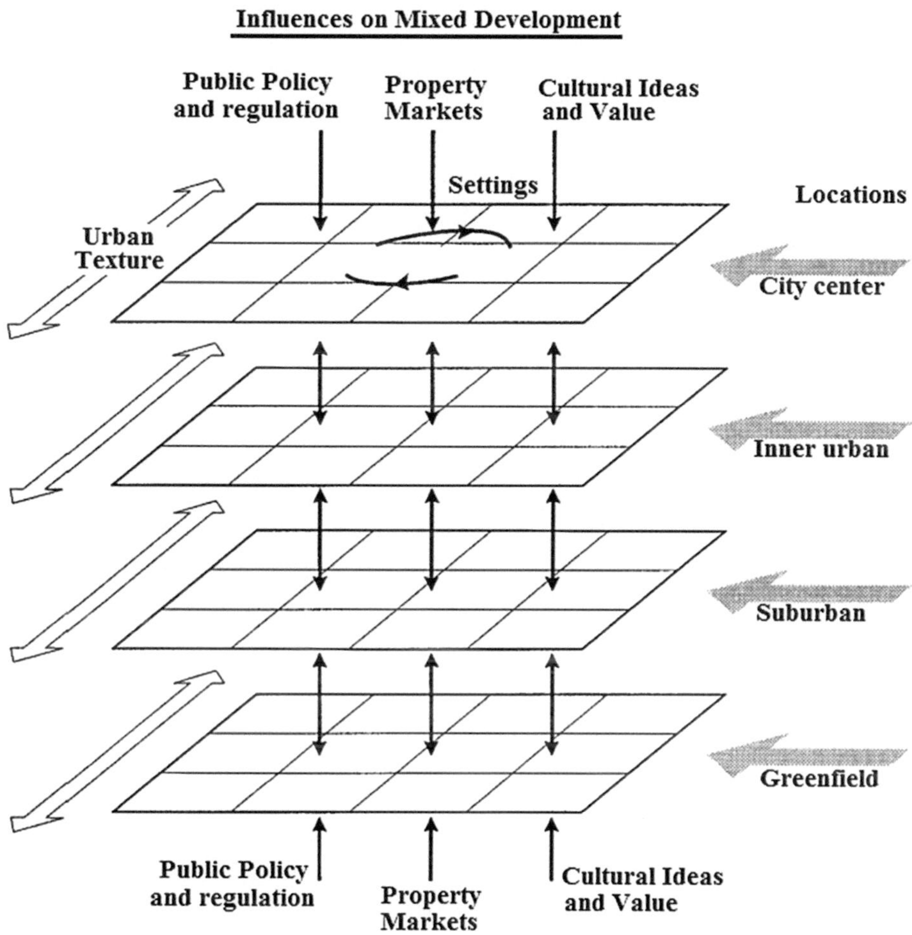

Fig. 3.9 Rowley's mixed land use development (Rowley 1996)

communities can be achieved through dense mixed land use, infill development, and proximity between the living and working places (TRB 2005). For instance, Gu et al. (2013) presented that the fuel consumption for transportation within the city could be significantly reduced by providing a mixture of living, working, and entertainment activities in one location. Gainza and Livert (2013) also observed the positive effect of high density and mixed land use development on the time and energy consumption of transportation. Ding (2004) and Song et al. (2013) stated that the close proximity of various land use types encourages walking and cycling behavior in a community. In addition, the close proximity of infrastructure and facilities (such as heat and power plants) to consumers (residential and commercial buildings) can reduce energy waste during transmission. Reusing energy wastes can increase energy consumption efficiency as well.

3.3.3 Urban Intensity

Urban intensity is related to the process of achieving city compactness. In fact, city intensification can be defined as the process to making an urban area more compact and sustainable. Infill development, brownfield revitalization, and mixed land use developments are examples of city intensification resulting in a positive density growth (Burton 2002; Lin and Yang 2006). Infill development can be supported by existing services and infrastructures, whereas rural development requires provision of various new services and utilities (water supply, electricity, road network, and so on). Burton (2002) identified three main tasks in implanting intensification: increase population, development, and land use diversity. Similar to the mixed land use development, an important aspect of urban intensity is related to the land use distribution pattern, which indicates the crowd and liability of an area. Higher urban intensity also reduces car dependency, conserve lands, and regenerate central parts of the city. According to the city compactness definition (self-dependency from outside), city intensification determines the degree of compactness of an urban area. Therefore, urban intensity can also be viewed in terms of the availability and accessibility of various required community facilities and services. In this manner, it supports the economic sustainability aspects as well (Burton 2002). Proper accessibility and availability of various required community facilities, such as health, educational, and shopping centers,

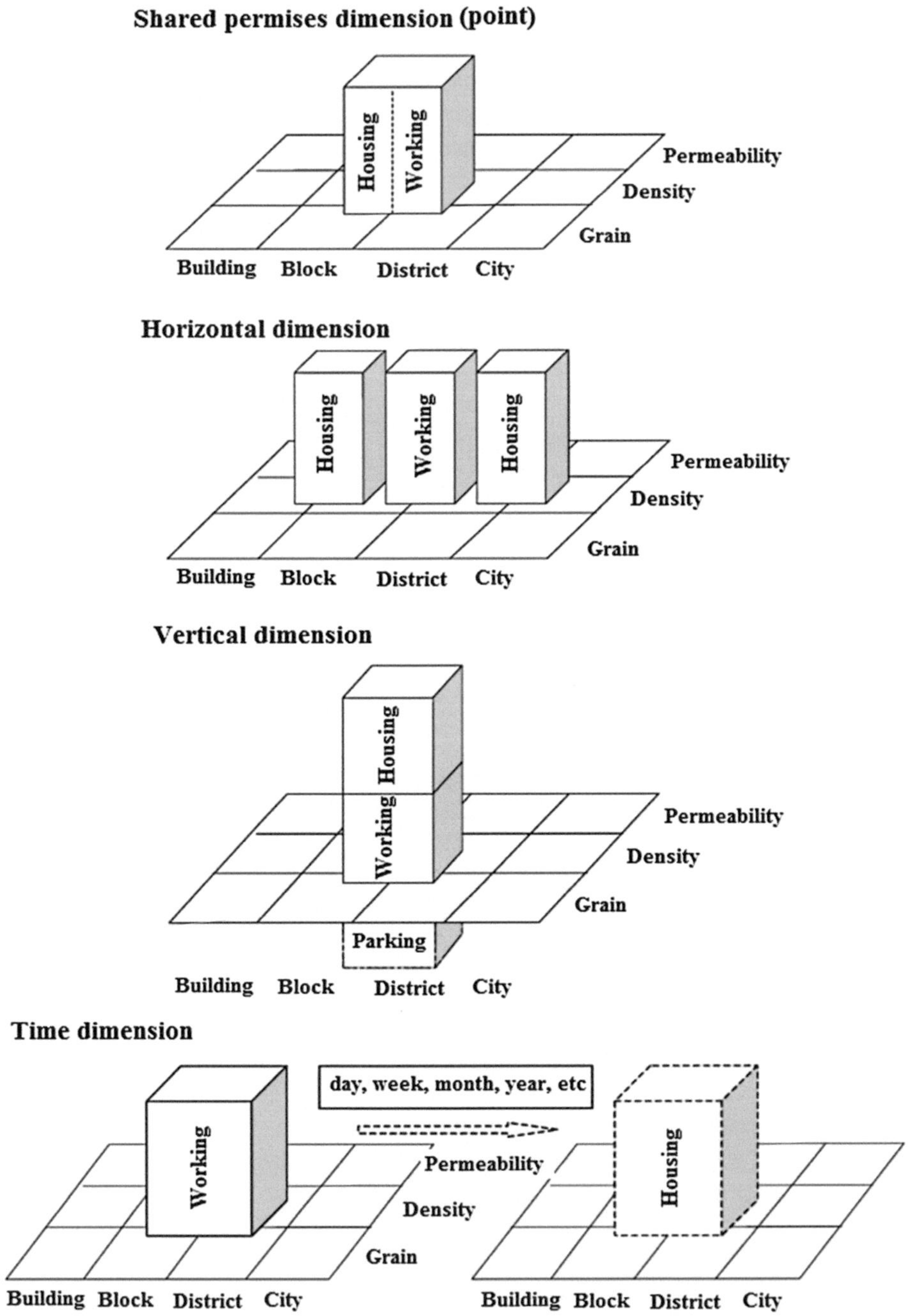

Fig. 3.10 Hoppenbrouwer and Louw mixed land use development (Hoppenbrouwer and Louw 2005)

to the residential and working areas decrease the dependence of the local residents on their own private vehicles. These facilities can be more efficiently provided and accessed in the compact urban form, thus reducing the cost of their provision and ensuring sustainability (Carruthers and Ulfarsson 2008; Arifwidodo 2012). Hence, urban intensity can be evaluated by considering the activeness, availability, proximity, quality, and quantity of each type of community facilities (such as health, educational, public transportation, point of interests, open space, and recreational facilities as well as job opportunity) with respect to the characteristics of the local residences and neighborhood. Detailed information on the local population is required to evaluate the local demands. Meanwhile, updated information on existing facilities, such as capacities, locations, and qualities, should be available. Transportation facility is one of the most important aspects to consider in these assessments. Evaluating the local population demands based on some local and/or standard guidelines, such as the one given by De Chiara (1990), which shown in Table 3.2 is important.

Compact urban development can be implemented in a variety of scales, from urban infill and central area

Table 3.2 Minimum population required for community facilities

Community facility	Minimum population	Community facility	Minimum population
Education		**Health**	
Kindergarten	500	Clink	10,000
Primary school	1800	Welfare center	25,000
Secondary school	5000	Hospital 100 beds	25,000
High school	9000	Public clinical center	35,000
Educational library	5000	Hospital 225 beds	50,000
Specialized college	50,000	Psychiatric hospital	50,000
College	100,000	Rehabilitation center	75,000
University	500,000	Hospital 340 beds	75,000
University (graduate studies)	1,000,000	Hospital 450 beds	100,000
Institutional		**Recreation**	
Post office	1200	Small playground	1000
Library	500	Restaurant	2000
Church	500	Local park	3000
Public city hall	5000	Play ground	5000
Fire station	10,000	Gym and fitness	10,000
Police station	10,000	Sport club	10,000
Waste management center	10,000	Museum	20,000
Employment		Theater	20,000
Institutional	10,000	Cinema	20,000
Services	10,000	Coffee shop	20,000
Light industry	10,000	Swimming pool	20,000
Heavy industry	50,000	Local TV station	20,000
Industrial park	100,000	City recreational park	500,000
Miscellaneous	50,000	Zoo	1,000,000
Commercial		**Transportation**	
Small shop	500	Private parking	100
Super market	2000	Workshop	2000
Bakery	3000	Public parking	15,000
Pharmacy	3000	Public bus services	20,000
Bank	5000	Taxi services	20,000
Shopping mall	20,000	Train services	50,000
Hotel	25,000	Local airport	70,000

revitalization to the creation of an entirely new development, such as the idea of Urban Villages in United Kingdom and New Urbanism in the United States (Zagorskas et al. 2007). In theory, urban villages are human scale, compact, of mixed land use, and of mixed tenure neighborhoods within a wider urban area, diverse open spaces, less car dependency, self-dependency in terms of residents' employment needs, and shopping, recreation, and community activities (Chung 2010).

Although numerous studies have been conducted on the advantages of compact city, some scholars still argue that the compact city concept is unrealistic and undesirable (Breheny et al. 1999). As an alternative, decentralized concentration based around a single city or groups of towns, may be acceptable. Like compact city, this kind of development is high density and a land use diverse settlement with clear boundaries (Williams et al. 2000). In this debate regarding urban density, some believe that high concentration makes it possible to achieve sustainability and contributes to global environmental preservation, while others are skeptical because they are concerned about the freedom of human nature, quality of life, and lifestyle (Zagorskas et al. 2007). For instance, Yang et al. (2012) stated that a proper clustered development in suburban zones helps maintain a shorter commuting duration. The linkage between the shorter commute and suburban development has been proven by other studies on Asian and European cities with similar density contexts (Alpkokin et al. 2008; Veneri 2009). In the case of the natural environment preservation in rural areas and the provision of green environments inside urban areas, some scholars believe that compact city preserves these green environments in suburban areas because of high-density and concentrated development (Gordon and Richardson 1997; Burton 2002). Meanwhile, the high land use diversity and high intensity promote social sustainability and improve quality of life by the provision and distribution of local parks and recreational facilities. In contrast, other studies concluded that high density urban pattern may decrease the amount of green environments and can threaten ecological variety and living environment because of the concentration of human settlements (Burton 2000; Van Der Waals 2000). In general, gradual and more concentrated development seems more effective in saving and protecting the rural environment through the utilization of the existing infrastructures and redevelopment of abandoned lands. Thus, compact urban development significantly affects the environmental aspect of urban sustainability.

Although the positive effects of compact city on social sustainability, such as improvements in neighborhood attraction, accessibility, public transportation, and social equity, are recognized clearly, some researchers also mentioned a few of the negative effects of high-density development, such as insufficient living space, higher expenditure because of congestion, and high crime rates and housing prices (Breheny 1996; Gordon and Richardson 1997; Burton 2000). These issues encourage moving from the concentrated development to low-density suburban areas, which continue the growth of sprawl development (Senior et al. 2004; Stewart 2005). In the case of negative health issues, the negative health effects of higher density living are likely to be the result of other aspects of the residence, such as being part of a segregated "ghetto", being located next to a polluted highway, and poor construction, rather than simply its high density.

Although a broad consensus on its merits exists, implementing a sustainable compact development is not an easy task. It is a complex and long-term process, which requires proper planning, a flexible law system, and a supportive government (Chinyio et al. 1998; CEC 2005). One of the main problems is the administrative fragmentation of urban regions, which usually consists of a national government, several provinces or states, and many municipalities from urban to rural areas. In this regard, Dieleman et al. (1999) listed a number of specific conditions that are responsible for the implementation of a compact city, such as:

- The strategic planning tradition,
- The municipal finance system,
- The mass production of social housing, and
- Land policy.

A powerful planning authority, which has the ability to govern the area for at least next 20 years, is required to develop a compact urban pattern or increase the existing compactness of a region. The authority should provide financial and planning supports to the required infrastructures and utilities, such as road network, rail systems, and water and electricity supply. Financial support is very important for the feasibility of the compact city policy. In addition to these powers and supports, having a wide consensus among the local residents on the policy of compact urban development and the advantages of this kind of urban form is important.

3.4 Compact Urban Development Versus Urban Sustainability

The concept of sustainable development has given a major stimulus to the question of the contribution of urban forms in reducing the energy consumption and pollution levels (Breheny et al. 1999). Thus, researchers and urban planners focused on the relationship between the forms and shape of the cities with urban sustainability. Although it is not simple and straightforward, urban forms can be significantly linked to urban sustainability. Several urban development paradigms for achieving urban sustainability exist. In this regard, compact city is one of the popular paradigms in urban application fields. Compact city with its characteristics, such as high density, land use diversity, and rural development containments, attempts to be effective in preventing the effects of urban sprawl. These properties are clearly compatible with urban sustainability. Nevertheless, illustrating a clear and straightforward relationship between compact city and urban sustainability is important. The advantages of compact urban development on urban sustainability might be hampered, neglected, and/or even proven detrimental to sustainable development. Some scholars and urban scientists examined and reviewed the concepts of compact city. For instance, Burton (2002) concluded that compact city paradigms provide both positive and negative effects on the social aspect of urban sustainability by assessing the compactness of several British cities. Meanwhile, regarding physical environment and development pattern of the urban areas, land use diversity and pedestrian-friendly streets are important to increase social interaction and provide a sense of

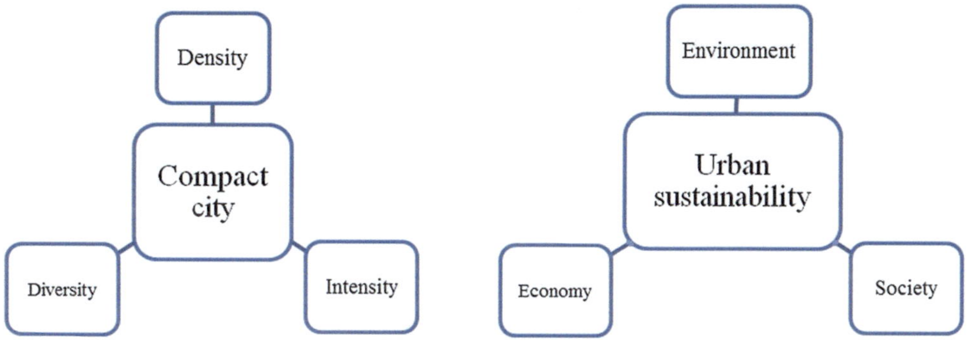

Fig. 3.11 Main indicators of compact city and sustainable development

community (Barton 2000). These properties have been applied in recent sustainable urban forms, such as New Urbanism and compact city, in which local residents have better opportunities for social interaction because of proximity (Nurul 2015). The assessment of the compact city and social issues can be implemented by investigating the compact form related parameters with social variables. For instance, the one-to-one assessment of urban density, diversity, and housing types related to the frequency of meeting, how well local residents know their neighbors, and how they interact with their neighbors (Nurul 2015).

Many concepts and indicators exist to define and measure compact city and urban sustainability. For instance, density is one of the major physical characteristics of compact city and can be evaluated in terms of various aspects, such as population, residential, and road densities. Meanwhile, social sustainability is one of the main aspects of sustainable development in the urban perspective that can be assessed in terms of accessibility, equity, security, and safety. Thus, the evaluation and investigation of the relationship between compact city and urban sustainability should be performed considering all the aspects and indicators of these two concepts. Nevertheless, directly evaluating the relationships among these aspects and indicators is not enough. The relationships could be comprehensively clarified if the extent to which the compact city paradigm affects sustainability could be examined in terms of concepts, rather than in terms of the component indicators (Lin and Yang 2006). Lin and Yang (2006) proposed a structural equation modeling to evaluate the relationship based on the concepts and indicators of compact city and urban sustainability. In fact, this model is able to analyze complex associations among various indicators by aggregating the capabilities of path and factor analysis.

Compact city is described by several concepts, such as free-standing, contained, autonomous, moderately sized, and self-contained (Scoffham and Vale 1996; Burton et al. 2003). However, some identical terms used in defining compact city in various literatures were explained in the previous section, namely, urban density, diversity, and intensity. Urban sustainability also involves various concepts and principles related to social, environmental, and economic issues (Fig. 3.11).

Most of the researchers usually evaluate these concepts based on a one-to-one indicators and indices correspondence with respect to several case studies. For instance, Lin and Yang (2006) evaluated this relationship by considering 44 indices of compact cities and 119 indices of sustainable development for 92 samples of cities in Taiwan (Fig. 3.12).

For instance, the environmental sustainability availability and accessibility of green and natural spaces for local residents are some of the main issues regarding the quality of life as a social issue. Thus, urban compactness pattern in terms of density, diversity, and intensity can be evaluated as an aspect of urban sustainability. The high and low densities of urban areas affect the amount of local parks and green spaces. Higher land use diversity, including recreational facilities and preserved natural environments, fulfill this aspect of urban sustainability. In contrast, single land use development, either fully residential or industrial land use, is implicated for less compactness and less sustainability. In the preservation of natural environment in rural areas, some scholars believe that compact city preserves these green environments in suburban areas because of high density and concentrated development (Gordon and Richardson 1997; Burton 2002). Meanwhile, convenient and proper public transportation facilities encourage traveling to the nearest natural and green environments, which indicates a sustainable intensifying urban pattern. Higher density and land use diversity encourage walking and cycling because of the availability and accessibility of various points of interests, while high intensified city with good quality of public transportation reduces private car dependency. These characteristics significantly affect fuel consumption, which is eventually an indication or measurement tool for the environmental aspect of urban sustainability. However, higher density can increase air pollution because of the concentration of activities. Thus, evaluation of air pollutants, such as PM_{10}, SO_2, NO_2, and Co, is good assessment tool for the environmental sustainability of a compact city.

In economical perspectives, high density and land use diversity increase economic production and efficiently

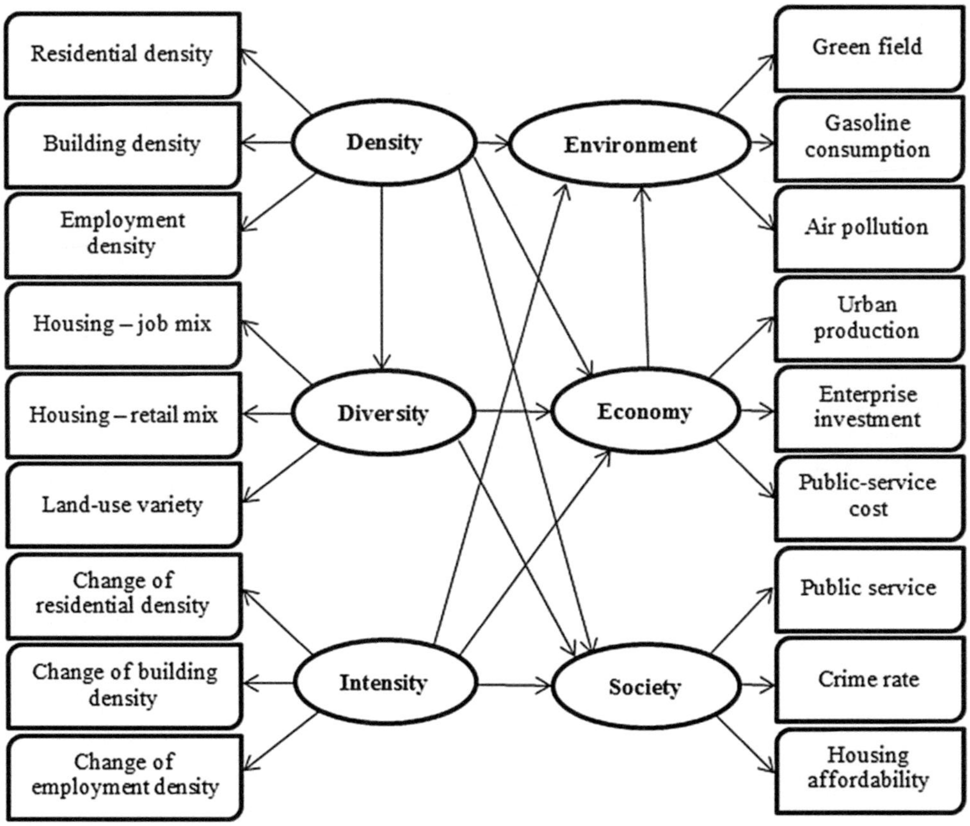

Fig. 3.12 Hypothetical path diagram for model specification (Lin and Yang 2006)

facilitate land use management. In addition, an intensified city with concentration of various facilities and services attracts enterprise investments. The assessment of these measures with respect to local residence is a suitable evaluation of the economic sustainability of a compact city because the amounts of urban production and investment are indications of sustainable development. Moreover, higher density and diversity increase the efficiency of infrastructures and utilities. Thus, an analysis regarding the expenditure per level of infrastructure supply can be a measure of public services efficiency.

In terms of social sustainability, a compact city affects various social issues because of its concentrated form and spatial distribution pattern of several community facilities based on the density, diversity, and intensity characteristics of the urban area. In general, higher density increases the concentration of community facilities, and consequently leads to a higher quality of life. Thus, the assessment of these facilities and services with respect to quality and quantity per capita of local residents can be used as measure of the sustainability of a compact city. Burton (2000) stated that an overly dense area can increase crime rate and reduce security and safety. In contrast, other researchers, like Elkin et al. (1991), believed that higher compactness increases safety by putting eyes on the street to deter wrongdoing. Hence, crime rate and personal site survey from local residents can be used in the evaluation of the safety aspect of urban compactness. Moreover, Burton (2000) estimated that a compact city usually increases housing prices because of the high demand and low supply of residential units. This growth reduces the amount of funds for expenditures on other aspects. Although they are more complicated and more parameters should be considered, housing price and affordability can be a used as measures of social sustainability of compact urban development.

As a conclusion and according to a comprehensive literature review in this field, compact urban development has a significant effect on various aspects of urban sustainability, which are either positive or negative effects. Lin and Yang (2006) and Thinh et al. (2002) found positive effects on the economic aspects in terms of production and enterprise investment. In contrast, several negative effects were found in terms of less green fields, higher crime rate, and increase in house and land prices. Nevertheless, these negative effects should not debilitate the concept of compact urban development because of its comprehensive and significant advantages with respect to urban sprawl development. Therefore, several complementary strategies are required to ensure the sustainability of compact urban development. These strategies should enhance the positive effects and attempt to mitigate the negative effects of compact urban pattern. For instance, several variables regarding the

Table 3.3 Current initiative towards compact city

Compact city paradigms	Site and details
Transit oriented development (TOD)	– KL central – Subang Jaya transit nodes residential areas – Sentul public housing – Mid Valley shopping mall
Mixed land use development	– Service apartment in KL – Subang Jaya – South city (Serdang)
Brownfield redevelopment	– KL central – KLCC – SOGO shopping center
High density development	– Pantai Dalam – Damansara residential neighborhood

demands and characteristics of local residents should be estimated and included in planning and designing a high-density development and in the general development of a compact city. These variables are as follows:

- The environmental capacity and proper green spaces;
- Community facilities and services, such as educational, recreational, and medical centers;
- Enough floor space area for living and recreational activities;
- The traveling and commuting demands and the capacity of road networks and public transportation;
- Enough security to control the crime rate.

Moreover, land and housing prices should be controlled to improve housing affordability. Meanwhile, the positive effects of compact city on urban system according to urban sustainability need to be enhanced and promoted. However, very precise and comprehensive assessment variables are required because a sustainable compact development is a long-term and very complex project. A high number of sampling units, accurate data, and powerful tools, such as remote sensing and geographical information system, are required for the assessment model implementation.

3.5 Compact Urban Development: Malaysian Perspectives

Malaysian government as documented in the 10th Malaysian Plan is committed to improve various aspects of urban areas to achieve more sustainable environments. This commitment involves efficient public transportation facilities, accessibility of various community facilities and services, security and safety, environmental conservation, and provision of green urban landscapes. One of the main current policies on urbanization in the 10th Malaysian Plan is to build an urban environment with high quality of life by building vibrant and livable cities and promoting compact and efficient cities (EPU 2010). In this plan, livability is defined as a city with vibrant, attractive, and secure environment for the local residents to live, work, and entertain/be entertained. This city has a proper governance, competitive economy, high quality of life, and sustainable environment. Promotion of mixed land use development to encourage living, working, and recreational activities within the same compact neighborhood is an example of a mechanism to increase livability. Meanwhile, zoning ordinance, urban growth boundaries, growth control regulations, and other development incentives are some mechanisms to encourage compact urban development. In addition, in National Physical Plan 2, concentrated development in core centers for higher efficiency and high value added growth is promoted. In this plan, achieving energy efficient compact city is one of the main policies of urban growth management. Table 3.3 and Fig. 3.13 present some of the local projects of the Kuala Lumpur city in the way of the sustainable aspect of compact urban development.

For instance, Kajang city has applied several development strategies regarding land use and development, landscape and biodiversity, security and safety, renewable energy, integrating transport and development, development accessibility, and urban design principles aspects. Specifically, the local planning authority of Kajang (JPBD) proposed several other strategies, as shown in Table 3.4. These strategies consisted of several aspects, such as mixed land use development, building design, housing design, sense of place, public transportation, neighborhood, and promotion of walking, cycling, and green environment.

Nurul (2015) explored the city compactness of Kuala Lumpur and Putrajaya cities and its effects on social sustainability. This aspect of urban sustainability was assessed with respect to the behavior of local residences within the physical environment. The assessment process was based on

Fig. 3.13 Initiative towards compact city; **a** Mid Valley shopping mall, **b** Sentul public housing, **c** Subang Jaya, **d** Kuala Lumpur conventional center, **e** Sogo shopping mall, and **g** South city plaza

quantitative approaches through questionnaire survey from local households. The results revealed the significant effects of land use diversity, pedestrian-friendly streets, density, and residential building types. Table 3.5 shows that people living in higher density neighborhoods (population and residential) have better social interaction with each other. Higher land use diversity provides opportunities for local residents to participate more within the neighborhood because of walking and cycling to and from local community facilities and services. Moreover, neighborhoods with terraced housing types have higher levels of social interaction than those with other types.

Table 3.4 City compactness strategies of Kajang city

Aspect	Criteria
Mixed land use development	– Encourage residents to live within the workplace – Concentration of activities in line with the centralized public transport networks (TOD) – Implemented a mixed development area and development of potential/infill site – Building design: variety of activities/functions in one building creating an effective vertical mixed land use
Advantages of public transportation	– Create a wide range of public transport modes – User-friendly public transport system (appropriate age group)
Housing design	– Various types of residential design according to location and needs – Residential types are developed to suit the compact city – Residence district, integrated with transport convenience lay
Sense of place	– Safe and active open space – Characterized commercial development to local community activities
Cycling and walking neighborhood	– Building design incorporates pedestrian-friendly features – Accessibility of public transport nodes for pedestrian/cycling – Safe network of pedestrian/cycling, and uninterrupted between the neighborhood and the city center
Environment preservation	– The green area is maintained – Create green corridor and blue part of the redevelopment potential

Table 3.5 Average level of social interaction based on city compactness variables (Nurul 2015)

Variable	Class	Regular social meeting	Daily chatting	Helping each other
Density	High	3.07	3.07	2.00
	Medium	2.72	2.82	1.81
	Low	2.39	2.51	1.56
Land use diversity	High	2.57	2.63	1.71
	Low	2.62	2.79	1.68
Housing type	Detached/semi detached	2.73	2.72	1.76
	Terraced	2.74	2.85	1.83
	Flat/apartment	2.42	2.54	1.56

References

Abdullahi, S., & Pradhan, B. (2015). Sustainable brownfields land use change modeling using GIS-Based weights-of-evidence approach. *Applied Spatial Analysis and Policy,* 1–18.

Abdullahi, S., Pradhan, B., & Jebur, M. N. (2015a). GIS-based sustainable city compactness assessment using integration of MCDM, Bayes theorem and RADAR technology. *Geocarto International, 30*(4), 365–387.

Abdullahi, S., Pradhan, B., Mansor, S., & Shariff, A. R. M. (2015b). GIS-based modeling for the spatial measurement and evaluation of mixed land use development for a compact city. *GIScience & Remote Sensing, 52*(1), 18–39.

Adolphson, M. (2010). Kernel densities and mixed functionality in a multicentred urban region. *Environment and Planning. B, Planning & Design, 37*(3), 550.

Alpkokin, P., Cheung, C., Black, J., & Hayashi, Y. (2008). Dynamics of clustered employment growth and its impacts on commuting patterns in rapidly developing cities. *Transportation Research Part A: Policy and Practice, 42*(3), 427–444.

Anderson, W. P., Kanaroglou, P. S., & Miller, E. J. (1996). Urban form, energy and the environment: A review of issues, evidence and policy. *Urban Studies, 33*(1), 7–35.

Angel, S., Parent, J., & Civco, D. (2007). *Urban sprawl metrics: An analysis of global urban expansion using GIS.* Paper presented at the Proceedings of ASPRS 2007 Annual Conference, Tampa, Florida May.

Arbury, J. (2005). *From urban sprawl to compact city: An analysis of urban growth management in Auckland.* Geography and Environmental Science, University of Auckland.

Arifwidodo, S. D. (2012). Exploring the effect of compact development policy to urban quality of life in Bandung, Indonesia. *City, Culture and Society, 3*(4), 303–311.

Badland, H., & Schofield, G. (2005). Transport, urban design, and physical activity: An evidence-based update. *Transportation Research Part D: Transport and Environment, 10*(3), 177–196.

Banister, D. (2012). Assessing the reality—Transport and land use planning to achieve sustainability. *Journal of Transport and Land Use, 5*(3), 1–14.

Barton, H. (2000). *Sustainable communities: The potential for eco-neighbourhoods*. Earthscan.

Bengston, D. N., Potts, R. S., Fan, D. P., & Goetz, E. G. (2005). An analysis of the public discourse about urban sprawl in the United States: Monitoring concern about a major threat to forests. *Forest Policy and Economics, 7*(5), 745–756.

Bhatta, B. (2010). *Analysis of urban growth and sprawl from remote sensing data*. Springer Science & Business Media.

Bhatta, B., Saraswati, S., & Bandyopadhyay, D. (2010). Urban sprawl measurement from remote sensing data. *Applied Geography, 30*(4), 731–740.

Bourne, L. S. (1996). Reurbanization, uneven urban development, and the debate on new urban forms. *Urban Geography, 17*(8), 690–713.

Breheny, M. J. (1996). Centrists, decentrists and compromisers: views on the future of urban form. *The Compact City: A sustainable urban form*, 13–35.

Breheny, M. J., Town, & Association, C. P. (1999). *The people: Where will they work?: Report of TCPA research into the changing geography of employment*. Town and Country Planning Association.

Burchell, R. W., & Mukherji, S. (2003). Conventional development versus managed growth: the costs of sprawl. *American Journal of Public Health, 93*(9), 1534–1540.

Burton, E. (2000). The compact city: Just or just compact? A preliminary analysis. *Urban Studies, 37*(11), 1969–2006.

Burton, E. (2002). Measuring urban compactness in UK towns and cities. *Environment and Planning B, 29*(2), 219–250.

Burton, E., Jenks, M., & Williams, K. (2003). *The compact city: A sustainable urban form?* Routledge.

Calthorpe, P. (1993). *The next American metropolis: Ecology, community, and the American dream*. Princeton Architectural Press.

Carruthers, J. I. (2002). The impacts of state growth management programmes: A comparative analysis. *Urban Studies, 39*(11), 1959–1982.

Carruthers, J. I., & Ulfarsson, G. F. (2008). Does smart growth'matter to public finance? *Urban Studies, 45*(9), 1791–1823.

CEC, C. o. t. E. C. (1992). *Sustainable mobility: Impact of transport on the environment* (Vol. COM 9246.). Brussels: Commission of the European Communities.

CEC, C. o. t. E. C. (2004). *Towards a thematic strategy on the urban environment* (Vol. 60). Commission of the European Communities.

CEC, C. o. t. E. C. (2005). *Green paper on energy efficiency or doing more with less*. Commission of the European Communities.

Cheng, J., & Masser, I. (2003). Modelling urban growth patterns: A multiscale perspective. *Environment and Planning A, 35*(4), 679–704.

Chinyio, E. A., Olomolaiye, P. O., Kometa, S. T., & Harris, F. C. (1998). A needs-based methodology for classifying construction clients and selecting contractors. *Construction Management & Economics, 16*(1), 91–98.

Chung, H. (2010). Building an image of Villages-in-the-City: A Clarification of China's Distinct Urban Spaces. *International Journal of Urban and Regional Research, 34*(2), 421–437.

Clark, T. A. (2013). Metropolitan density, energy efficiency and carbon emissions: Multi-attribute tradeoffs and their policy implications. *Energy Policy, 53,* 413–428.

Dadi, D., Azadi, H., Senbeta, F., Abebe, K., Taheri, F., & Stellmacher, T. (2016). Urban sprawl and its impacts on land use change in Central Ethiopia. *Urban Forestry & Urban Greening, 16,* 132–141.

De Chiara, J. (1990). *Time-saver standards for building types*. New York: McGraw-Hill.

Deal, B., & Schunk, D. (2004). Spatial dynamic modeling and urban land use transformation: A simulation approach to assessing the costs of urban sprawl. *Ecological Economics, 51*(1), 79–95.

DETR, D. o. t. E., Transport and the Regions. (1998). *The use of density in urban planning*. The Stationery Office, London: Department of the Environment, Transport and the Regions.

Dieleman, F., Dijst, M. J., & Spit, T. (1999). Planning the compact city: The Randstad Holland experience. *European Planning Studies, 7*(5), 605–621.

Dieleman, F., & Wegener, M. (2004). Compact city and urban sprawl. *Built Environment, 30*(4), 308–323.

Ding, C. (2004). Urban spatial development in the land policy reform era: Evidence from Beijing. *Urban Studies, 41*(10), 1889–1907.

Doi, K. (2005). Multiagent land-use and transport model for the policy evaluation of a compact city. *Environment and Planning B: Planning and Design, 32,* 485–504.

Duany, A., Plater-Zyberk, E., & Speck, J. (2001). *Suburban nation: The rise of sprawl and the decline of the American dream*. Macmillan.

Elkin, T., McLaren, D., & Hillman, M. (1991). *Reviving the City: Towards sustainable urban development*. Friends of the Earth London.

EPU. (2010). *Tenth Malaysian plan 2011–2015*. Putrajaya: Economic Planning Unit.

Ewing, R. (1997). Is Los Angeles-style sprawl desirable? *Journal of the American Planning Association, 63*(1), 107–126.

Ewing, R., Pendall, R., & Chen, D. (2003). Measuring sprawl and its transportation impacts. *Transportation Research Record: Journal of the Transportation Research Board, 1831,* 175–183.

Force, U. T., & Rogers, R. G. (1999). *Towards an urban renaissance*. Spon London.

Frenkel, A., & Ashkenazi, M. (2008). The integrated sprawl index: Measuring the urban landscape in Israel. *The Annals of Regional Science, 42*(1), 99–121.

Frumkin, H., Frank, L., & Jackson, R. J. (2004). *Urban sprawl and public health: Designing, planning, and building for healthy communities*. Island Press.

Gainza, X., & Livert, F. (2013). Urban form and the environmental impact of commuting in a segregated city, Santiago de Chile. *Environment and Planning B: Planning and Design, 40*(3), 507–522.

Galster, G., Hanson, R., Ratcliffe, M. R., Wolman, H., Coleman, S., & Freihage, J. (2006). Wrestling sprawl to the ground-defining and measuring an elusive concept. *US Geological Survey professional paper, 1726*, 55–58.

Garreau, J. (1991). *Edge city: Life on the frontier*. New York.

Gillham, O. (2002). *The limitless city: A primer on the urban sprawl debate*. Island Press.

Gordon, P., & Richardson, H. W. (1997). Are compact cities a desirable planning goal? *Journal of the American Planning Association, 63*(1), 95–106.

Grant, J. (2002). Mixed use in theory and practice: Canadian experience with implementing a planning principle. *Journal of the American planning association, 68*(1), 71–84.

Grant, J. (2006). *Planning the good community: New urbanism in theory and practice* (Vol. 9). Taylor & Francis.

Gu, Z., Sun, Q., & Wennersten, R. (2013). Impact of urban residences on energy consumption and carbon emissions: An investigation in Nanjing, China. *Sustainable Cities and Society, 7,* 52–61.

Hayek, U. W., Jaeger, J. A., Schwick, C., Jarne, A., & Schuler, M. (2011). Measuring and assessing urban sprawl: What are the remaining options for future settlement development in Switzerland for 2030? *Applied Spatial Analysis and Policy, 4*(4), 249–279.

Herndon, J. D. (2011). *Mixed-use development in theory and practice: Learning from Atlanta's mixed experiences*. USA: Georgia Tech Library. Accessed at: http://hdl.handle.net/1853/40790

Hillman, M. (1996). In favour of the compact city. *The Compact City: A sustainable urban form*, 36–44.

Hoppenbrouwer, E., & Louw, E. (2005). Mixed-use development: Theory and practice in Amsterdam's Eastern Docklands. *European Planning Studies, 13*(7), 967–983.

Irwin, E. G., & Bockstael, N. E. (2007). The evolution of urban sprawl: Evidence of spatial heterogeneity and increasing land fragmentation. *Proceedings of the National Academy of Sciences, 104*(52), 20672–20677.

Jaeger, J. A., Bertiller, R., Schwick, C., Cavens, D., & Kienast, F. (2010). Urban permeation of landscapes and sprawl per capita: New measures of urban sprawl. *Ecological Indicators, 10*(2), 427–441.

Kelly-Schwartz, A. C., Stockard, J., Doyle, S., & Schlossberg, M. (2004). Is sprawl unhealthy? A multilevel analysis of the relationship of metropolitan sprawl to the health of individuals. *Journal of Planning Education and Research, 24*(2), 184–196.

Kropp, W. W., & Lein, J. K. (2013). Scenario analysis for urban sustainability assessment: A spatial multicriteria decision-analysis approach. *Environmental Practice, 15*(02), 133–146.

Le Goix, R. (2005). Gated communities: Sprawl and social segregation in Southern California. *Housing studies, 20*(2), 323–343.

Lin, J., & Yang, A. (2006). Does the compact-city paradigm foster sustainability? An empirical study in Taiwan. *Environment and Planning B: Planning and Design, 33*(3), 365.

Livingstone, K., & Authority, G. L. (2003). *Housing for a compact city*. Greater London authority.

Macfarlane, G. S., Garrow, L. A., & Mokhtarian, P. L. (2015). The influences of past and present residential locations on vehicle ownership decisions. *Transportation Research Part A: Policy and Practice, 74,* 186–200.

Milan, B. F., & Creutzig, F. (2016). Municipal policies accelerated urban sprawl and public debts in Spain. *Land Use Policy, 54,* 103–115.

Mindali, O., Raveh, A., & Salomon, I. (2004). Urban density and energy consumption: A new look at old statistics. *Transportation Research Part A: Policy and Practice, 38*(2), 143–162.

Morris, A. E. J. (2013). *History of urban form before the industrial revolution*. Routledge.

Mumford, L. (1961). *The City in History: Its origins and transformations, and its prospects*. Harcourt, Brace & World.

Musakwa, W., & Van Niekerk, A. (2013). Implications of land use change for the sustainability of urban areas: A case study of Stellenbosch, South Africa. *Cities, 32,* 143–156.

Naess, P., & Sandberg, S. L. (1996). Workplace location, modal split and energy use for commuting trips. *Urban Studies, 33*(3), 557–580.

Neuman, M. (2005). The compact city fallacy. *Journal of planning education and research, 25*(1), 11–26.

Newman, P., & Kenworthy, J. (1999). *Sustainability and cities: Overcoming automobile dependence*. Island Press.

Newton, P., Tucker, S., & Ambrose, M. (2000). Housing form, energy use and greenhouse gas emissions. *Achieving sustainable urban form* (pp. 74–83).

Norman, J., MacLean, H. L., & Kennedy, C. A. (2006). Comparing high and low residential density: Life-cycle analysis of energy use and greenhouse gas emissions. *Journal of Urban Planning and Development, 132*(1), 10–21.

Nozzi, D. (2003). *Road to ruin: An introduction to sprawl and how to cure it*. Greenwood Publishing Group.

Nurul, W. M. R. W. (2015). Compact urban form for sociability in urban neighbourhoods. *International Journal of Social Science and Humanity, 5*(10), 822.

Rabianski, J., Gibler, K., Tidwell, O. A., & Clements, J. S., III. (2009). Mixed-use development: A call for research. *Journal of Real Estate Literature, 17*(2), 205–230.

Rowley, A. (1996). Mixed-use development: Ambiguous concept, simplistic analysis and wishful thinking? *Planning Practice and Research, 11*(1), 85–98.

Schäfer, A. (2009). *Transportation in a climate-constrained world*. MIT Press.

Schwanke, D. (2003). *Mixed-use development handbook*. Urban Land Inst.

Schwarz, N. (2010). Urban form revisited—Selecting indicators for characterising European cities. *Landscape and Urban Planning, 96* (1), 29–47.

Scoffham, E., & Vale, B. (1996). How compact is sustainable—How sustainable is compact. *The Compact City: A sustainable urban form*, 66–73.

Senior, M., Webster, C., & Blank, N. (2004). Residential preferences versus sustainable cities: Quantitative and qualitative evidence from a survey of relocating owner-occupiers. *Town Planning Review, 75*(3), 337–357.

Song, Y., & Knaap, G.-J. (2004). Measuring the effects of mixed land uses on housing values. *Regional Science and Urban Economics, 34*(6), 663–680.

Song, Y., Merlin, L., & Rodriguez, D. (2013). Comparing measures of urban land use mix. *Computers, Environment and Urban Systems, 42,* 1–13.

Song, Y., & Rodríguez, D. A. (2005). The measurement of the level of mixed land uses: a synthetic approach. *Environment and Planning B*.

Southworth, F. (2001). On the potential impacts of land use change policies on automobile vehicle miles of travel. *Energy Policy, 29*(14), 1271–1283.

Stewart, J. (2005). *Room to move? Reconciling housing consumption aspirations and land-use planning*. Paper presented at the Social Market Foundation seminar The Wrong Kind of Housing.

Stone, B., Jr., Mednick, A. C., Holloway, T., & Spak, S. N. (2007). Is compact growth good for air quality? *Journal of the American Planning Association, 73*(4), 404–418.

Stone, B., Jr., & Rodgers, M. O. (2001). Urban form and thermal efficiency: How the design of cities influences the urban heat island effect. *Journal of the American Planning Association, 67*(2), 186–198.

Thinh, N. X., Arlt, G., Heber, B., Hennersdorf, J., & Lehmann, I. (2002). Evaluation of urban land-use structures with a view to sustainable development. *Environmental Impact Assessment Review, 22*(5), 475–492.

Tian, G., Ewing, R., White, A., Hamidi, S., Walters, J., Goates, J., et al. (2015). Traffic generated by mixed-use developments: Thirteen-region study using consistent measures of built environment. *Transportation Research Record: Journal of the Transportation Research Board, 2500*, 116–124.

Torrens, P. M., & Alberti, M. (2000). Measuring sprawl. *Centre for Advanced Spatial Analysis, University College, London* (Working paper no. 27).

TRB. (2005). *Does built environment influence physical activity? Examining the evidence* (Special Report). Washington, DC: Transportation Research Board.

Van Der Waals, J. (2000). The compact city and the environment: A review. *Tijdschrift voor economische en sociale geografie, 91*(2), 111–121.

Veneri, P. (2009). Urban polycentricity and the social costs of commuting. *Growth and Change: A Journal of Urban and Regional Policy, 41*(3), 403–429.

Weng, Y.-C. (2007). Spatiotemporal changes of landscape pattern in response to urbanization. *Landscape and Urban Planning, 81*(4), 341–353.

Wheeler, S. M. (2000). Planning for metropolitan sustainability. *Journal of Planning Education and Research, 20*(2), 133–145.
White, R. R. (1994). Strategic decisions for sustainable development in the Third World. *Third World planning review, 16*(2), 103.
Williams, K. (1999). Urban intensification policies in England: Problems and contradictions. *Land Use Policy, 16*(3), 167–178.
Williams, K., Jenks, M., & Burton, E. (2000). *Achieving sustainable urban form*. Taylor & Francis.
Yang, J., Shen, Q., Shen, J., & He, C. (2012). Transport impacts of clustered development in Beijing: Compact development versus overconcentration. *Urban Studies, 49*(6), 1315–1331.
Yeh, A., & Li, X. (2001). A constrained CA model for the simulation and planning of sustainable urban forms by using GIS. *Environment and Planning B: Planning and Design, 28*(5), 733–753.
Zagorskas, J., Burinskienė, M., Zavadskas, E., & Turskis, Z. (2007). Urbanistic assessment of city compactness on the basis of GIS applying the COPRAS method. *Ekologija, 53*(2), 55–63.

Part II
Practice/Implementation

Urban Sprawl Assessment

4

Abubakr A.A. Al-sharif, Biswajeet Pradhan and Saleh Abdullahi

4.1 Background

Urban form is related to the size, shape, and intensity of human settlements and the spatial distribution of various land use categories and urban features. Urban forms are described by several key variables, such as density, shape, degree of dispersal or concentration, and quality of public transport infrastructures.

Many studies show that urban form and pattern significantly affect urban sustainability, especially social and environmental sustainability. Therefore, assessing and controlling the current and future patterns of urban areas are essential in achieving urban sustainability paradigms. In this regard, mapping urban expansion can reveal the abstracted and simplified changes in urban regions (Abed and Kaysi 2003). Change detection in urban area applications has recently shifted from basic detection to the measurement of patterns, analysis of processes and patterns of urban expansion, and quantification of urban change. However, urban systems analysis covers a varied spectrum of scientific studies, such as political and historical processes (Flint 2002), urban crime analysis (Ceccato et al. 2002), site selection analysis (Abdullahi et al. 2014), urban population estimation (Zhang 2003), urban heat island research (Rigo and Parlow 2007), merging urban ecology and socioeconomics (Zhang et al. 2006), land use/cover evaluation (Yuan 2008), urban change analysis and the growth modeling (Park et al. 2011; Al-shalabi et al. 2013), compactness and sprawl assessment (Abdullahi et al. 2015; Al-sharif and Pradhan 2015) and so on.

Urban expansion is quantified by determining the amount of built-up areas between two periods (Bhatta 2009). As a process or pattern, sprawl quantification is a challenging subject that lacks a clear universal definition, and sprawl cannot be easily modeled or quantified (Bhatta 2010). As an alternative to urban sprawl, urban expansion modeling allows researchers to quantify the amount of areas that have been transformed for urban land use and to identify urban sprawl based on their judgment (Angel et al. 2007). However, this concept makes the urban sprawl phenomenon indistinct, thereby discouraging researchers from measuring urban sprawl. Although previous studies have measured urban expansion forms, they have several limitations in capturing urban sprawl characteristics. Such measurement processes also produce outcomes that cannot be easily interpreted because of the limitations and inaccuracy of the input information, including remote sensing data (Paolini et al. 2006; Bhatta et al. 2010b). Bhatta et al. (2010a) mentioned that urban sprawl could be assessed in absolute and relative scales. Relative assessments measure many elements that can be compared across various regions within a city, different metropolises, or different periods. Relative urban measurements allow analysts to decide whether the study area is sprawling or not, while absolute assessments clearly distinguish compact cities from sprawled ones. However, most urban sprawl measurement methods are relative measures that can be considered as indicators of urban sprawl. Therefore, defining a clear threshold is a crucial and challenging step in the absolute measurement of urban sprawl. Some researchers have proposed rules for defining such a threshold, but these rules are unclear to other urban experts. Urban relative measures may not provide wise conclusions on urban sprawl, and the defined threshold that is applied in one city may not be reliably applied in others.

Several statistics and spatial metrics have been proposed for evaluating and measuring urban sprawl. Spatial metrics are numeric figures that determine spatial patterns of land cover classes, land cover patches, or whole landscape mosaics of a geographic zone (McGarigal and Marks 1995). These metrics, known as landscape metrics, are used in landscape ecology to describe and identify main ecological relationships, such as adjacency and connectivity (Yeh and Huang 2009; Ramachandra et al. 2013). Apart from

B. Pradhan (ed.), *Spatial Modeling and Assessment of Urban Form*,
DOI 10.1007/978-3-319-54217-1_4

landscape ecology, the assumptions and approaches of landscape metrics can be used in various environments, such as urban areas (Pham et al. 2011). These spatial metrics have significant applications in identifying and quantifying urban growth, sprawl, and fragmentation (McGarigal and Marks 1995; McGarigal et al. 2002). These metrics are classified into three types. Class metrics are calculated for each class in the landscape, patch metrics are calculated for each patch in the landscape, and landscape metrics are calculated for the entire patch mosaic. Many spatial metrics in the literature are based on these three types, as shown in Table 4.1 (McGarigal and Marks 1995).

An important question in this field is the following: *what are the appropriate spatial metrics for urban growth/sprawl analysis and measurement?* Several researchers have defined eight dimensions of land use patterns to quantify urban sprawl (Galster et al. 2006). Angel et al. (2007) proposed numerous landscape metrics for evaluating, manifesting, and characterizing urban sprawl, but did not recommend any sprawl threshold for distinguishing growth from sprawl. Moreover, the results that are derived from applied metrics are confusing and difficult to comprehend because some metrics may contradict one another.

Many researchers have established multiple indices through descriptive statistical and geographic information system (GIS) analysis to measure urban sprawl (Hasse 2004). These urban indices analyze various aspects, such as employment, resource consumption, population, living quality, and architecture aesthetics. The most-used indices include spatial configuration (accessibility, fragmentation, and proximity), density (employment, residential, and population densities), growth rate (built-up area and population), land use efficiency, and per capita land consumption (Sutton 2003; Jiang et al. 2007). However, the following questions still demand straight answers: (1) What is the per capita consumption of land in sprawling cities? and (2) What is the built-up area expansion rate in non-sprawling cities?

Torrens (2008) suggested that urban sprawl should be analyzed and measured at multiple scales and should cover gross and net lands. To provide a clear insight into sprawl, Torrens examined this concept at the metropolitan area, intra-urban, and land parcel levels using 42 measures. However, the complexity of such methodology resulted in confusing outcomes because of the various metrics and scales used. Jiang et al. (2007) suggested 13 integrated geospatial indices for measuring urban sprawl in the Beijing metropolis. Such an approach required minimal interpretation effort yet required extensive inputs of multi-temporal data, such as GDP, population, land use master planning, land use maps, highway maps, floor area ratio, and city center maps. Given that developing countries have insufficient temporal data, most of the proposed indices are difficult to derive. Jiang et al. also did not propose a clear threshold for characterizing sprawl. However, the applied temporal analysis is a valuable technique for comparing various cities and zones within a metropolis or the status of an urban area at different periods. The number of metrics to be applied presents another problem in urban sprawl measurement. Some researchers have proposed and compared various metrics for urban expansion analysis. However, such comparisons did not yield a standard set of urban spatial metrics for measuring urban sprawl (Alberti and Waddell 2000; Herold et al. 2003). Given that several spatial metrics are correlated and produce redundant information, some of these metrics cannot quantify different patterns. Urban analysts must use metrics that are relatively independent of one another to produce a reliable measure of urban sprawl and to achieve a meaningful detection of the urban landscape structure. Many metrics are often necessary to describe and characterize urban landscape because a single measure cannot cover everything (Turner et al. 2001). However, different spatial metrics may also produce varied conclusions (Herold et al. 2003)

The spatial resolution of remotely sensed data presents another challenge in urban analysis and sprawl assessment. Several metrics, such as spatial heterogeneity and patch analysis, are dependent on spatial resolution. For example, separate objects may appear falsely compact and may be wrongly merged together. On the one hand, a greater spatial resolution corresponds to an improved urban sprawl interpretability. On the other hand, a very high spatial resolution leads to high object diversity that can produce unexpected problems, such as an additional number of patches or increased heterogeneity.

In the multi-temporal analysis of urban sprawl, using different resolutions will also render resolution-dependent metrics unusable. The intensity of annual urban expansion and the density of built-up areas efficiently describe the sprawl characteristics of rapidly changing and low-density areas, but these metrics have weak ability in identifying the specific spatial patterns of urban growth and sprawl. These spatial metrics are not explicitly common; for instance, the expansion of built-up areas to the growth of households in a city. This category of urban metrics measures what is present without referring to a particular location on the landscape.

The entropy method of Shannon is a very popular technique that determines urban sprawl by integrating GIS into remotely sensed data (Yeh and Xia 2001; Kumar et al. 2007; Ramachandra et al. 2013). Relative entropy is used to measure entropy values ranging from 0 to 1. According to Yeh and Li (2001), given that the entropy method can evaluate the distribution of a geographical phenomenon, one can determine the degree of urban sprawl change by measuring the entropy difference between two time instances. The entropy method is also more spatial, robust, and static than other methods (Bhatta et al. 2010a). Many studies show

Table 4.1 Main urban landscape metric from 'FragStats' software

Metrics	Patch	Class	Landscape
Area-edge	Patch area		
	Radius of gyration		
	Patch perimeter	Largest patch index	
	–	Total edge	
	–	Edge density	
	–	Total area	
	–	Percentage of land scape	–
Shape	Perimeter area ratio		
	Shape index		
	Fractal dimension index		
	Related circumscribing circle		
	Contiguity index		
	–	Perimeter area fractal dimension	
Core area	Core area		
	Number of core area		
	Core area index		
	–	Total core area	
	–	Core area percent of landscape	–
	–	Number of disjunction core area	
	–	Disjunction core area density	
Contrast	Edge contrast index		
	–	Contrast weighted edge density	
	–	Total edge contrast index	
Aggregation	Euclidean nearest neighbor distance		
	Proximity index		
	Similarity index		
	–	Number of patches	
	–	Patch density	
	–	Landscape division index	
	–	Splitting index	
	–	Effective mesh size	
	–	Aggregation index	
	–	Clumpiness	Contagion
	–	Landscape shape index	
	–	Normalized LSI	Patch cohesion index
Diversity	–	–	Patch richness
	–	–	Patch richness density
	–	–	Relative patch richness
	–	–	Shannon's diversity index
	–	–	Simpson's diversity index
	–	–	Shannon's evenness
	–	–	Simpson's evenness

that this method outperforms other spatial dispersal statistics, such as Moran's coefficient, which depends on the shape, size, and number of zones used to compute the statistics (Tsai 2005). Relative entropy also alleviates the scale effect of the modifiable areal unit problem (Yeh and Xia 2001; Bhatta et al. 2010a).

In the case of a well-planned city, future urban expansion is often modeled and planned using a highly advanced approach. Given that urban growth is generally restricted within clearly defined borderlines, real urban growth cannot exceed the planned expansion, and the extent to which the observed urban expansion meets the expected or planned expansion needs to be evaluated. Pearson's chi-square statistics (degree of freedom) measures and compares the observed expansion with the expected expansion (Almeida et al. 2005). However, given that this technique considers urban expansion both as a process and pattern, chi-square cannot distinguish the urban expansion patterns from its processes. The chi-square and entropy methods are different urban growth measures that may be unrelated to each other. Bhatta et al. (2010a) proposed a degree-of-goodness measure to assess the degree of relation and compactness between observed and planned expansions. However, this measure is a new technique in urban analysis and assessment that requires further study and evaluation because of its limitations.

4.2 Case Study: Methodological Process of Urban Sprawl Assessment for Tripoli Metropolitan Area, Libya

As a case study, this section assesses and analyzes the spatiotemporal patterns of urban expansions in the Tripoli metropolitan area based on the urban sprawl assessment concept. Urban expansion and sprawl are assessed and investigated as a pattern and process using Urban Expansion Intensity Index (UEII), population and urban expansion proportions, landscape metrics, entropy model, and degree of freedom model.

Libya lies along the Mediterranean coast and stretches deep into the Saharan region. Although Libya mostly consists of rocky plains and sandy seas, a narrow band of fertile lowlands stretches across the northern edge of the country. Nearly three-fourth of the Libyan population resides in urban areas that occupy only 1.5% of the country's land area along the coast. As the capital city of Libya, the Tripoli agglomeration has the largest concentration of population and economic activities not only in the Tripoli region but also in the entire country. Therefore, this area has a very important role in the socioeconomic development of the country (UPA 2009). The study area is located along the Mediterranean coast in the northwestern part of Libya between longitudes 12°54′ 04″ E and 13°26′ 38″ E and between latitudes 32°36′ 18″ N and 32°54′ 17″ N, and occupies a total land area of approximately 1143.73 km^2. The Tripoli metropolitan area includes the districts of Tripoli Center, Hey Alandalus, Tajoura, Janzur, Kaser Ben Ghashir, Alswani, AinZara, AbuSlim, and Suq Aljumma (Fig. 4.1).

Regarding urban sprawl, the Urban Planning Agency reported in 2009 that citizens generally demanded better housing and larger land to build their houses. Those areas that can accommodate such demands are generally situated in the peripheries of urban areas and mostly comprise agricultural land within the agglomeration boundary. Sprawl can easily occur in areas without careful land use planning and with cheap personal transportation. Sprawl is not a land use category, but a settlement type that may take several forms. In Tripoli, sprawl may be observed in housing, industries, commercial activities, and services that are mixed with the old agricultural landscape. Sprawl may be either legal or illegal, but the category in itself is very extensive and all-encompassing. Economy is also an important factor in the spread of suburban sprawl. The lands near the city center have very high prices, while those located far away from the center have very modest prices. Numerous industrial enterprises are not welcomed in Tripoli, and many young couples cannot afford a site or a house in the city. However, couples with relatives who own farmlands may build and expand such lands to meet their increasing economic and family needs.

Tripoli was selected for this case study because of its significant growth over the last decades, but the urban growth patterns of this city were never analyzed in the literature. Tripoli serves many functions, such as the political center; the economic, industrial, and service capital; and the communicator of Libya with other countries. Despite the urban plans of the Libyan government, Tripoli has witnessed a rapid yet haphazard urban growth over the past decades. Corruption, political unrest, and economic conditions may have affected urban planning and subsequently resulted in massive urban sprawl. Therefore, the main concerns for Tripoli include its fast and uncontrolled urban expansion and the conversion of fertile green lands and environmental reserve areas, both of which have resulted in socioeconomic and physical problems.

4.2.1 Data Collection and Preprocessing

Remotely sensed and GIS-prepared data were used in this study (Table 4.2). Unfortunately, the available data are limited in quantity and the images cannot be collected for equal periods.

Several approaches have been developed and used for the preparation, processing, and extracting of information from remotely sensed data. Moreover, the selection of algorithms or methods to be applied depends on the objective of the study. For this study, the ENVI and ARC/INFO GIS software packages were used for image processing, generating classified land cover/land use maps, spatial analysis, and map production.

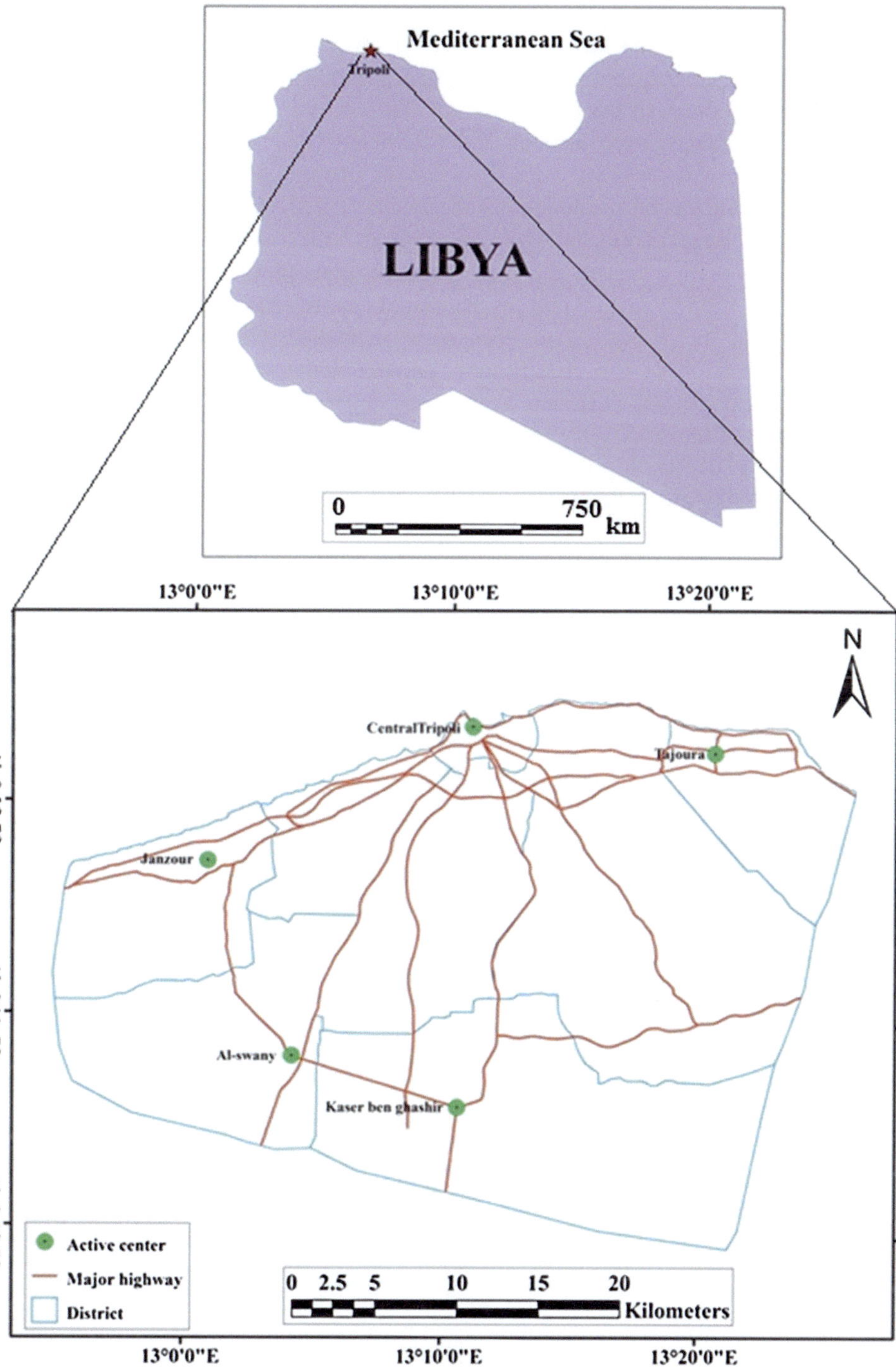

Fig. 4.1 The location map of Tripoli metropolitan (Libya)

Table 4.2 Collected data and sources

Data	Source of data
Landsat image 1984 (30 m resolution)	Biruni remote sensing center
Landsat image 1996 (30 m resolution)	Biruni remote sensing center
Spot 5 image 2002 (2.5 m resolution)	Biruni remote sensing center
Spot 5 image 2010 (5 m resolution)	Libyan centre for remote sensing and space science
Roads network (shape file)	Urban planning agency, Libya
Population data census	General information authority, Libya
Digital contour map (5 m interval)	Biruni remote sensing center

The collected images were standard products that were radiometrically and geometrically corrected. However, image supplying agencies adopt different standards, thereby resulting in low accuracy of the image overlay. To address this problem, the images were rectified and georeferenced during preprocessing to achieve a highly accurate image overlay. The varying spatial resolutions of the images used in this study were corrected by resampling the high-resolution images to match the low-resolution ones. Given that the resampling process reduces the spatial details, the pixel sizes of the images were unchanged to avoid changing the precision of the classification process with various radiometric spectral and spatial resolutions.

A vector map of the Tripoli metropolitan area was used for clipping images. The maximum likelihood supervised classification method was applied to the images during the classification process. All images were then classified by selecting precise samples (78 polygons) as training area samples to present different classes for each individual image. Three classes, namely, built-up (impervious surfaces), non-built-up (agriculture), and restricted or excluded areas, were investigated. The classified images were then resampled to the same spatial resolution (30 m × 30 m), in which each map contains 1,816,750 cells. The pixel size was selected to avoid reducing the spatial details of the images. Therefore, resampling was conducted after the image classification.

For modeling input, thematic raster maps of all variables were prepared and calculated in the Arc-Info GIS environment, and were presented in raster maps with grid cell sizes of 30 m × 30 m as shown in Figs. 4.2, 4.3, 4.4 and 4.5. The independent input data included the following:

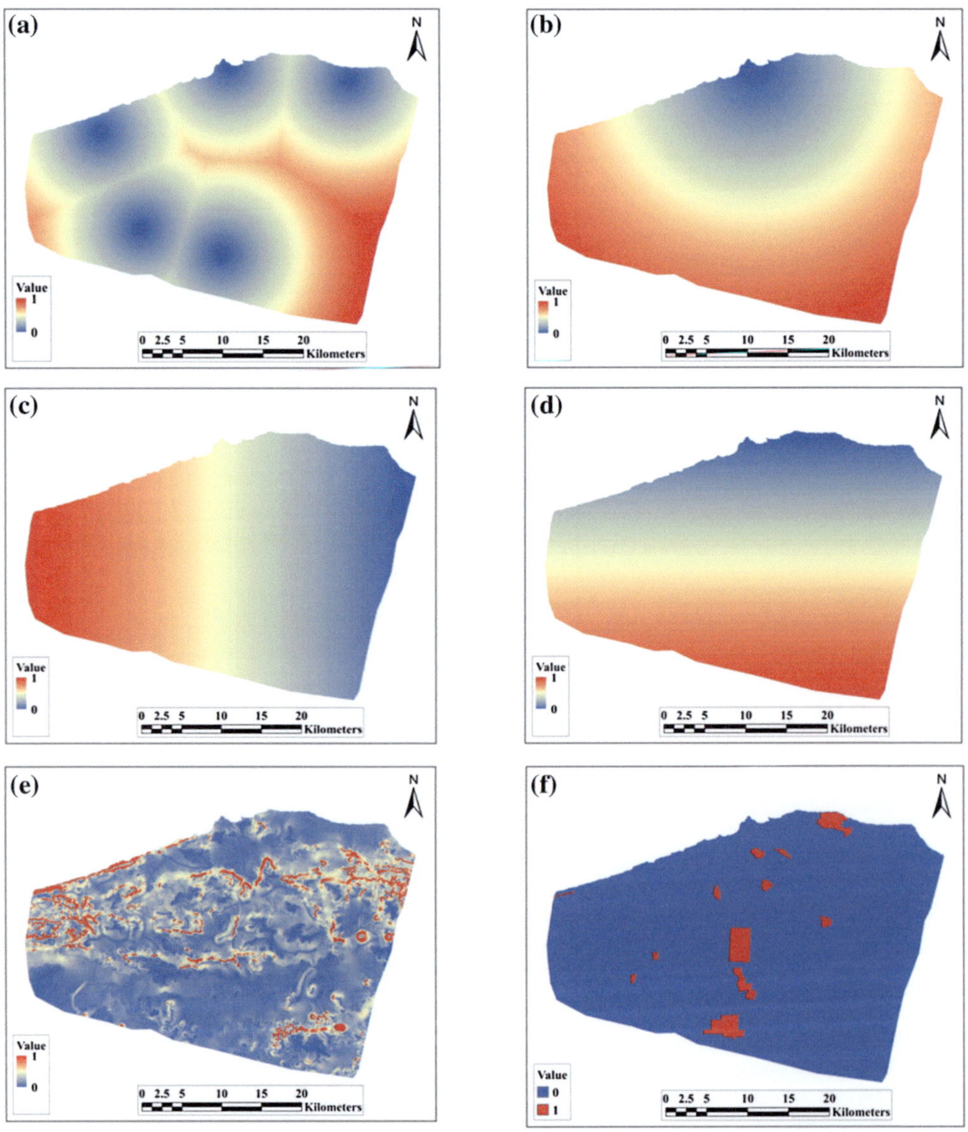

Fig. 4.2 Thematic raster maps of independent variables: **a** Distance to active economy centers, **b** Distance to CBD, **c** Easting coordinate, **d** Northing coordinate, **e** Slope, **f** Restricted areas, **g** Distance to nearest urbanized area, **h** Population density, **i** Distance to educational area, **j** Urban area, **k** Distance to roads, **l** Distance to coastal area

Fig. 4.2 (continued)

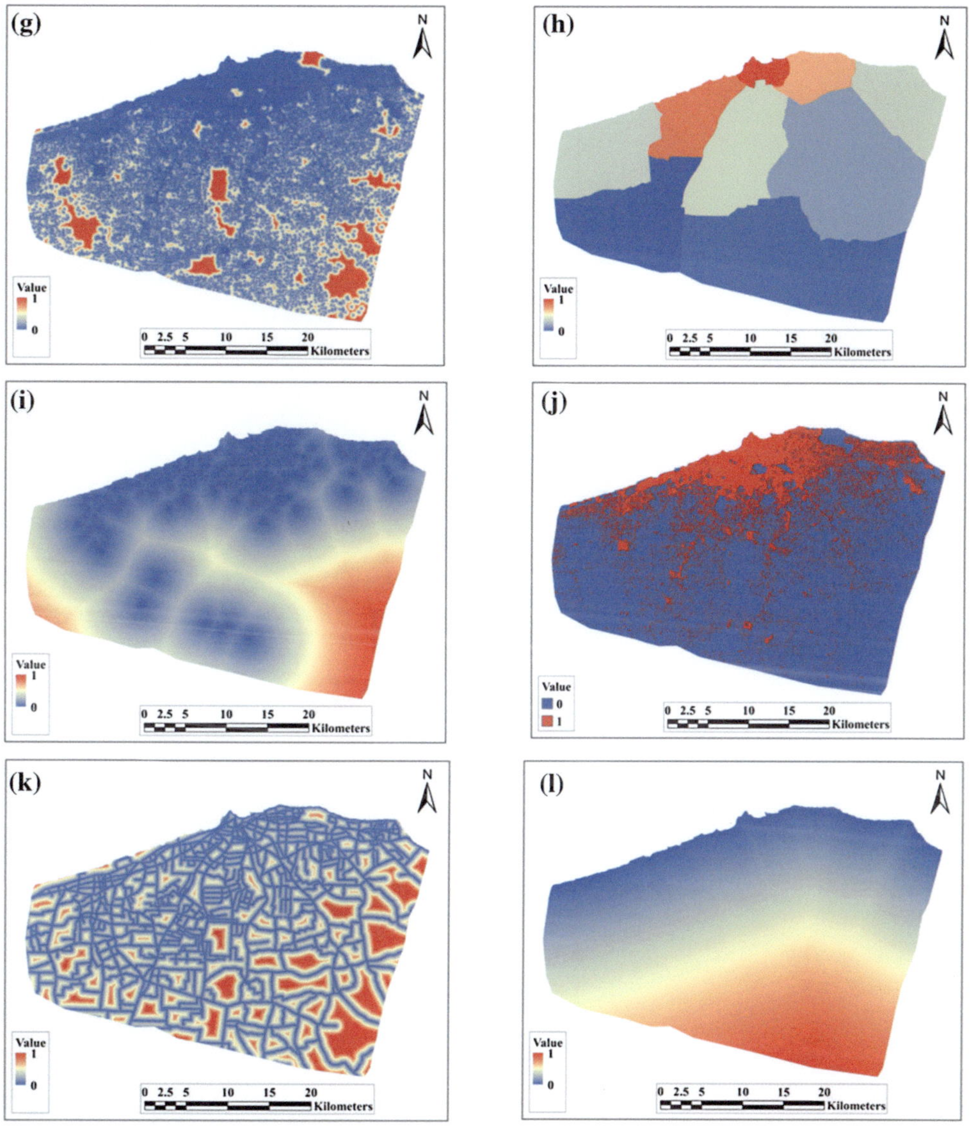

- Distance to active economy centers
- Distance to central business districts (CBDs)
- Easting coordinate
- Northing coordinate
- Slope
- Restricted areas
- Distance to nearest urbanized area
- Population density
- Distance to educational area
- Urban area
- Distance to roads
- Distance to coastline.

All prepared data were converted to ASCII and IDRISI formats for further analysis and simulation using the IBM SPSS Statistics 20, IDRISI-Selva, and FRAGSTATS software.

4.2.2 Zones Division to Assess Urban Expansion and Sprawl

Most urban application processes and projects are based on zonal format. Urban sprawl assessments cover many different zone divisions, such as circular buffer zones for the city center, buffer zones for roads, eight-pie sections to represent eight directions, designed transects along the axes of urban expansions, and zone division based on administrative boundaries (Bhatta et al. 2010a; Sarvestani et al. 2011; Yue et al. 2013). Given that the entropy method does not depend on the type and number of divisions or the zoning manners of the study area, we applied two approaches to assess the urban expansion patterns of the study area. The first approach is based on the administrative boundaries of districts in the study area (Fig. 4.1), while the second approach divides the study area into five

Fig. 4.3 Urban expansion from 1984 to 2002

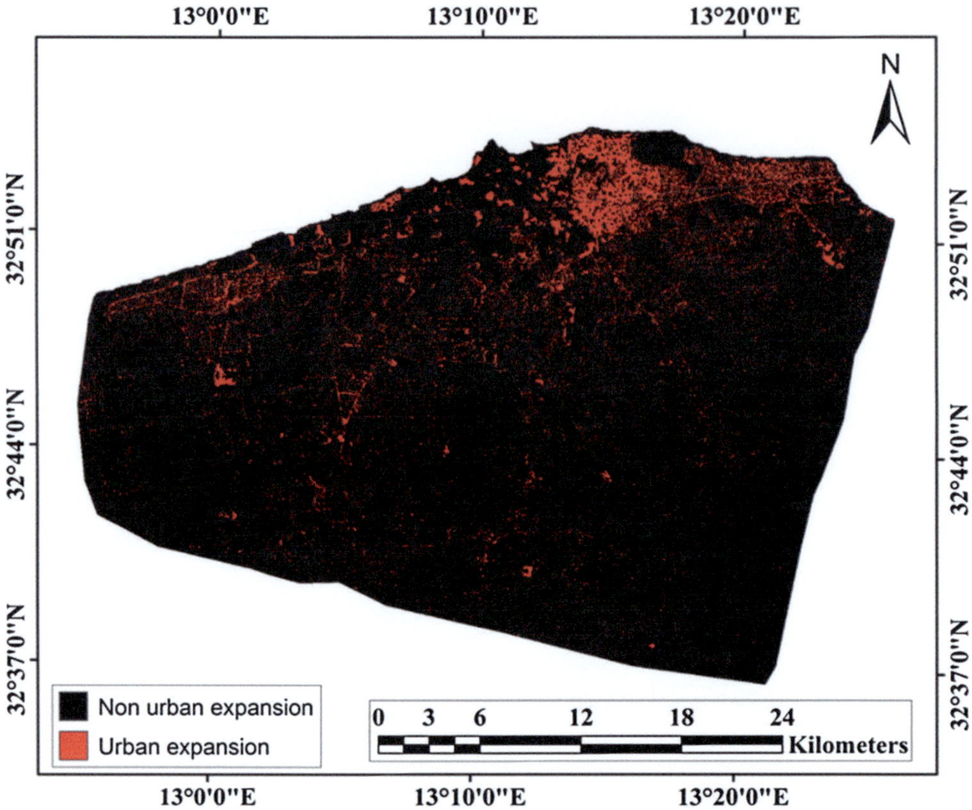

Fig. 4.4 Urban expansion from 1996 to 2002

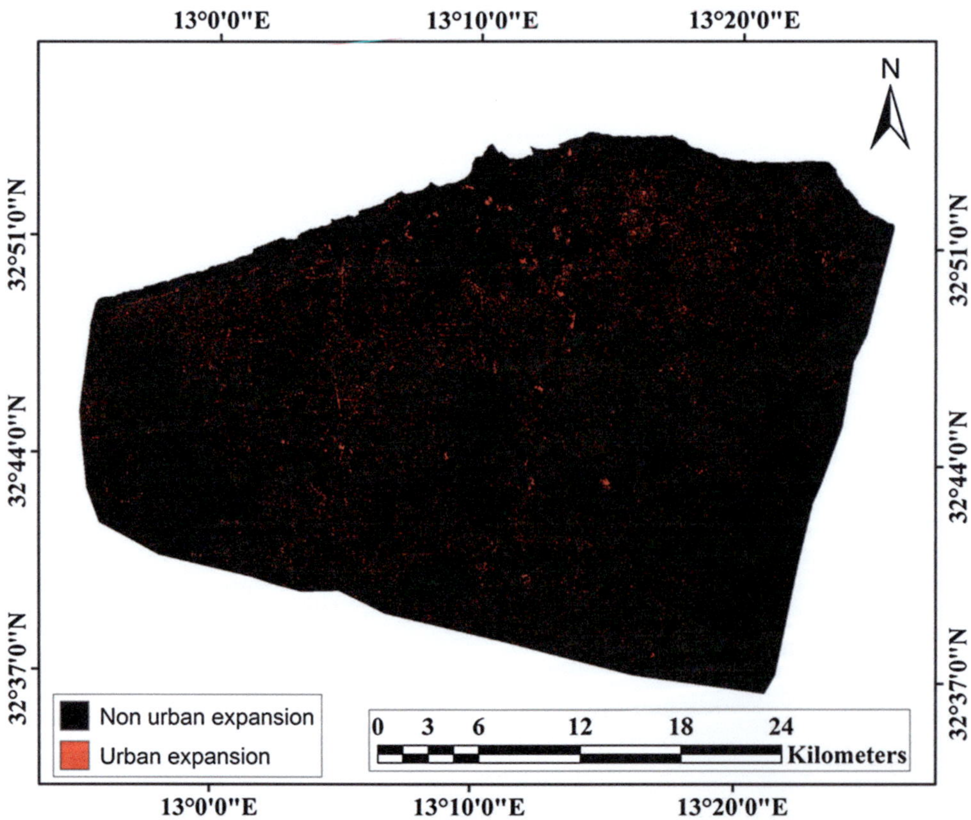

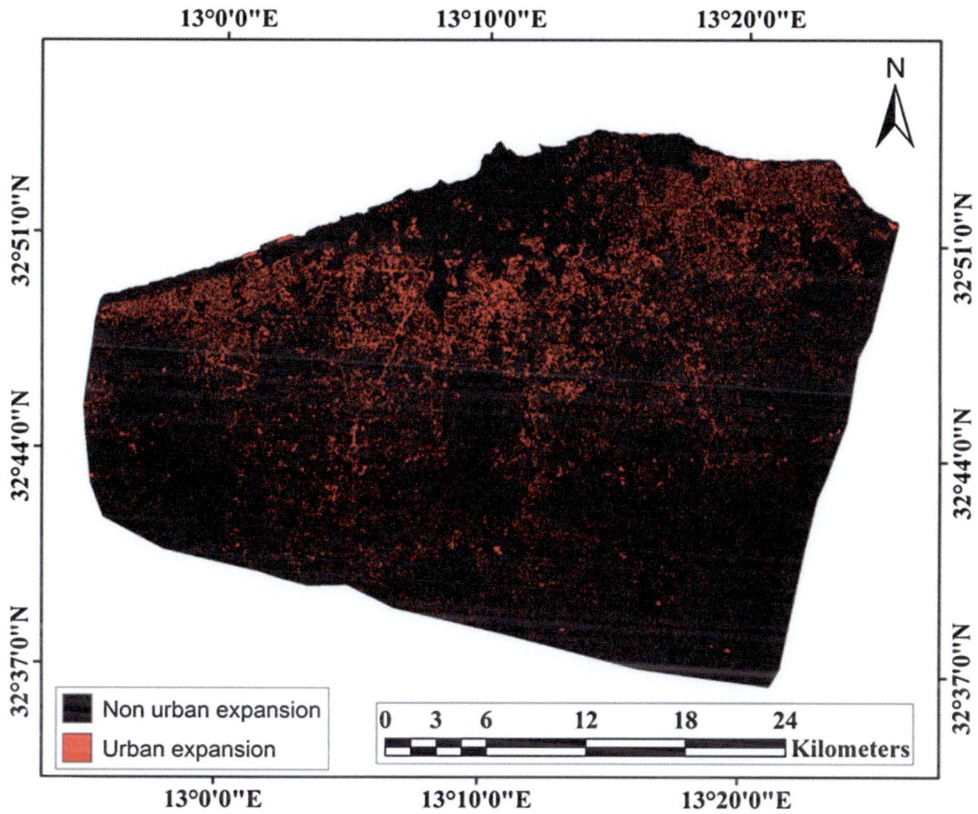

Fig. 4.5 Urban expansion from 2002 to 2010

pie sections to consider and assess urban sprawl direction. These sections are then divided into multiple zones to consider the effect of distance from CBD and to determine the sprawl in each zone, as shown in Fig. 4.6.

The aforementioned approach provides additional details on the urban growth process and its patterns in the entire study area, for each zone, and at different periods. However, the central point of Tripoli, which matches the CBD and represents the starting point of the urbanization process along the history of the study area, had already been determined. Consequently, a 51-zone vector map of the study area was used to clip the classified imageries and to divide the area into 51 zones (Fig. 4.6). The urban growth and built-up area for all zones and for each temporal point were calculated with the respective zone borders by multiplying pixel size by the number of pixels in each zone.

The following sections present different approaches for spatially analyzing and assessing urban development. These techniques identify whether the urban expansion process and its patterns can be considered either as sprawl or growth. These assessments provide a clear understanding of urban growth/sprawl and highlight urban sprawl from different perspectives.

4.2.3 Built-up Area and Population

The urbanization process (urban sprawl) is among the major significant drivers of land cover/use change and is mainly associated with population growth (Barredo et al. 2003; Weng 2007). Rapid population growth is also considered as the main factor in urban sprawl and urban spatial problems (Maktav et al. 2005; Bhatta 2009; Sarvestani et al. 2011). Urban sprawl is directly related to socioeconomic information and population figures. For example, unexpected population growth, which is associated with unplanned development activities, will result in uncontrolled urban sprawl with poor infrastructure and economic performance (Sarvestani et al. 2011). Unplanned development and urban sprawl occur when the percentage of built-up areas exceeds that of population growth (Barnes et al. 2001; Soffianian et al. 2010). Therefore, the population in an area is an important metric for measuring the urban sprawl process. The proportion of total built-up areas in the total population is a simple acceptable measure for identifying and quantifying urban sprawl (Sharma et al. 2012; Sandhya Kiran and Joshi 2013).

In the case study, the amounts of built-up lands for districts in four time instances were obtained by clipping

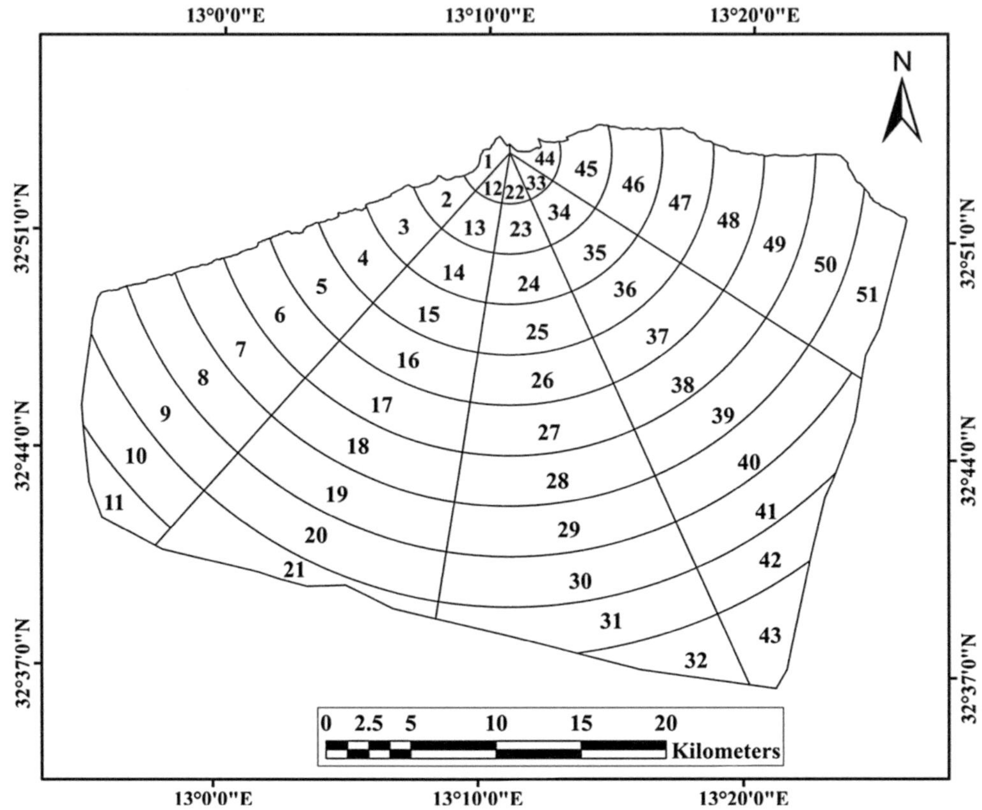

Fig. 4.6 Zone divisions of the study area

the final classified images using the provided vector map of districts. The quantity of pixels in each district was multiplied by grid size. The imaging time of some images deviated from the population census dates (1984, 1995, and 2006). To address this problem, the population data were interpolated to match the RS data dates. For the 2010 population, the annual population growth rate from 1995 to 2006 was assumed constant. Therefore, the 2010 population was computed based on the 2006 population. Table 4.3 shows the observed and interpolated population figures for Tripoli. The built-up area and population data were used to assess and investigate the urban sprawl process and its spatial patterns. The relationships among built-up area, population, and urban sprawl were analyzed further to provide a clear understanding of the urbanization in the study area.

Table 4.3 Total population and total area (km^2) of Tripoli districts

District	Year				District area
	1984	1996	2002	2010	
Central Tripoli	132,505	128,270	129,130	130,354	16.62
Suq Aljumma	81,378	103,207	118,124	131,650	44.03
Tajoura	59,575	93,852	110,281	133,677	82.93
AinZara	145,562	184,619	206,501	237,662	256.03
Kasr Ben Ghashir	49,348	66,782	76,829	83,137	274.62
AbuSlim	201,886	212,156	221,280	234,273	137.53
Hey Alandlus	218,317	240,051	240,454	241,029	68.91
Janzour	87,331	126,593	144,369	169,681	109.20
Alswani	13,340	17,396	19,842	23,324	153.86
Total population	**989,242**	**1,172,928**	**1,266,810**	**1,384,787**	**1,143.73**

Source General information authority, Libya

4.2.4 Observed and Theoretical Expected Built-up Area Expansion

The observed expansion must be compared with the forecasted urban expansion to understand the divergence of urban growth. However, unlike developed countries, developing countries usually lack clear urban plans or estimations of urban expansion. Equation (4.1) was used to calculate the theoretical expected urban growth for all zones in each period as follows (Almeida et al. 2005):

$$G_i^e = \frac{G_t^z \times G_t^t}{G}, \quad (4.1)$$

where G_i^e is the expected urban growth in the zone, G_t^t is the total growth in three periods in one zone, G_t^z is the total growth in 51 zones in one period, and G is the total growth of the study area in all periods.

4.2.5 Urban Expansion Intensity Index (UEII)

In the urbanization process, the expansion differs per region and direction because of the policy on urban driving factors and their spatial effects. Such factors include road network, population density, slope, and economics. The differences in expansion are referred to as the preference of urban growth. In this process, the UEII was employed to assess and analyze quantitatively the differences in urban spatial expansion as well as to recognize the preference of urban growth in a certain period (Ren et al. 2013).

The UEII reflects the future direction and potential of urban expansions as well as compares the speed or intensity of urban land use change in different periods. UEII values ranging from 0 to 0.28, 0.28 to 0.59, 0.59 to 1.05, 1.05 to 1.92, and >1.92 indicate slow, low-speed, medium-speed, high-speed, and very high-speed development, respectively. The UEII for the entire study area, each temporal span, and each zone is calculated as follows:

$$\mathrm{UEII}_{it} = \left[\frac{\mathrm{ULA}_{i,b} - \mathrm{ULA}_{i,a}}{t}\right] / \mathrm{TLA}_i \times 100, \quad (4.2)$$

where UEII_{it} is the annual average UEII of the ith zone in period t; $\mathrm{ULA}_{i,a}$ and $\mathrm{ULA}_{i,b}$ are the quantities of the built-up area at periods a and b in the (ith) spatial zone, respectively; and TLA_i is the total area of the (ith) spatial zone.

4.2.6 Shannon's Entropy Model

The entropy technique of Shannon is widely used to study the urban sprawl phenomenon (Ramachandra et al. 2013). As a favorable measure of spatial dispersion or concentration, the entropy model can be used to analyze and assess any geographical variable, reveal the configuration and orientation of spatial patterns, and investigate spatial variables within a GIS (Yeh and Xia 2001; Sudhira et al. 2004; Kumar et al. 2007). Previous urban researches have used this technique to analyze the urban patterns and identify the urban sprawl of a specific area for a specified period, but do not consider sprawl direction, sprawl variation, and distance to CBD. The level of urban sprawl is represented by the entropy value, which ranges from 0 to $\log_e(n)$. A value of 0 indicates compact distribution in an urban area, while a value nearer to $\log_e(n)$ indicates dispersed distribution. In other words, high entropy values indicate sprawl occurrence (Yeh and Xia 2001; Sandhya Kiran and Joshi 2013).

In this study, the absolute value of entropy H_n was determined as follows:

$$H_n = \sum_i^n P_i \log_e\left(\frac{1}{P_i}\right), \quad (4.3)$$

where P_i is the percentage of the variable in the ith district (i.e., percentage of built-up lands in each district) obtained by dividing the proportion of built-up lands in the ith district by the total proportion of built-up areas in all districts, and n indicates the number of districts (n = 9).

The relative entropy as demonstrated in Eq. (4.4) can be obtained by dividing the calculated absolute Shannon's entropy by $\log_e(n)$. The relative entropy value always varies between 0 and 1, with a value of 0 denoting compact distribution and values near 1 indicating dispersed distribution.

$$H_n = \sum_i^n P_i \log_e\left(\frac{1}{P_i}\right) / \log_e(n) \quad (4.4)$$

To evaluate the urban sprawl process, the change rate of urban dispersion (change in sprawl) between two periods was calculated as follows:

$$\Delta H_n = H_n(t_2) - H_n(t_1), \quad (4.5)$$

where ΔH_n denotes the variation of entropy values between two periods, $H_n(t_1)$ is the relative entropy at time (t_1), and $H_n(t_2)$ is the relative entropy at time (t_2).

Using the proposed zone division approach, we recalculated the relative entropy values using Eq. (4.4), where P_i is the probability or percentage of the variable occurring within zone i (i.e., percentage of urban area in the ith zone determined by the urban area in the ith zone divided by the zone area), and n is the total number of zones (n = 51).

4.2.7 Degree of Freedom Model

Pearson's chi-square was used to check the degree of freedom between pairs of variables and to describe the same class of land cover change (Almeida et al. 2005; Bhatta et al. 2010a). In this study, the degree of freedom for urban growth was calculated as follows:

$$\frac{(\text{Observed Growth} - \text{Expected rowth})^2}{\text{Expected Growth}} \quad (4.6)$$

Pearson's chi-square statistics estimates the freedom or degree of variation between the observed and expected urban growth. The chi-square statistics for each time period was computed using Eq. (4.7) as follows:

$$D_i^t = \sum_i^n D_i^z, \quad (4.7)$$

where D_i^t is the degree of freedom of growth in the ith period, and D_i^z is the degree of freedom of growth for the ith zone in the same period.

The degree of freedom for each zone was computed using Eq. (4.8) as follows:

$$D_i^z = \sum_i^n D_i^t \quad (4.8)$$

The overall degree of freedom of the study area was calculated by summing the degrees of freedom across all periods or zones. The lower limit of the chi-square was 0, which indicates that the observed and expected growth values are equal.

4.2.8 Landscape Metrics

Numerous landscape metrics were developed, tested, and used for landscape structure and composition analysis in the last three decades (Turner et al. 1989; Yeh and Huang 2009). Generally, the urban sprawl process changes and modifies the landscape compositions over time by increasing landscape fragmentation and generating small urban patches. In this study, six landscape metrics analyses were performed to investigate and analyze the spatiotemporal patterns of urban sprawl in the study area and to assess sprawl from different prospective. The applied landscape metrics, which measure clumpiness, aggregation, complexity, and level of dispersion of urban area classes in the study area landscape (Taubenböck et al. 2009), include edge density (ED), largest patch index (LPI), shape index (SHAPE), landscape shape index (LSI), patch density (PD), and Simpson's evenness index (SIEI). For urban landscape analysis, the FRAGSTATS Version 4 (McGarigal et al. 2002) statistical package was employed to calculate all landscape quantitative measures as shown in Table 4.4.

Table 4.4 Description of landscape metrics used to investigate and quantify urban sprawl patterns in Tripoli (McGarigal et al. 2002)

Landscape metrics	Description
Edge density (ED)	$\text{ED} = \frac{E}{A}(10,000)$ $\text{ED} \geq 0$, without limit (in hectares) E = total length (m) of edge in landscape A = total landscape area (m^2)
Largest patch index (LPI)	$\text{LPI} = \frac{\text{MAX}(a_{ij})}{A}(100)$, $0 < \text{LPI} \leq 100\%$ a_{ij} = area (m^2) of patch ij A = total landscape area (m^2)
Shape index (SHAPE)	SHAPE equals patch perimeter (m) divided by the square root of patch area (m^2), adjusted by a constant to adjust for a square standard $\text{SHAPE} = \frac{0.25P_{ij}}{\sqrt{a_{ij}}}$ SHAPE ≥ 1, without limit SHAPE = 1 when the patch is square and increases without limit as patch shape becomes more irregular
Landscape shape index (LSI)	$\text{LSI} = \frac{0.25E}{\sqrt{A}}$, LSI ≥ 1, without limit E = total length (m) of edge in landscape; includes the entire landscape boundary and some or all background edge segments A = total landscape area (m^2) LSI increases without limit as landscape shape becomes more irregular and/or as the length of edge within the landscape increases
Patch density (PD)	$\text{PD} = \frac{N}{A}(10,000)(100)$, $\text{PD} > 0$, constrained by cell size N = total number of patches in the landscape A = total landscape area (m^2)
Simpson's evenness index (SIEI)	$\text{SIEI} = \frac{1-\sum_{i=1}^{m} P_i^2}{1-\left(\frac{1}{m}\right)}$, $0 \leq$ SIEI ≤ 1 P_i = proportion of the landscape occupied by patch type (class) i m = number of patch types (classes) present in the landscape, excluding the landscape border if present

4.3 Results and Discussion of Urban Sprawl Assessment for the Tripoli Metropolitan Area (Libya)

This section presents the results of investigating and assessing urban sprawl using the applied approaches, and then discusses these findings in detail with respect to the study area.

4.3.1 Urbanized Area and Urban Growth

The satellite images were classified into non-built-up and built-up areas for the four temporal dates. Figure 4.7 shows the abstracted and simplified visual maps of urban extents at specific periods for the study area. The classified maps were assessed using the confusion matrix method. Real ground reference polygons were compared with the classified output maps to assess accuracy. The classified maps for 1984, 1996, 2002, and 2010 obtained accuracy values of 91, 93.2, 95.7, and 94% as well as Kappa coefficient values of 0.89, 0.93, 0.94, and 0.93, respectively. Figure 4.7 shows that the urban expansions of Tripoli have different signatures. Specifically, some zones have very dense built-up areas, while wide open spaces are present between urbanized areas. In some regions, the edges between urban and nonurban areas are very clear, whereas the two classes are very near each other in other areas. The signatures of the urbanization process in each district differ across periods and behaviors. The infill of the non-built-up areas between previously urbanized areas, which leads to an increased compactness level, can also be observed. Based on the classified images, the study area demonstrates dispersed growth, especially over the last decade.

Quantitative measures that sum up the urban growth properties of the study area are necessary to illustrate different urban patterns, compare districts/zones, and identify how the districts/zones transform over time. The amount of built-up areas was used as a primary quantitative measure and a direct indicator of urban development patterns. The percentages of built-up areas were computed for each district and for all classified images as shown in Fig. 4.8.

Figure 4.8 shows the concentration percentages of built-up areas in each district. A noticeably large concentration rate of built-up areas was observed in Central Tripoli, Suq Aljumma, Hey Alandalus, and AbuSlim. Central Tripoli and the adjacent zones particularly had an extremely high concentration rate. The urban growth rate in these districts was also lower than that of other districts because of built-up area saturation. These results indicate the controlled urban growth in these districts.

The percentages of built-up areas in Tajoura, Janzour, AinZara, Kaser Ben Ghashir, and Alswany were lower than 20% in 1984, 1996, and 2002. However, these districts had high urban expansion ratios between 2002 and 2010. The percentages of urban areas in these districts remained lower than 50%, which indicated the higher availability of urban growth lands in these districts than in other districts. The urban expansion rates in these districts almost doubled in 2010, which is highly alarming because of the high urban growth ratio and clear dispersion growth patterns. The overall urban expansion rate of the study area continuously increased, especially over the last decade during which the growth rate exceeded 40% of the urban extent in 2002. These findings suggest the urgent need to control the dispersed urban sprawl by applying a suitable urban plan and a wise urban policy; otherwise, the situation will worsen.

The proposed division map of the study area (Fig. 4.6) is used to calculate the amount of built-up areas in each zone and to highlight the effect of direction and distance to CBD on urban expansion quantities in four periods (1984, 1996, 2002, and 2010). The column graphs of the built-up areas in each zone (Fig. 4.9) can help identify the quantity of changing built-up areas in each zone. Figure 4.9 provides clear basic information on the built-up areas in Tripoli and how these areas change over time in different zones and directions. Those zones located very near the CBD had very low urban development, while those zones that included urban fringes witnessed the highest urban expansion. This dramatic urban area increase was recorded in 2010. The increasing distance between CBDs and urban fringes decreases urban development because a low urban development expands in scattered mode. The expansion rate of urbanized areas increases dramatically along the history of the study area. Tripoli witnessed a very high urban expansion ratio over the past decade. Given that these findings reflect the rapid increase of urban expansion, the study area requires further analysis.

4.3.2 Population and Built-up Area Growth Rates

Population growth results in the expansion of built-up areas, that is, the expansion rate of built-up areas is interrelated to population growth rate. Therefore, urban sprawl can be determined by assessing both population and built-up area growth rates. Figure 4.10 illustrates the growth ratio of built-up areas and population in three periods. The expansion rate of built-up areas was constantly higher than the population growth rate, which was

Fig. 4.7 Extent of built-up area in different years: **a** 1984, **b** 1996, **c** 2002, and **d** 2010

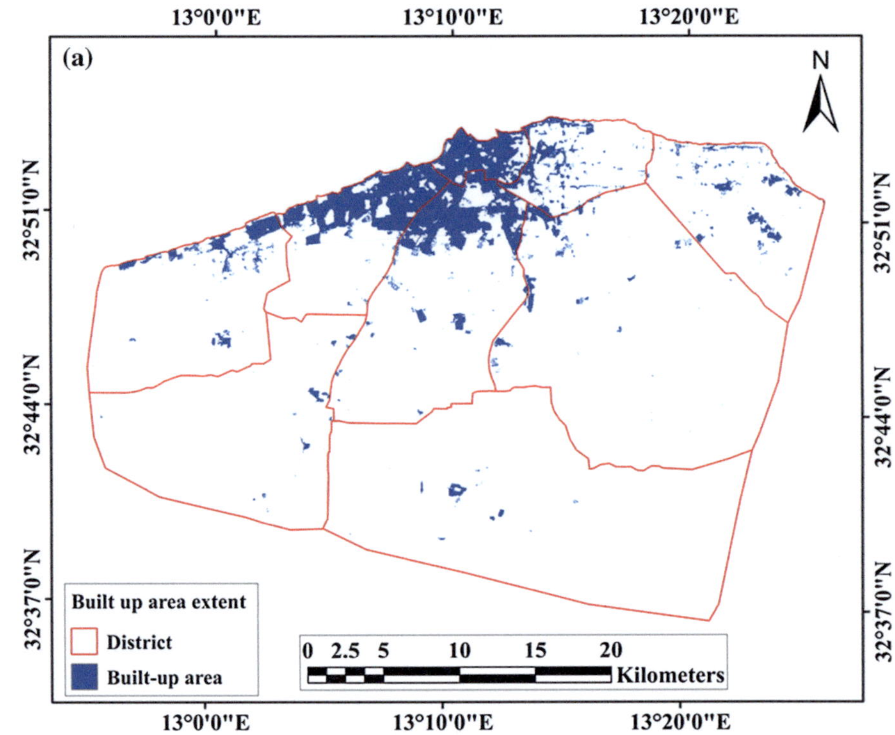

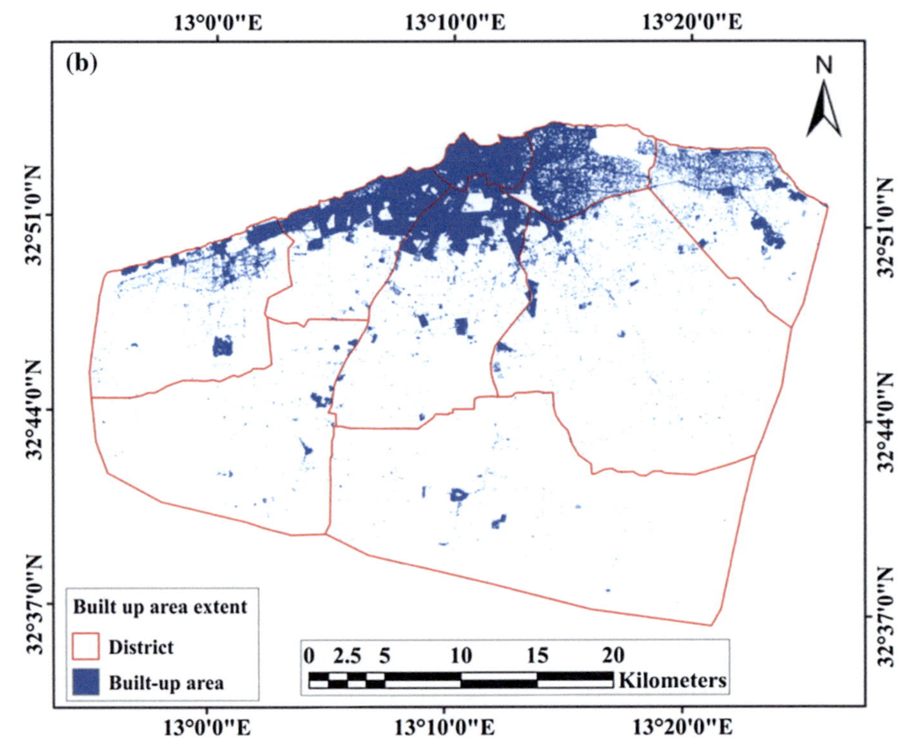

extremely high in the last decade. This finding contradicts that of Acioly and Davidson (1996), who found that cities in developing countries were becoming more compact despite their decreasing population growth rate. In the case of Tripoli, even if the study area demonstrated a tendency toward compactness between 1996 and 2002, the metropolitan area showed a general tendency toward high dispersion, which is an indicator of higher urban sprawl.

Fig. 4.7 (continued)

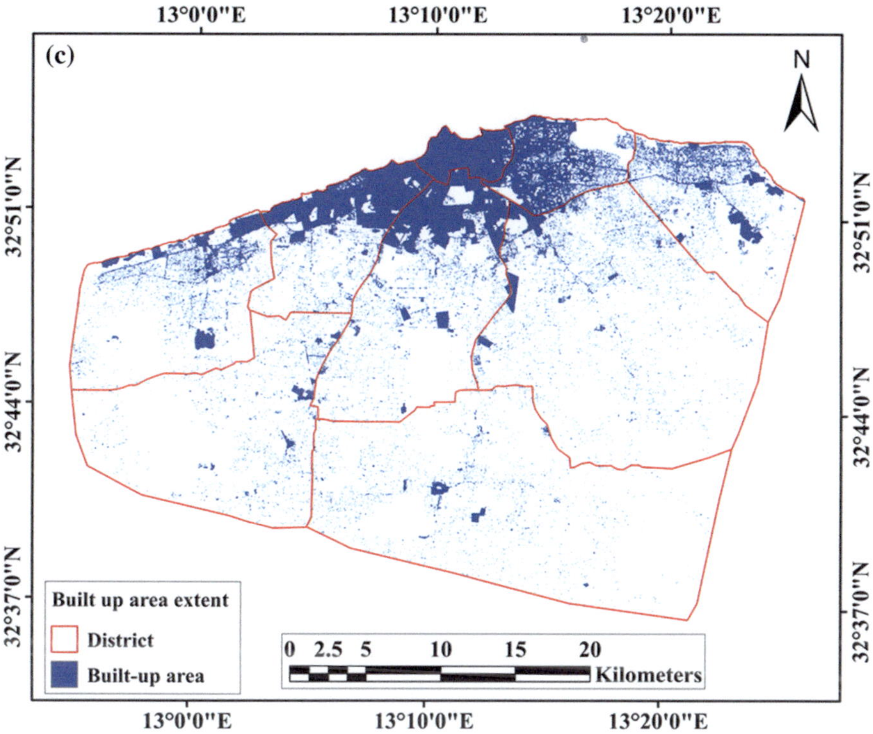

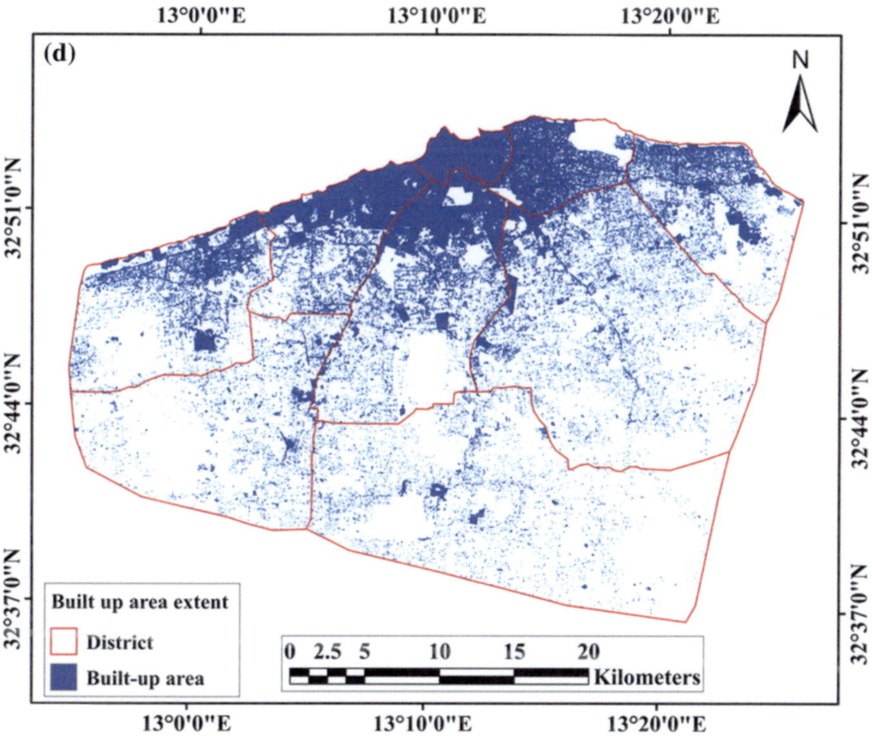

4.3.3 Relationship Between Built-up Area and Population Proportions

The sprawl of urbanization patterns can be measured and quantified as the percentage of built-up areas and population in a zone. District population comprises the non-built-up lands in the same district. As an indicator of built-up land, the percentages of built-up areas and population in each district were related and used as sprawl measures instead of population data. These percentages were computed by

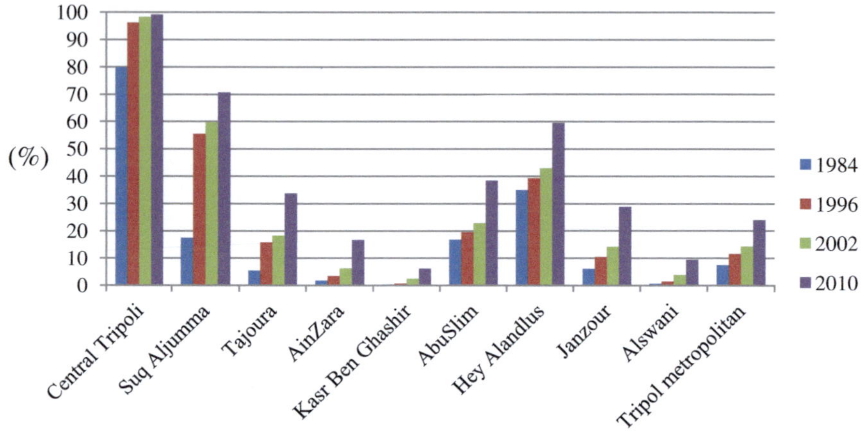

Fig. 4.8 Percentage of built-up areas in each district and the overall study area in different time periods

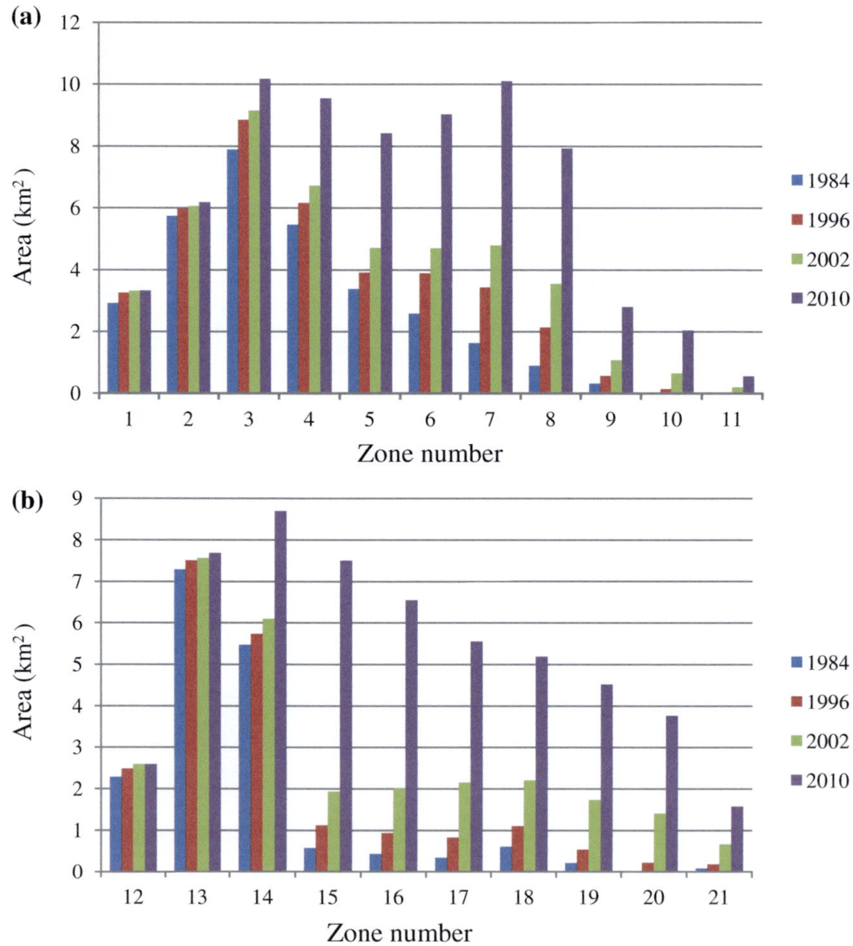

Fig. 4.9 Built-up area in each zone in different years (in km^2): **a** Zones within the first growth direction, **b** Zones within the second growth direction, **c** Zones within the third growth direction, **d** Zones within the fourth growth direction, and **e** Zones within the fifth growth direction

dividing the amount of built-up areas and population in each district by the overall built-up area and population of the Tripoli metropolitan area, respectively. The interchangeable relationship between urban growth and population was evaluated by subtracting the population ratio from the built-up area ratio in each district. Figure 4.11 shows that the differences range between −1 and 1, where a value of 0 indicates moderate conditions.

Higher positive values indicate a higher built-up area consumption per capita, which in turn indicates better environment and urban services. Meanwhile, higher negative values indicate population crowding, which may produce negative effects at the social, economic, and urban levels. Figure 4.11 shows that the Central Tripoli district has a more compact urban growth pattern than the other districts. Hey Alandalus, Suq Aljumma, AbuSlim, and Tajoura

Fig. 4.9 (continued)

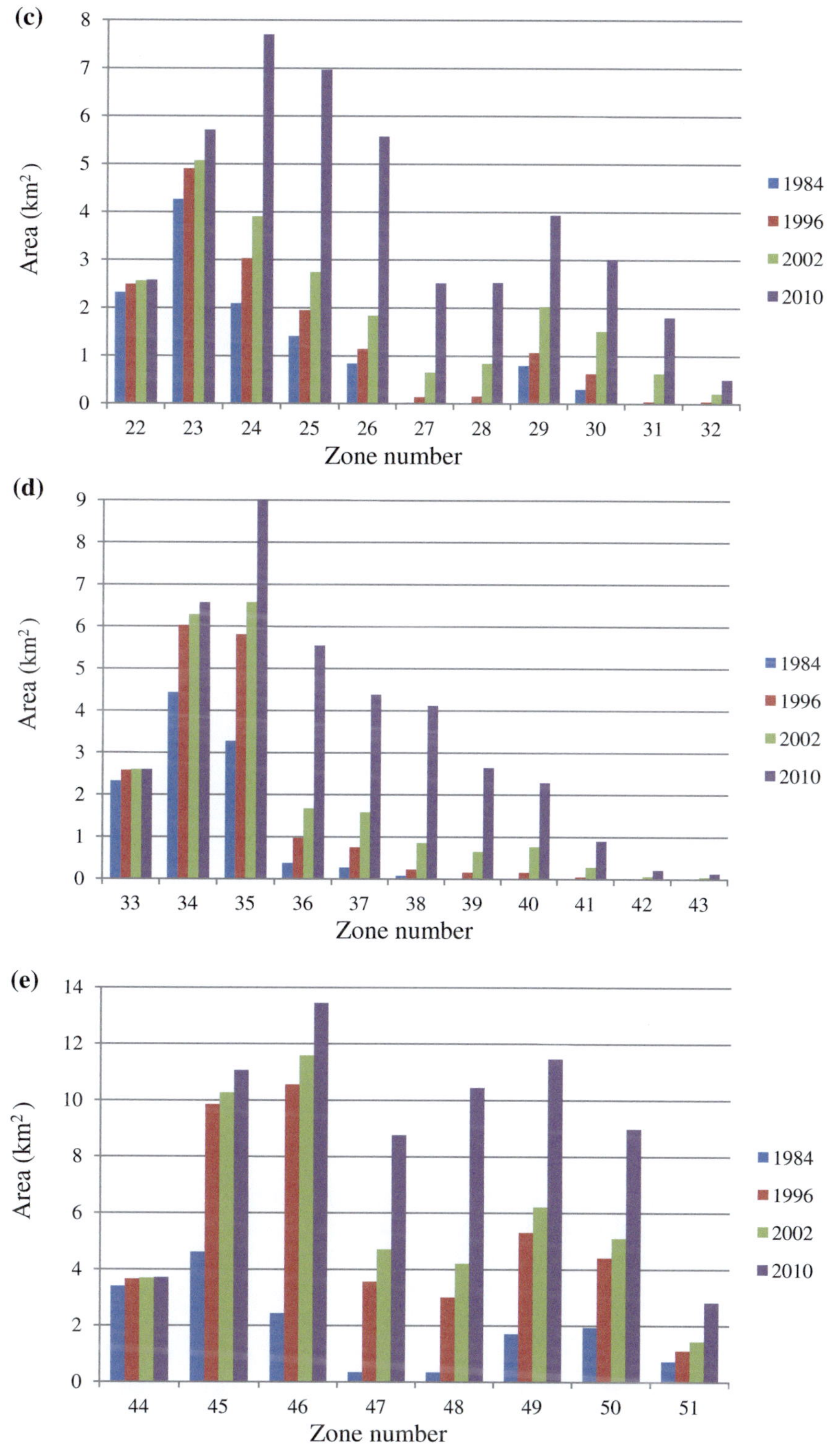

demonstrated high urban land consumption before 1996, but such consumption decreased thereafter, indicating the increasing compactness of these districts. In comparison, AinZara, Kaser Ben Ghashir, Janzour, and Alswani demonstrated very low urban land consumption between 1984 and 1996, but showed a remarkable increase in their percentage of built-up areas between 2002 and 2010, thereby reflecting the gradual increase of the urban sprawl pattern.

Land absorption rate evaluates and measures urban expansion and sprawl as a process based on the relationship

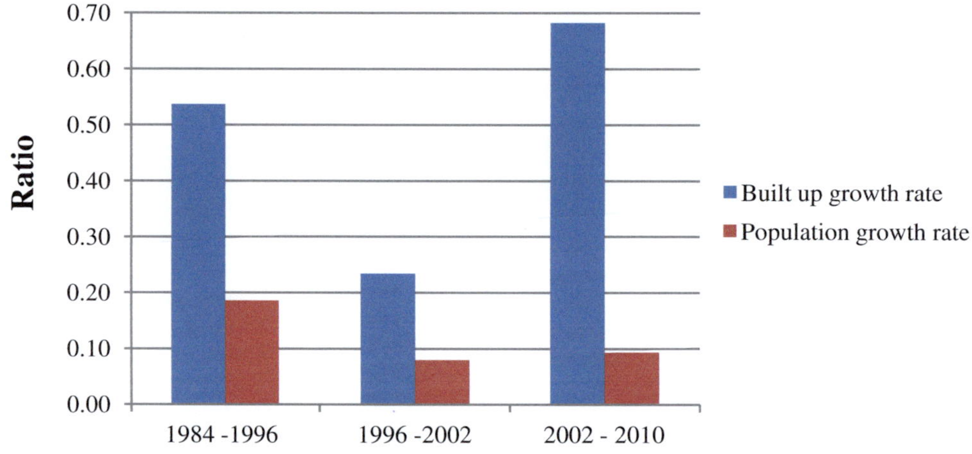

Fig. 4.10 Growth ratios of population and built up area

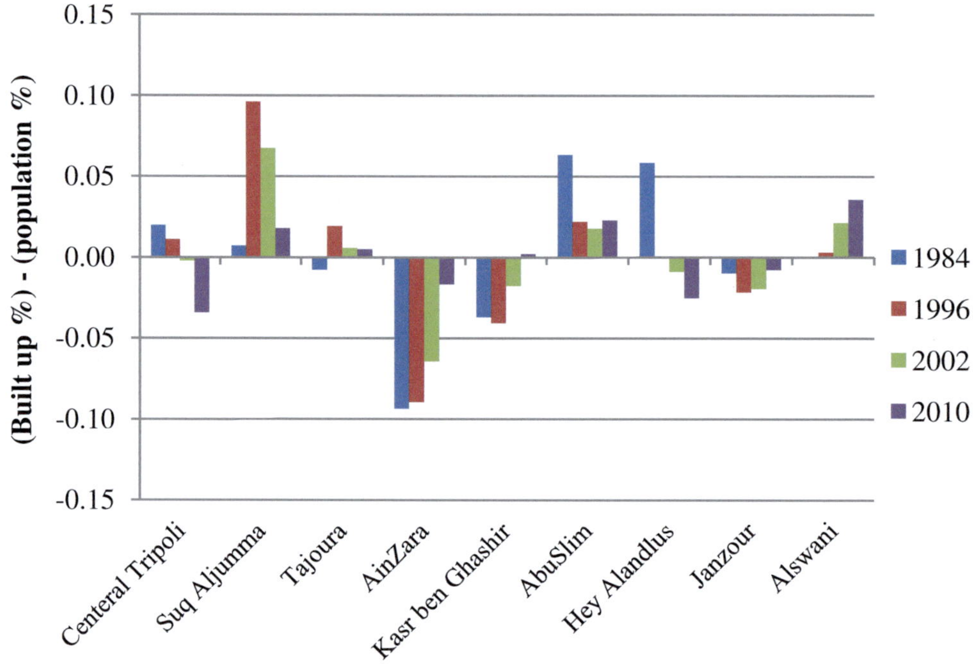

Fig. 4.11 Percentage of built-up area versus percentage of district population

between built-up area and population. The land absorption rate technique applied in this research was based on the evaluation of changes in built-up area and population data within a defined period. Land absorption rate was calculated as follows:

$$\text{Land absorption rate} = \frac{A_2 - A_1}{P_2 - P_1}, \tag{4.9}$$

where P_1 and P_2 denote the population for the first and second periods, and A_1 and A_2 denote the quantity of built-up areas for the first and second periods.

Figure 4.12 presents the land absorption rate analysis results. The Central Tripoli and Suq Aljumma districts showed a moderate level of urban land absorption rate across their urban expansion history. In other words, these districts faced a compacted urban growth process than an urban sprawl because of their controlled growth and vertical urban expansion. However, the other seven districts witnessed increasing urban land absorption rates especially in the last decade, which indicated their rapid uncontrolled urban expansion (i.e., increase of urban sprawl). These findings reveal the deteriorating situation of the urban process in the study area, thereby warranting the attention of urban planners. Hey Alandalus had a remarkable urban land absorption rate that reflects high sprawl as an urban process. However, given that assessing urban expansion as a pattern may yield different results (Fig. 4.11), urban growth must be assessed both as a process and pattern to obtain a comprehensive understanding of urban expansion.

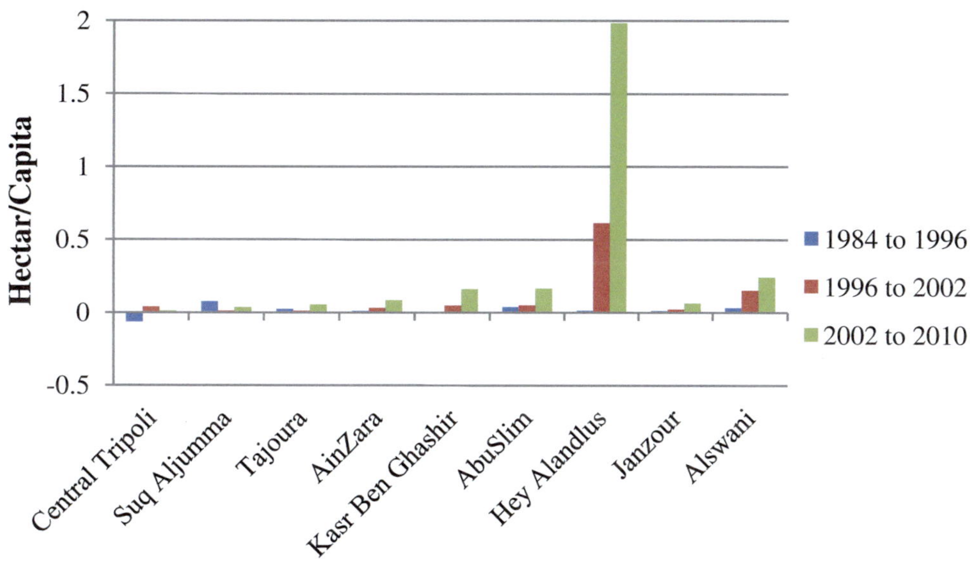

Fig. 4.12 Land absorption rate in each district

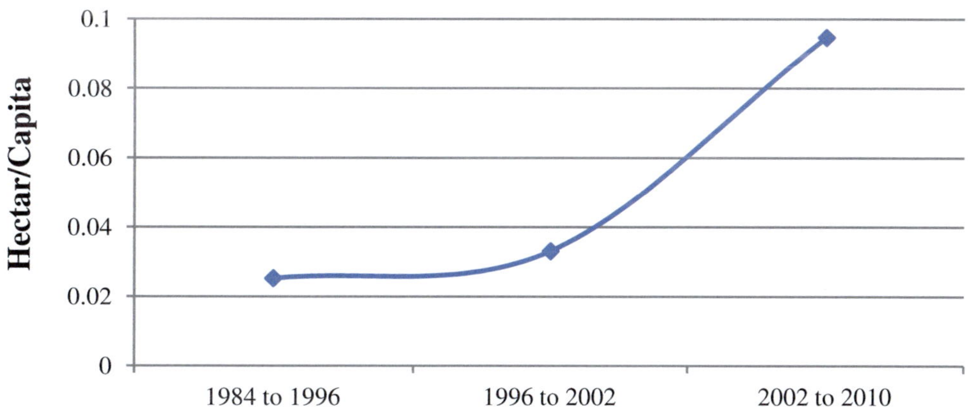

Fig. 4.13 Overall land absorption rate of the study area

The results in Figs. 4.11 and 4.12 complement each other and can collectively explain the relationship between urbanization and population. Evaluating the urban growth process of the Tripoli metropolitan area (Fig. 4.13) reveals a large increase in the urban land absorption rates in the study area between 2002 and 2010, thereby indicating a high overall sprawl.

4.3.4 Detected and Expected Theoretical Urban Expansion

The divergence of urban growth for each zone and temporal period can be easily identified (Fig. 4.14) by subtracting the calculated theoretical expected urban growth from the observed growth. Positive values indicate that the actual growth is higher than expected, while negative values indicate that the actual growth is lower than expected. The magnitude of the difference also reflects the level of variance. The observed urban expansion in several zones (especially at built-up area fringes) significantly deviated from the expectations, and such deviation remained associated with urban growth and continued to increase over time. Higher deviations reflect the freedom and independence of the urban expansion process, such that a high deviation indicates that the studied variable is independent from other similar variables. These findings indicate the occurrence of a clear urban sprawl in most zones far from the CBD in all directions, especially between 1996 and 2010.

4.3.5 Intensity of Urban Expansion and Sprawl

Table 4.5 shows that the study area has an expansion intensity index of 0.66, which is considered moderate urban expansion speed. However, its UEII increased dramatically from 0.35 in 1984–1996 to 1.28 in 2002–2010, thereby indicating an alarming increase in urban sprawl occurrence.

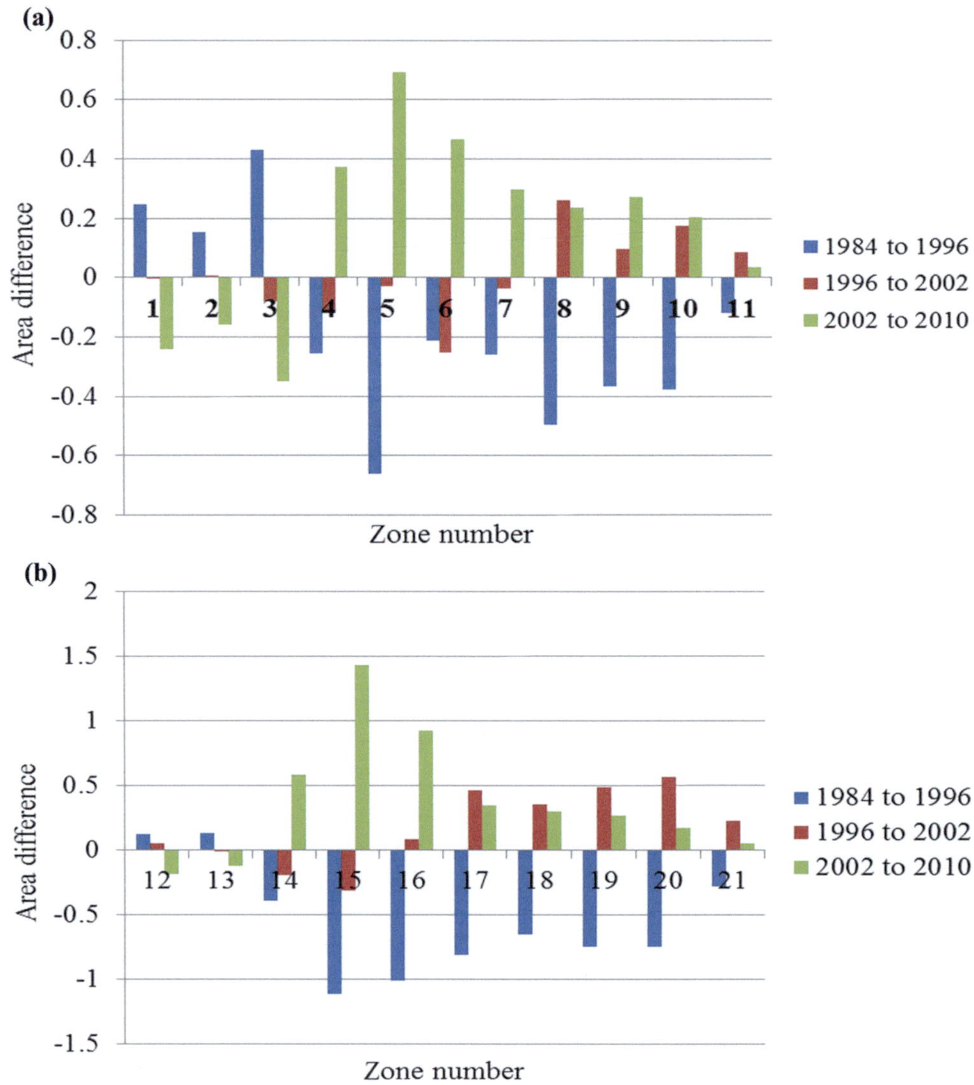

Fig. 4.14 Difference between observed and expected built up area growth (in km^2): **a** in zones within the first direction, **b** in zones within the second direction, **c** in zones within the third direction, **d** in zones within the fourth direction, and **e** in zones within the fifth direction

Figure 4.15 shows that by having the lowest UEIIs, zones near the CBD have relatively stable expansion rates. Meanwhile, zones that included built-up area fringes had very high UEIIs, thereby suggesting their high sprawled urban expansion. However, the UEIIs in most urban expansion directions of the study area decreased after exceeding the urban area fringes, thereby indicating a low urban expansion.

4.3.6 Shannon's Entropy of Urban Expansion Based on District Boundaries

Shannon's entropy method was applied to identify, quantify, and measure the occurrence of urban sprawl in the study area based on the boundaries of its nine districts. Table 4.6 and Fig. 4.16 show that the entropy values are always higher than the middle point of $\log_e(n)$ (i.e., 1.099). A very high entropy value also was recorded in 2010. Therefore, the Tripoli metropolitan area has an overall high dispersed urban expansion or urban sprawl.

Figure 4.17 shows the tendency of the urban expansion process. The study area experienced a clearly progressing urban sprawl process, and the positive change of ΔH_n indicates the increasingly uncontrolled dispersal pattern of the urbanization process. These findings highlight the urgent need for sustainable urban growth control and planned urban management. Moreover, cities in developing countries do not necessarily become more compact with decreasing population growth rate.

Instead of predicting the occurrence of urban expansion, urban planners need to prioritize the measurement of urban

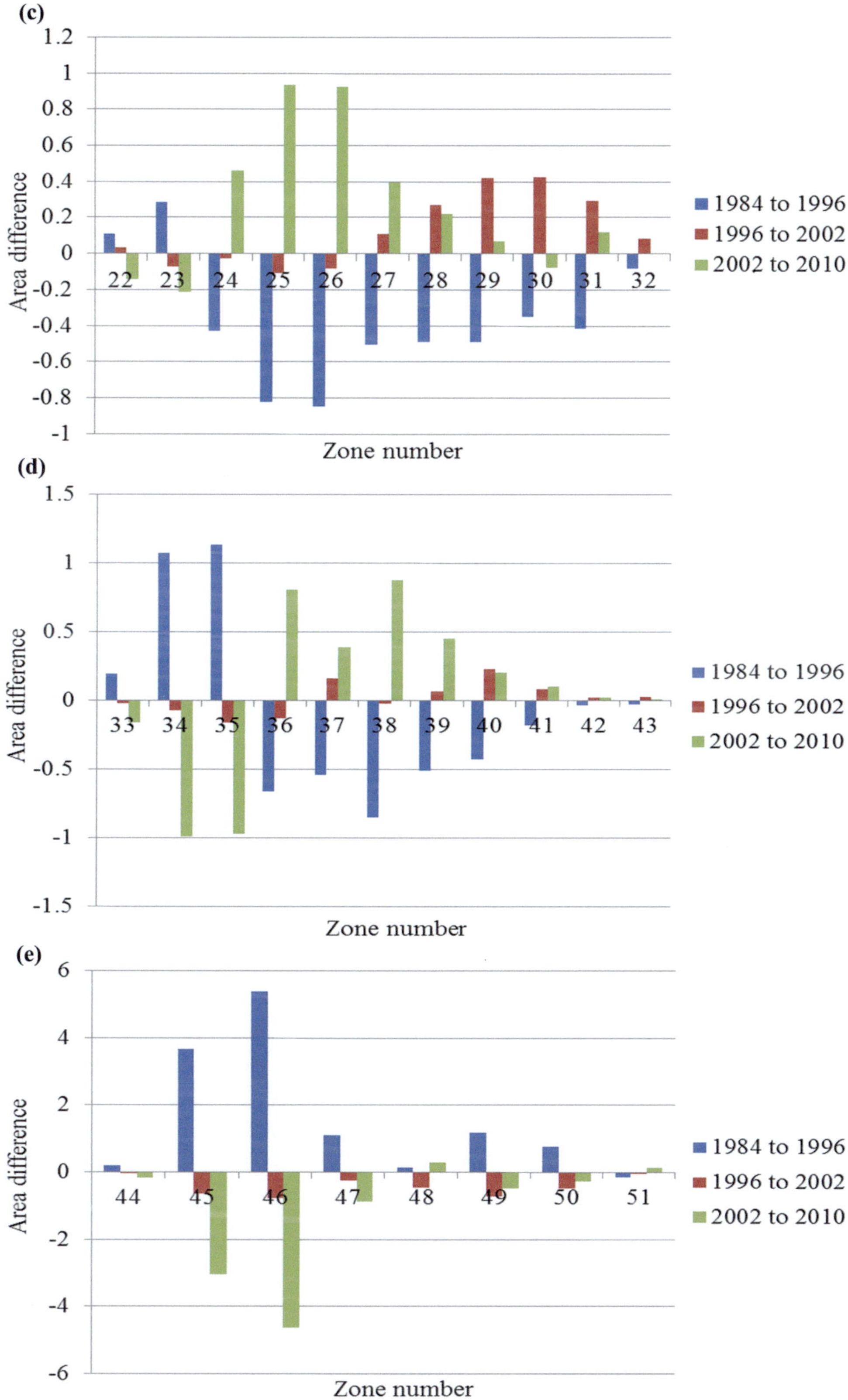

Fig. 4.14 (continued)

Table 4.5 Urban expansion intensity index of time periods

Time period	UEII
1984–1996	0.35
1996–2002	0.45
2002–2010	1.28

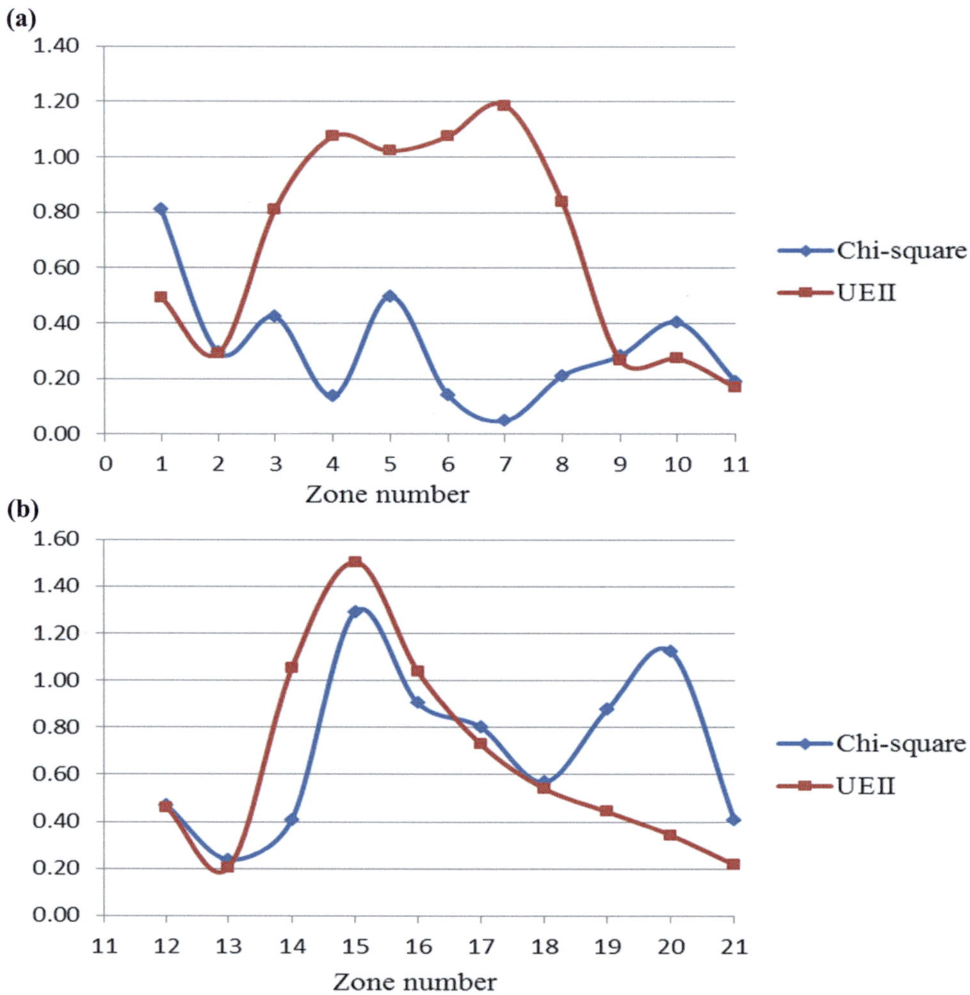

Fig. 4.15 Variation of UEII and Degree of freedom in different growth directions: **a** in the first direction, **b** in the second direction, **c** in the third direction, **d** in the fourth direction, and **e** in the fifth direction

growth and the identification of urban requirements to be accomplished in preparation for future urban demands. Shannon's entropy model can guide the identification and measurement of the likely changes that may result from urban history. This model can also be applied to each district for a more specific level of research. Given that different urban growth patterns may result from the varying intensities of compactness of each district, a single policy for the entire metropolitan area will not have the same degree of efficiency as that for each district.

4.3.7 Shannon's Entropy of Urban Expansion and Effect of Direction and Distance to CBD

The relative Shannon's entropy technique was applied based on the proposed 51-zone division (Fig. 4.6). The effects of expansion direction and distance to CBD in the study area were examined. The analysis results are presented as follows.

Table 4.7 shows that the overall relative entropy values in all years are much larger than the half-way point (i.e., 0.5).

Fig. 4.15 (continued)

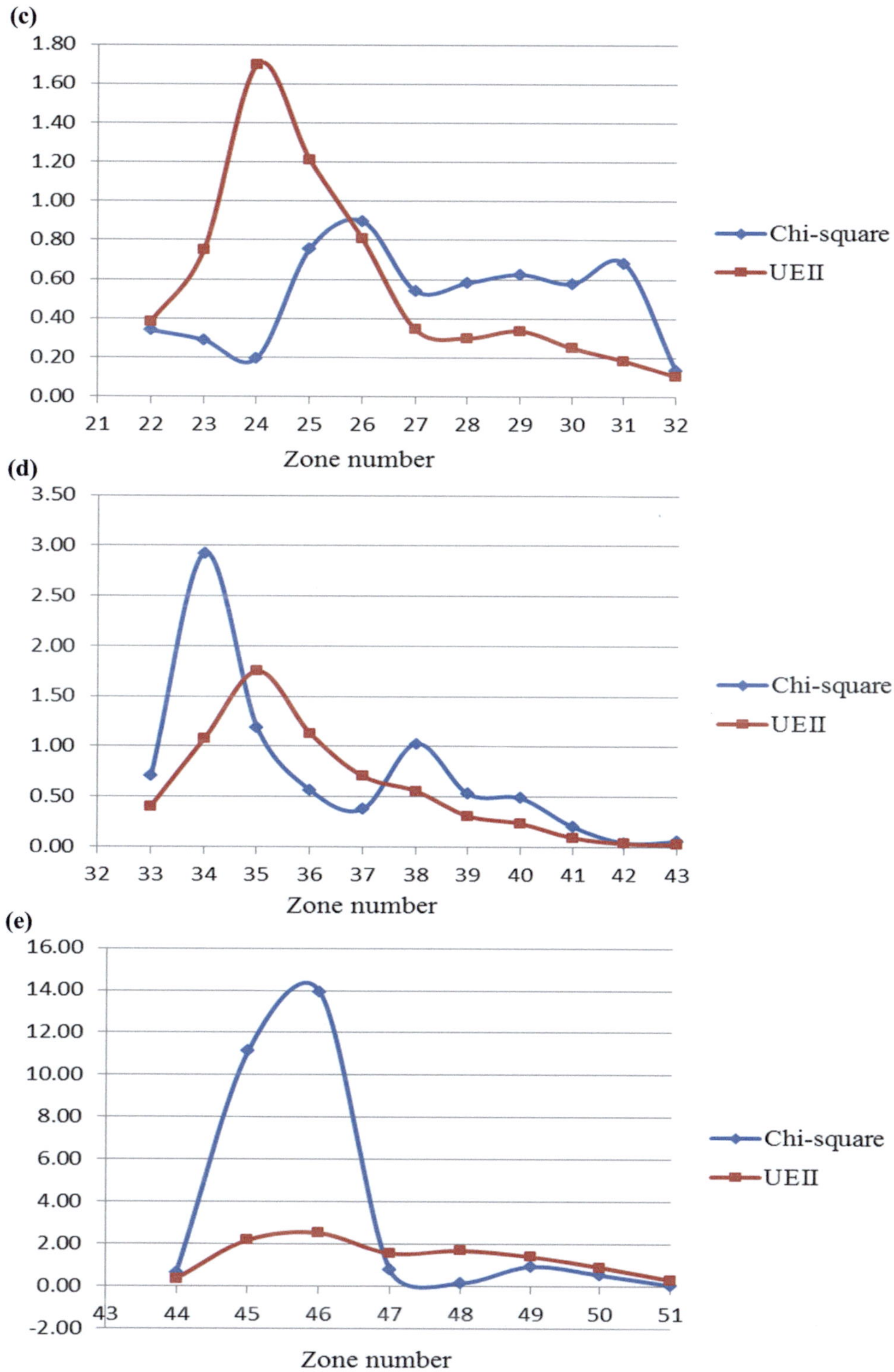

Table 4.6 Shannon's entropy values for different time periods

	1984	1996	2002	2010	$\log_e(n)$	$\log_e/2$
Absolute entropy value	1.478	1.628	1.737	1.936	2.197	1.099
Relative entropy value	0.673	0.741	0.791	0.881	–	–

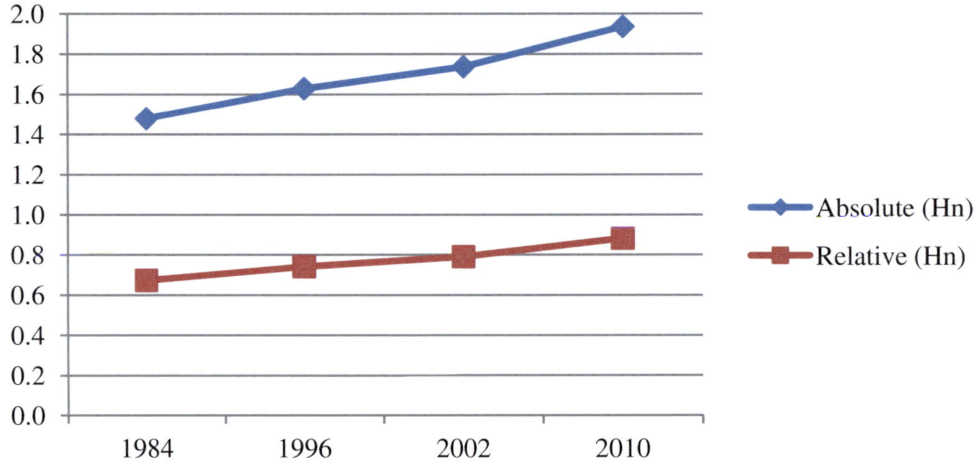

Fig. 4.16 Absolute and relative Shannon's entropy values in different time periods

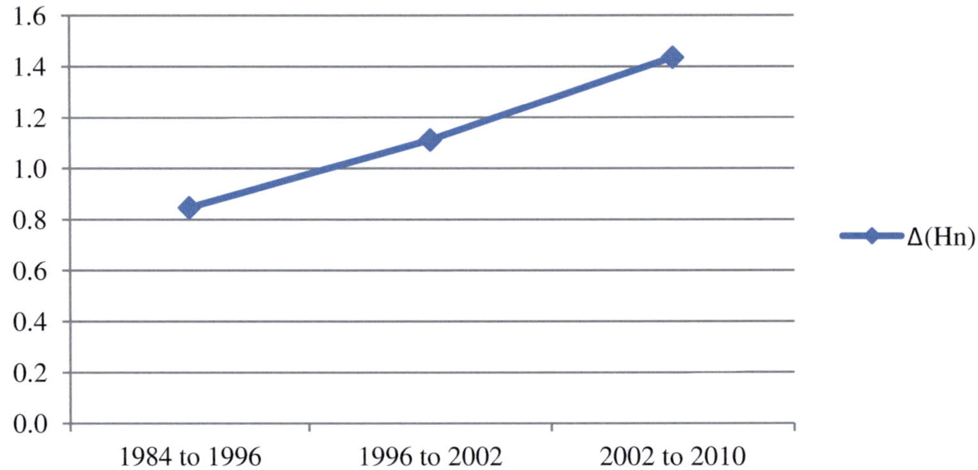

Fig. 4.17 Shannon's entropy change rate values

Table 4.7 Overall relative Shannon's entropy of study area in different years

Year	1984	1996	2002	2010
Entropy (H_n)	0.74	0.79	0.83	0.90

Therefore, the urban expansion in Tripoli is sprawling, and the sprawling trend is increasing. In this case, the general urban growth process requires further control and a clear urban planning policy.

To understand and assess the urban sprawl process in each zone and direction across the history of the study area (1984–2010), the historical relative entropy values were subtracted from the recent relative entropy values in each zone. Higher positive values indicate an increasing urban sprawl and a highly dispersed urban growth, while higher negative values indicate a decreasing sprawl and increased compactness in a zone or crowded urban area. Figure 4.18 clearly shows the variation of relative entropy and spatiotemporal urban sprawl behaviors in the study area. The change rate of relative entropy values in the first direction were negative in Zones 1, 2, 3, and 4 across all three periods, but the opposite situation was observed in Zones 5–11, which demonstrated positive change rates (i.e., sprawl increase). The highest sprawl growth rate was clearly observed between 2002 and 2010, especially in Zones 5–8. In the second direction, Zones 12–14 became more compact, while the remaining zones in the second direction witnessed increased sprawl rates at all times. A dramatic sprawl increase was also recorded in Zones 15–17. However, in the third direction, Zones 22 and 23 became more compact along with time. The other third-direction zones almost had similar sprawl increase rates between 1996 and 2002, but Zones 24–27 witnessed a remarkable sprawl increase

Fig. 4.18 Variation of the sprawl in different growth directions with the respect to each zone: **a** in the first direction, **b** in the second direction, **c** in the third direction, **d** in the fourth direction, and **e** in the fifth direction

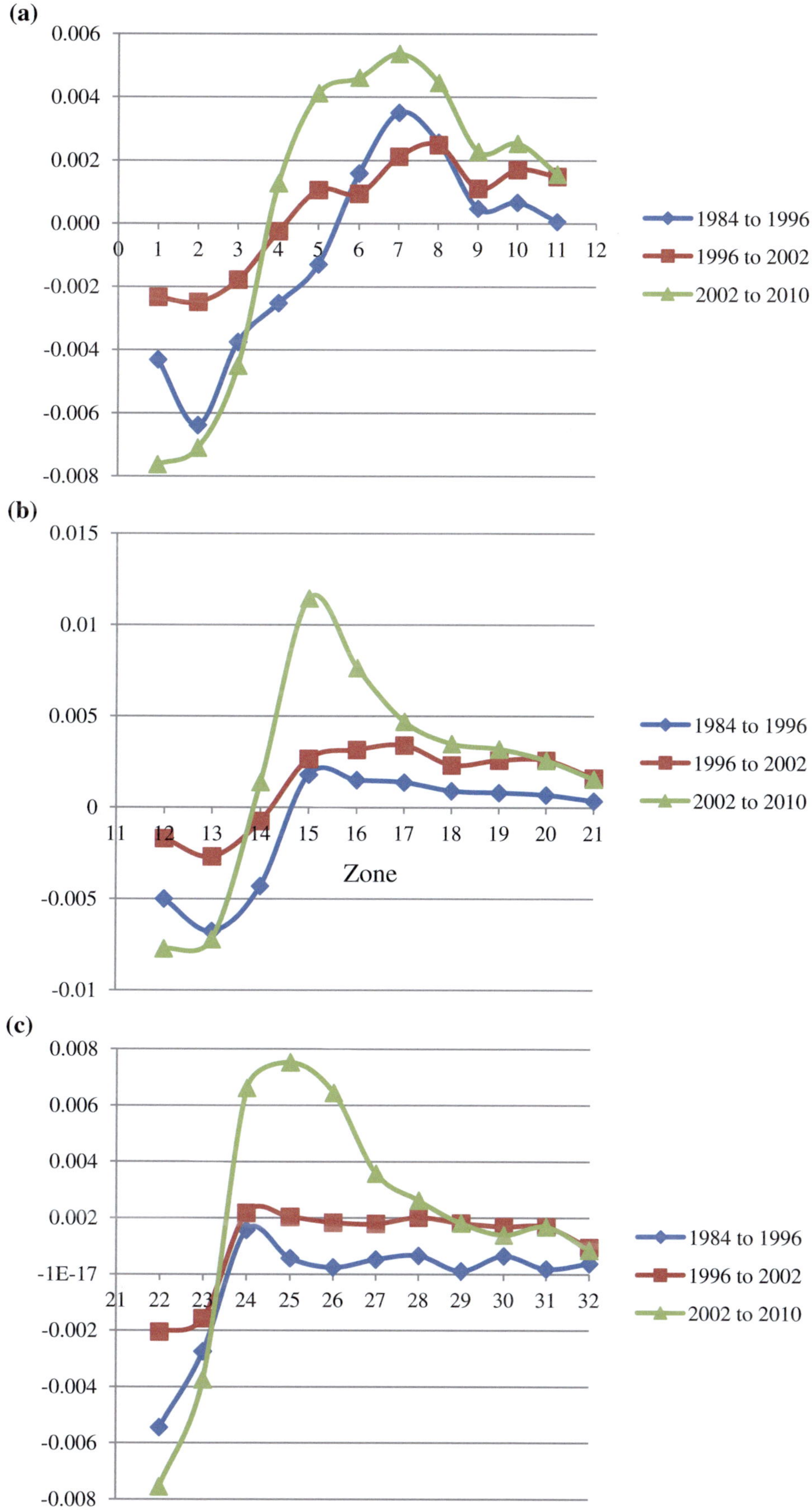

Fig. 4.18 (continued)

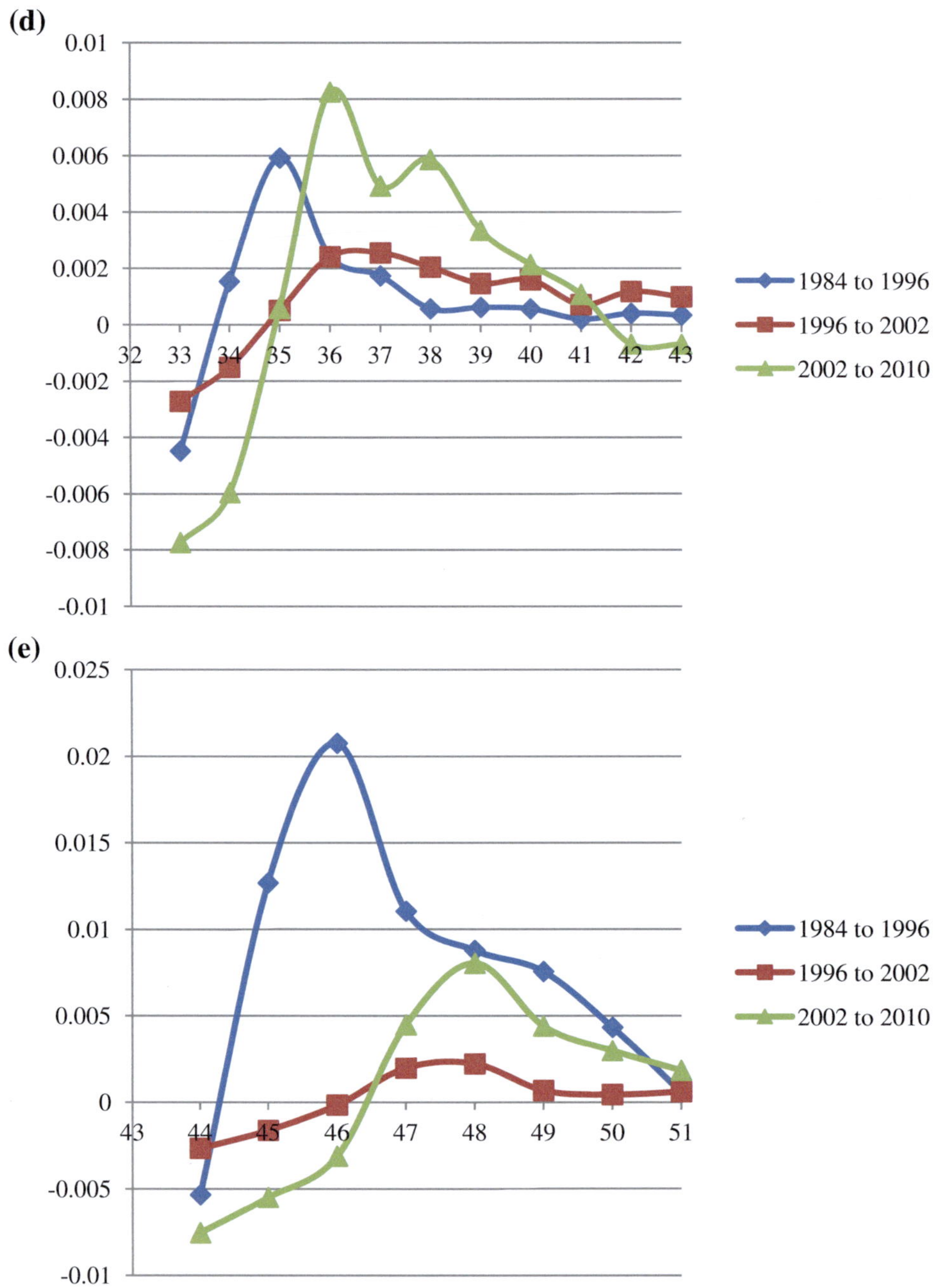

between 2002 and 2010. The urban sprawl trends in the fourth direction are generally similar to those in the second and third directions. However, Zones 34–36 showed the largest entropy value between 1984 and 1996, and this value gradually decreased in the zones located further away from the CBD. Nonetheless, the fifth direction demonstrated a different sprawl tendency than the other four directions. Specifically, all zones within the fifth direction recorded the highest urban sprawl rate between 1984 and 1996 except Zone 44, which was adjacent to the CBD. This rate of sprawl evidently decreased to its lowest point for Zones 47–51 between 1996 and 2002. From 2002, the change rate of entropy value increased yet remained lower than the values recorded between 1984 and 1996. The entire study area showed a similar urban sprawl trend, but the analysis results illustrate that CBD-adjacent zones have the lowest relative entropy change rate, especially between 2002 and 2010. Therefore, these urban zones have a high compactness. Given that the urban sprawl rate increases along with increasing distance from the CBD, zones with high compactness, including the urban fringes, have an extremely high sprawl increase rate.

Thereafter, the sprawl rate decreased and reached a very low value in those zones near the border of the study area. The effect of direction on the urban sprawl process in Tripoli slightly differed in the first, second, third, and fourth directions, and only the fifth direction showed a different fluctuating sprawl change. The sprawl rate reached its highest value between 1984 and 1996, declined to its lowest value between 1996 and 2002, and increased again between 2002 and 2010, but remained lower than that recorded during the first observed period. Therefore, the study area has clearly experienced a general urban sprawl trend in the majority of its directions and zones.

However, these urban sprawls differ from one another and are influenced by sprawl direction and distance from CBD. This section explained how the relative Shannon's entropy model could be implemented to investigate the sprawl of urbanization in the Tripoli metropolitan area and demonstrated the relationships among urban sprawl, sprawl direction, and distance from CBD in different periods to provide clear and specific spatiotemporal descriptions of such phenomena.

4.3.8 Degree of Freedom of Urban Expansion

The degree of freedom of the urbanization process was estimated to offer another perspective toward urban sprawl. Pearson's chi-square method assesses the deviation of actual urban expansion from the planned or expected growth, with a high deviation indicating the occurrence of urban sprawl. Table 4.8 shows a high degree of freedom across all three periods. The study area has an extremely high overall degree of freedom of 52.41, thereby indicating a very high difference in its observed and expected urban expansion. Figure 4.15 shows the varying degrees of freedom in each zone, with Zones 46 and 42 demonstrating the highest and lowest degrees of freedom. A higher degree of freedom generally suggests the need for consistency in planning, managing, and controlling urban growth. A zone with high degrees of freedom may experience an unbalanced growth over time, whereas a period with high degrees of freedom indicates a high inter-zone inconsistency in urban growth. Nonetheless, a high degree of freedom cannot be directly considered as sprawl, but as disparity in urban growth.

4.3.9 Landscape Metrics and Urban Sprawl Detection

Figures 4.19, 4.20, 4.21, 4.22, 4.23 and 4.24 show the rapid urban expansions in the Tripoli metropolitan area between 1984 and 2010. The presence of sprawl was identified and assessed quantitatively using the different definitions of urban sprawl and the analysis results from the applied landscape metrics.

As shown in Figs. 4.19, 4.20, 4.21, 4.22, 4.23 and 4.24, a synoptic analysis of the implemented spatial metrics offers an overall picture of urban sprawl spatial patterns. The SIEI measure illustrates that the overall diversity level of the study area increased between 1984 and 2010, and this same measure clearly increased between 2002 and 2010. The landscape metrics of PD and ED increased remarkably in the entire studied landscape. Such large increases reflect the increasing number and irregular formations of isolated urban patches as well as indicate a high urban fragmentation and the increasing trend of the overall urban sprawl process. However, the overall LPI metric continuously decreased since 1984, making this decrease the largest to be recorded in the last decade (i.e., the largest urban patch in the study area becomes progressively small) and another indicator of increased urban sprawl.

The landscape metrics of LSI and SHAPE continuously increased according to the urban expansion history of the study area. Such an increase indicates irregularity in the urban area, that is, Tripoli may be currently facing an unplanned urban growth. Figures 4.19, 4.20, 4.21, 4.22, 4.23 and 4.24 evaluate the urban sprawl patterns in each district to identify which district has a higher sprawl level. The PD analysis results reveal that Central Tripoli has a decreasing PD with the smallest decrease rate being recorded between 2002 and 2010, thereby increasing the compactness of this zone. Hey Alandalus and Janzour also showed decreasing urban patch densities in 2010, while the other seven districts showed different behaviors and witnessed high occurrences of dispersed urban clusters. The LPI analysis results demonstrated that Central Tripoli manifested an increased LPI at all times. In 1984, Suq Aljumma and Hey Alandalus recorded the highest LPI, but such value declined between 1996 and 2002 and then increased again after 2002, thereby reflecting the high urban compactness in

Table 4.8 Degree of freedom of time periods

Time period	Degree of freedom (D_i^t)
1984–1996	10.77
1996–2002	5.22
2002–2010	36.43

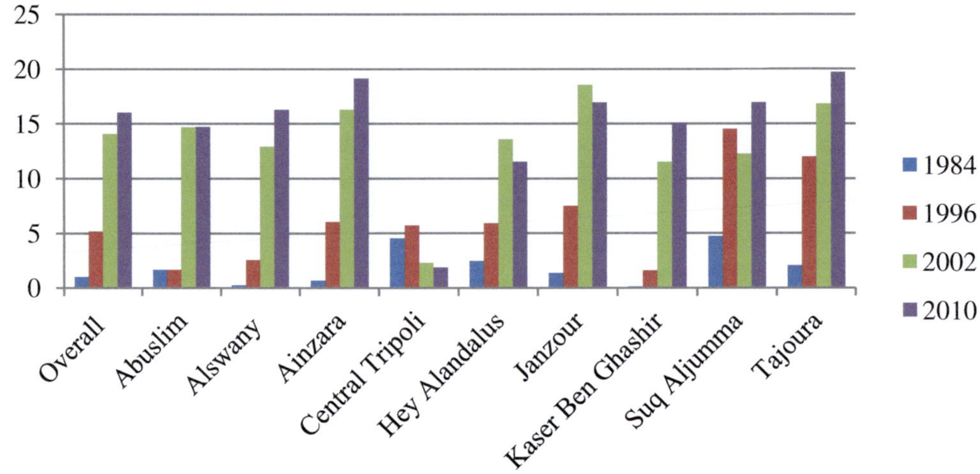

Fig. 4.19 Variation of PD measure in different time periods

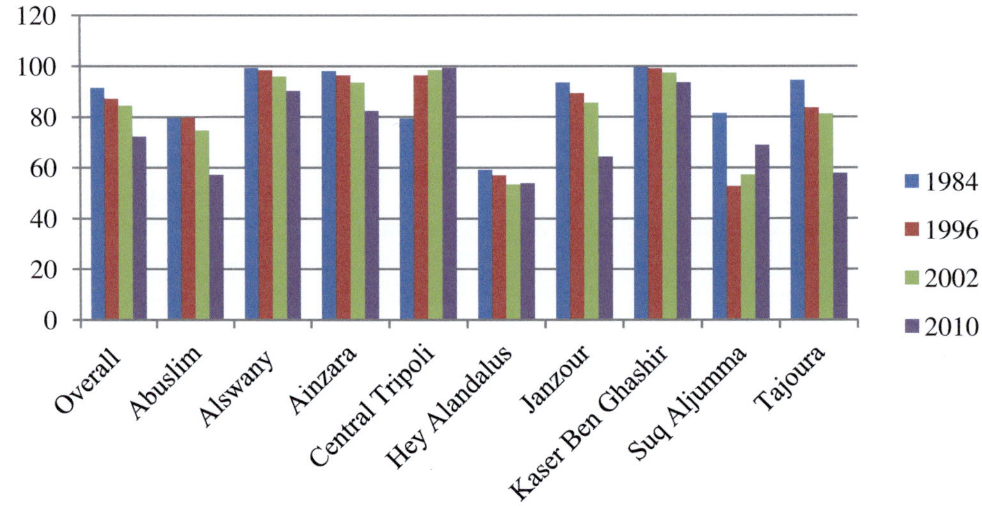

Fig. 4.20 Variation of LPI measure in different time periods

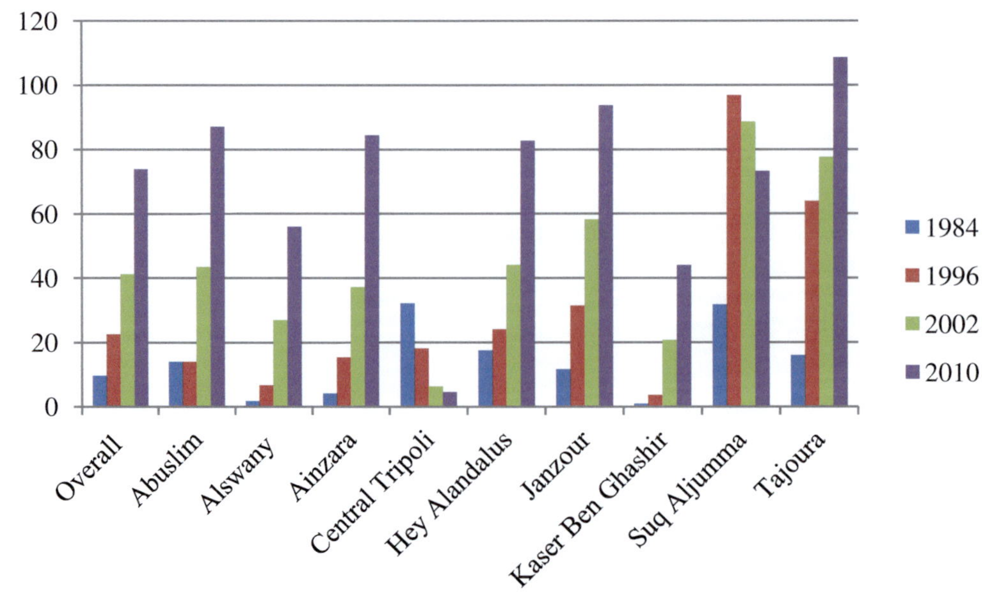

Fig. 4.21 Variation of ED measure in different time periods

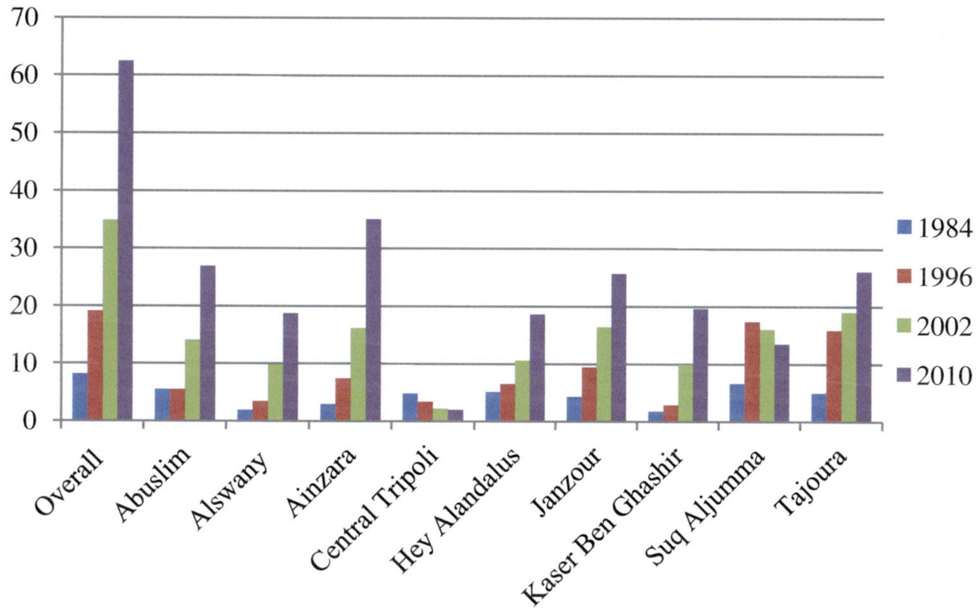

Fig. 4.22 Variation of LSI measure in different time periods

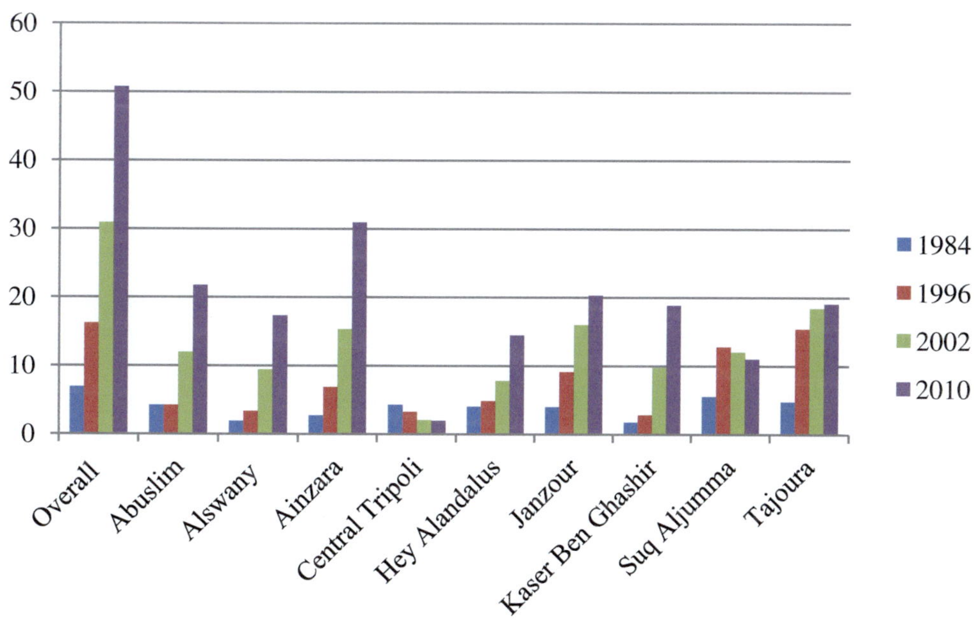

Fig. 4.23 Variation of SHAPE measure in different time periods

these districts. The other districts faced a dramatic decrease of LPI values. Therefore, the occurrence of urban sprawl is clear and easily detectable.

ED, LSI, and SHAPE confirm the presence of highly fragmented complex urban patches in Hey Alandalus, AbuSlim, Tajoura, AinZara, Janzour, Kaser Ben Ghashir, and Alswany. These metrics gradually increased between 1984 and 1996, but their trends significantly differed after 1996. The remarkable increase in the ED, LSI, and SHAPE values indicated deteriorating urban expansion patterns and uncontrolled sprawl presence. The SIEI measure also recorded extremely high diversity values in most districts especially in 2010, except in Central Tripoli, Suq Aljumma, and Hey Alandalus. This result confirms the findings of other landscape metrics analyses and can be used as strong evidence of urban sprawl.

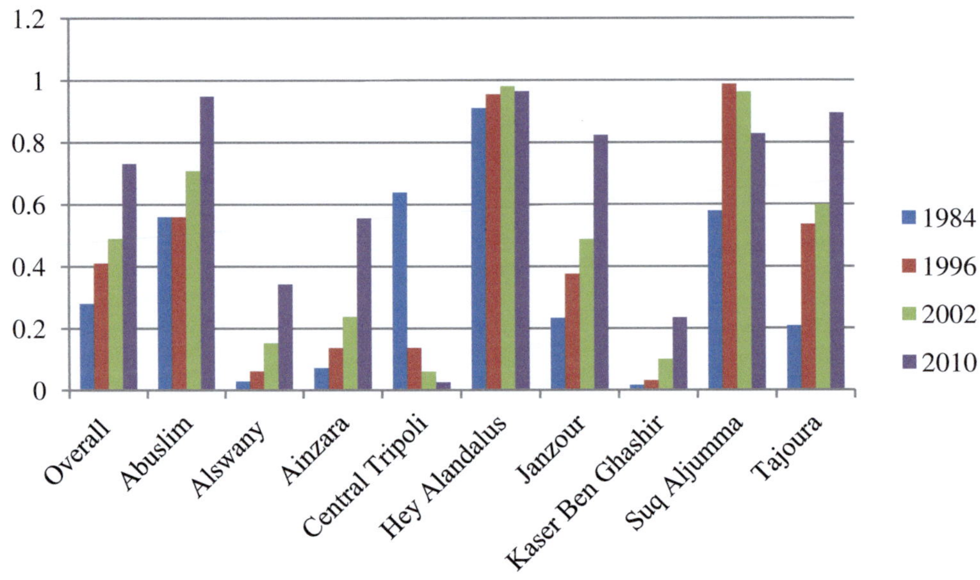

Fig. 4.24 Variation of SIEI measure in different time periods

4.4 Conclusion

This chapter utilized remotely sensed data and GIS to analyze the urban expansion process and its patterns in Tripoli. This study focused on the spatial patterns and extents of urban growth change in the Tripoli metropolitan area from 1984 to 2010, and the findings could be used to direct the urban plans and urbanization policies for Tripoli. Urban planners need to measure urban expansions and determine the urban requirements to be accomplished in preparation for future urban demands. The identified models can also be used to guide the identification and measurement of changes that may result from urban history.

This study successfully highlighted and discovered the spatiotemporal urban expansions patterns in the Tripoli metropolitan area. The overall urban development of Tripoli increased remarkably after 2002 when the growth rate of urbanized lands exceeded 40% of the existing urban areas. Some districts, such as Central Tripoli, Suq Aljumma, and Hey Alandalus, had obvious concentrations of built-up areas and low growth ratios. However, the urban growth process in the study area requires a wise urban management plan to control the urgent demands of urban lands. Given that the districts of Tripoli had different urban growth ratios and urban land concentrations, more than one urban development policy should be implemented because a single urban policy for the whole study area will not be as effective as that for all area districts.

The findings are summarized as follows:

- The advantages and significance of the analytical tools are based on the combinations of population and built-up area data for urban expansion evaluations. Such a simple analysis tool is a very good method for identifying and measuring urban sprawl.
- The land absorption rate is generally increasing, which indicates that the urban area growth rate exceeds the population growth rate. This result signals the occurrence of urban sprawl. The last decade also witnessed extreme levels of consumption rate.
- To increase the compactness of the Tripoli metropolitan area, the quantity of urban lands absorbed by population increase should be decreased by adopting vertical urban growth plans.
- The observed urban expansion in most zones of Tripoli obviously deviated from the expected theoretical urban growth (especially at built-up area fringe zones). This outcome reflects the uncontrolled urbanization and the independence of urban expansion.
- The UEII reflected the probable future direction and potentials of urban development, and compared the speed or intensity of urban land use change across different periods. The intensity and speed of urban growth increased across the urban history of the study area, and the intensity remarkably increased after 2002 (i.e., very rapid growth). The urbanization processes in most

directions showed inconsistent growth intensities (i.e., differences in uncontrolled growth). However, the overall urban intensity index indicated a moderate urban expansion process, which serves as a favorable indicator of the possibility of accommodating and managing probable future expansions before the occurrence of a deteriorating situation.

- Urban sprawl was identified and quantified by applying Shannon's entropy method to assess Tripoli urban growth (as a process and a pattern) at different analysis levels (periods, zones, and directions). The two approaches used in each zone division showed an increased overall sprawl in the study area in all periods. The 51-zone division approach provided better insights into the urban expansion process, illustrated the urban sprawl variation in each direction and zone, and highlighted the effect of distance to CBD.
- The urban growth in the study area showed an extremely high overall degree of freedom. The highest level of freedom was observed between 2002 and 2010. The varying degrees of freedom across the divided zones showed the dispersed unbalanced urban growths and emphasized the need for consistency in planning.
- The assessment landscape metrics quantitatively assessed the dispersion, aggregation, diversity, complexity, and shape of urban areas. All the applied metrics demonstrated clear and uncontrolled fragmented urban pattern and urban sprawl (i.e., increased number of non-uniform dispersed small urban patches).

These findings demonstrated the urban dispersion and sprawl for the Tripoli metropolitan area (i.e., the sprawl increases with time). Furthermore, these findings show that cities in developing countries require highly compact development scenarios to achieve sustainable urban development principles.

References

Abdullahi, S., Mahmud, A. R. B., & Pradhan, B. (2014). Spatial modelling of site suitability assessment for hospitals using geographical information system-based multicriteria approach at Qazvin city, Iran. *Geocarto International, 29*(2), 164–184.

Abdullahi, S., Pradhan, B., & Jebur, M. N. (2015). GIS-based sustainable city compactness assessment using integration of MCDM, Bayes theorem and RADAR technology. *Geocarto International, 30*(4), 365–387.

Abed, J., & Kaysi, I. (2003). Identifying urban boundaries: application of remote sensing and geographic information system technologies. *Canadian Journal of Civil Engineering, 30*(6), 992–999.

Acioly, C., & Davidson, F. (1996). *Density in urban development* (Vol. 8). Lund University, Lund Centre for Habitat Studies.

Alberti, M., & Waddell, P. (2000). An integrated urban development and ecological simulation model. *Integrated Assessment, 1*(3), 215–227.

Almeida, C., Monteiro, A. M. V., Câmara, G., Soares-Filho, B. S., Cerqueira, G. C., Pennachin, C. L., et al. (2005). GIS and remote sensing as tools for the simulation of urban land-use change. *International Journal of Remote Sensing, 26*(4), 759–774.

Al-shalabi, M., Billa, L., Pradhan, B., Mansor, S., & Al-Sharif, A. A. (2013). Modelling urban growth evolution and land-use changes using GIS based cellular automata and SLEUTH models: The case of Sana'a metropolitan city, Yemen. *Environmental Earth Sciences, 70*(1), 425–437.

Al-sharif, A. A., & Pradhan, B. (2015). A novel approach for predicting the spatial patterns of urban expansion by combining the chi-squared automatic integration detection decision tree, Markov chain and cellular automata models in GIS. *Geocarto International*, 1–24 (ahead-of-print).

Angel, S., Parent, J., & Civco, D. (2007). *Urban sprawl metrics: An analysis of global urban expansion using GIS*. Paper presented at the Proceedings of ASPRS 2007 Annual Conference, Tampa, Florida May.

Barnes, K. B., Morgan J. M., III., Roberge, M. C., & Lowe, S. (2001). *Sprawl development: Its patterns, consequences, and measurement* (pp. 1–24). Towson: Towson University. Available online at http://chesapeake.towson.edu/landscape/urbansprawl/download/Sprawl_white_paper.pdf. Last accessed February 13, 2008.

Barredo, J. I., Kasanko, M., McCormick, N., & Lavalle, C. (2003). Modelling dynamic spatial processes: Simulation of urban future scenarios through cellular automata. *Landscape and Urban Planning, 64*(3), 145–160.

Bhatta, B. (2009). Analysis of urban growth pattern using remote sensing and GIS: A case study of Kolkata, India. *International Journal of Remote Sensing, 30*(18), 4733–4746.

Bhatta, B. (2010). *Analysis of urban growth and sprawl from remote sensing data*. Springer Science & Business Media.

Bhatta, B., Saraswati, S., & Bandyopadhyay, D. (2010a). Quantifying the degree-of-freedom, degree-of-sprawl, and degree-of-goodness of urban growth from remote sensing data. *Applied Geography, 30* (1), 96–111.

Bhatta, B., Saraswati, S., & Bandyopadhyay, D. (2010b). Urban sprawl measurement from remote sensing data. *Applied Geography, 30*(4), 731–740.

Ceccato, V., Haining, R., & Signoretta, P. (2002). Exploring offence statistics in Stockholm city using spatial analysis tools. *Annals of the Association of American Geographers, 92*(1), 29–51.

Flint, C. (2002). The theoretical and methodological utility of space and spatial statistics for historical studies: the Nazi party in geographic context. *Historical Methods: A Journal of Quantitative and Interdisciplinary History, 35*(1), 32–42.

Galster, G., Hanson, R., Ratcliffe, M. R., Wolman, H., Coleman, S., & Freihage, J. (2006). Wrestling sprawl to the ground-defining and measuring an elusive concept. *US Geological Survey Professional Paper* (1726), 55–58.

Hasse, J. (2004). A geospatial approach to measuring new development tracts for characteristics of sprawl. *Landscape Journal, 23*(1), 52–67.

Herold, M., Goldstein, N. C., & Clarke, K. C. (2003). The spatiotemporal form of urban growth: Measurement, analysis and modeling. *Remote Sensing of Environment, 86*(3), 286–302.

Jiang, F., Liu, S., Yuan, H., & Zhang, Q. (2007). Measuring urban sprawl in Beijing with geo-spatial indices. *Journal of Geographical Sciences, 17*(4), 469–478.

Kumar, J. A. V., Pathan, S., & Bhanderi, R. (2007). Spatio-temporal analysis for monitoring urban growth—A case study of Indore city. *Journal of the Indian Society of Remote Sensing, 35*(1), 11–20.

Maktav, D., Erbek, F., & Jürgens, C. (2005). Remote sensing of urban areas. *International Journal of Remote Sensing, 26*(4), 655–659.

McGarigal, K., Cushman, S. A., Neel, M. C., & Ene, E. (2002). FRAGSTATS: Spatial pattern analysis program for categorical maps.

McGarigal, K., & Marks, B. J. (1995). *Spatial pattern analysis program for quantifying landscape structure*. Gen. Tech. Rep. PNW-GTR-351. US Department of Agriculture, Forest Service, Pacific Northwest Research Station.

Paolini, L., Grings, F., Sobrino, J. A., Jiménez Muñoz, J. C., & Karszenbaum, H. (2006). Radiometric correction effects in Landsat multi-date/multi-sensor change detection studies. *International Journal of Remote Sensing, 27*(4), 685–704.

Park, S., Jeon, S., Kim, S., & Choi, C. (2011). Prediction and comparison of urban growth by land suitability index mapping using GIS and RS in South Korea. *Landscape and Urban Planning, 99*(2), 104–114.

Pham, H. M., Yamaguchi, Y., & Bui, T. Q. (2011). A case study on the relation between city planning and urban growth using remote sensing and spatial metrics. *Landscape and Urban Planning, 100* (3), 223–230.

Ramachandra, T., Bharath, H., & Sowmyashree, M. (2013). Analysis of spatial patterns of urbanisation using geoinformatics and spatial metrics. *Theoretical and Empirical Researches in Urban Management, 8*(4), 5–24.

Ren, P., Gan, S., Yuan, X., Zong, H., & Xie, X. (2013). Spatial expansion and sprawl quantitative analysis of mountain city built-up area. *Geo-Informatics in Resource Management and Sustainable Ecosystem* (pp. 166–176). Springer.

Rigo, G., & Parlow, E. (2007). Modelling the ground heat flux of an urban area using remote sensing data. *Theoretical and Applied Climatology, 90*(3–4), 185–199.

Sandhya Kiran, G., & Joshi, U. B. (2013). Estimation of variables explaining urbanization concomitant with land-use change: A spatial approach. *International Journal of Remote Sensing, 34*(3), 824–847.

Sarvestani, M. S., Ibrahim, A. L., & Kanaroglou, P. (2011). Three decades of urban growth in the city of Shiraz, Iran: A remote sensing and geographic information systems application. *Cities, 28* (4), 320–329.

Sharma, L., Pandey, P. C., & Nathawat, M. (2012). Assessment of land consumption rate with urban dynamics change using geospatial techniques. *Journal of Land Use Science, 7*(2), 135–148.

Soffianian, A., Nadoushan, M. A., Yaghmaei, L., & Falahatkar, S. (2010). Mapping and analyzing urban expansion using remotely sensed imagery in Isfahan, Iran. *World Applied Sciences Journal, 9* (12), 1370–1378.

Sudhira, H., Ramachandra, T., & Jagadish, K. (2004). Urban sprawl: Metrics, dynamics and modelling using GIS. *International Journal of Applied Earth Observation and Geoinformation, 5*(1), 29–39.

Sutton, P. C. (2003). A scale-adjusted measure of "urban sprawl" using nighttime satellite imagery. *Remote Sensing of Environment, 86*(3), 353–369.

Taubenböck, H., Wegmann, M., Roth, A., Mehl, H., & Dech, S. (2009). Urbanization in India—Spatiotemporal analysis using remote sensing data. *Computers, Environment and Urban Systems, 33*(3), 179–188.

Torrens, P. M. (2008). A toolkit for measuring sprawl. *Applied Spatial Analysis and Policy, 1*(1), 5–36.

Tsai, Y.-H. (2005). Quantifying urban form: Compactness versus 'sprawl'. *Urban Studies, 42*(1), 141–161.

Turner, M. G., Gardner, R. H., & O'neill, R. V. (2001). *Landscape ecology in theory and practice* (Vol. 401). Springer.

Turner, M. G., O'Neill, R. V., Gardner, R. H., & Milne, B. T. (1989). Effects of changing spatial scale on the analysis of landscape pattern. *Landscape Ecology, 3*(3–4), 153–162.

UPA (2009). Tripoli agglomeration plan. In *3rd generation planning project.* Libyan Government, Libya.

Weng, Y.-C. (2007). Spatiotemporal changes of landscape pattern in response to urbanization. *Landscape and Urban Planning, 81*(4), 341–353.

Yeh, C.-T., & Huang, S.-L. (2009). Investigating spatiotemporal patterns of landscape diversity in response to urbanization. *Landscape and Urban Planning, 93*(3), 151–162.

Yeh, A., & Li, X. (2001). A constrained CA model for the simulation and planning of sustainable urban forms by using GIS. *Environment and Planning B: Planning and Design, 28*(5), 733–753.

Yeh, A., & Xia, L. (2001). Measurement and monitoring of urban sprawl in a rapidly growing region using entropy. *Photogrammetric Engineering and Remote Sensing, 67*(1), 83–90.

Yuan, F. (2008). Land-cover change and environmental impact analysis in the Greater Mankato area of Minnesota using remote sensing and GIS modelling. *International Journal of Remote Sensing, 29*(4), 1169–1184.

Yue, W., Liu, Y., & Fan, P. (2013). Measuring urban sprawl and its drivers in large Chinese cities: The case of Hangzhou. *Land Use Policy, 31*, 358–370.

Zhang, B. (2003). Application of remote sensing technology to population estimation. *Chinese Geographical Science, 13*(3), 267–271.

Zhang, Y., Yang, Z., & Li, W. (2006). Analyses of urban ecosystem based on information entropy. *Ecological Modelling, 197*(1), 1–12.

5 Urban Compactness Assessment

Saleh Abdullahi and Biswajeet Pradhan

5.1 Background

A compact city is well recognized as one of the most sustainable urban forms. However, the most important step in implementing this objective for a specific urban area is to evaluate the existing compactness to realize the current situations of urban form. The concept of compactness is related to the form and arrangement of urban features in urban areas such as spatial design and distribution of land use categories or transportation network. This definition relates urban compactness only to physical properties of an urban system. However, urban compactness is also related to human activities and the living behavior of the local residents of an urban region. City compactness is evaluated through self-sufficiency and independence from external forces (Burton 2002). It can also be defined as a measure for evaluating the travel behavior of residents of a community to meet their daily requirements, such as working, shopping, entertaining, and others. Compactness assessment is useful in land use planning to calculate land development demands, site capacity evaluation, and individual site development. Burton (2002) summarized the advantages of compactness assessment based on three main reasons:

(1) to assist research on the effects of compactness, and thus to guide policy;
(2) to enable the measurement of progress toward sustainability; and
(3) to be used as a planning tool.

However, measuring city compactness by assessing not only various aspects of the compact city but also its relation to urban sustainability is not an easy task. Therefore, no standard and comprehensive method is available for this assessment, and every study evaluates city compactness based on the objective of the study and data availability. For instance, measuring urban density and land use diversity is usually based on the census tracts, which vary in size and resolution. Therefore, the assessments are not comprehensive and reliable enough because the results may differ according to various zoning manners, cell sizes, and types of input data. In addition, urban areas are evaluated usually in terms of zonal or district level, and most of the previous studies evaluate city compactness in a zonal manner. However, evaluations based on the predefined zoning are not sufficiently reliable and comprehensive. Any changes in the boundary of the zones disturb the area of the zones, which is very important especially in density evaluations, such as population, build-up, and residential densities. Furthermore, on the one hand, these changes affect the ownership of the community facilities and services of each zone. Thus, the result varies depending on the zoning manner. On the other hand, in large-scale regions, such as countries, urban compactness is generally measured based on the cellular concept and the concentration of built-up cells in a specific area, as discussed by Li and Yeh (2004), who assessed the urban compactness using entropy and compactness index methods. Mubareka et al. (2011) introduced a composite index that characterizes urban expansion patterns to describe the degree of compactness of European urban lands. In addition, principal component and cluster analysis were applied to build the composite index. Thinh et al. (2002) presented a measuring compactness method based on GIS raster analysis and used gravitation approach for built-up grid surfaces of 500 m 500 m of the study area. Meanwhile, city compactness in addition to urban built-up density (which is an implication of physical compactness) consists of other various aspects related to functional compactness, which reveals valuable and useful information about the existing condition of cities (Turskis et al. 2006; Zagorskas et al. 2007). Therefore, the evaluation of compactness independent of any assumptions and local guidelines is effective and can be considered as a flexible and global approach.

B. Pradhan (ed.), *Spatial Modeling and Assessment of Urban Form*,
DOI 10.1007/978-3-319-54217-1_5

Statistical analysis is the main approach of these studies in calculating city compactness. Evaluating city compactness by applying common statistical techniques to measure, for example, mixed land use development can only indicate the land use richness of a local neighborhood (Bhat and Gossen 2004; Van Eck and Koomen 2008; Manaugh and Kreider 2013). However, the distribution pattern, which depends on the adjacency and relationships among various land use categories, can be evaluated only through spatial and mapping-based approaches (Abdullahi et al. 2015b).

A compact city is described according to several concepts, such as free-standing, contained, autonomous, moderately sized, and self-contained (Scoffham and Vale 1996; Burton et al. 2003). However, among various studies, some identical terms are used in defining a compact city, namely, urban density, diversity, and intensity. The high density of compact development refers to the concentration of activities (such as living, entertainment, and business) and settlements (such as residential, commercial, and industrial). The high land use diversity of a compact city refers to the mixture of several activities and land use in horizontal or vertical dimensions. On the one hand, this aspect includes the proximity of various land use types, which are the destination of local residents' daily transportation, such as living, working, educational, recreational, and other locations. On the other hand, city intensity is the process of achieving urban compactness related to the proper distribution of facilities and services, brownfield redevelopment, urban regeneration, and others. (Abdullahi et al. 2015a).

5.1.1 Urban Density

Density is considered as an important parameter of urban development. High-density development is widely proven to be an effective task for achieving urban sustainability (Carruthers 2002; Arifwidodo 2012). For example, Carruthers and Ulfarsson (2008), Burton (2002), and Regan (2000) discussed the advantages of high-density development, high residential density, and high population density, respectively. Urban density has the following measurement aspects:

(1) Population density: The number of people within a standard spatial unit is usually referred to as population density. In general, population density can be measured by the number of habitants divided by the total area of region of interest (Lin and Yang 2006). In a more sophisticated way, population density is measured by the number of habitants divided by the area of urban development land or the number of habitants divided by only the built-up areas of the region of interest (residential, commercial, and community facilities without open spaces such as recreational fields), with a unit of people per hectare. In the case of the evaluation of population with respect to housing availability, the number of habitants is divided by the total number of residential units in the region of interest.
(2) Residential density: Residential density is normally calculated by the number of residential units divided by the total built-up land or total residential and commercial land, with a unit of unit per hectare.
(3) Building density: Building density usually is calculated by the area of floor space divided by the total built-up land. This quantity has no unit.
(4) Employment density: This quantity is measured by the number of employees divided by the total built-up land, with a unit of people per hectare.

Other density measurement tools are available, such as floor area ratio or construction density (area of building for all floors divided by area of land parcel) and per capita spaces, which are normally calculated and considered in very specific cases. The aforementioned indicators measure the density of major urban activities and built-up properties. The numerators represent the volume of activities, and the denominators represent the available area of land for the corresponding activities (Lin and Yang 2006). High values of these indicators indicate high levels of city compactness. However, in identifying specific values for urban sustainability or compactness, a question is still raised regarding the appropriate values (Burton 2002).

5.1.2 Land Use Diversity

The measurement of land use diversity is an important challenge in terms of evaluating and implementing this type of development to achieve a compact city. Several studies are conducted on mixed land use measurement and assessment. For instance, Turskis et al. (2006) and Zagorskas et al. (2007) used Bayes theorem and complex proportional assessment, respectively, to evaluate the sustainable city compactness of Kaunas, Lithuania. Mixed land use development was assessed by measuring the ratio of (a) population and working places, and (b) population and object of attraction. The distribution level of nonresidential areas within residential areas was evaluated in this manner. Burton (2002) measured the land use diversity of 25 English cities relative to social sustainability with very detailed input data such as the number of key facilities for every 100 residents, number of newsagents for every 10,000 residents, variation in the number of facilities per postcode, and other data. Hoppenbrouwer and Louw (2005) discussed mixed land use development through the use of typology concept for the Eastern Docklands in Amsterdam. This typology entailed four aspects: urban functions, dimension, urban scale, and urban texture. The proposed typology provided a clear view of mixed land use properties and facilitated systematic analysis. Song and Knaap (2004) evaluated the effect of mixed land use development on residential property value using regression and statistical analyses. Manaugh and Kreider (2013) proposed a new technique called "land use interaction" based on the contiguity of different land use types. However, this technique is sensitive to street networks spatially located in different land types. In the next step, the results of this technique are compared with those of land use diversity measurement through an entropy index. This technique is commonly used in the biodiversity measurement, ecology, and communication fields. It can be traced backed to the work of Shannon (2001). However, this method only focuses on the diversity of land uses without considering the land's distribution and proximity, which can be a drawback in urban planning applications. Song and Knaap (2004) reviewed different GIS-based quantitative measures of mixed land use and tested the effectiveness of such measures according to proximity to residential buildings, proportion of nonresidential land, entropy method, and balance between jobs and population. Bhat and Gossen (2004) measured land use diversity on the basis of the weekend recreational-type choice. Only three land use categories (residential, commercial/industrial, and other types of land use) were considered in this measurement. Van Eck and Koomen (2008) evaluated land use diversity using entropy theorem and Simpson's diversity index S. This technique is based on the probability that two different polygons within a pixel or a zone have different functions. Musakwa and Van Niekerk (2013) differentiated mixed land use from land use frequency and reported that mixed land use is a more reliable indicator for capturing socioeconomic costs associated with urban sustainability. Song et al. (2013) examined several common measures of mixed land use, such as Atkinson index, balance, entropy, and Herfindahl–Hirschman indices, to identify the measures' strengths and weaknesses. However, most of these techniques are applicable only to two land use types.

In summary, regardless of the quantity assessment of land use categories through the various aforementioned methods, mixed land use assessment techniques should be capable of evaluating spatial aspects as well because the distribution pattern of various land use types and their relations with one another are important factors as well.

5.1.3 Urban Intensity

Urban intensity deals with the crowd and livability of an area. This process can be evaluated through the increase in population density, development, and land use diversity. Thus, urban intensity can be assessed through the activeness, availability, proximity, quality, and quantity of each type of community facility (such as health, educational, public transportation, point of interests, open space, and recreational facilities as well as job opportunity) with respect to the characteristics of the local residents and neighborhoods. Lin and Yang (2006) measured city intensification through three observed variables: change in residential density (from 1991 to 2001), change in building density, and change in employment density. Therefore, 10 years of changes in such variables in terms of measuring density can be considered in evaluating city compactness and consequently increase the process of compactness. On the one hand, in evaluating the local demands, detailed information on the local population is required. On the other hand, up-to-date information on existing facilities, such as capacities, locations, and qualities, should be available. Evaluating the local population demands based on some local and/or standard guidelines such as those proposed by De Chiara (1990) is important (Table 5.1).

5.2 Methodological Process of Urban Compactness Assessment with Kajang City (Malaysia) as Case Study

This section presents the methodological process of city compactness assessment of Kajang City based on main compact city paradigms for four temporal land use maps of this city (2004, 2008, 2012, and 2015). Kajang is a city located in the eastern part of Selangor province in the southwestern region of Peninsular Malaysia. This city is located 21 km away from Kuala Lumpur, which is the capital city of Malaysia, and covers an area of 60 km^2. Moreover, it has a population of 300,000 (2010). The current population of Kajang has grown rapidly in the past few years. The eastern part of this region is mainly occupied by agricultural and forest lands. Agricultural land is a high-proportion land use category in this region. Central and peripheral parts of the city are mainly occupied by community facilities and residential buildings. However, commercial buildings such as shopping malls have higher growth in the city center than the other categories. Industrial areas are mainly located in the central west of the city.

The local planning authority of Kajang (JPBD) has proposed several other strategies in addition to general development strategies, especially to increase city compactness. These strategies consisted of several aspects such as mixed land use development, building design, housing design, sense of place, public transportation, neighborhood, and promotion of walking, cycling, and green environment, as listed in the following:

Table 5.1 Minimum population required for various community facilities (De Chiara 1990)

Community facility	Minimum population	Community facility	Minimum population
Education		*Health*	
Kindergarten	500	Clink	10,000
Primary school	1800	Hospital 100 beds	25,000
Secondary school	5000	Welfare center	25,000
High school	9000	Hospital 225 beds	50,000
University	500,000	Hospital 340 beds	75,000
University (graduate studies)	1,000,000	Hospital 450 beds	100,000
Institutional		*Recreation*	
Post office	1200	Local park	3000
Library	500	Play ground	5000
Fire station	10,000	Gym and fitness	10,000
Police station	10,000	Cinema	20,000
Waste management center	10,000	Swimming pool	20,000
Commercial		*Employment*	
Super market	2000	Institutional	10,000
Pharmacy	3000	Services	10,000
Bank	5000	Light industry	10,000
Shopping mall	20,000	Heavy industry	50,000
Hotel	25,000	Industrial park	100,000

(1) Mixed land use development
- Residents encouraged to live within the workplace
- Concentration of activities in accordance with the centralized public transport networks (TOD)
- Implementation of a mixed development area and development of potential/in-fill site
- Building design: variety of activities or functions in one building, creating an effective vertical mixed land use

(2) Advantages of public transportation
- Wide range of public transport modes created
- User-friendly public transport system (appropriate age group)

(3) Housing design
- Various types of residential design according to location and needs
- Residential types developed to suit the compact city
- Residential district integrated with convenient transportation facilities

(4) Sense of place
- Safe and active open space
- Commercial development characterized by local community activities

(5) Cycling and walking neighborhood
- Building design incorporated with pedestrian-friendly features
- Accessibility of public transport nodes for pedestrians/cyclists
- Safe network for pedestrians/cyclists and uninterrupted access between the neighborhood and the city center

(6) Green and environment preservation
- Maintenance of green area
- Creation of green corridor and blue part of the redevelopment potential.

These strategies required several analyses and processing tools such as site suitability analysis, readiness analysis, evaluation of land development potential, accessibility analysis, network analysis, and other methods that can be accomplished through several GIS mapping and processing tools. Regardless of those related to planning and design (such as those in housing and building design, creation and promotion of walking and cycling environment), others related to general urban sustainability (such as preservation of natural and green environment, less car dependency thus less carbon emission, and promotion of public transportation) are involved in the discussion in this section.

As previously explained regarding the drawbacks of zonal assessment, this study evaluated the city compactness of Kajang City first with the predefined zoning of the district proposed by the local planning authority and second without the zonal format. In the second phase, instead of quantifying the availability of facilities, services, and other statistical measurements within a specific border, proximity analysis using Euclidean distance theory was applied. In this manner,

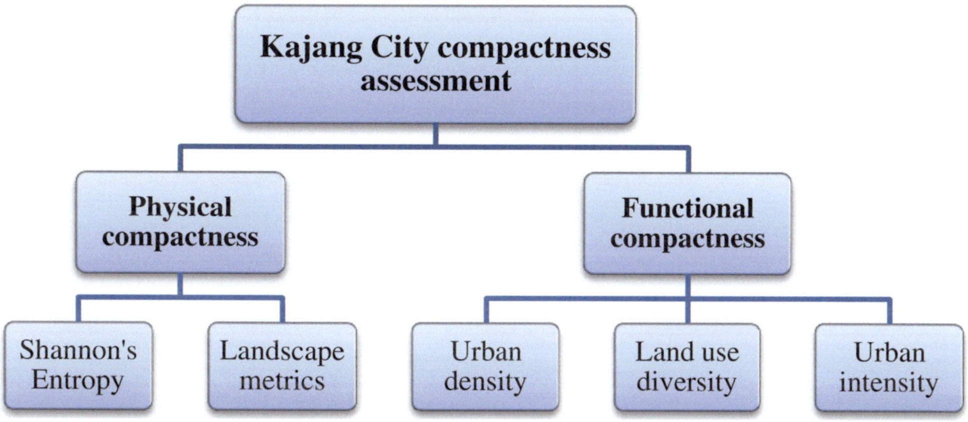

Fig. 5.1 General flowchart of city compactness assessment

the city compactness of every pixel of the study area was evaluated based on the entire study area. In addition, to improve most of the compactness assessment studies in the literature, which were based on only the concentration of built-up areas, this study evaluated city compactness in two main phases (Fig. 5.1):

(1) Physical compactness
(2) Functional compactness.

5.2.1 Physical City Compactness

Physical compactness refers to the physical appearance and spatial configuration of urban areas (built-up areas), determining whether they areas look clustered or dispersed. In this assessment, structural metrics measure the physical composition or configuration of the patch mosaic without explicit reference to an urban functional process. This perspective actually accounts for the shape complexity of cities without considering the details about functional, activity, and land use pattern perspectives. In general, concentrated or clustered areas imply high compactness. By contrast, sprawl development changes urban landscapes over time by increasing fragmentation and generating several small urban patches.

5.2.1.1 Landscape Metrics Measurements

For this purpose, at the first stage, various landscape metrics were used to measure urban compactness and to investigate the spatiotemporal trend of urban growth and land use changes by using the popular FRAGSTATS (McGarigal and Marks 1995) software. These metrics were applied to all four land use maps (2004, 2008, 2012 and 2015) in two sampling strategies. The first one was applied with no sampling strategy, and the second run was performed with predefined zoning districts proposed by the local planning authority (Fig. 5.2).

The main logic behind these metrics is to compute several statistics for each patch, land use type (class level), and the landscape as a whole. Some of these metrics quantify landscape composition, whereas others quantify landscape configuration. Thus, understanding each metric in terms of the aspect of landscape pattern being quantified is especially important. Patch indices represent small fragmentation in the landscape regardless of its type and class. Class indices represent the spatial distribution and pattern within a landscape of a single patch type, whereas landscape indices represent the spatial pattern of the entire landscape mosaic, considering all patch types simultaneously.

Some of these metrics, which measure clumpiness, aggregation, complexity, level of dispersion, diversity, and other attributes, have direct relations to physical compactness and can play effective roles in assessing city compactness (Li et al. 2008). Thus, in addition to the urban metrics applied in the previous section, several other metrics particularly suitable for compactness assessment were used in this process (Table 5.2).

Landscape configuration should be evaluated in terms of the complexity of patch shapes at different levels. Most of these shape metrics are based on perimeter–area relationships. Thus, shape index (SHAPE) was used to measure the complexity of patch shapes compared with a standard shape (square) of the same size, therefore mitigating the size dependency of the perimeter–area ratio.

An alternative approach in shape assessment, which has a closer relation to compactness assessment, is based on the ratio of the patch area to the area of the smallest circumscribing circle. This index measures the overall patch elongation of the study area. A highly convoluted but narrow patch will have a low related circumscribing circle index (CIRCLE) because of the relative compactness of the patch, but a narrow and elongated patch will have a high related circumscribing square index. This index may be particularly useful for distinguishing patches that are linear (narrow) and elongated.

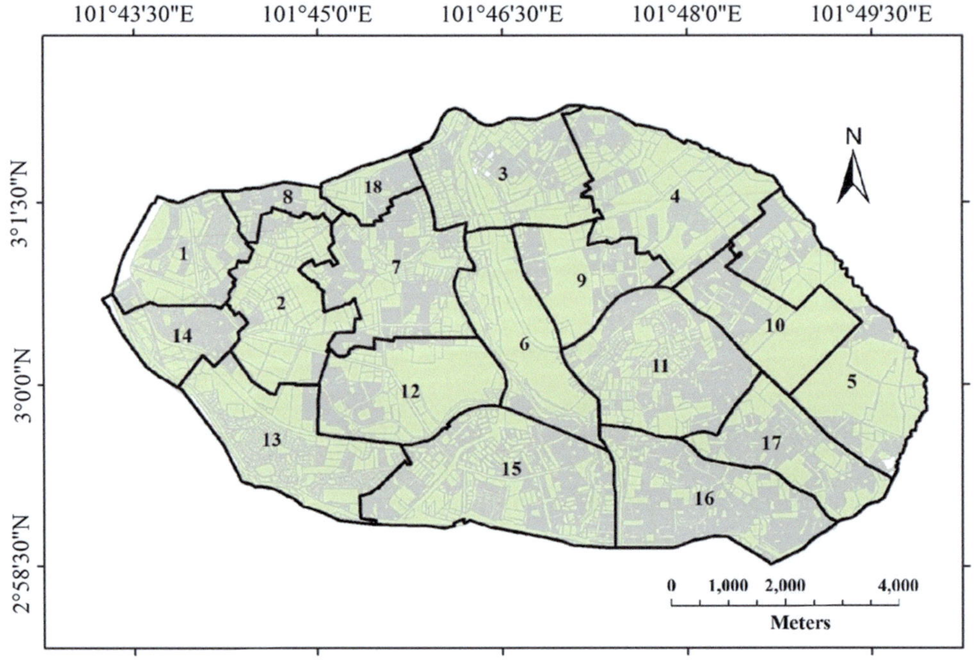

Fig. 5.2 Kajang zoning district sampling

Edge (perimeter) metrics explain landscape configuration, although they are not spatially explicit. Total edge is the measure of the total perimeter of a patch type in a class or all patches in a landscape. Thus, edge density was used to standardize edges to a per unit area to ensure comparability among patches with varying sizes. In addition, landscape shape index (LSI) was used to compare a simple geometric shape with a number of edges and that with no internal edges.

Number of patches (NP) and patch density (PD) provide information regarding the subdivision aspects of aggregation. However, in this case, considering the patch class (land use type) is very important. Thus, high patch density in a specific area does not necessarily indicate high compactness. High built-up patch density implies high compactness. By contrast, high agricultural field patch density does not imply city compactness.

Area metrics evaluate landscape composition instead of configuration. The area of each landscape mosaic is the main and fundamental information about the landscape. However, the extension of the patch is also very important, in addition to the size of a patch. Thus, the radius of gyration metric can measure the extent of patches. It evaluates the distance that a patch extends its reach within the landscape. If all other variables remain constant, then the larger patch has a higher radius of gyration. Similarly, in holding the area constant, when the patch causing a high radius of gyration is more extensive, a less compact composition landscape results. This metric can be considered as a measure of the average distance that an organism can move within a patch before encountering the patch boundary from a random starting point. This measure provides the ability to distinguish the distribution of area among patches within the study area.

The compactness of the largest path index evaluates landscape fragmentation. A regularly shaped landscape with a few patches gains high CLPI value (Huang et al. 2007). Compact zones usually have high population and built-up density and hence low land consumption and fragmented urban zones. Consequently, a high value of CLPI represents high city compactness.

Aggregation metrics deal with the tendency of landscape patch types to be spatially aggregated, that is, occur in large, aggregated, or contagious distribution. These metrics also evaluate the landscape texture in terms of dispersion, interspersion, subdivision, and isolation. The main metrics related to city compactness under the aggregation concept are aggregation index (AI), Euclidean nearest-neighbor distance (ENN), and proportion of like adjacencies (PLADJ). High aggregation index values such as PLADJ show aggregated patch types with large and compact shapes (McGarigal and Marks 1995).

Regardless of physical configuration and composition, the land use diversity of Kajang City can be evaluated through the FRAGSTATS software. In this measurement, richness and evenness were the main components of evaluation. Richness refers to number of land use types, thereby illustrating the compositional components, and evenness refers to the distribution of areas among different land use types, therefore illustrating the structure and pattern of diversity. Several metrics are used for diversity measurements, such as Shannon's diversity index (SHDI), Simpson's diversity index (SIDI), Shannon's evenness index (SHEI), and Simpson's evenness index (SIEI).

Table 5.2 Detail information about landscape metrics used to assess city compactness of Kajang City

Landscape metrics	Abbreviation	Descriptions
Shape index	SHAPE	SHAPE equals patch perimeter (m) divided by the square root of patch area (m^2), adjusted by a constant to adjust for a square standard $\text{SHAPE} = \frac{0.25P_{ij}}{\sqrt{a_{ij}}}$ $\text{SHAPE} \geq 1$, without limit SHAPE = 1 when the patch is square and increases without limit as patch shape becomes more irregular
Landscape shape index	LSI	$\text{LSI} = \frac{0.25E}{\sqrt{A}}$ $\text{LSI} \geq 1$, without limit E = total length (m) of edge in landscape; includes the entire landscape boundary and some or all background edge segments A = total landscape area (m^2)
Edge density	ED	$\text{ED} = \frac{E}{A}(10{,}000)$ $\text{ED} \geq 0$ E = total length (m) of edge in landscape A = total landscape area (m^2)
Largest patch index	LPI	$\text{LPI} = \frac{\text{MAX}(a_{ij})}{A}(100)$, $0 < \text{LPI} \leq 100\%$ a_{ij} = area (m^2) of patch ij A = total landscape area (m^2)
Related circumscribing circle	CIRCLE	$\text{CIRCLE} = 1 - \left(\frac{a_{ij}}{a_{ij}}\right)$, $0 \leq \text{CIRCLE} \leq 1$ without limit
Patch density	PD	$\text{PD} = \frac{N}{A}(10{,}000)(100)$, $\text{PD} > 0$, constrained by cell size N = total number of patches in the landscape A = total landscape area (m^2)
Simpson's evenness index	SIEI	$\text{SIEI} = \frac{1-\sum_{i=1}^{m} P_i^2}{1-\left(\frac{1}{m}\right)}$, $0 \leq \text{SIEI} \leq 1$ P_i = proportion of the landscape occupied by patch type (class) i m = number of patch types (classes) present in the landscape, excluding the landscape border if present
Aggregation index	AI	$\text{AI} = \left(\frac{g_{ij}}{\max g_{ij}}\right)(100)$, $0 \leq \text{AI} \leq 100$ With percent unit
Proportion of like adjacencies	PLADJ	$\text{PLADJ} = \left(\frac{\sum_{i=1}^{m} g_{ij}}{\sum_{i=1}^{m}\sum_{k=1}^{m} g_{ij}}\right)(100)$, $0 \leq \text{PLADJ} \leq 100$ With percent unit
Radius of gyration	GYRATE	$R_g^2 = \frac{1}{2N^2}\sum_{k=1}^{N}(r_i - r_j)^2$

5.2.1.2 Shannon's Entropy

Although the concept of Shannon's entropy was already used in the first stage in the case of land use diversity assessment (in the FRAGSTATS software), in the second phase, Shannon's entropy model was used independently for all four land use maps to evaluate their physical compactness. In fact, because of this model's capability of spatial dispersion and concentration measurements, it is widely used in urban sprawl studies (Ramachandra et al. 2013). This model is applied to understand the growth and land use changes of Kajang City, that is, whether city compactness is increased or decreased. This goal provides an insight into the development trend in various zones of the city. Thus, Shannon's entropy was applied based on Kajang zoning districts. This application means that the entropy value was computed based on the built-up areas of each zone. The level of sprawl and/or compactness is recognized by the entropy value, which ranges from zero to $\log_e(n)$. In this calculation, compact zones are assigned with a value near zero, and dispersed zones are assigned with a value near $\log_e(n)$.

5.2.2 Functional City Compactness

Functional compactness can be defined as an attribute that explicitly measures the landscape pattern in a manner that is functionally relevant to the organism or process under consideration. Functional metrics require additional parameterization prior to their calculation, such that the same metric can return multiple values depending on the user's specifications, which are more about the availability of facilities, land use diversity, land use pattern, and others. This section evaluates the compactness of Kajang City by considering various details on land use pattern and distribution, urban densities (population, residential, and built-up), and self-sufficiency and independence from the outside. This kind of functional city compactness assessment is quite new and is a complex research. In the case of physical compactness assessment, only satellite data may be sufficient in performing the analysis and obtaining the results. However, for functional assessments, several layers of data consisting of very detailed information on land use maps, population data, road network maps, public transportation facilities, job availability (in some cases), and other information are required. In addition, various statistical and other methodological processes are required to manipulate and evaluate this large amount of data and information to assess city compactness. In fact, no steady and standard approaches are used in this field, and each study has assessed the city compactness of a specific region based on the data availability and the research objective.

In this study, city compactness was evaluated based on three main indicators (urban density, urban intensity, and land use diversity) of compact urban development (Fig. 5.1). Several thematic maps of Kajang City were used to define these compactness indicators.

- **Population data**: Population data were used to create a population density map, which is one of the most important variables for city compactness.
- **Land use map**: All four available land use maps were used:
  - To extract various land use types (residential, commercial, industrial, and others)
  - To extract the location of various available community facilities (health, educational, point of attraction)
  - To extract the location of job opportunities (commercial, institutional, industrial, and others)
  - To extract the location of existing parks and open spaces to be used as recreational facilities
  - To extract residential areas and create a residential density map
  - To extract built-up areas for building density assessment
  - To evaluate land use diversity of the study area.
- **Public transportation facilities map**: These data such as train and bus station locations are required in evaluating transportation facilities.
- **Road network map**: These data are needed to evaluate the role of road networks in city compactness as well as to perform road density assessment.

As shown in Fig. 5.3, compactness assessment was implemented for zonal and without zonal bases.

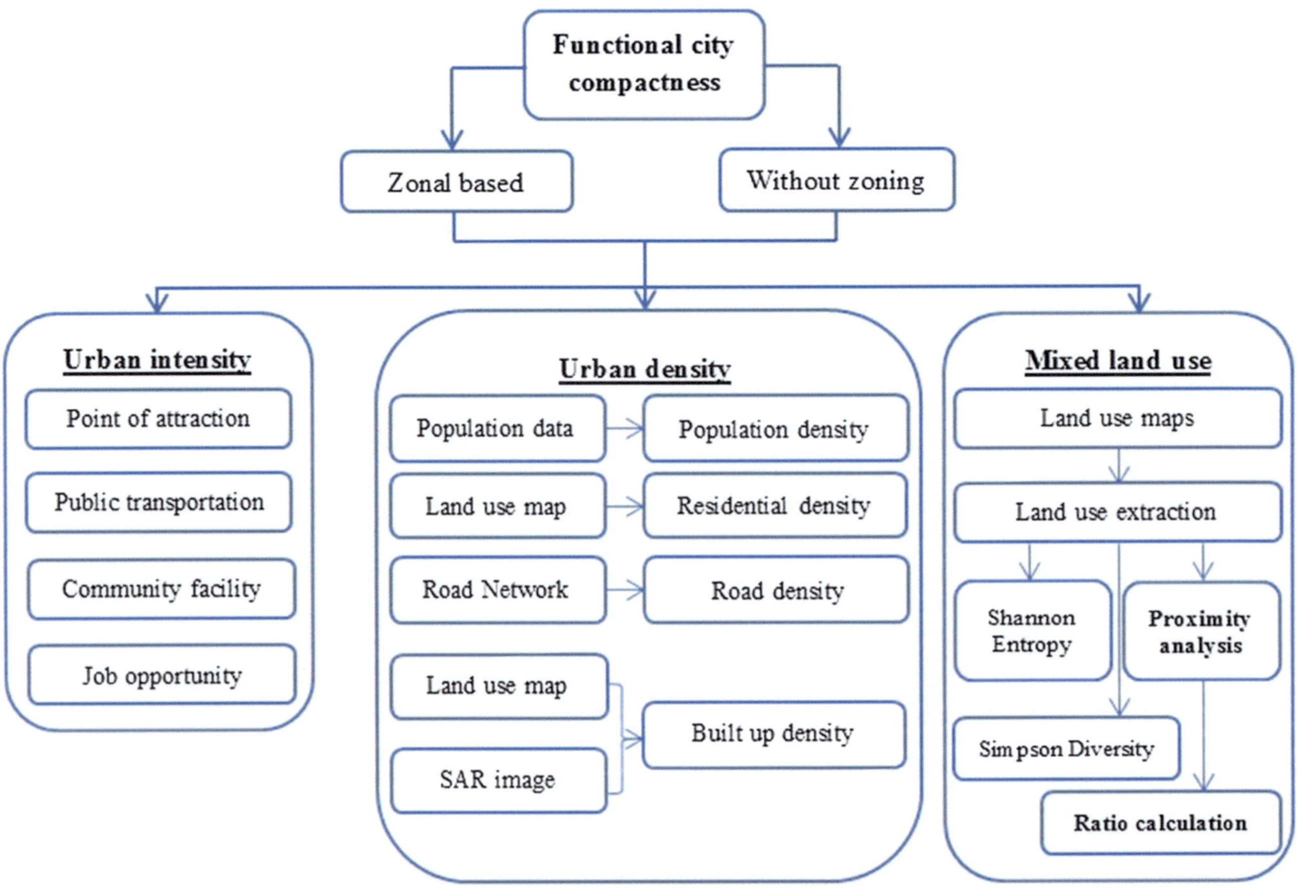

Fig. 5.3 Functional city compactness assessment flowchart

For zonal basis, all the aspects were evaluated with respect to the area and properties of each zone, whereas for the second case, various aspects were evaluated independently and with respect to the entire Kajang City area.

5.2.2.1 Urban Density Assessment

As mentioned, determining the best value for various urban density calculations to achieve better quality of life remains a challenging issue (Burton 2002). Several values can be assumed as optimum according to the local situation and the objective of the studies. For instance, 225–300 is the optimum value of housing density according to earth environmental scientists (Burton 2002). However, because this study's main concern is land use pattern and distribution analysis, and because a strong agreement exists on high density instead of low density from the point of view of land conservation, it assumes that higher values of urban densities are more compact and sustainable than lower densities.

(i) **Population density**: The available data for population were for the years 2000 and 2010, with populations of 250,000 and 300,000, respectively. In matching the population data with land use maps, the annual rate of population growth per year was assumed according to the available data. Thus, the population assigned for each land use map was calculated and used for population density analysis.

The evaluation of population density in the case of zonal basis was calculated according to persons per hectare in built-up areas, as shown in Eq. 5.1.

$$\text{Population density of zone } i = \frac{\text{Population of zone } i}{\text{Build up area of zone } i} \quad (5.1)$$

In case of density assessment without zoning basis, the moving-window method using kernel density was applied to create a raster-based (1 m × 1 m) population density surface. The advantages of this density surface processing compared with conventional zoning-based computation are consistency of results, higher accuracy of population distribution impression, and ability of further analytical processing (Adolphson 2010).

(ii) **Residential density**: Residential areas of Kajang City mainly consist of single-story, double-story, multistory, and high-rise buildings. The available land use maps collected from Kajang planning authority included information on the number of residential units. For the zonal density assessment, residential density was calculated in three aspects for all four land use maps:

$$\text{Residential density of zone } i = \frac{\text{Number of residential units of zone } i}{\text{Built up area of zone } i} \quad (5.2)$$

$$\text{Residential density of City (Built−up based)} = \frac{\text{Total number of residential units}}{\text{Total built up area of Kajang}} \quad (5.3)$$

$$\text{Residential density of City (Overall)} = \frac{\text{Total number of residential units}}{\text{Total area of Kajang City}}. \quad (5.4)$$

In this manner, the zonal aspect of residential growth can be assessed, and the total growth of the residential area of Kajang City can be achieved. In the case without zonal basis, areas and the number of residential units were used to create raster-based residential density surface through moving windows.

(iii) **Road density**: Road density was also evaluated with respect to zonal basis for each land use map through the following equations:

$$\text{Roaddensity of zone } i = \frac{\text{Road lenght of zone } i}{\text{Total area of zone } i} \quad (5.5)$$

$$\text{Roaddensity Kajang City} = \frac{\text{Total road lenght of Kajang}}{\text{Total area of Kajang}} \quad (5.6)$$

In the case without zonal basis, road length was used to create a raster-based road density surface through moving windows.

(iv) **Building density**: Researchers are mainly concerned with population density when they analyze urban density. However, building density is also an important aspect of urban density. In the case of environmental conservation, the built-up areas destroy agricultural and forest lands. High built-up and residential densities not only have benefit of land savings but also reduce energy consumption as well as increase affordable housing. For building density assessment, land use categories such as residential, commercial, industrial, and some facilities and infrastructures were considered as the built-up areas. By contrast, open spaces, water bodies, agricultural and green fields, local parks, and recreational areas were omitted from analysis. Thus, building density was calculated based on the following equations:

$$\text{Building density of zone } i = \frac{\text{Built up area of zone } i}{\text{Total area of zone } i} \tag{5.7}$$

$$\text{Building density Kajang City} = \frac{\text{Total Built up area of Kajang}}{\text{Total area of Kajang}}. \tag{5.8}$$

5.2.2.2 Mixed Land Use Measurements

The existence of several land use categories in a neighborhood reduces car dependency and encourages walking and cycling behavior. Thus, finding the right balance between residential and nonresidential development is important. Mixed land use can be considered as horizontal and vertical development. However, because this study focuses on neighborhood planning and especially employs land use pattern analysis, horizontal mixed land use development is the main research concern. Measuring mixed land use development is a challenging topic in city compactness assessment. Although previous studies have measured land use diversity in individual cities with respect to zonal properties, no consistent indicators consider the statistical as well as spatial distribution of various land use types especially for comparing different regions of interest. Therefore, the current study proposes a new approach to fulfill these limitations.

At the initial stage, as two common methods used in this field, Shannon's entropy and Simpson's diversity index are applied to the study area, and then proximity analysis is proposed and explained in detail.

(i) **Shannon's entropy**: In addition to sprawl development analysis, Shannon's entropy has been the most widely used index for measuring land use diversity especially in the biodiversity, ecology, and communication fields (Manaugh and Kreider 2013). In urban application, entropy can be applied to various scales, such as urban versus rural levels, or to evaluate mixing among various urban land use categories. Thus, in this case, an entropy index was used to evaluate the land use diversity of Kajang City:

$$\text{Entropy} = \frac{-\sum\left(A_{ij}\ln A_{ij}\right)}{\ln N_j}, \tag{5.9}$$

where A_{ij} is the percentage of land use i in land use map j, and N_j is the number of land use categories in land use map j. After each land use type was extracted separately, the total area of each land use type and the percentage occupied with respect to the total study area were calculated. Thereafter, through the entropy equation, the relative and absolute entropy for all four land use maps were calculated. Absolute entropy ranges from zero to $\ln(m)$, in which m is the number of land use categories. Relative entropy ranges from zero to one. In both cases, zero represents single land use, and $\ln(m)$ and 1 represent mixed land use development. Entropy and other measures from this family are sensitive to the distribution of the size of all land uses within a raster cell (Manaugh and Kreider 2013).

(ii) **Simpson's diversity index**: This index is based on the probability that two random places in a raster cell have different states or categories. This index is calculated through following equation:

$$S = 1 - \sum_{i=1}^{n} P_i^2, \tag{5.10}$$

where P_i is the percentage of the total study area occupied by each land use category. This index ranges from zero to $1 - (1/m)$, where m is the number of land use categories. Through this index, land use diversity can be further measured as effective function richness $1/(1 - S)$, which ranges from 1 to m. Furthermore, the proportionality index can also be calculated through $S/(1 - (1/\text{m}))$, ranging from zero to one. Similar to entropy measures, Simpson's index is sensitive to the distribution of the size of land use types in a grid cell. Simpson's index is more sensitive to larger land use types and less sensitive to smaller types.

These two methods as well as the other mixed land use measurement approaches are mainly based on the number of land use categories or land use type richness of the landscape, but they do not consider the spatial distribution of land use types and their relationships.

(iii) **Proximity analysis**: The main advantage of proximity analysis is the consideration of a distribution pattern of various land use categories with respect to one another. Thus, unlike most of the existing methods evaluating land use diversity only quantitatively, proximity analysis in addition to quantitative results, which show the level of diversity, produce a graphical representation through a raster-based map that shows the mixed land use pattern of the landscape. Usually, mixed land use developments are assessed with respect to the entire landscape or to zonal and regional bases. These results are not consistent and are not a fair representation of the entire landscape because highly mixed development of only a small part of the region may result in high diversity of the land use of the entire landscape. Therefore, a raster-based format of representation provides the ability to evaluate each small pixel of the landscape because each pixel can represent one value regarding land use diversity level.

In addition, raster-based format benefits the user in case of further analytical processes, especially in overlaying analysis. However, the proximity analysis method can also be modified to evaluate land use diversity based on zonal or whole landscape level.

Similar to Shannon's Entropy and Simpson's diversity index assessment, all land categories were extracted as a separate layer. The selection of land use to be included in the process depends on the objective of the study. One may use only a specific land use type (residential vs. nonresidential) or all land uses and land covers of the landscape (including bodies of water). In this study, among all existing land use types in Kajang City, only five main land uses are included, which are the daily (or weekly) destinations of the local residents. Therefore, these land uses affect traffic load and local transportation (residential, commercial, industrial, community, and recreational facilities). Through the selection of these land use types, the land use diversity of a neighborhood can be measured accurately to achieve sustainable city compactness.

Thereafter, through Euclidean distance theory, proximity analysis was conducted for each selected land use separately. This process created a raster-based layer for each land use type that covers its spatial extent with respect to the corresponding land use polygon in the center. Then, the classification process was performed to assign every pixel to one class of proximity (Fig. 5.4). Several classification schemes

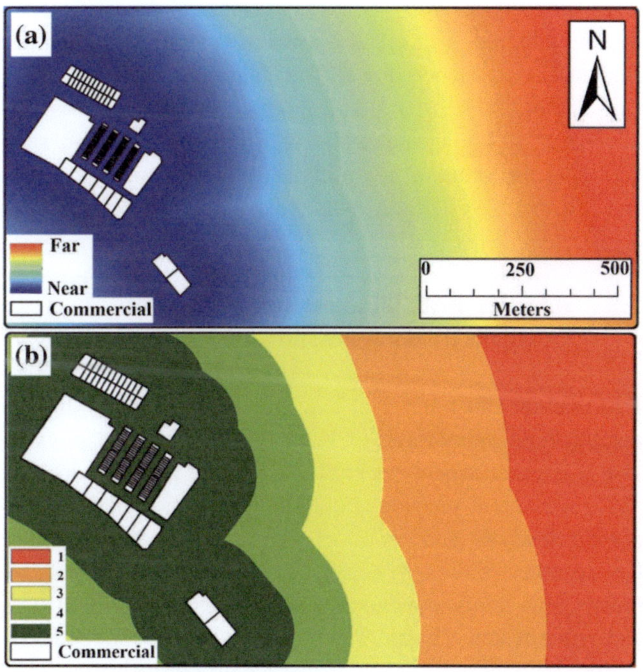

Fig. 5.4 Proximity analysis, **a** Euclidean distance analysis, **b** classified image (5 as nearest and 1 as farthest)

are available according to the requirements of the study, such as manual, equal interval, defined interval, quantile, natural breaks, and others. However, a standard classification for all land use types is important for consistent results.

All classified land use map layers were overlaid to summarize and create an overall proximity result of all land use types. This process produced a graphical representation with a raster-based format of the land use diversity level of each pixel of the landscape. This process was conducted four times for each land use map of Kajang City (2004, 2008, 2012, and 2015). Finally, the following equation was developed to evaluate the land use diversity of the entire landscape:

$$\mathrm{LD}_{\mathrm{Prox.}} = \sum_{i=n}^{m} P_i . V_i, \tag{5.11}$$

where $\mathrm{LD}_{\mathrm{Prox}}$ is land use diversity using the proximity concept; P_i is the proportional percentage of the landscape area (pixels) with corresponding V_i value; and V_i is the level of proximity of each pixel with respect to other land use categories, which is assessed from the overlay process that creates the overall proximity results. V_i ranges from n (number of the land use categories) to m (number of land use categories times number of proximity classes). In this study, the value of n is 5 (representing five land use types: commercial, recreational, residential, industrial, and community facilities), and m is 25, which is computed from 5 (number of land use types) × 5 (number of distance classes). Similarly, $\mathrm{LD}_{\mathrm{Prox}}$ is found in the range of n to m for each landscape. In this case, the entire landscape (100%) assigned V_i as n and the rest of V_i as zero; the $\mathrm{LD}_{\mathrm{Prox}}$ value of landscape was n. This case represented a landscape with one land use type or a single land use development neighborhood. By contrast, if all landscape pixels obtain a value of m for V_i, then the $\mathrm{LD}_{\mathrm{Prox}}$ value of landscape is m. This case represents a highly mixed land use development of a landscape in which all land use types are distributed at the proximity of one another properly.

$\mathrm{LD}_{\mathrm{Prox}}$ can be further modified and standardized in the rage of zero to one through the following equation:

$$\mathrm{LD}_{\mathrm{Prox.}} = \frac{\left(\sum_{i=n}^{m} P_i \cdot V_i\right) - n}{m - n}. \tag{5.12}$$

In this formula, $\mathrm{LD}_{\mathrm{Prox}}$ is zero if a landscape has a single land use type or one if a landscape has highly mixed land uses. These processes were applied for all four available land use maps of Kajang City.

Finally, the proximity analysis was further modified to assess the land use diversity on a zonal basis. For this process, from raster-based proximity maps, the value of $\mathrm{LD}_{\mathrm{Prox.}}$ for each zone of Kajang City was calculated separately. In considering a number of pixels in each zone, a ranking process was performed to evaluate and extract the least and highly mixed land use zones. Evaluation results were standardized to range from zero as least mixed to one as a highly mixed developed zone. In this manner, each zone of Kajang City can be evaluated for land use diversity within the selected period of time (from 2004 to 2015).

5.2.2.3 Urban Intensity Measurements

For urban intensity measurements, this study emphasized self-sufficiency from external forces and resources. Thus, the distribution pattern of community facilities and other resources of Kajang City as a whole and zonal basis were evaluated. In conducting this process, we assessed the availability, proximity, quality, and quantity of various facilities and resources with respect to the characteristics of local residents and neighborhoods. The facilities and resources involved in this process are enumerated as follows:

- Health: clinic, hospital, welfare center, and others
- Education: kindergarten, primary school, secondary school, high school, university, and others
- Open spaces and recreational facilities: local park, playground, stadium, golf course, swimming pool, and green and natural environment
- Public transportation facilities: bus and train stations
- Commercial buildings: shopping malls and commercial services
- Infrastructure and utilities: water supply, electricity, and waste management centers
- Other resources such as post offices, security, religious centers, points of attraction, and institutional and industrial use related to job opportunities.

The availability of these facilities and resources were extracted from land use maps. Information on various public attraction points, such as megamalls, markets, places of worship, and others was obtained from a recently developed specialized plan for the dislocation of these places.

In performing this analysis, the same concept and equation of proximity analysis were used to evaluate each pixel of the study area with respect to urban intensity. However, at this stage, more complex considerations were involved in the process, and the analysis was not solely based on the proximity concept. For instance, population data included detailed information on age, gender, ethnic, religion, and race. Thus, the required facilities were evaluated based on these characteristics (De Chiara 1990). Educational facilities were evaluated by considering the availability and proximity of kindergarten, primary and secondary schools, and others with respect to the number of children and young people in various age groups. Health facilities were assessed according to the availability of clinics, general hospitals, specialized hospitals, and welfare centers with respect to the population especially the number of elderly. The number and types of places of worship were evaluated according to the number of people belonging to each religion. The availability and proximity of public transportation stations, shopping malls, and recreational facilities were assessed with respect to the population and location of living and working places.

In addition to population requirements, urban intensity assessment can be further modified by considering different weights for each variable (facilities and resources) of landscape. This weighting concept is application dependent. For instance, one may want to prioritize educational facilities, or health and clinical resources are more important in health studies. The common approach for these weighting processes is multi-criteria decision making (MCDM) using expert knowledge (Saaty 1980). Therefore, the priority values can be applied to the urban intensity assessments as coefficients for each variable (facilities and resources) during the overlay process of the created proximity layers. This ability means that after proximity layers are created for all variables and during the calculation of V_i (level of proximity of each pixel with respect to selected variables), a predefined coefficient or weight should be inserted to each proximity layer to represent the importance level of each variable. Hence, when the modified V_i is used in the equation, it already bears the weights or priorities of the variables with respect to one another. However, to avoid bias in the city compactness analysis and to go through the detailed processing of conventional MCDM, this study assumed constant weights for all variables.

In the case of zonal assessment, all of these evaluations were conducted with respect to each zone's properties and characteristics. Hence, for all four land use maps of Kajang City (2004, 2008, 2012, and 2015), this process was performed as a whole and with zonal basis. Finally, in standardizing the value of urban intensity in the range of zero to one, Eq. 5.12 was used.

5.3 Results and Discussion of Urban Compactness Assessment for Kajang City

This section illustrates and discusses the results of the city compactness assessment of Kajang City based on physical and functional aspects as explained in the previous section. This assessment provides baseline information and guidelines for analysis and proposes a compact land use pattern. Unlike other studies, to achieve comprehensive and reliable results, this study conducted compactness assessment with and without zonal basis. The zonal basis was conducted on the predefined zoning of the district proposed by the local planning authority of Kajang City. In the second phase, instead of the assessments quantified within specific borders, all measurements were based on the total extent of the study area. In this manner, evaluating the city compactness of every cell (with any cell size) of Kajang City was feasible.

5.3.1 Physical Compactness

In the case of physical compactness, the physical composition and spatial configuration of the city (without the details and functional purposes) were assessed in two main stages: landscape metric analysis and Shannon's entropy compactness assessment. In general, in both cases, concentration and building clusters are considered to result in high compactness.

5.3.1.1 Landscape Metrics Analysis

Overall, the results of urban metric analysis demonstrated urban spatial patterns. Several urban metrics were listed to evaluate landscape structure, composition, and configuration of the study area. However, a few of them show direct and effective relation to city compactness.

SHAPE measures the complexity of patch shapes compared with a standard shape (square) of the same size (Fig. 5.5). A significant irregularity in urban development can be observed from 2008 to 2012, which shows the reduction in urban compactness as well. However, a constant trend occurred during the first and last time periods. In the case of zonal basis, SHAPE did not present noticeable differences among various zones, but because largest shape index (LSI) was based on the number of edges, this metric shows more significant differences than others. Urban growth, especially the conversion of large open spaces and agricultural areas to small residential and commercial parcels, increases the number of edges. Consequently, as shown in the LSI (zonal) graph, the value of LSI increases (from 2004 to 2015) especially for zones with high growth potential because of the large amount of open spaces (zone numbers 4 and 5). Highly developed zones such as 11, 15, 16, and 17 generally have very high LSI value because of their proximity to the CBD and train station, but undergo gentle growth during the selected period.

The CIRCLE metric evaluated the overall patch elongation of the study area. A high value of this metric indicates low compactness because of the narrow and elongated pattern. The result of this metric illustrates a growth of narrow and elongated patches from 2008 to 2012, indicating a decrease in compactness similarly obtained in SHAPE metrics (Fig. 5.6).

Edge density was applied to evaluate edges in a per unit area to compare the parcel growth and changes in various years. The continuous growth of this metric in Kajang City indicates the growth and changes especially from 2004 to 2012 and constant and gradual changes during the last period. Undeveloped zones generally have lower values of ED but higher growth in contrast to developed zones with higher value of ED but more gentle growth during the 11-year period (Fig. 5.7).

Radius of gyration was used to measure the compactness of the study area regarding the extent of patches. More extensive patches indicate higher radius of gyration and less compactness. This assessment revealed the average distance of various land use growth extended within the study area. Thus, the growth of various land uses, especially the conversion of large parcels to small parcels, reduces this extension from 2004 to 2012, as shown in Fig. 5.8. By contrast, similar to other analyses, insignificant development growth from 2012 to 2015 stopped the reduction of GYRATE value during this period. Interestingly, zones with a longitudinal shape, such as zone numbers 6, 13, and 12, have high GYRATE, and the zones with a square shape have low value for this metric. This finding is due to the existence of linear land use types or parcels such as river in zone 6, infrastructure line in zone 12, lengthy road in zone 13, and others. Therefore, the dominance of the small circle- and square-shaped parcels, such as single-family residential buildings in a zone, causes a low GYRATE value and in some cases indicates high compactness.

Largest patch index (LPI) is a measure of the size of the patches with respect to the total area of study. With a constant area size (Kajang City in 4 year land use maps), the map with the largest patch has the highest LPI. Residential and road networks have the largest coverage in Kajang City. Therefore, the growth of these land use types from 2004 to 2012 caused an increase in LPI during this period (Fig. 5.9). The reduction of significant development growth in last time period resulted in a constant LPI value during this time. However, in the case of zonal basis, zone number 5 with large agricultural fields had a very high LPI in 2004. However, because of agricultural conversion to residential use, this value decreased significantly. The compactness of the largest path index has an inverse presentation of LPI. Compact zones with low fragmentations and small parcels have high LPI.

The PLADJ metric was used to evaluate the study area in terms of dispersion, interspersion, subdivision, and isolation. Similar to other metrics, this metric shows that the development growth from 2004 to 2012 slightly reduced the compactness of Kajang City (Fig. 5.10). However, this situation has changed since 2012, and compactness is increasing every year. A similar result was observed in the case of zonal assessment.

Various diversity indices, such as SHDI, SIDI, SHEI, and SIEI were used to evaluate the land use diversity, evenness, and richness of Kajang City. All these metrics show a reduction in variables from 2004 to 2008, a significant

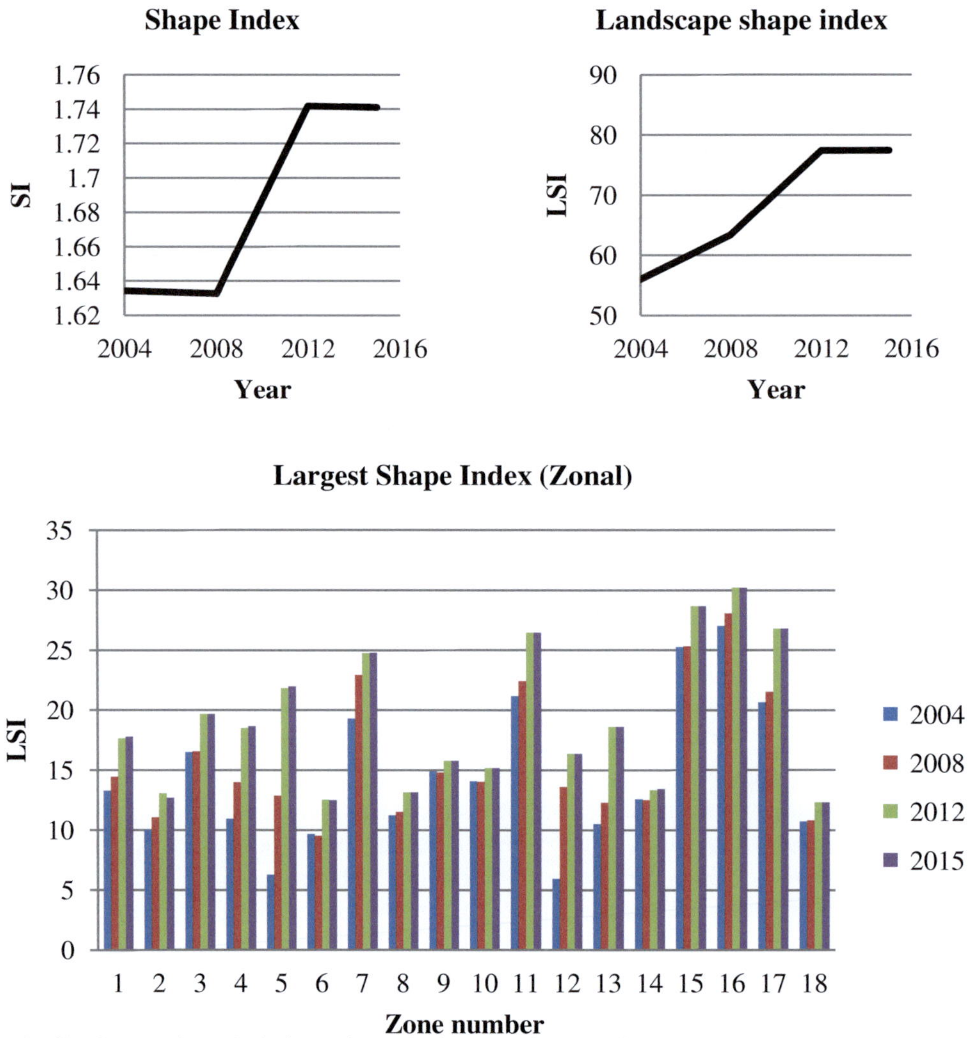

Fig. 5.5 SHAPE and LSI urban metric analysis for zoning and without zoning sampling

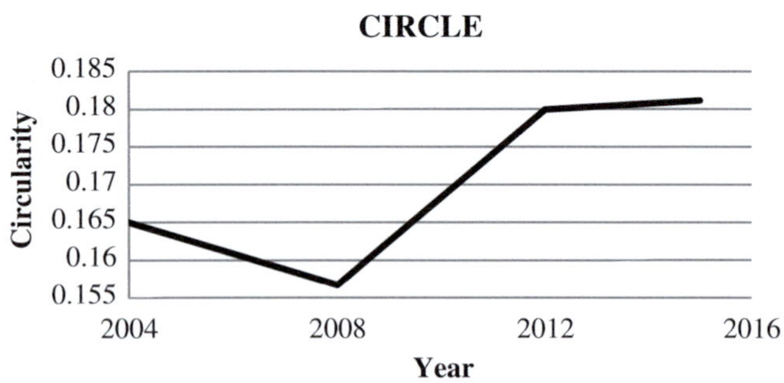

Fig. 5.6 CIRCLE landscape metric

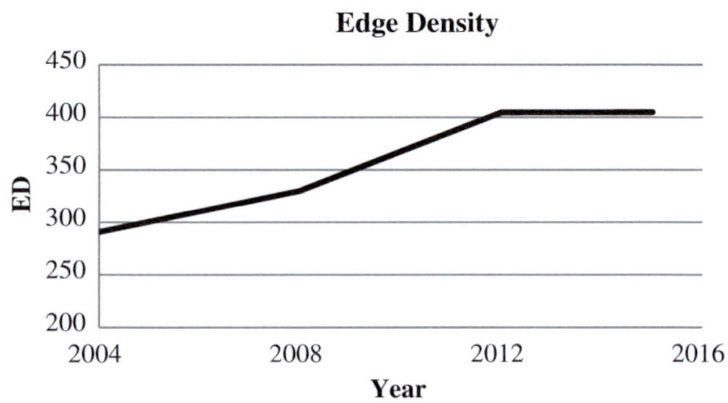

Fig. 5.7 Edge density assessment for zonal and non-zonal sampling

growth during 2008–2012, and again a slight reduction in the last time period (Fig. 5.11). In the case of zonal basis, no significant difference can be observed from various metrics. However, in zone numbers 4, 5, 6, and 12, because of the large proportion of open spaces and high potential for new development growth, an increase in land use diversity can be noticed. By contrast, because of saturated development in developed zones, such as 15, 16, and 17, no significant difference in land use diversity and richness can be observed.

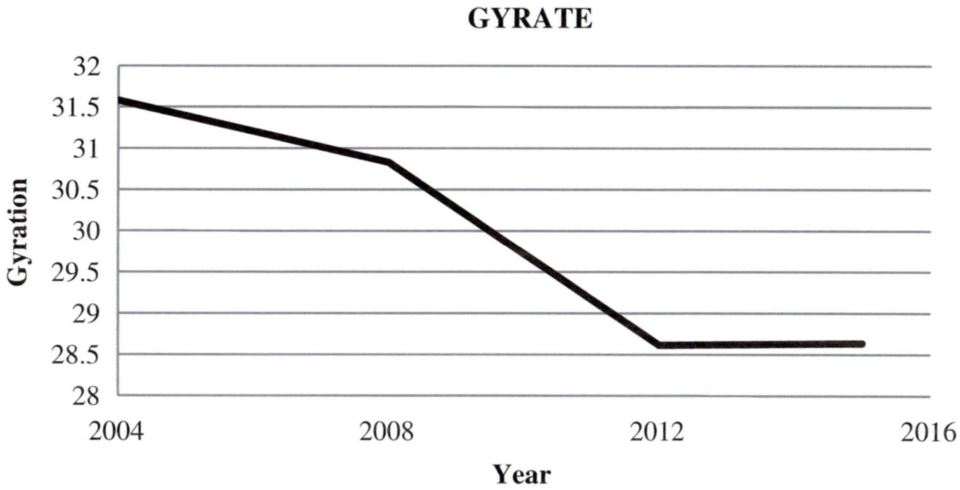

Fig. 5.8 GYRATE metric assessment for zonal and non-zonal sampling

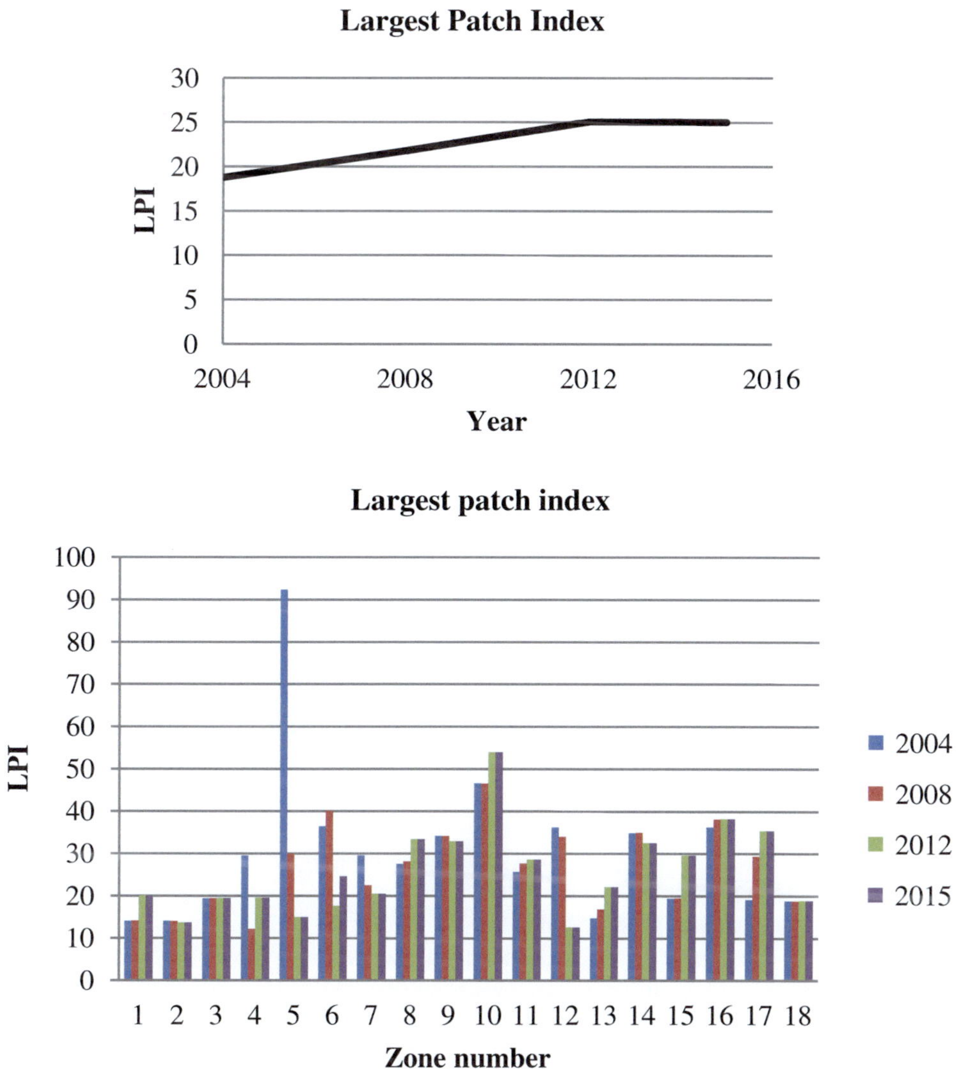

Fig. 5.9 Largest patch index metric assessment for zonal and non-zonal sampling

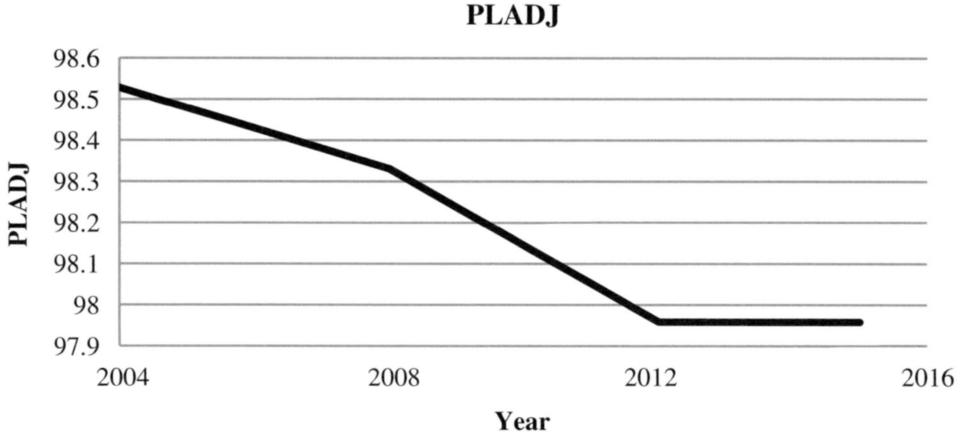

Fig. 5.10 PLADJ metric assessment for zonal and non-zonal sampling

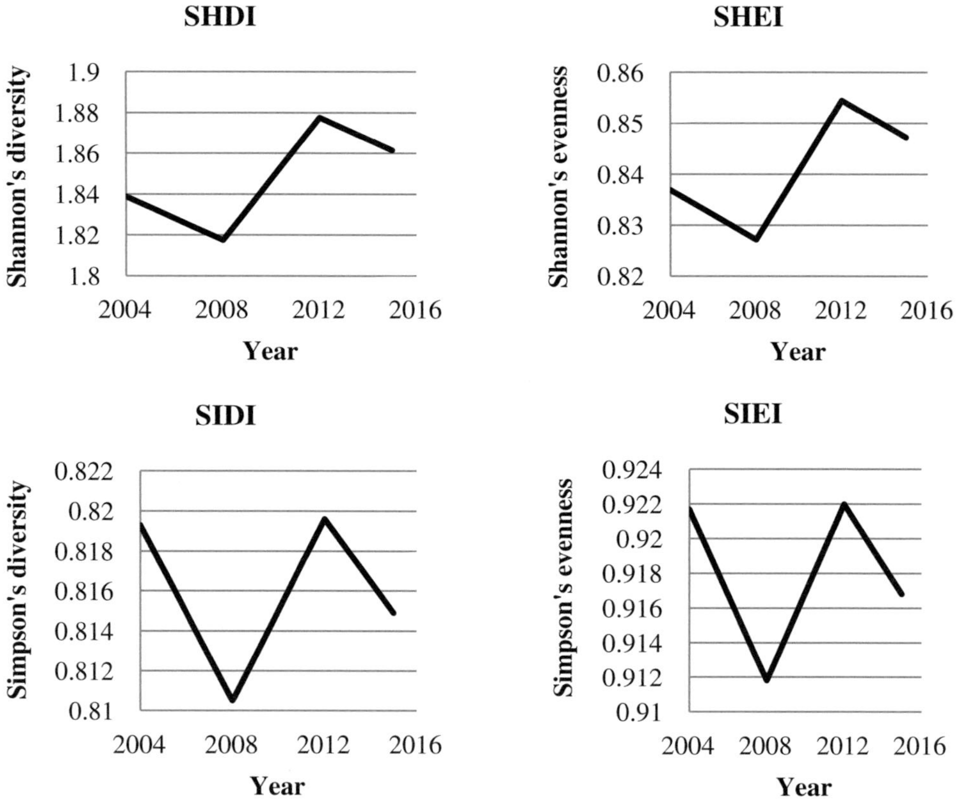

Fig. 5.11 SHDI, SHEI, SIDI and SIEI metrics assessment for Kajang City

5.3.1.2 Shannon's Entropy Analysis

With the capability of spatial dispersion and concentration assessment, Shannon's entropy was selected to evaluate physical city compactness. The first step was to extract the built-up area of each zone of all available land use maps (Table 5.3). At this stage, all land uses consisting of buildings such as residential, commercial, industrial, and some community facilities were extracted, and open spaces, local parks, playgrounds, road networks, bodies of water, and agricultural fields were omitted from analysis.

Thereafter, according to the given equation of Shannon's entropy, the percentage of built-up area in each zone with respect to the total built-up area of the study area was evaluated. Next, the natural logarithmic of each reverse percentage value was calculated. The sum of the products of these two values is the absolute Shannon's entropy (Tables 5.4 and 5.5). The relative entropy value for each land use map was also calculated based on Ln(18), where 18 is the number of involved zones.

The values of absolute and relative entropy for all available land use maps are summarized in Table 5.6, and the trend is shown in Fig. 5.12. Absolute entropy ranges from 0 to 2.89 [Ln(18)], and relative entropy ranges from zero to one. In both cases, a value near zero indicates

Table 5.3 Built-up area of each zone with respect to each land use maps (m^2)

Zone no.	Zone area	2004	2008	2012	2015
1	2,325,979.27	756,376	1,094,408	1,105,887	1,166,005
2	3,435,020.57	1,324,999	1,374,403	1,492,114	1,919,694
3	3,540,824.40	1,986,365	2,083,046	2,005,682	2,009,607
4	4,400,541.10	607,289	825,789	1,384,052	1,460,032
5	4,758,501.54	67,777	217,058	1,599,746	1,735,128
6	3,365,709.71	577,630	985,316	1,166,769	1,166,769
7	3,961,381.07	1,350,321	1,730,100	1,948,818	2,055,010
8	683,832.01	362,280	431,539	396,732	396,732
9	2,423,413.60	1,410,871	1,447,033	1,705,289	1,705,289
10	2,639,943.46	1,914,270	2,002,309	2,068,949	2,069,113
11	4,519,842.86	2,285,209	2,388,225	2,805,926	2,805,926
12	3,639,339.89	1,530,970	1,628,733	1,777,444	1,855,723
13	3,315,233.81	1,784,536	1,749,210	2,199,413	2,205,245
14	1,221,672.72	675,053	698,666	783,517	763,477
15	4,944,197.80	2,399,841	2,642,736	2,852,269	2,852,269
16	3,940,328.57	1,729,385	1,899,801	2,073,160	2,104,777
17	2,716,453.26	980,095	999,783	1,264,633	1,359,098
18	705,482.03	353,992	367,976	346,860	346,860
Total	56,537,697.67	22,097,259	24,566,131	28,977,260	29,976,754

Table 5.4 Shannon's entropy calculation for land use map 2004 and 2008

Zone no.	2004		2008	
	Proportion of built-up	Prop. × Ln(1/prop.)	Proportion of built-up	Prop. × Ln(1/prop.)
1	0.03	0.11551295	0.04	0.138600297
2	0.06	0.16873598	0.06	0.161314936
3	0.09	0.216562867	0.08	0.209230914
4	0.03	0.098777889	0.03	0.114048251
5	0.00	0.017749919	0.01	0.041783481
6	0.03	0.095262619	0.04	0.128996115
7	0.06	0.170803872	0.07	0.186854134
8	0.02	0.067395577	0.02	0.070999365
9	0.06	0.175662224	0.06	0.166806299
10	0.09	0.211905414	0.08	0.204343297
11	0.10	0.234650383	0.10	0.226593212
12	0.07	0.184955252	0.07	0.179909281
13	0.08	0.203211595	0.07	0.188135871
14	0.03	0.106568276	0.03	0.10124577
15	0.11	0.241105452	0.11	0.239847395
16	0.08	0.19938821	0.08	0.197946018
17	0.04	0.138186557	0.04	0.130296911
18	0.02	0.066224492	0.01	0.062928358

Table 5.5 Shannon's entropy calculation for land use map 2012 and 2015

Zone no.	2012		2015	
	Proportion of built-up	Prop. × Ln(1/prop.)	Proportion of built-up	Prop. × Ln(1/prop.)
1	0.04	0.124638291	0.04	0.126292204
2	0.05	0.152743352	0.06	0.175996752
3	0.07	0.184842473	0.07	0.181171344
4	0.05	0.14527214	0.05	0.147186181
5	0.06	0.159916109	0.06	0.164926843
6	0.04	0.129342131	0.04	0.126349459
7	0.07	0.181536202	0.07	0.183732955
8	0.01	0.0587488	0.01	0.057238778
9	0.06	0.16670669	0.06	0.163077392
10	0.07	0.188455711	0.07	0.184521791
11	0.10	0.226081198	0.09	0.221717299
12	0.06	0.171218454	0.06	0.17222998
13	0.08	0.195698022	0.07	0.191974461
14	0.03	0.097623712	0.03	0.093478614
15	0.10	0.228202758	0.10	0.223820547
16	0.07	0.188693812	0.07	0.18650236
17	0.04	0.136675737	0.05	0.140258912
18	0.01	0.05297172	0.01	0.051597902

Table 5.6 Absolute and relative Shannon's entropy values for all land use maps

Shannon's entropy	2004	2008	2012	2015
Absolute entropy	2.713	2.750	2.789	2.792
Relative entropy	0.939	0.951	0.965	0.966
Time period	2004–2008	2008–2012		2012–2105
ΔH_n	0.012	0.014		0.001

compactness, and a value near 2.89 and 1 indicates sprawled development. Hence, in general, according to these results, Kajang City physically seems to have low compactness. Evaluating the trend of entropy value shows that although very small differences can be seen, in both cases, Kajang City's compactness decreased during the selected period. In subtracting the later time entropy value from the earlier time (ΔH_n), the tendency of urban growth process can be achieved. A positive value of ΔH_n in all three cases indicates the sprawl's growing process. However, a significant reduction of this value from 2012 to 2015 indicates changes in the development pattern, which decreases sprawl growth.

An important issue in urban growth analysis is the measurement of growth and the determination of the urban requirements to be accomplished in preparation for future urban demands. Shannon's entropy model was used to guide the identification and measurement of the change that is likely to happen given the tendency of urban history to persist. Thus, these findings highlighted the necessity of a sustainable urban form with higher urban density, especially built-up density, to reduce land consumption and consequently preserve agricultural and green environments.

5.3.2 Functional Compactness

The functional compactness of the study area was assessed with respect to the actual activities in the neighborhoods and especially to the planning and development of land use pattern and distribution. This assessment required detailed land use maps and the consideration of other variables to calculate various aspects of compact development with respect to land use pattern. Functional compactness assessment was conducted based on three main compactness indicators: urban density, urban intensity, and land use diversity.

5.3.2.1 Urban Density

In the case of urban density, four main variables were considered: population, built-up, residential, and road density.

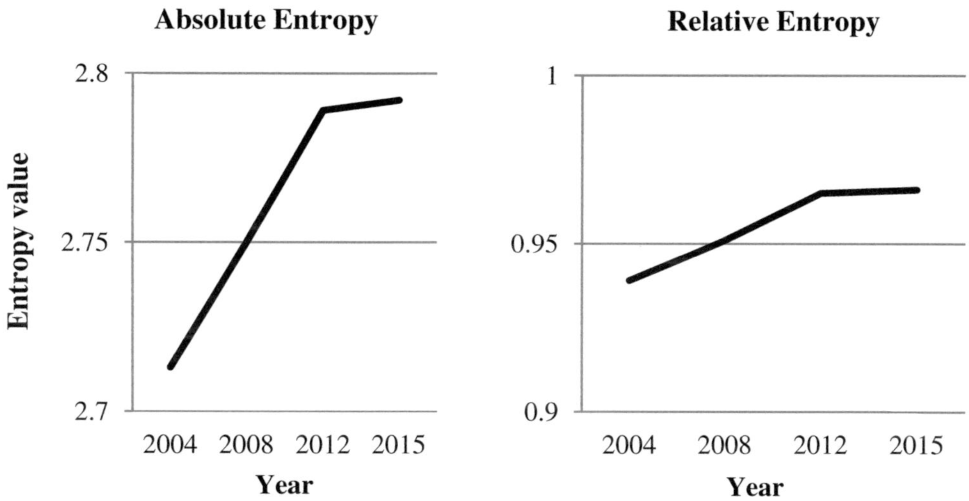

Fig. 5.12 Absolute and relative Shannon's entropy values in different time periods

Table 5.7 shows the population density calculation based on the built-up area of each zone. Figure 5.2 depicts the zoning districts of Kajang City. Eastern zones (zone numbers 4 and 5) are mainly covered by agricultural fields, and a smaller built-up area can be observed in these zones. Thus, although these zones have a small population, they have high population density. However, through the significant growth of the built-up area during the selected period (2004 to 2015) and the slight population growth, the population densities of these two zones have decreased significantly. The most populated zones are those near the main train station (KTM commuter) of Kajang City (zone numbers 11, 15, and 16). These zones are already saturated with a large population and many built-up areas. Hence, the population density value also remained almost constant from 2004 to 2015. Two small zones located in the northern borders of Kajang City (zone numbers 8 and 18) also have high population densities because of several high-rise buildings with large populations. The proximity of these northern and western zones to other regions and city centers of Malaysia affects the development pattern of these zones regardless of the characteristics and conditions of Kajang City's development pattern. Industrial

Table 5.7 Calculation of population density for each zone (person/ha)

Zones		2004		2008		2012		2015	
No.	Area (ha)	Built-up area	Density	Built-up area	Density	Built-up area	Density	Built-up area	Density
1	232.60	75.64	131	109.44	97	110.59	104	116.60	103
2	343.50	132.50	20	137.44	21	149.21	21	191.97	17
3	354.08	198.64	103	208.30	105	200.57	118	200.96	123
4	440.05	60.73	106	82.58	84	138.41	54	146.00	54
5	475.85	6.78	191	21.71	64	159.97	10	173.51	9
6	336.57	57.76	104	98.53	65	116.68	60	116.68	63
7	396.14	135.03	263	173.01	221	194.88	210	205.50	209
8	68.38	36.23	217	43.15	196	39.67	228	39.67	240
9	242.34	141.09	84	144.70	88	170.53	80	170.53	84
10	263.99	191.43	95	200.23	98	206.89	101	206.91	107
11	451.98	228.52	81	238.82	83	280.59	76	280.59	80
12	363.93	153.10	14	162.87	14	177.74	13	185.57	14
13	331.52	178.45	39	174.92	43	219.94	36	220.52	38
14	122.17	67.51	76	69.87	79	78.35	75	76.35	80
15	494.42	239.98	175	264.27	171	285.23	169	285.23	177
16	394.03	172.94	250	189.98	244	207.32	239	210.48	247
17	271.65	98.01	217	99.98	228	126.46	193	135.91	188
18	70.55	35.40	274	36.80	283	34.69	322	34.69	337

Table 5.8 Total population density of Kajang City with respect to built-up area and total Kajang City area (person/ha)

	2004		2008		2012		2015	
	Area (ha)	Population density	Area (ha)	Density	Area (ha)	Density	Area (ha)	Density
Based on built-up area	2210	122	2457	118	2898	107	2998	109
Based on Kajang City area	5653.7	48	5653.7	52	5653.7	55	5653.7	58

zones such as zone numbers 2 and 12 have many industrial buildings and a small resident population, which resulted in low population density in these zones and in some cases a reduction in this value during the selected period. The river and water bodies and their buffer zones in the central area of Kajang City (zone number 6) resulted in a small built-up area in this zone, but development growth caused the reduction of population density in this zone.

Table 5.8 assessed the total population density based on the built-up area and total area of Kajang City for each available land use map. Obviously, because the population of this region has grown during the 4-year period and the area of Kajang City is constant, the population density of this region increases every year. Elaborate and reliable evaluation can be achieved by calculating the density based on the built-up area of the region because this assessment considers the urban structural growth as well. Thus, because of the significantly higher growth of built-up area than the population from 2004 to 2015, the population density decreases every year. However, a slight growth in density can be observed between 2012 and 2015, which can be due to the reduced growth in building development.

In addition to the tabular quantitative assessment of population, Fig. 5.13 illustrates a raster-based map of population density assessment to present the cellular-based analytical model.

The residential density of Kajang City was evaluated by calculating the number of residential units with respect to the built-up area. Thereafter, this assessment was used to produce a raster-based presentation through kernel density analysis for all land use maps. Finally, the maps were classified into three main classes of low (1) to high (3) residential density to create standardized map layers.

As shown in Fig. 5.14, the area near public transportation and the zones with saturated development pattern have high residential density. By contrast, as expected, eastern zones

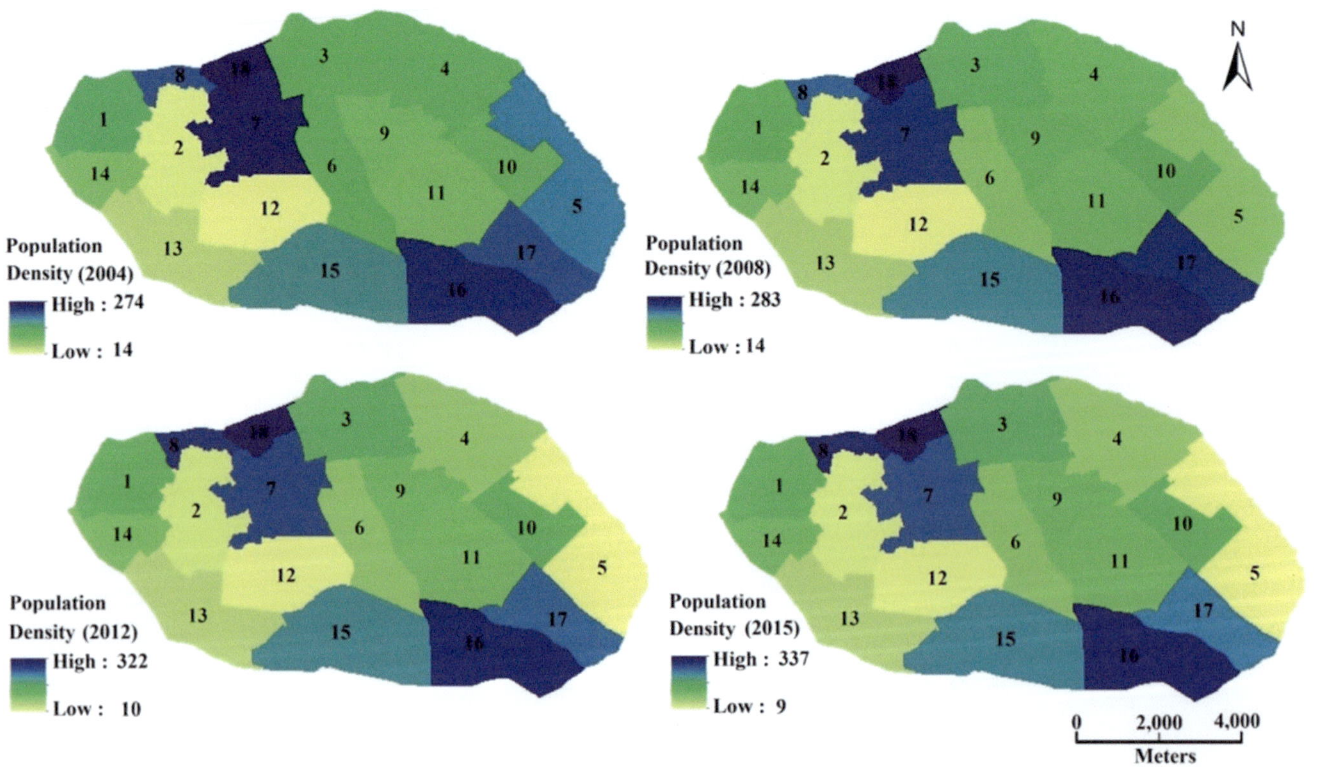

Fig. 5.13 Graphical presentation of population density for four years (person per hectare)

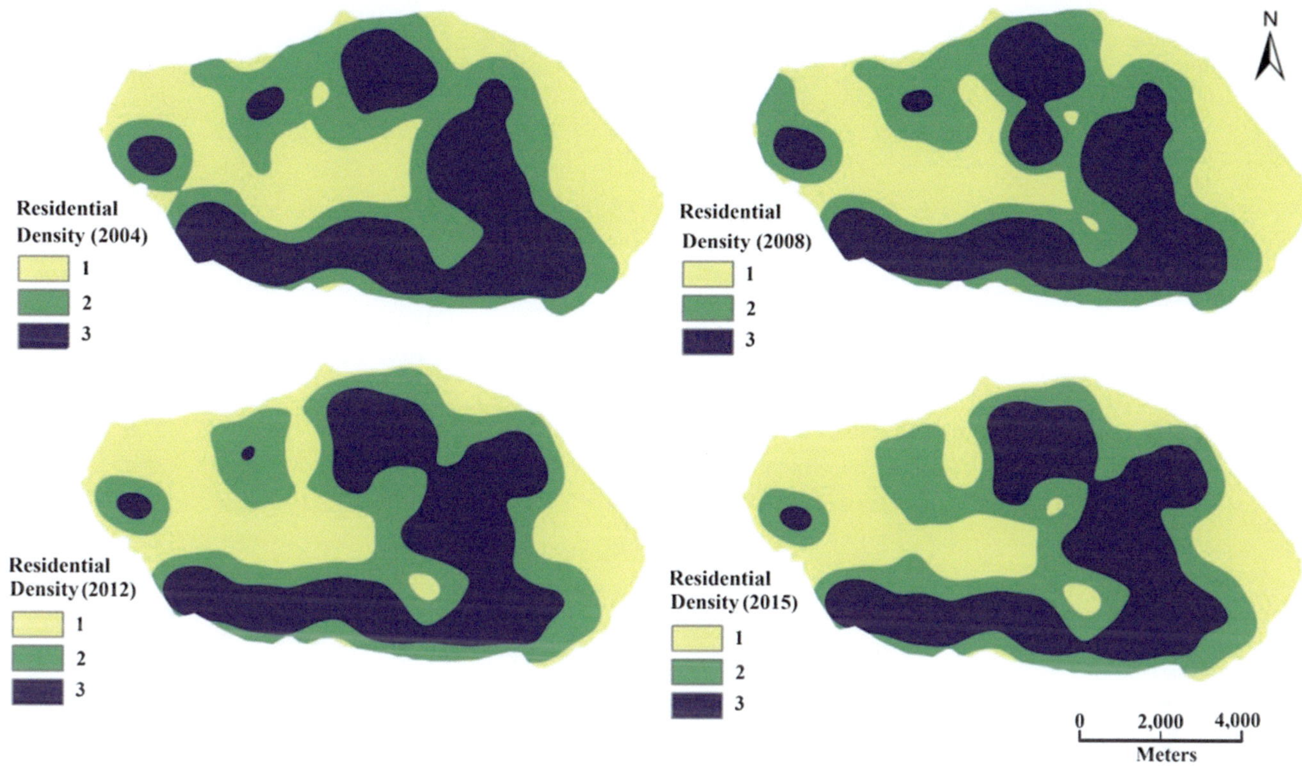

Fig. 5.14 Kernel density analysis for evaluation of residential density of Kajang City

(agricultural covered areas) and industrial areas (central west) have very low residential density. Although at first glance, no significant difference can be observed among the four land use maps, significant changes can be extracted through cellular-based investigation.

Road density was calculated based on the length of the existing road networks with respect to the total area of each zone and of Kajang City (Table 5.9). In the case of road density, the size and the development situation of the zones are the effective factors. Zone numbers 8, 14, and 18, as the smallest zones, and zone numbers 15, 16, and 17, as the most populated and developed zones, have very high road densities. By contrast, zone numbers 4 and 5, as the least developed zones, have the lowest road densities. Industrial zones (2 and 12) have short road lengths because of large parcel size, thereby having low road density value. Similar to highly developed zones, these zones have slightly grown from 2004 to 2015. In the case of the whole Kajang City perspective, this city has grown in road density from 2004 to 2012 and has maintained a constant density value in the last period. Similar to the findings of the previous section (urban growth analysis), no significant difference in development pattern can be observed between 2012 and 2015.

The graphical illustration of road density (Fig. 5.15) shows that the growth of road development in the eastern parts and central west of the city can be observed during the 11-year period. However, insignificant differences can be seen in the south and northwest, which were already developed properly.

Building density was evaluated by considering the built-up area with respect to the area of each zone. Table 5.10 shows the calculation of building density. The last row of this table shows the total area of Kajang City (5653.7 ha), the total built-up area for each land use map, and the corresponding density value.

Similar to other assessments, developed zones (zone numbers 3, 10, 14, 15, 16, and 18,) have insignificant growth in the built-up area because of development saturation. By contrast, significant growth can be observed in undeveloped zones with large proportions of open spaces. The main building growth and development of Kajang City happened from 2008 to 2012 (Fig. 5.16).

These assessments show that in saturated zones, a reduction of growth in density can be observed because of the low potential for growth in these areas. Thus, in the case of land consumption and eventually green environment preservation objective, focusing on the areas with high potential for new development and growth is appropriate. In addition, neighborhoods and zones along the border of the study area are affected by some external and outside forces from other regions, whether urban and rural areas, over the city boundaries. Hence, unlike internal zones, which are

Table 5.9 Road density calculation based on area of each zone (meter/hectare)

Zone		Road length (m)				Road density (m/ha)			
No.	Area (ha)	2004	2008	2012	2015	2004	2008	2012	2015
1	232.6	36,012	32,897	37,279	37,249	154.8	141.4	160.2	160.1
2	343.5	24,846	26,524	28,647	28,647	72.33	77.21	83.40	83.40
3	354.0	44,716	42,903	46,384	46,384	126.2	121.1	131.0	131.0
4	440.0	27,015	36,567	50,044	50,044	61.39	83.10	113.7	113.7
5	475.8	14,517	36,690	64,373	64,373	30.51	77.10	135.2	135.2
6	336.5	33,504	21,865	24,516	24,516	99.5	64.9	72.8	72.8
7	396.1	63,449	75,810	72,066	72,066	160.1	191.3	181.9	181.9
8	68.3	14,885	15,211	16,010	16,010	217.6	222.4	234.1	234.1
9	242.3	37,659	36,679	37,910	37,910	155.4	151.3	156.4	156.4
10	263.9	30,012	29,978	32,118	32,118	113.6	113.5	121.6	121.6
11	451.9	75,751	75,776	85,263	85,263	167.6	167.6	188.6	188.6
12	363.9	16,803	38,919	41,417	41,417	46.17	106.9	113.8	113.8
13	331.5	26,433	31,676	44,684	44,684	79.7	95.5	134.7	134.7
14	122.1	27,518	24,537	20,260	20,260	225.2	200.8	165.8	165.8
15	494.4	89,980	88,073	97,343	97,343	181.9	178.1	196.8	196.8
16	394.0	94,823	96,895	100,743	100,743	240.6	245.9	255.6	255.6
17	271.6	52,917	54,992	63,966	63,966	194.8	202.4	235.4	235.
18	70.5	14,489	11,862	13,950	13,950	205.3	168.1	197.7	197.7
Sum	5653.7	725,327	777,849	876,969	876,939	128.2	137.5	155.1	155.1

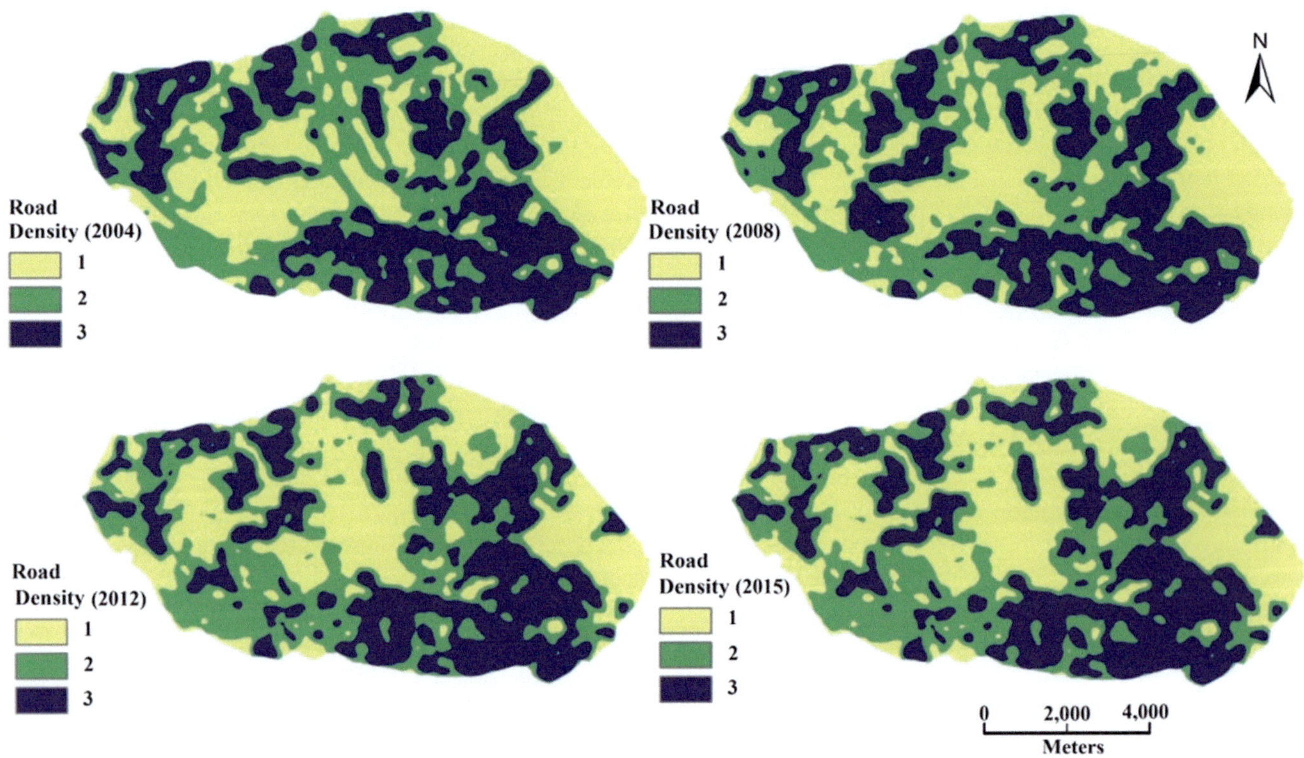

Fig. 5.15 Kernel density analysis for evaluation of road density of Kajang City

Table 5.10 Building density calculation based on area of each zone (ha/ha)

Zones		2004		2008		2012		2015	
No.	Area (ha)	Built-up	Density	Built-up	Density	Built-up	Density	Built-up	Density
1	232.60	75.64	0.325	109.44	0.470	110.59	0.475	116.60	0.501
2	343.50	132.50	0.385	137.44	0.400	149.21	0.434	191.97	0.558
3	354.08	198.64	0.561	208.30	0.588	200.57	0.566	200.96	0.567
4	440.05	60.73	0.138	82.58	0.187	138.41	0.314	146.00	0.331
5	475.85	6.78	0.014	21.71	0.045	159.97	0.336	173.51	0.364
6	336.57	57.76	0.171	98.53	0.292	116.68	0.346	116.68	0.346
7	396.14	135.03	0.340	173.01	0.436	194.88	0.492	205.50	0.518
8	68.38	36.23	0.529	43.15	0.631	39.67	0.580	39.67	0.580
9	242.34	141.09	0.582	144.70	0.597	170.53	0.703	170.53	0.703
10	263.99	191.43	0.725	200.23	0.758	206.89	0.783	206.91	0.783
11	451.98	228.52	0.505	238.82	0.528	280.59	0.620	280.59	0.620
12	363.93	153.10	0.420	162.87	0.447	177.74	0.488	185.57	0.509
13	331.52	178.45	0.538	174.92	0.527	219.94	0.663	220.52	0.665
14	122.17	67.51	0.552	69.87	0.571	78.35	0.641	76.35	0.624
15	494.42	239.98	0.485	264.27	0.534	285.23	0.576	285.23	0.576
16	394.03	172.94	0.438	189.98	0.482	207.32	0.526	210.48	0.534
17	271.65	98.01	0.360	99.98	0.368	126.46	0.465	135.91	0.500
18	70.55	35.40	0.501	36.80	0.521	34.69	0.491	34.69	0.491
Sum	5653.77	2210	39.08	2457	43.451	2898	51.253	2998	53.021

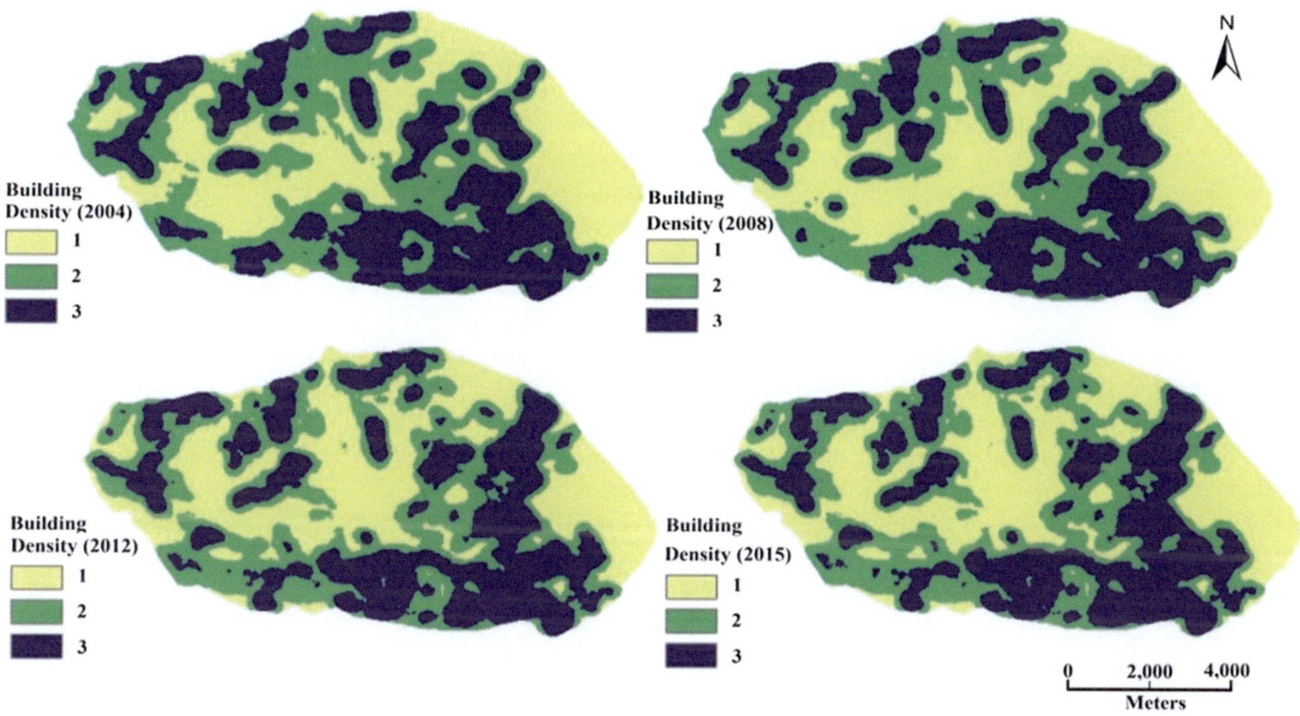

Fig. 5.16 Kernel density analysis for evaluation of building density of Kajang City

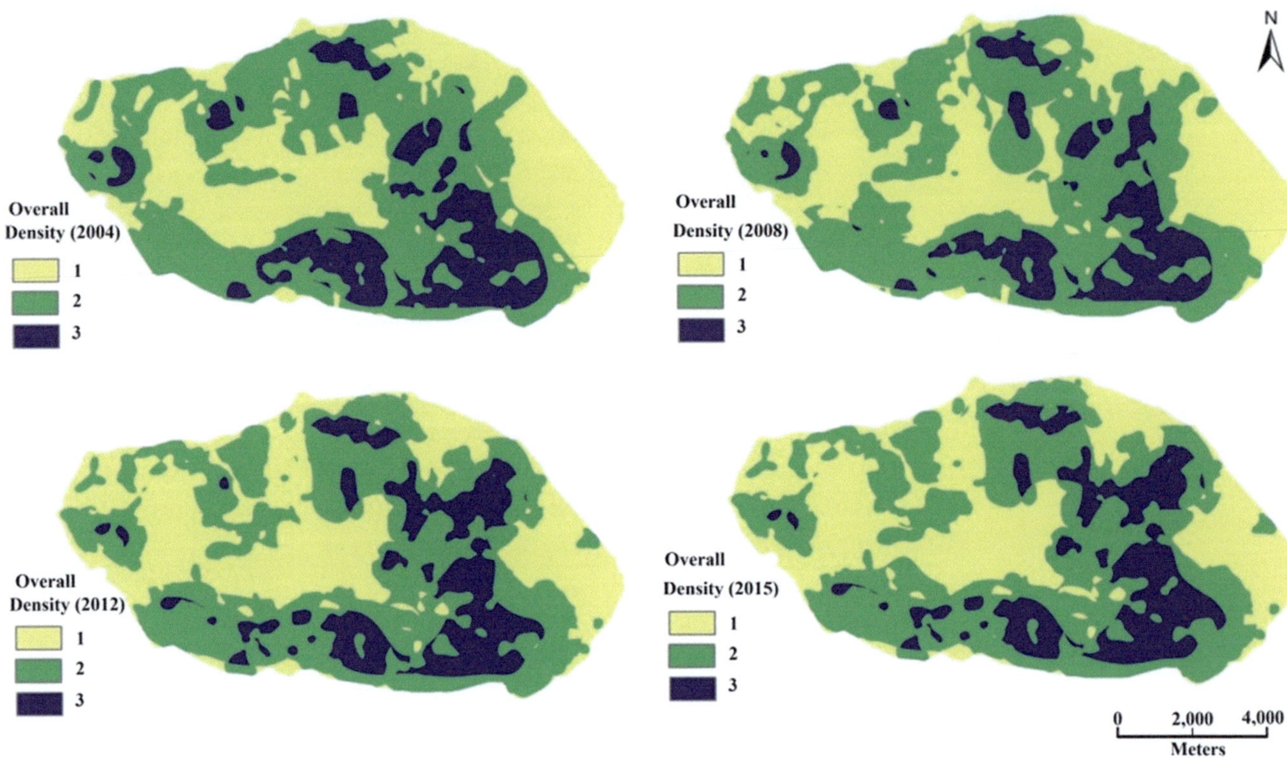

Fig. 5.17 The overall urban density of the study area

mainly dependent on city characteristics, boundary zones are affected by other parameters. Therefore, considering these issues and viewing the study area from wide perspectives are important.

Finally, creating one output map by aggregating all created variables in urban density was required to show the overall density of the study area. Figure 5.17 shows the overall density for all available land use maps. These maps are reclassified into three classes to standardize them when they are aggregated with other compactness indicators.

5.3.2.2 Mix Land Use Assessment

The evaluation of land use diversity or land use richness is an important task in compact city modeling and urban sustainability analysis. This study evaluated mixed land use development using two common methods in this field and one proposed approach based on proximity and relationship among various land use categories.

Originally, Shannon's entropy was the most common index in measuring land use diversity especially in the biodiversity field. In this study, in evaluating the land use diversity of Kajang City, the proportion of each land use category that covered the study area was calculated. Shannon's entropy is based on the sum of the natural logarithms of the land use percentages. Absolute entropy ranges from zero to Ln(m), where m is the number of land use categories. A value near zero indicates low land use diversity, and a value near 2.19 [Ln(9)] refers to high land use diversity. As shown in Tables 5.11 and 5.12, the value of absolute entropy for all years is approximately 1.8, which shows that in general, Kajang City has a high level of land use diversity.

Absolute entropy can be modified to relative entropy and to change the range of the values from zero to one. Similarly, zero and one indicate low diversity and high diversity, respectively. In this case, in all years, relative entropy values near 1 indicate the high land use richness of Kajang City. However, as shown in Fig. 5.18, a slight reduction in diversity was observed from 2004 to 2008, a noticeable growth was seen from 2008 to 2012, and again a reduction was observed during the last period.

In addition to Shannon's entropy, Simpson's diversity index was used to confirm the trend of land use diversity of Kajang City through the frequent and common methods (Table 5.13). In this method, the percentage of land use coverage is an important component. In this assessment, three indices are calculated: Simpson's diversity, which ranges from zero to $1 - (1/m)$ (thus, from 0 to 0.88); effective function richness, which ranges from zero to m (thus, from 0 to 9); and proportionality index, which ranges from zero to one. In all cases, zero is assigned for low land use diversity, and 0.88, 9, and 1 are assigned for high land use diversity. Similar to Shannon's entropy, Simpson's diversity index shows that Kajang City has high mixed land use development. Furthermore, the same trend of increase

Table 5.11 Mixed land use assessment using Shannon's entropy for year 2004 and 2008 land use maps

Land use	2004			2008		
	Area (m^2)	Proportion	Prop. × Ln (prop.)	Area (m^2)	Proportion	Prop. × Ln (prop.)
Agr.	8,820,560	0.16	−0.2897	5,103,330	0.09	−0.2170
Com.	984,740	0.02	−0.0705	1,107,992	0.02	−0.0770
Open	14,443,095	0.26	−0.3485	14,289,551	0.25	−0.3475
Hous.	11,160,288	0.20	−0.3202	12,926,373	0.23	−0.3373
Inds.	4,650,866	0.08	−0.2054	4,847,926	0.09	−0.2105
Infra	915,870	0.02	−0.0667	1,048,338	0.02	−0.0739
Fac.	3,928,078	0.07	−0.1852	3,968,596	0.07	−0.1864
Trans.	11,321,348	0.20	−0.3219	12,932,886	0.23	−0.3373
Water	332,694	0.01	−0.0302	332,545	0.01	−0.0301
Sum	56,557,539	Sum	−1.838	56,557,537	Sum	−1.817
	Absolute entropy		1.838	Absolute entropy		1.817
	Relative entropy		0.836	Relative entropy		0.827

Note Agr. agriculture; *Com.* commercial; *Open* Open spaces; *Hous.* housing; *Inds.* industry; *Infra* infrastructure; *Fac* facility; *Tran.* transportation

Table 5.12 Mixed land use assessment using Shannon's entropy for year 2012 and 2015 land use maps

Land use	2012			2015		
	Area (m^2)	Proportion	Perc. × Ln(perc.)	Area (m^2)	Proportion	Perc. × Ln(perc.)
Agr.	5,621,719	0.10	−0.2294	4,210,981	0.07	−0.1933
Com.	1,744,100	0.03	−0.1072	1,757,851	0.03	−0.1078
Open	7,951,358	0.14	−0.2758	8,382,406	0.15	−0.2829
Hous.	14,757,211	0.26	−0.3505	15,327,836	0.27	−0.3536
Inds.	5,605,319	0.10	−0.2290	6,002,177	0.11	−0.2380
Infra	1,528,637	0.03	−0.0975	1,529,810	0.03	−0.0976
Fac.	4,530,455	0.08	−0.2022	4,530,455	0.08	−0.2022
Trans.	14,386,034	0.25	−0.3482	14,386,124	0.25	−0.3482
Water	432,858	0.01	−0.0372	432,858	0.01	−0.0372
Sum	56,557,691	Sum	−1.877	56,560,498	Sum	−1.861
	Absolute entropy		1.877	Absolute entropy		1.861
	Relative entropy		0.854	Relative entropy		0.847

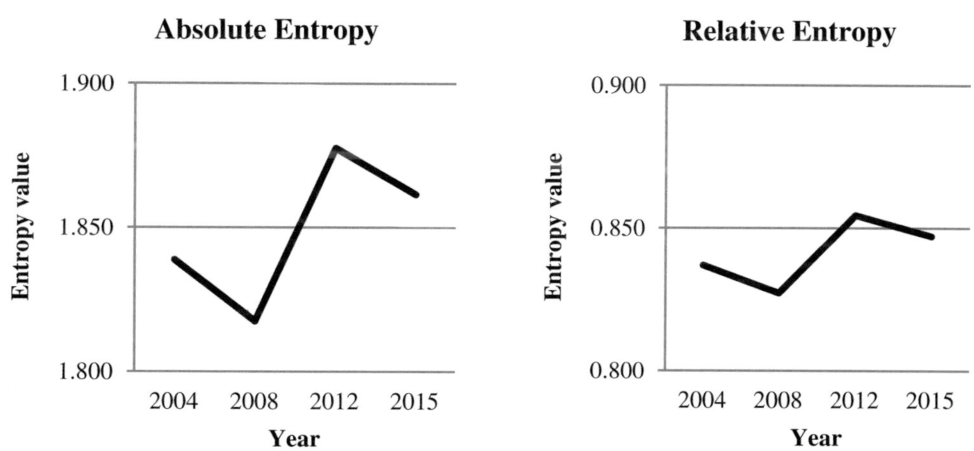

Fig. 5.18 The trend of land use diversity of Kajang City using absolute and relative entropy techniques

Table 5.13 Mixed land use assessment using Simpson's diversity index for all land use maps

	2004		2008		2012		2015	
Land use	Proportion	Pro.^{2}	Pro.	Pro.^{2}	Pro.	Pro.^{2}	Pro.	Pro.^{2}
Agr.	0.16	0.024	0.09	0.0081	0.10	0.0098	0.07	0.0055
Com.	0.02	0.0003	0.02	0.0003	0.03	0.0009	0.03	0.0009
Open	0.26	0.065	0.25	0.0638	0.14	0.0197	0.15	0.0219
Hous.	0.20	0.038	0.23	0.0522	0.26	0.0680	0.27	0.0734
Inds.	0.08	0.006	0.09	0.0073	0.10	0.0098	0.11	0.0112
Infra	0.02	0.0002	0.02	0.0003	0.03	0.0007	0.03	0.0007
Fac.	0.07	0.004	0.07	0.0049	0.08	0.0064	0.08	0.0064
Trans.	0.20	0.040	0.23	0.0522	0.25	0.0647	0.25	0.0646
Water	0.01	0.00003	0.01	0.00,003	0.01	0.00006	0.01	0.00006
Sum	1.00	0.180	1.00	0.1895	1.00	0.1804	1.00	0.1850
Simpson's diversity		0.819		0.8104		0.8196		0.8149
Effective function richness		5.533		5.2760		5.5431		5.4032
Proportionality index		0.9216		0.9117		0.9220		0.9167

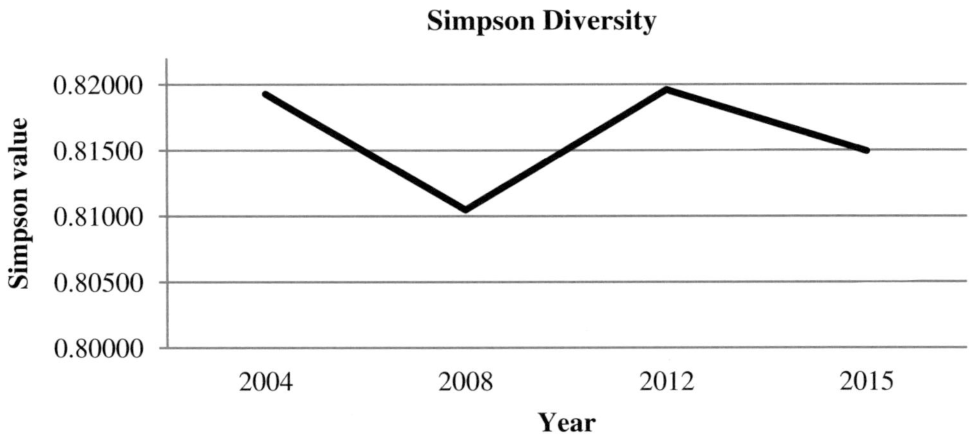

Fig. 5.19 The trend of land use diversity of Kajang City using Simpson's diversity index

and decrease in diversity can be observed in these assessments (Fig. 5.19 and Table 5.13).

Both models of diversity assessments seem to provide similar results. However, realizing that both methods are based only on land use richness of the study area without considering spatial pattern and distribution and the relationship among various land use types is important. Thus, a new approach was proposed, in which the spatial patterns and distribution of various land use types in addition to land use richness are considered. This process is based on the proximity analysis among existing land use types. However, only five main urban land use types that affected local transportation were selected. Euclidean distance analysis was applied for these five land use types separately, and each layer was classified to assign every pixel of the study area to one class of proximity (1–5, in which 1 means farthest and 5 means nearest). Five classified layers had five classes. Therefore, the aggregated proximity map assigned all pixels with proximity values ranging from 5 to 25 (Table 5.14). The number and percentage of pixels bearing each proximity value are estimated. Finally, the sum of the products of the percentage and proximity value for each land use map indicates the level or degree of land use diversity of that map. This result also ranges from 5, when 100% of the study area obtains a value of 5 (minimum proximity value) and 0 for reset, to 25, when 100% of the study area obtains a value of 25 (maximum proximity value) and 0 for reset. Through the proposed standardization approach, this assessment can result in the range of 0 (minimum land use diversity) to 1 (maximum land use diversity).

Tables 5.14 and 5.15 present the quantitative assessment of this analysis for all land use maps without zoning consideration. Therefore, the produced results are the value for whole Kajang City. The standardized values of land use diversity for all land use maps are found in the middle range (0.5) from 0 to 1, indicating that although Kajang City has a high value of land use richness, the pattern of the existing categories (especially those mainly related to human daily

Table 5.14 Land use diversity assessment using proposed proximity analysis for 2004 and 2008

2004				2008			
Proximity value	No. of cells	Proportion	Prop. × value	Proximity value	No. of cells	Proportion	Prop. × value
5	1,620,920	0.0287	0.14	5	1,709,934	0.0302	0.15
6	1,379,331	0.0244	0.15	6	961,740	0.0170	0.10
7	878,105	0.0155	0.11	7	584,402	0.0103	0.07
8	877,831	0.0155	0.12	8	992,917	0.0176	0.14
9	1,614,007	0.0286	0.26	9	1,586,691	0.0281	0.25
10	1,936,101	0.0343	0.34	10	1,989,770	0.0352	0.35
11	2,494,687	0.0441	0.49	11	2,901,680	0.0513	0.56
12	3,454,480	0.0611	0.73	12	3,336,293	0.0590	0.71
13	3,853,974	0.0682	0.89	13	3,540,586	0.0626	0.81
14	4,161,219	0.0736	1.03	14	3,703,359	0.0655	0.92
15	4,700,199	0.0832	1.25	15	4,522,871	0.0800	1.20
16	5,475,487	0.0969	1.55	16	5,815,099	0.1029	1.65
17	5,907,560	0.1045	1.78	17	6,234,250	0.1103	1.87
18	5,633,348	0.0997	1.79	18	6,069,473	0.1074	1.93
19	5,041,381	0.0892	1.69	19	5,170,999	0.0915	1.74
20	3,638,416	0.0643	1.29	20	3,761,407	0.0665	1.33
21	2,349,010	0.0415	0.87	21	2,276,630	0.0402	0.84
22	1,158,134	0.0204	0.45	22	1,051,611	0.0186	0.41
23	329,949	0.0058	0.13	23	284,922	0.0050	0.12
24	33,505	0.0006	0.01	24	43,009	0.0008	0.02
Sum	56,537,644	1	15.075	Sum	56,537,643	1	15.184
Standardized LD_{Prox}			0.504	Standardized LD_{Prox}			0.509

activities) are not well -distributed in the entire city. However, this value increased slightly during the selected period, which indicates the good distribution of various land use types in Kajang City.

Tables 5.14 and 5.15 also show that land use diversity in 2004 and 2008 reached a maximum of 24, but the maximum value of diversity in 2012 and 2015 reached 25. However, all years have a constant reduction and almost 0 percentages for maximum proximity values (near the value of 25).

Figure 5.20 depicts the overall land use diversity trend of Kajang City during the selected period. As shown in the percentage–proximity value graph, the percentage of minimum proximity value (the value of 5) decreased from 2008 to 2012 and even became 0 in 2015 (minimum value starts from 6). In addition, the percentage with middle proximity values (near the value of 15) increased from 2004 to 2015. This figure also shows the growth of the standardized LD_{Prox} from 2004 to 2015.

In addition to quantitative assessments, the proximity analysis method produces a graphical illustration that shows the overall land use diversity condition in the entire study area. As shown in Fig. 5.21, the eastern sides, which are mainly covered by agricultural fields, have a minimum value of land use diversity (near 5). This condition is changing, and new developments in these areas increase the level of mixed developments. The level of diversity value in the northern parts are decreasing yearly mainly because of the increase in residential growth in these areas (as shown in the residential density assessment). Central areas along train stations always keep a high level of diversity because of the proximity to the CBD. Among the four land use maps, the diversity in 2012 looks better distributed and more sustainable than others, especially in 2004, which shows a few mixed land use centers. Notably, the trend of high mixed land use development is moving from north to south because of the potential of these areas for the redevelopment of brownfield sites. The growth of land use diversity values from 2004 to 2015 can also be noticed in the legends of this figure.

Finally, as shown in Fig. 5.22, these maps were reclassified into three classes, as performed for overall urban density maps, to create a standard aggregation process.

Mixed land use development assessment using proximity analysis was also conducted using zoning format to rank the land use diversity of the study area on a zonal basis. For this process, the same proximity values for each pixel of land use

Table 5.15 Land use diversity assessment using proposed proximity analysis for 2012 and 2015

2012				2015			
Proximity value	No. of cells	Proportion	Prop. × Value	Proximity value	No. of cells	Proportion	Prop. × Value
5	312,362	0.0055	0.03	6	163	0.0000	0.00
6	755,125	0.0134	0.08	7	465,687	0.0082	0.06
7	890,062	0.0157	0.11	8	1,252,867	0.0222	0.18
8	1,283,098	0.0227	0.18	9	1,978,436	0.0350	0.31
9	2,353,747	0.0416	0.37	10	2,496,131	0.0441	0.44
10	2,073,939	0.0367	0.37	11	2,367,881	0.0419	0.46
11	2,589,843	0.0458	0.50	12	3,611,271	0.0639	0.77
12	3,244,631	0.0574	0.69	13	4,224,172	0.0747	0.97
13	3,922,450	0.0694	0.90	14	5,213,258	0.0922	1.29
14	4,496,600	0.0795	1.11	15	6,217,185	0.1100	1.65
15	5,177,448	0.0916	1.37	16	6,327,274	0.1119	1.79
16	5,649,717	0.0999	1.60	17	6,480,122	0.1146	1.95
17	5,807,342	0.1027	1.75	18	5,659,448	0.1001	1.80
18	5,740,249	0.1015	1.83	19	4,502,222	0.0796	1.51
19	4,944,520	0.0875	1.66	20	2,894,632	0.0512	1.02
20	3,472,511	0.0614	1.23	21	1,635,725	0.0289	0.61
21	2,213,026	0.0391	0.82	22	840,591	0.0149	0.33
22	1,101,154	0.0195	0.43	23	302,112	0.0053	0.12
23	412,311	0.0073	0.17	24	58,285	0.0010	0.02
24	90,018	0.0016	0.04	25	10,213	0.0002	0.00
25	7522	0.00	0.00	Sum	56,537,675	1.00	15.295
Sum	56,537,675	1.00	15.245	Standardized LD_{Prox}			0.515
Standardized LD_{Prox}			0.512				

maps were used. Unlike the previous case in which whole pixels of the study area were considered, this case involved proximity values calculated for each zone separately. Large zones obtain higher land use diversity values than small zones because of the greater number of pixels. Thus, in this assessment, the result does not reveal the actual level of mixed development. To avoid this bias in the result, land use diversity values were calculated based on the area of each zone, in which the estimated values of diversity were divided by the number of pixels of the corresponding zones. This process not only provided a reliable result but also standardized the values of diversity of the zones.

In addition, similar results can be obtained in considering the percentage of the amount of pixels bearing a specific diversity value with respect to the corresponding zone area. Tables 5.16 and 5.17 show the calculations for each zone of the land use maps. The first two columns indicate the zone number and the number of pixels in each zone. The next column for each land use year shows the total diversity values obtained from the sum of the diversity values of all pixels of the corresponding zone. The next column shows the average diversity values, which are the calculated diversity values based on the area of each zone. Finally, the average diversity values are standardized from zero to one, in which zero means minimum and one indicates maximum level of diversity condition. By considering the area of each zone, we can expect that small zones (8, 9, 10, 14, and 18) obtain high values of diversity. However, in fact, land use richness and distribution pattern is the main aspect of this assessment. For instance, zone number 10 has a small area with moderate land use diversity value. Regardless of the area considerations, the eastern zones have very low mixed land use development because of the high coverage of agricultural fields. However, similar to previous analysis results, these areas developed slightly during the selected period. By contrast, highly developed and saturated zones (3, 9, 11, 15, 16, and 17), especially those located in the central parts or near the CBD, have high values of diversity in all land use maps. The western zones of the study area (Fig. 5.23) encountered a growth in diversity value from 2004 to 2012 and almost a constant situation in the last period. In general, a very slight reduction in diversity values from 2012 to 2015 indicates better and more efficient mixed land use development in 2012.

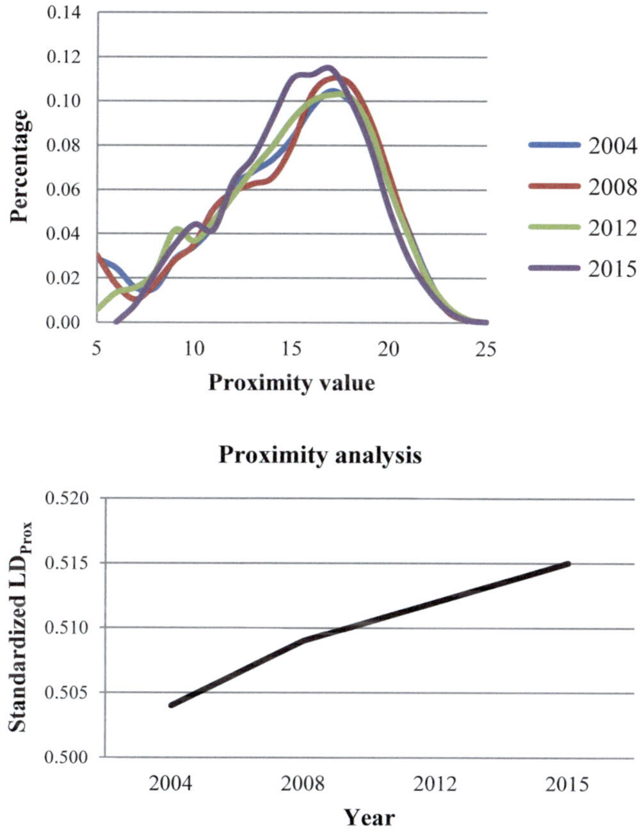

Fig. 5.20 Graphical presentation of quantitative assessment of proximity analysis

Figure 5.23 depicts the graphical presentation of land use diversity using these assessments. This figure shows the reduction and growth of mixed development in the northern and southern parts of the city, respectively. The similarity of diversity values in 2004 and 2008 (Tables 5.16) can be observed in this figure as well, resulting in the same color for all corresponding zones. This figure shows the reduction of the mixed development in zone number 3 because of the increase in residential development and in zone number 2 because of industrial growth. Although the growth of diversity was revealed from previous analysis (without zonal basis), zone numbers 5 and 10 in this part maintain an almost constant condition.

A review of the results of zonal and without zonal-based analyses shows that in the case of zonal basis, although the method attempted to remove the bias, borders and areas of the zones affected the results. In addition, non-zonal basis provided more detailed and accurate results during the selected period than zonal basis.

5.3.2.3 Urban Intensity Assessment

Urban intensity was evaluated with respect to the availability, quality, and proximity of various community facilities and services, such as health, educational, recreational, security, and others in Kajang City. Overall, seven variables with five proximity classes were considered. Thus, as shown in Tables 5.18 and 5.19, the minimum intensity value is 7, and the maximum intensity value can reach 35 ($7 \times 5=35$). Thereafter, the number and percentage of pixels bearing each intensity value were calculated. Finally, the sum of the products of percentage and intensity values presents the urban intensity level for each land use map. The output values can vary from 7, as the least intensified city, to 35, as the most intensified city. As shown in Tables 5.18 and 5.19, all land use maps obtained almost the same value near 20. However, Fig. 5.24 shows a proper presentation with more details regarding the proportion of pixels belonging to various intensity values. This figure reveals that although all land use maps have constant intensity value in general, subtle differences have occurred during the selected period. The development growth from 2008 to 2012 reduced the number of pixels (from 3 to 1%) with a low intensity value of 10. However, the development growth from 2012 to 2015 showed a reduced intensity value of 25 near 20 for more than 6% of the pixels of the study area. In addition, the 2004 and 2012 land use maps have the highest and lowest number of pixels with intensity value of 35, respectively. Through the standardization of the total intensity value in the range of 0–1, all the land use maps obtained a value near 0.5. All these quantitative results showed that in general, the recent development pattern of Kajang City did not improve the intensification properties of this region.

The graphical illustration of urban intensity assessment reveals more information on the spatial pattern of the city intensification condition of Kajang City (Fig. 5.25). In all land use maps, the most intensified part of the city is near Kajang main train station (KTM) and a few small centers near the central and central east parts. However, the trend of intensification growth in the southern parts of the city during selected periods can be observed clearly. Similar to other assessments, agricultural (eastern parts) and industrial areas have very low intensity values. In addition, because of the proximity of the northern parts to other regions and cities, these parts have high intensity values from 2004 to 2012 but have reduced values in the last period. Finally, these maps were also classified into three main classes to aggregate with other compactness indicators in a standard format.

In evaluating urban intensity for zonal-based analysis, the value of urban intensity for each zone was calculated separately. In this part, similar to land use diversity assessment, the total intensity value for each zone was calculated, and an average intensity value was evaluated with respect to the area of the corresponding zone. Tables 5.20 and 5.21 present these assessments. Similar to the values in the non-zonal basis, the average intensity values range from 7 to 35. In these tables, the values range from 11.33 to 27.52 for 2004, 11.14 to 27.44 for 2008, 15.13 to 27.27 for 2012, and 14.75

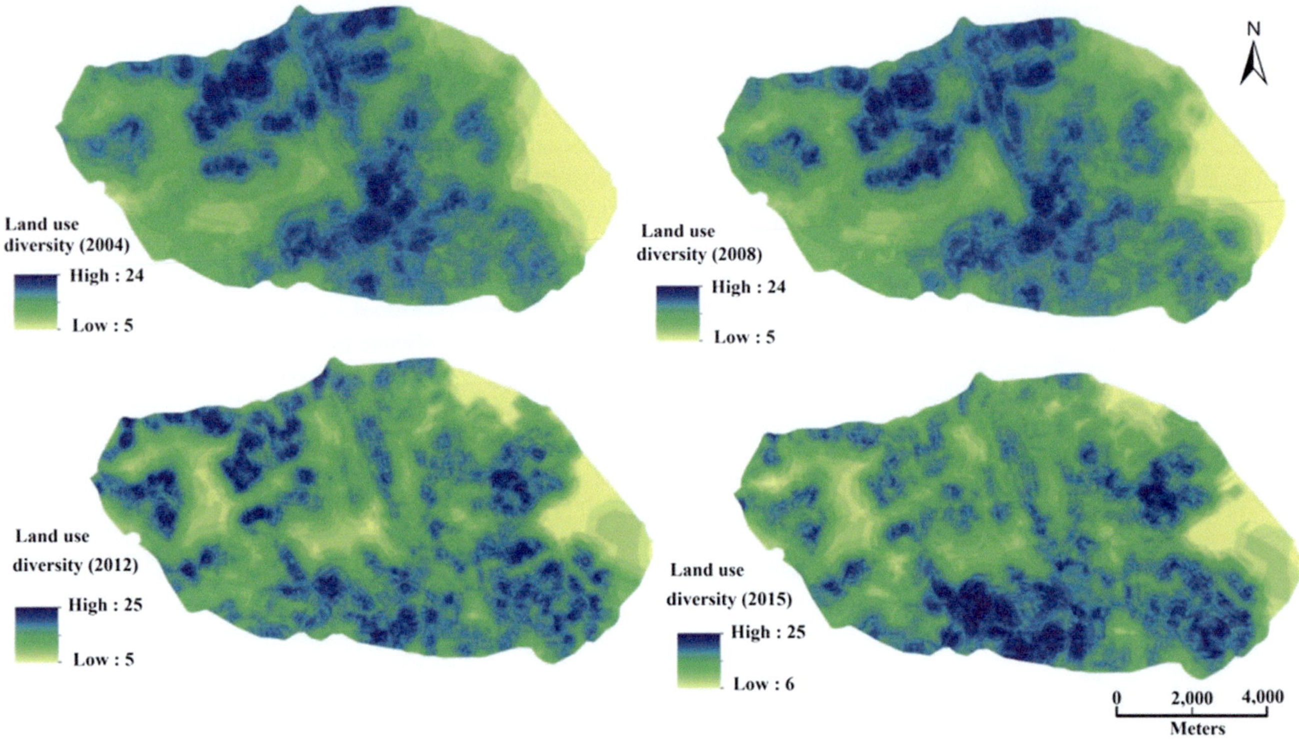

Fig. 5.21 Graphical presentation of land use diversity condition of Kajang City

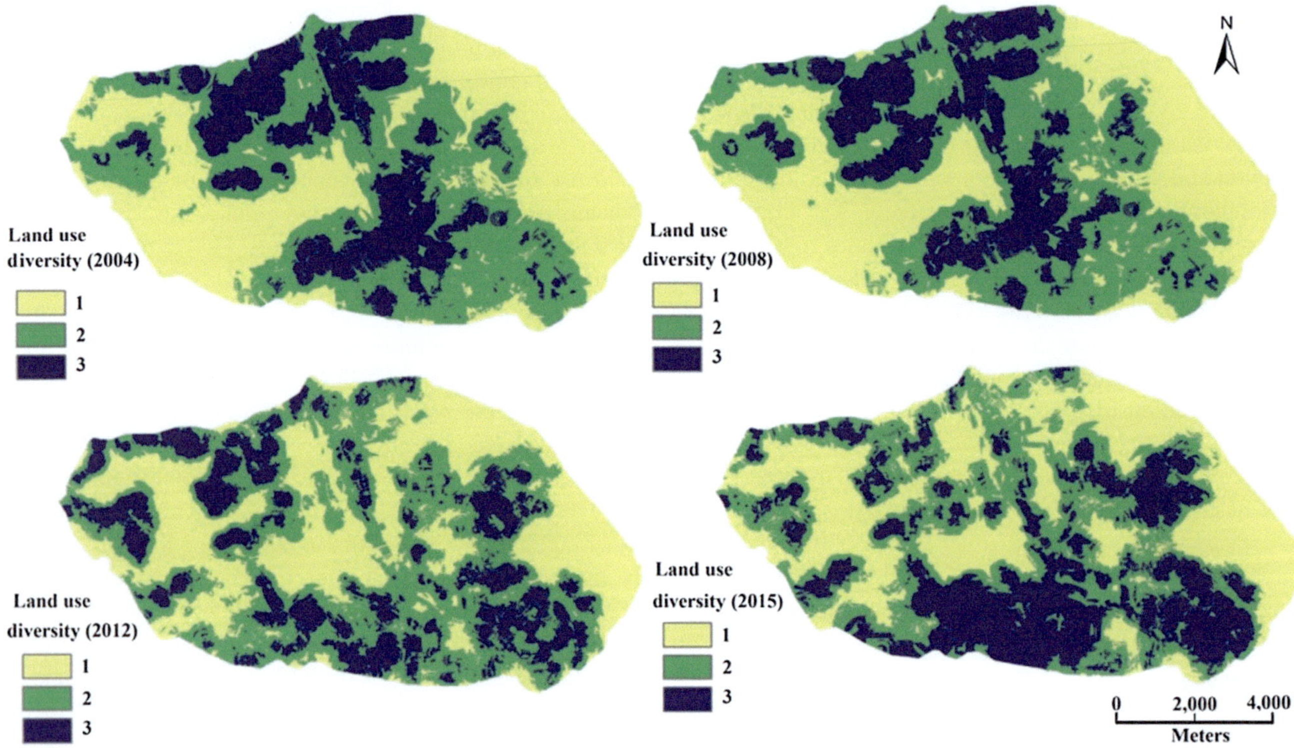

Fig. 5.22 Reclassified land use diversity maps

Table 5.16 Calculation of land use diversity using zonal basis for years 2004 and 2008

Zoning		2004			2008		
Zone no.	No. of pixels	Total diversity values	Average diversity value	Standard diversity value	Total diversity values	Average diversity value	Standard diversity value
1	2,326,007	32,999,018	14.19	0.75	33,714,063	14.49	0.75
2	3,435,010	49,900,009	14.53	0.77	49,625,469	14.45	0.75
3	3,540,793	66,614,513	18.81	0.99	65,753,063	18.57	0.96
4	4,400,533	58,018,979	13.18	0.69	60,950,742	13.85	0.72
5	4,758,531	35,871,846	7.54	0.40	39,850,738	8.37	0.43
6	3,365,698	55,990,584	16.64	0.88	55,409,128	16.46	0.85
7	3,961,364	74,650,678	18.84	0.99	76,482,357	19.31	1.00
8	683,864	12,578,968	18.39	0.97	12,527,335	18.32	0.95
9	2,423,400	41,383,232	17.08	0.90	42,305,207	17.46	0.90
10	2,639,950	30,914,338	11.71	0.62	30,680,128	11.62	0.60
11	4,519,894	79,355,614	17.56	0.93	78,833,651	17.44	0.90
12	3,639,343	46,204,856	12.70	0.67	47,227,113	12.98	0.67
13	3,315,292	39,306,048	11.86	0.62	39,056,839	11.78	0.61
14	1,221,664	19,389,050	15.87	0.84	19,053,076	15.60	0.81
15	4,944,235	87,064,714	17.61	0.93	86,250,016	17.44	0.90
16	3,940,104	65,486,401	16.62	0.88	65,022,508	16.50	0.85
17	2,716,433	43,033,847	15.84	0.83	43,062,035	15.85	0.82
18	705,560	13,386,617	18.97	1.00	12,558,379	17.80	0.92

Table 5.17 Calculation of land use diversity using zonal basis for years 2012 and 2015

Zoning		2012			2015		
Zone no.	No. of pixels	Total diversity values	Average diversity value	Standard diversity value	Total diversity values	Average diversity value	Standard diversity value
1	2,326,007	37,046,180	15.93	0.85	32,760,443	14.08	0.72
2	3,435,010	44,048,620	12.82	0.68	42,498,078	12.37	0.63
3	3,540,793	54,275,753	15.33	0.82	50,394,125	14.23	0.73
4	4,400,533	53,407,691	12.14	0.65	57,988,973	13.18	0.67
5	4,758,531	58,088,113	12.21	0.65	58,269,929	12.25	0.63
6	3,365,698	47,328,916	14.06	0.75	52,996,910	15.75	0.80
7	3,961,364	69,031,651	17.43	0.93	63,312,623	15.98	0.82
8	683,864	12,842,586	18.78	1.00	11,498,758	16.81	0.86
9	2,423,400	38,006,942	15.68	0.84	38,528,809	15.90	0.81
10	2,639,950	35,471,646	13.44	0.72	36,265,385	13.74	0.70
11	4,519,894	73,458,779	16.25	0.87	72,807,876	16.11	0.82
12	3,639,343	48,910,123	13.44	0.72	52,778,576	14.50	0.74
13	3,315,292	52,118,020	15.72	0.84	52,682,525	15.89	0.81
14	1,221,664	20,567,261	16.84	0.90	18,619,108	15.24	0.78
15	4,944,235	89,044,868	18.01	0.96	96,842,139	19.59	1.00
16	3,940,104	66,107,919	16.78	0.89	66,264,749	16.82	0.86
17	2,716,433	49,908,051	18.37	0.98	49,541,142	18.24	0.93
18	705,560	12,249,024	17.36	0.92	10,689,723	15.15	0.77

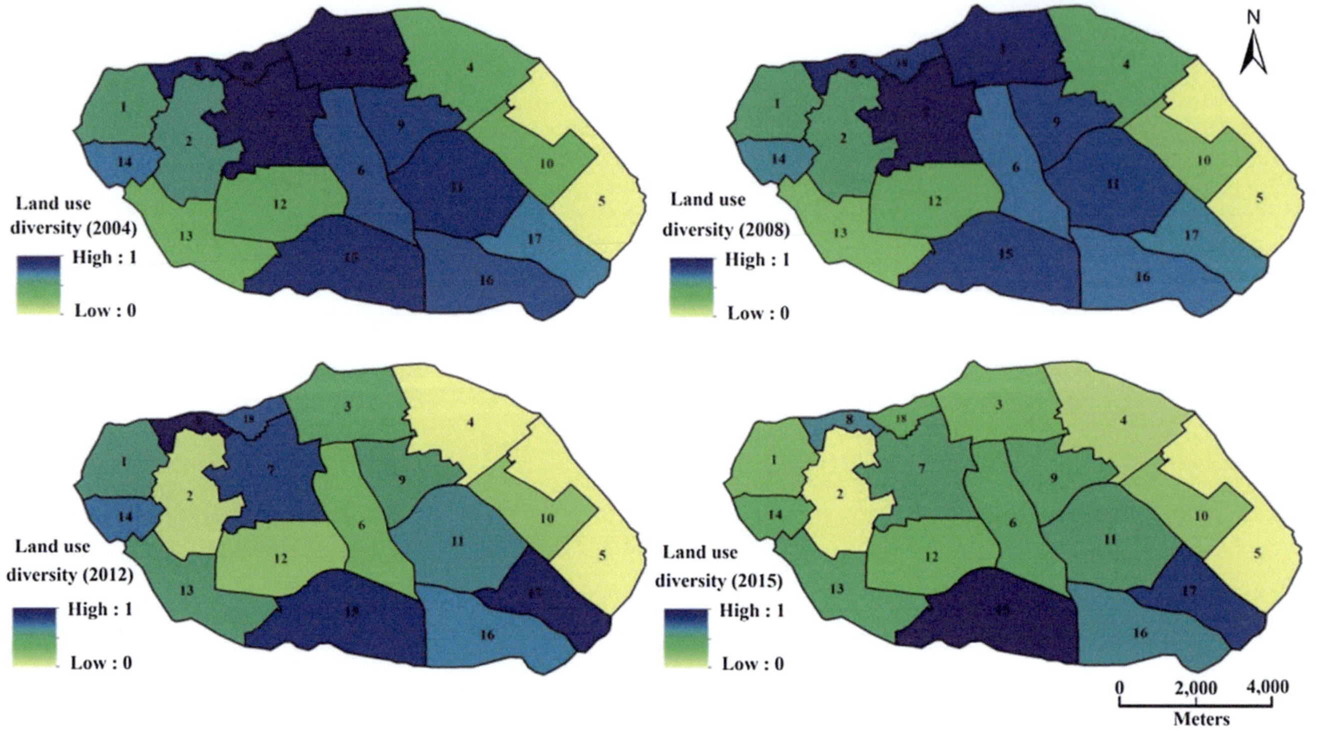

Fig. 5.23 Graphical presentation of land use diversity on zonal basis

to 27.29 for 2015. The minimum value of urban intensity increased until 2012 and then decreased in 2015, but no significant changes can be observed in the maximum value. In general, the intensity of most of the zones increased from 2004 to 2012, but an inverse situation can be observed in the last period. Zone number 11 in 2004 and 2008 and zone number 16 in the last two years are the most intensified zones. By contrast, zone number 13 in 2004 and 2008, zone number 12 in 2012, and zone number 2 in the last year are the least intensified zones.

Figure 5.26 shows that the central zones (from north to south) near the CBD and main public transportation are the most intensified zones from 2004 to 2008. However, the intensity of the northern zones decreased in the last two time periods, and only the center and southern zones intensified. The growth of residential areas in zone numbers 3 and 4 and some parts of zone numbers 2, 7, and 12 disturbed the balance of the available facilities for local residents, thereby reducing the intensity values of these zones. We conclude that the availability of transportation facilities is one of the main aspects of urban intensity because this facility attracts a high population. Moreover, more facilities and services will be automatically provided to the population.

5.4 Evaluation of Trend of Compactness (ToC) of Kajang City

After the evaluation of all land use maps based on several compactness indicators (functional compactness, urban densities, urban intensities, and land use diversity), all the derived maps from each concept were added using the following equation to extract the overall compactness of Kajang City:

$$\mathrm{DoC}_i = \sum \mathrm{UD}_i + \sum \mathrm{UI}_i + \mathrm{LD}_i \tag{5.13}$$

where DoC_i is the degree of compactness of land use map i (i = 2004, 2008, 2012, and 2015), $\sum \mathrm{UD}_i$ is the sum of all urban density values, $\sum \mathrm{UI}_i$ is the sum of all urban intensity values, and LD_i is the land use diversity of the corresponding land use map. The range of DoC depends on the number of land use categories involved in the process, the number of urban intensity variables involved in the process, and the classification schemes of urban density calculations. For instance, if five land use types (with five proximity classes, as explained in detail in the section discussing land use diversity assessment by proximity method), seven urban

Table 5.18 Urban intensity assessment for years 2004 and 2008

2004				2008			
Intensity value	No. of pixels	Perc.	Perc. × value	Intensity value	No. of pixels	Perc.	Perc. × value
7	154,239	0.0027	0.0191	7	170,952	0.003	0.0212
8	946,359	0.0167	0.1339	8	890,312	0.015	0.1260
9	1,705,251	0.0302	0.2715	9	1,509,206	0.026	0.2403
10	1,656,322	0.0293	0.2930	10	1,460,346	0.025	0.2583
11	1,570,460	0.0278	0.3056	11	1,397,996	0.024	0.2720
12	1,340,877	0.0237	0.2846	12	1,292,439	0.022	0.2743
13	1,171,238	0.0207	0.2693	13	1,412,750	0.025	0.3249
14	1,771,205	0.0313	0.4386	14	1,509,135	0.026	0.3737
15	1,733,058	0.0307	0.4598	15	1,435,426	0.025	0.3809
16	2,062,117	0.0365	0.5836	16	2,361,553	0.041	0.6684
17	2,648,479	0.0468	0.7964	17	2,837,114	0.050	0.8532
18	2,458,926	0.0435	0.7829	18	2,516,739	0.044	0.8013
19	2,860,107	0.0506	0.9613	19	2,820,065	0.049	0.9478
20	2,954,778	0.0523	1.0454	20	3,094,296	0.054	1.0947
21	3,264,092	0.0577	1.2125	21	3,614,929	0.063	1.3428
22	3,262,492	0.0577	1.2696	22	3,649,307	0.064	1.4202
23	3,592,930	0.0636	1.4618	23	3,851,785	0.068	1.5671
24	3,448,539	0.0610	1.4640	24	3,318,983	0.058	1.4090
25	3,404,175	0.0602	1.5054	25	3,605,192	0.063	1.5943
26	3,147,155	0.0557	1.4474	26	3,026,454	0.053	1.3919
27	2,728,190	0.0483	1.3030	27	2,727,747	0.048	1.3028
28	2,303,141	0.0407	1.1407	28	2,203,508	0.039	1.0914
29	1,844,676	0.0326	0.9463	29	1,551,145	0.027	0.7957
30	1,239,087	0.0219	0.6576	30	1,159,475	0.020	0.6153
31	952,091	0.0168	0.5221	31	987,877	0.017	0.5417
32	683,094	0.0121	0.3867	32	638,964	0.011	0.3617
33	624,924	0.0111	0.3648	33	614,129	0.010	0.3585
34	662,401	0.0117	0.3984	34	652,056	0.011	0.3922
35	341,359	0.0060	0.2113	35	221,863	0.003	0.1374
Sum	56,531,762	1	20.94	Sum	56,531,743	1	20.96
Standardized value			0.498	Standardized value			0.499

intensity variables (with 5 proximity classes), and five classes are considered for density analysis, then the DoC ranges from 13 to 65. In avoiding this complexity, using Eq. 5.13 is better when all three variables are reclassified properly and then inserted into the equation. In this manner, a proper range can be obtained in the final DoC value. However, during analysis, we observed that any type of classification, especially with the low number of classes, tends to generalize and reduce the precision and accuracy of the results. Thus, this study preferred the values of compactness indicators in original format as they resulted from the analysis. Nevertheless, in considering the results whether using classified or original format, because the minimum and maximum values of the obtained results can be estimated easily, the DoC value can be standardized in the range from zero to one by Eq. 5.14:

$$\mathrm{DoC_{Std}} = \frac{\mathrm{DoC_{Org}} - \mathrm{DoC_{min}}}{\mathrm{DoC_{max}} - \mathrm{DoC_{min}}}, \tag{5.14}$$

where $\mathrm{DoC_{std}}$ is the standardized degree of compactness ranging from zero to one, $\mathrm{DoC_{org}}$ is the original DoC obtained from Eq. 5.13, and $\mathrm{DoC_{min}}$ and $\mathrm{DoC_{max}}$ are the minimum and maximum values of DoC, respectively, which depend on the number of land use categories, number of urban intensity variables, and classification schemes of

Table 5.19 Urban intensity assessment for years 2012 and 2015

2012				2015			
Intensity value	No. of pixels	Perc.	Perc. * value	Intensity value	No. of pixels	Perc.	Perc. * value
7	159,230	0.0028	0.0197	7	94,390	0.0017	0.0117
8	486,520	0.0086	0.0688	8	441,263	0.0078	0.0624
9	855,714	0.0151	0.1362	9	311,288	0.0055	0.0496
10	755,197	0.0134	0.1336	10	673,616	0.0119	0.1191
11	995,587	0.0176	0.1937	11	964,965	0.0171	0.1877
12	1,620,114	0.0287	0.3439	12	1,604,466	0.0284	0.3406
13	1,948,314	0.0345	0.4480	13	1,666,267	0.0295	0.3831
14	2,007,889	0.0355	0.4972	14	2,191,713	0.0388	0.5427
15	2,231,711	0.0395	0.5921	15	2,697,762	0.0477	0.7158
16	2,463,466	0.0436	0.6972	16	3,223,692	0.0570	0.9123
17	2,650,375	0.0469	0.7969	17	3,297,611	0.0583	0.9916
18	3,025,803	0.0535	0.9633	18	3,511,526	0.0621	1.1180
19	3,330,002	0.0589	1.1191	19	3,666,327	0.0648	1.2321
20	3,261,876	0.0577	1.1539	20	3,603,821	0.0637	1.2749
21	3,460,370	0.0612	1.2853	21	3,112,589	0.0551	1.1561
22	3,728,655	0.0660	1.4509	22	3,157,238	0.0558	1.2286
23	3,563,046	0.0630	1.4495	23	2,845,378	0.0503	1.1575
24	3,295,255	0.0583	1.3988	24	3,012,846	0.0533	1.2790
25	3,217,651	0.0569	1.4228	25	2,905,951	0.0514	1.2850
26	2,993,741	0.0530	1.3768	26	2,574,135	0.0455	1.1838
27	2,604,871	0.0461	1.2440	27	2,464,231	0.0436	1.1768
28	2,159,250	0.0382	1.0694	28	2,152,529	0.0381	1.0660
29	1,603,468	0.0284	0.8225	29	1,781,913	0.0315	0.9140
30	1,307,988	0.0231	0.6941	30	1,456,752	0.0258	0.7730
31	1,303,188	0.0231	0.7146	31	883,429	0.0156	0.4844
32	872,192	0.0154	0.4937	32	771,797	0.0137	0.4368
33	453,357	0.0080	0.2646	33	780,072	0.0138	0.4553
34	160,423	0.0028	0.0965	34	485,919	0.0086	0.2922
35	21,502	0.0004	0.0133	35	203,269	0.0036	0.1258
Sum	56,536,755	1	20.96	Sum	56,536,755	1	20.96
Standardized value			0.499	Standardized value			0.498

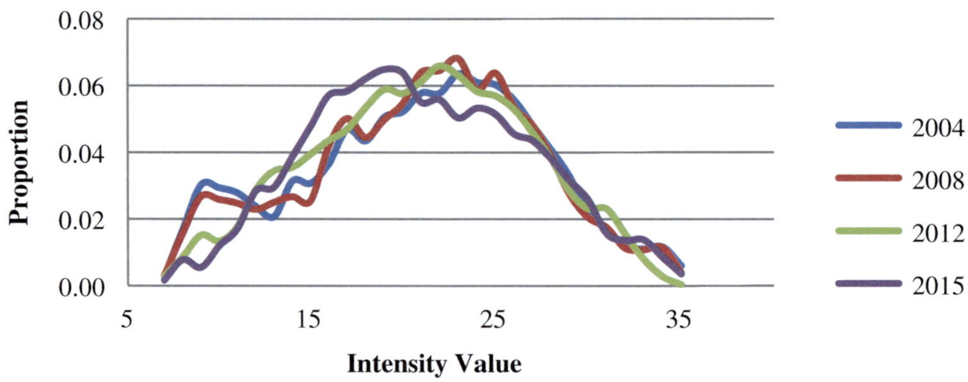

Fig. 5.24 Urban intensity value compare to percentage of pixels

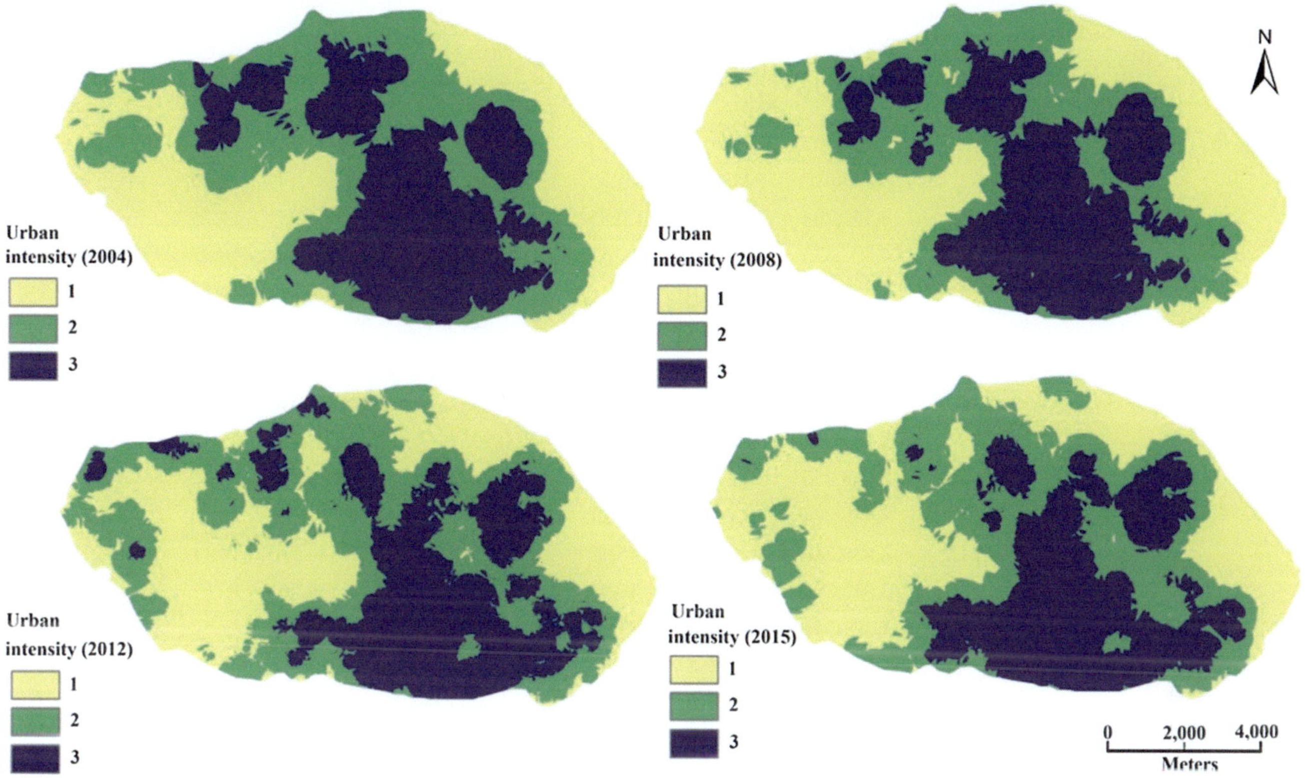

Fig. 5.25 Reclassified urban intensity maps

Table 5.20 Calculation of urban intensity using zonal basis for years 2004 and 2008

Zoning		2004			2008		
Zone no.	No. of pixels	Total intensity values	Average intensity values	Standard intensity values	Total intensity values	Average intensity values	Standard intensity values
1	2,326,007	43,069,889	18.52	0.67	41,888,091	18.01	0.66
2	3,435,010	61,228,219	17.82	0.65	59,130,535	17.21	0.63
3	3,540,793	84,815,430	23.95	0.87	82,513,691	23.30	0.85
4	4,400,533	81,818,293	18.59	0.68	82,772,695	18.81	0.69
5	4,758,531	60,399,277	12.69	0.46	66,868,970	14.05	0.51
6	3,365,698	85,865,218	25.51	0.93	85,001,315	25.26	0.92
7	3,961,364	93,085,157	23.50	0.85	94,866,407	23.95	0.87
8	683,864	14,662,368	21.44	0.78	14,390,097	21.04	0.77
9	2,423,400	61,874,806	25.53	0.93	62,046,315	25.60	0.93
10	2,639,950	56,088,283	21.25	0.77	56,138,333	21.26	0.77
11	4,519,894	124,371,838	27.52	1.00	124,042,678	27.44	1.00
12	3,639,343	53,702,271	14.76	0.54	56,420,088	15.50	0.56
13	3,315,292	37,546,213	11.33	0.41	36,942,288	11.14	0.41
14	1,221,664	21,471,467	17.58	0.64	20,775,173	17.01	0.62
15	4,944,235	124,164,341	25.11	0.91	122,239,297	24.72	0.90
16	3,940,104	103,073,397	26.16	0.95	102,327,936	25.97	0.95
17	2,716,433	60,525,852	22.28	0.81	61,094,067	22.49	0.82
18	705,560	15,835,698	22.44	0.82	15,393,758	21.82	0.80

Table 5.21 Calculation of urban intensity using zonal basis for years 2012 and 2015

Zoning		2012			2015		
Zone no.	No. of pixels	Total intensity values	Average intensity values	Standard intensity values	Total intensity values	Average intensity values	Standard intensity values
1	2,326,007	45,620,593	19.61	0.72	40,929,353	17.60	0.64
2	3,435,010	52,613,128	15.32	0.56	50,683,177	14.75	0.54
3	3,540,793	73,825,646	20.85	0.76	69,936,051	19.75	0.72
4	4,400,533	77,301,126	17.57	0.64	81,716,041	18.57	0.68
5	4,758,531	77,003,014	16.18	0.59	76,374,019	16.05	0.59
6	3,365,698	78,194,127	23.23	0.85	83,854,740	24.91	0.91
7	3,961,364	82,566,959	20.84	0.76	76,428,703	19.29	0.71
8	683,864	15,610,499	22.83	0.84	14,269,472	20.87	0.76
9	2,423,400	61,060,364	25.20	0.92	61,627,164	25.43	0.93
10	2,639,950	56,278,165	21.32	0.78	56,893,748	21.55	0.79
11	4,519,894	118,444,033	26.21	0.96	117,791,025	26.06	0.95
12	3,639,343	55,051,720	15.13	0.55	58,644,273	16.11	0.59
13	3,315,292	54,724,738	16.51	0.61	55,279,563	16.67	0.61
14	1,221,664	23,349,405	19.11	0.70	21,405,533	17.52	0.64
15	4,944,235	125,709,871	25.43	0.93	133,433,648	26.99	0.99
16	3,940,104	107,437,276	27.27	1.00	107,530,150	27.29	1.00
17	2,716,433	65,474,094	24.10	0.88	64,788,790	23.85	0.87
18	705,560	14,766,602	20.93	0.77	13,202,600	18.71	0.69

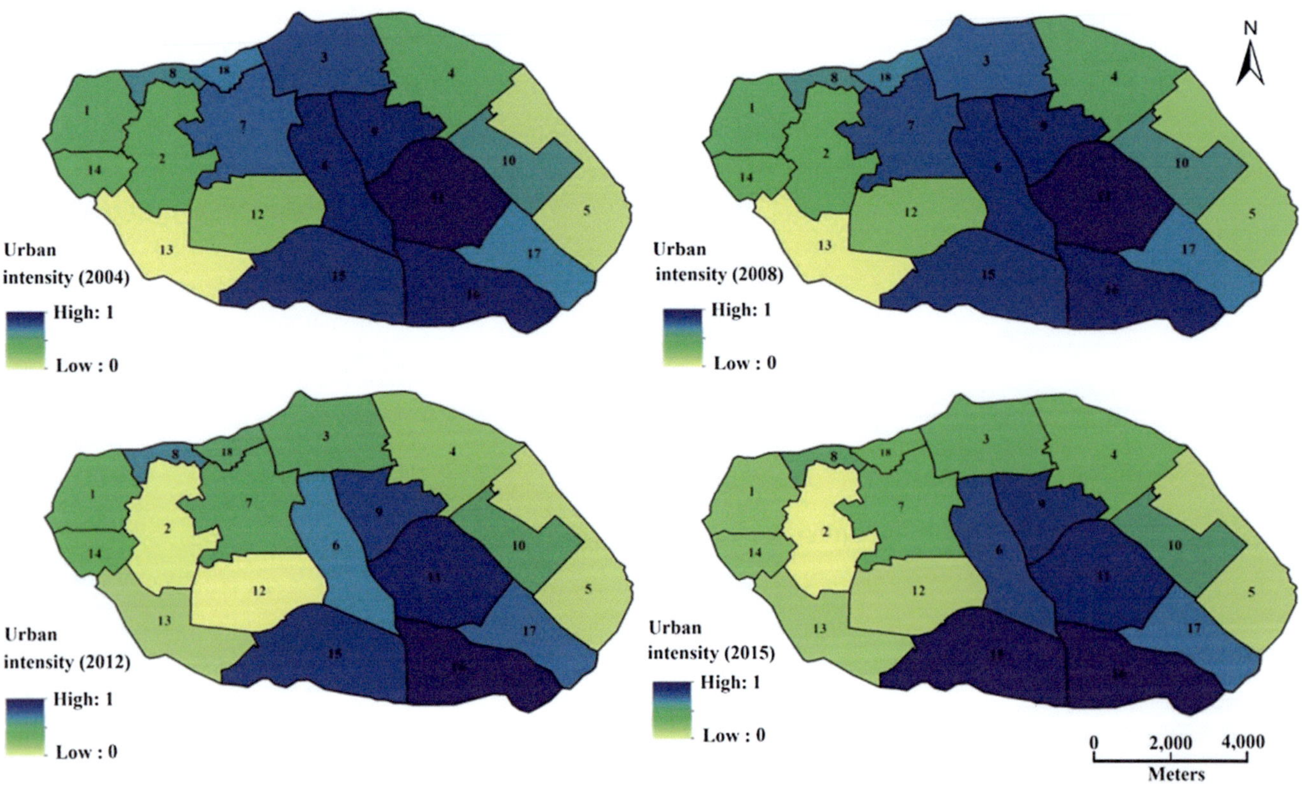

Fig. 5.26 Graphical presentation of urban intensity on zonal basis

urban densities calculations, as explained in the preceding parts of this paper.

After evaluating the DoC for all land use maps, we extracted and evaluated the ToC. One of the main contributions of this study is to propose the concept of ToC in urban planning. After numerous studies on urban growth analysis and change detection in urban behavior, the ToC provides a clear perspective regarding city compactness and eventually sustainable urban development. In general, the ToC reveals the trend of city compactness (degree of compactness) for each pixel of the study area based on a given period, to show whether each pixel in a specific neighborhood encountered growth and/or decrease in city compactness. ToC provides straightforward and valuable information for urban planners and decision makers to have good judgment on the existing situation and future trend of urban growth based on the urban sustainability. Thus, decision makers are able to propose more effective solutions and make the best decisions.

The ToC can be evaluated at various scales, whether regional or country basis, city basis, zonal basis, and finally pixel basis (according to the pixel size of the system analysis). After the analysis scale is selected, the assessment undergoes various aspects of the system regarding location and effective variables that cause the increase or decrease of the DoC. These variables mainly arise from city compactness indicators and their various aspects such as urban densities (built-up, population, and others), urban intensities, and land use diversity. However, other external factors that can also affect the DoC of a specific landscape directly or indirectly can be included in the ToC evaluation. In this manner, the modeling and simulation of ToC becomes very useful because they demonstrate actions such as locating a new shopping mall or providing a new educational facility in a specific neighborhood, which can affect city compactness.

In general, for a given period, the ToC of a specific landscape can be evaluated by the following equation:

$$\mathrm{ToC} = \mathrm{DoC}_{(t+1)} - \mathrm{DoC}_{(t)} \tag{5.15}$$

where ToC is the trend of city compactness, $\mathrm{DoC}_{(t+1)}$ is the degree of compactness at a later time, and $DoC_{(t)}$ is the degree of compactness at an earlier time of the landscape.

The result of ToC from this equation is based on the scale of the analysis (regional, zonal, and cellular). Moreover, because of the spatial basis of ToC, this result is presented in a range format. Thus, a visual illustration of the result provides a clear observation and easy interpretation regarding the ToC. The spatial basis of ToC or the raster-based presentation of the results provides the ability to use this evaluation for further analysis and processing.

Therefore, for this study, the ToC concept was used to evaluate the trend of compactness of Kajang City for all pairs of land use maps from 2004 to 2015. Finally, this study attempted to evaluate the reasons for this trend and to extract the main causative factors.

5.5 Results and Discussion of ToC

After assessing all land use maps based on compactness indicators and their variables, we create a map that illustrates the overall compactness of each available land use map. These maps were created by aggregating all three compactness indicators (urban density, urban intensity, and land use diversity). Uncertainty is always present in discussing the effect of the classification scheme on the final results. Therefore, to avoid this issue, we produced the results twice. First, the overall compactness was produced by the aggregation of all reclassified maps created from compactness indicators. Second, the overall compactness was produced by the aggregation of grayscale maps of compactness indicators. In this manner, the output maps can be evaluated without the effect of classification processes. These maps can be aggregated through any kind of weighting technique (AHP, expert choice, and others) to involve them in the analysis with different priorities. This method is a simple weighting process and raises questions about the ranking and priorities given by the experts. Therefore, it was ignored in this study. Hence, to generalize and keep the applicability of the process for other study areas, an equal priority was assumed for all indicators. Nevertheless, by giving a different priority for each indicator, various scenarios can be defined based on the objective of the research and characteristics of the local neighborhood of interest.

Figure 5.27 illustrates the overall compactness of the reclassified results of the compactness indicators maps. As mentioned and presented, each map was reclassified into three classes: 1 as lowest and 3 as highest value of density, intensity, and diversity. Therefore, in aggregating these three maps for each year, the overall compactness maps range from 3 to 9 in terms of DoC. Figure 5.28 depicts the overall compactness of Kajang City in the original grayscale format (without classification). Thus, each pixel has an actual value calculated from the aggregation of the three compactness indicators. These assessment results range from 13 to 65 (DoC). However, in the legend of these maps, the standardized range from zero to one is displayed as well. Eastern agricultural areas are the most noticeable areas with minimum DoC because of the low number of urban structures and facilities. However, this region is gradually developing, and the increase in DoC can be observed during the selected period. Next, the central west region, which is mainly occupied by industrial use, can be noticed as a less compact area. A few people are attracted to live in this area because

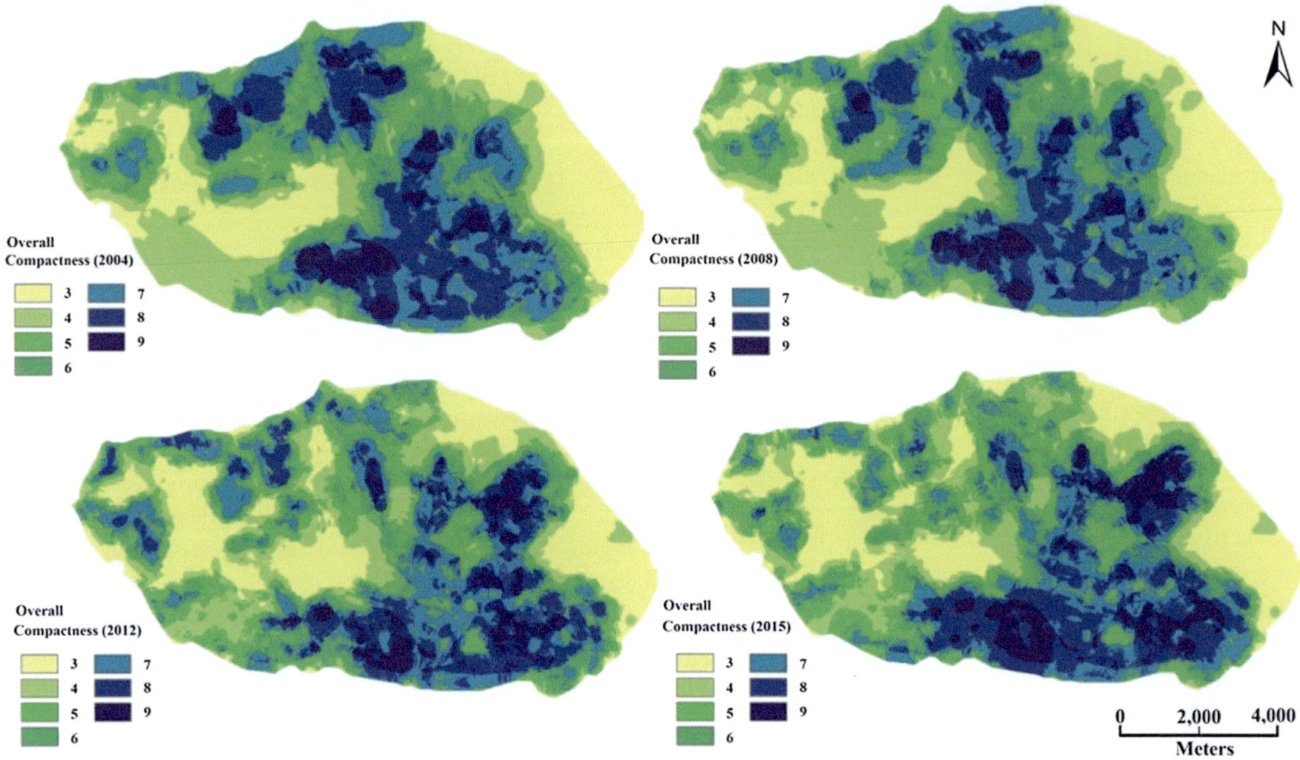

Fig. 5.27 Overall reclassified compactness of Kajang City

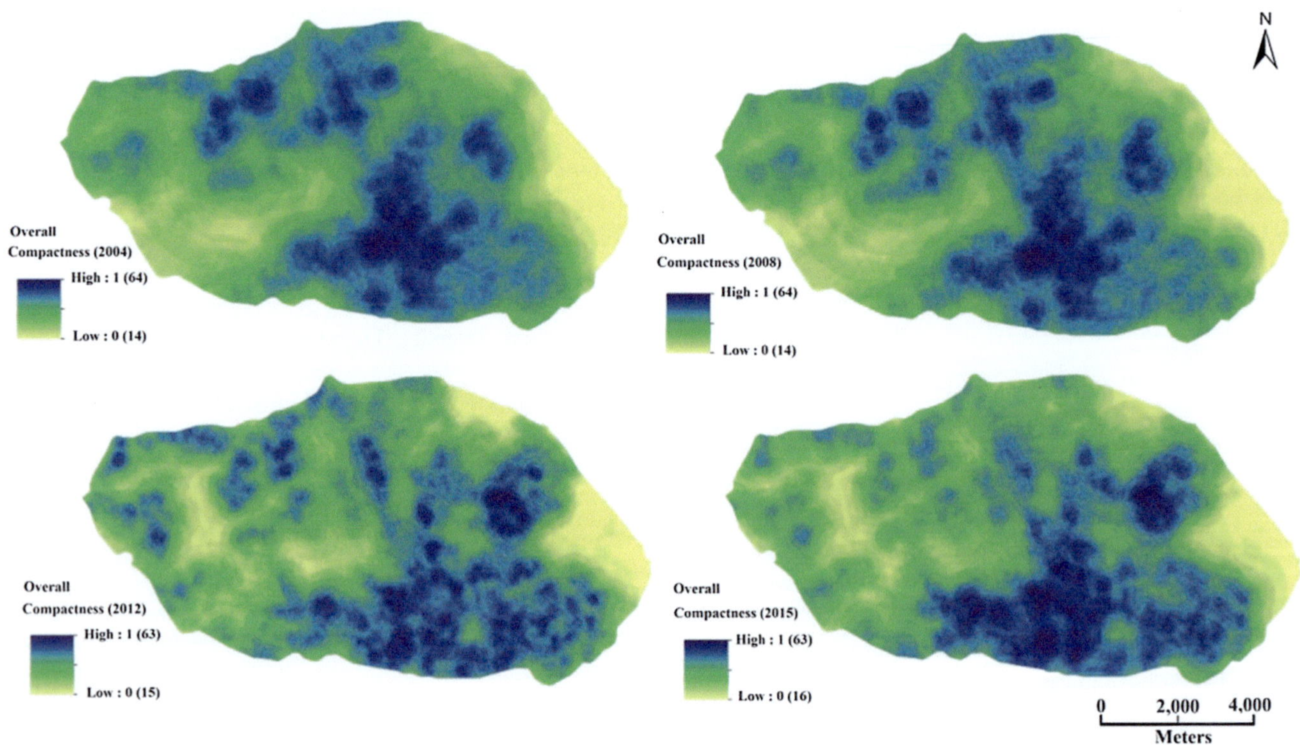

Fig. 5.28 Overall compactness maps of Kajang City without classification process

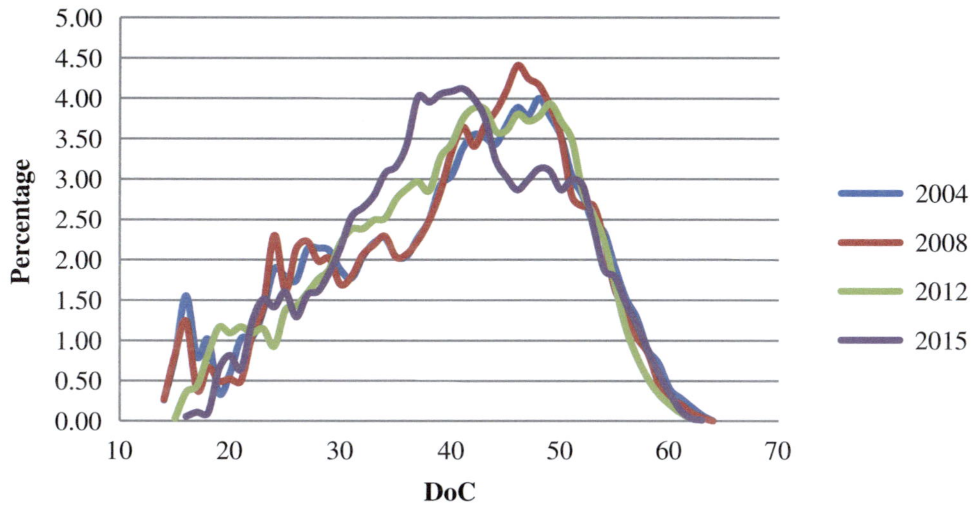

Fig. 5.29 Graphical presentation of compactness value with respect to area percentage

of the industrial buildings. Thus, few facilities and services are developed in this area. Consequently, because of the low population density, residential density, land use diversity, and urban intensity, this area is assigned as a less compact area. As shown in the figure, unlike the DoC of the agricultural region, the DoC of the industrial area decreased from 2004 to 2015 because of the growth of industrial land use especially through existing brownfield sites. The most stable compact areas are the central and southern parts of the city. The availability of the train station is the main role of compactness in the southern region. This train links Kajang City to other parts and the city centers of Kuala Lumpur and neighboring provinces. This accessibility can be the most attractive point for residential, commercial, recreational, and community facility uses. Consequently, for most of the compactness variables (urban density, intensity, and land use diversity), these areas have high priority and value. As presented in the figure, a high level of DoC expands through the east and west of Kajang City during the selected period. In addition, this area is gradually being distributed from a mono compact center to several small centers through the east and west. The legend of this figure shows that the minimum DoC increased two units from 2008 to 2015, but the maximum DoC decreased from 64 to 63 in 2012. However, these maps show that the overall ToC during the available time periods cannot be extracted. Figure 5.29 depicts the trend of DoC for this period. The DoC value of a large proportion of the study area decreased from 50 to 40 during 2008 to 2015. However, a slight increase in DoC can be observed from 2008 to 2012. Tables 5.22 and 5.23 present the quantitative calculation of compactness assessment, which finally shows a unique value of compactness for each land use map.

In these tables, the percentage of pixels bearing a DoC value is calculated, and finally the overall DoC for each land use map is evaluated by the sum of the product of percentage and DoC value. Although Fig. 5.29 shows that the overall compactness decreased especially in 2015, the actual DoC of each map shows that the compactness of Kajang City increased gradually from 2004 to 2015 (Fig. 5.30).

Figure 5.31 illustrates the spatial location of growth and loss of compactness from 2004 to 2008, 2008 to 2012, and 2012 to 2015, and the overall changes from 2004 to 2015. This figure shows that from 2004 to 2008, some points in the center, one point in the northeast, and one point in the east (agricultural areas) have significant growth in compactness. In this period, no significant reduction in compactness was observed. However, from 2008 to 2012, a substantial reduction in compactness occurred in the center and northern region of Kajang City. Interestingly, the growth of compactness is mainly found along borders, north to west to south, and a few points in the eastern region. Insignificant changes in compactness level from 2012 to 2015 are mainly due to the similarity of these two maps and one period being shorter than the other. Finally, the overall changes from 2004 to 2015 are shown in the lower right map (ToC from 2004 to 2015), which summarizes all the other periods. In general, a few small points in the west and two large areas in the east of Kajang City are the main developed areas that experienced growth in compactness. However, a significantly large area in the north and center of the city has experienced reduction in compactness during the recent development patterns.

Table 5.22 Quantitative assessment of overall compactness for years 2004 and 2008

2004				2008			
DoC	No. of cells	Percentage	Perc. × DoC	DoC	No. of cells	Percentage	Perc. × DoC
14	5831	0.26	0.04	14	6033	0.27	0.04
15	18,130	0.80	0.12	15	18,609	0.82	0.12
16	35,117	1.55	0.25	16	27,950	1.24	0.20
17	18,034	0.80	0.14	17	8590	0.38	0.06
18	22,856	1.01	0.18	18	15,075	0.67	0.12
19	7703	0.34	0.06	19	11,177	0.49	0.09
20	13,240	0.59	0.12	20	11,788	0.52	0.10
21	23,078	1.02	0.21	21	11,171	0.49	0.10
22	23,457	1.04	0.23	22	24,112	1.07	0.23
23	30,921	1.37	0.31	23	32,104	1.42	0.33
24	42,604	1.88	0.45	24	52,065	2.30	0.55
25	40,520	1.79	0.45	25	36,564	1.62	0.40
26	39,423	1.74	0.45	26	48,277	2.14	0.56
27	48,155	2.13	0.58	27	50,336	2.23	0.60
28	48,421	2.14	0.60	28	44,806	1.98	0.55
29	47,795	2.11	0.61	29	45,536	2.01	0.58
30	42,376	1.87	0.56	30	38,292	1.69	0.51
31	40,081	1.77	0.55	31	40,573	1.79	0.56
32	46,077	2.04	0.65	32	46,597	2.06	0.66
33	49,972	2.21	0.73	33	49,413	2.19	0.72
34	51,236	2.27	0.77	34	51,875	2.29	0.78
35	46,149	2.04	0.71	35	45,960	2.03	0.71
36	46,196	2.04	0.74	36	46,756	2.07	0.74
37	51,351	2.27	0.84	37	50,600	2.24	0.83
38	56,360	2.49	0.95	38	56,332	2.49	0.95
39	66,120	2.93	1.14	39	64,466	2.85	1.11
40	68,948	3.05	1.22	40	75,646	3.35	1.34
41	76,596	3.39	1.39	41	82,331	3.64	1.49
42	80,309	3.55	1.49	42	76,951	3.40	1.43
43	79,438	3.51	1.51	43	83,049	3.67	1.58
44	77,572	3.43	1.51	44	86,893	3.84	1.69
45	83,076	3.68	1.65	45	92,855	4.11	1.85
46	87,935	3.89	1.79	46	99,680	4.41	2.03
47	85,601	3.79	1.78	47	96,008	4.25	2.00
48	90,479	4.00	1.92	48	94,066	4.16	2.00
49	85,362	3.78	1.85	49	88,149	3.90	1.91
50	79,692	3.53	1.76	50	79,516	3.52	1.76
51	68,638	3.04	1.55	51	62,992	2.79	1.42
52	63,078	2.79	1.45	52	60,232	2.66	1.39
53	55,356	2.45	1.30	53	60,395	2.67	1.42
54	52,959	2.34	1.27	54	50,683	2.24	1.21
55	43,130	1.91	1.05	55	36,885	1.63	0.90
56	34,437	1.52	0.85	56	31,439	1.39	0.78

(continued)

Table 5.22 (continued)

2004				2008			
DoC	No. of cells	Percentage	Perc. × DoC	DoC	No. of cells	Percentage	Perc. × DoC
57	28,887	1.28	0.73	57	23,582	1.04	0.59
58	20,071	0.89	0.52	58	19,113	0.85	0.49
59	16,154	0.71	0.42	59	10,471	0.46	0.27
60	9126	0.40	0.24	60	6362	0.28	0.17
61	6494	0.29	0.18	61	4734	0.21	0.13
62	4021	0.18	0.11	62	2421	0.11	0.07
63	1679	0.07	0.05	63	1026	0.05	0.03
64	110	0.00	0.00	64	7	0.00	0.00
Sum	2,260,351	100	40.04	Sum	2,260,543	100	40.16

Table 5.23 Quantitative assessment of overall compactness for years 2012 and 2015

2012				2015			
DoC	No. of cells	Percentage	Perc. × DoC	DoC	No. of cells	Percentage	Perc. × DoC
15	695	0.03	0.00	16	1260	0.06	0.00
16	8011	0.35	0.06	17	2411	0.11	0.01
17	9842	0.44	0.07	18	2221	0.10	0.01
18	18,682	0.83	0.15	19	14,480	0.64	0.12
19	26,293	1.16	0.22	20	18,420	0.81	0.16
20	24,683	1.09	0.22	21	14,444	0.64	0.13
21	26,422	1.17	0.25	22	27,734	1.23	0.26
22	24,694	1.09	0.24	23	34,157	1.51	0.34
23	25,829	1.14	0.26	24	31,961	1.41	0.33
24	20,956	0.93	0.22	25	36,199	1.60	0.40
25	31,087	1.38	0.34	26	29,175	1.29	0.33
26	32,269	1.43	0.37	27	35,334	1.56	0.42
27	35,826	1.58	0.43	28	36,407	1.61	0.45
28	39,772	1.76	0.49	29	41,573	1.84	0.53
29	42,509	1.88	0.55	30	48,495	2.15	0.64
30	49,766	2.20	0.66	31	57,249	2.53	0.78
31	53,827	2.38	0.74	32	59,857	2.65	0.84
32	53,945	2.39	0.76	33	63,286	2.80	0.92
33	56,384	2.49	0.82	34	69,475	3.07	1.04
34	56,878	2.52	0.86	35	71,347	3.16	1.10
35	62,147	2.75	0.96	36	77,646	3.43	1.23
36	65,141	2.88	1.04	37	90,952	4.02	1.48
37	67,056	2.97	1.10	38	89,302	3.95	1.50
38	64,586	2.86	1.09	39	91,736	4.06	1.58
39	73,925	3.27	1.28	40	92,447	4.09	1.63
40	77,794	3.44	1.38	41	93,155	4.12	1.68
41	84,797	3.75	1.54	42	90,274	3.99	1.67
42	87,753	3.88	1.63	43	84,580	3.74	1.60
43	87,351	3.86	1.66	44	73,076	3.23	1.42

(continued)

Table 5.23 (continued)

2012				2015			
DoC	No. of cells	Percentage	Perc. × DoC	DoC	No. of cells	Percentage	Perc. × DoC
44	81,141	3.59	1.58	45	68,359	3.02	1.36
45	81,446	3.60	1.62	46	64,794	2.87	1.31
46	85,945	3.80	1.75	47	67,626	2.99	1.40
47	84,012	3.72	1.75	48	70,732	3.13	1.50
48	85,421	3.78	1.81	49	70,085	3.10	1.51
49	89,022	3.94	1.93	50	64,809	2.87	1.43
50	83,906	3.71	1.86	51	67,778	3.00	1.52
51	78,635	3.48	1.77	52	66,014	2.92	1.51
52	64,199	2.84	1.48	53	54,312	2.40	1.27
53	57,816	2.56	1.36	54	42,399	1.88	1.012
54	50,582	2.24	1.21	55	40,349	1.78	0.98
55	38,217	1.69	0.93	56	32,946	1.46	0.81
56	26,048	1.15	0.65	57	26,947	1.19	0.67
57	18,027	0.80	0.45	58	19,759	0.87	0.50
58	11,946	0.53	0.31	59	13,329	0.59	0.34
59	7527	0.33	0.20	60	7834	0.35	0.20
60	4618	0.20	0.12	61	3015	0.13	0.081
61	2312	0.10	0.06	62	739	0.03	0.02
62	737	0.03	0.02	63	233	0.01	0.00
63	235	0.01	0.01	Sum	2,260,712	100	40.27
Sum	2,260,712	100	40.23				

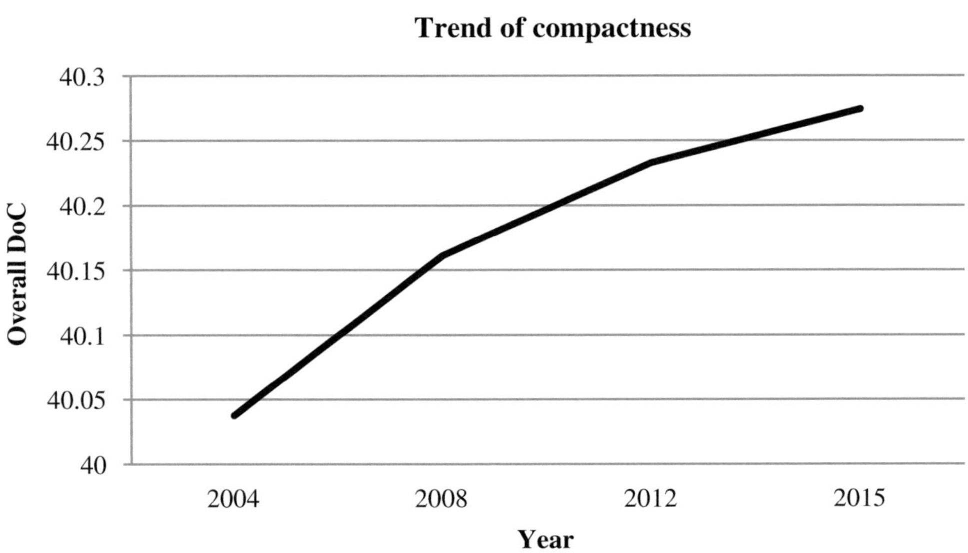

Fig. 5.30 Trend of DoC during selected time period

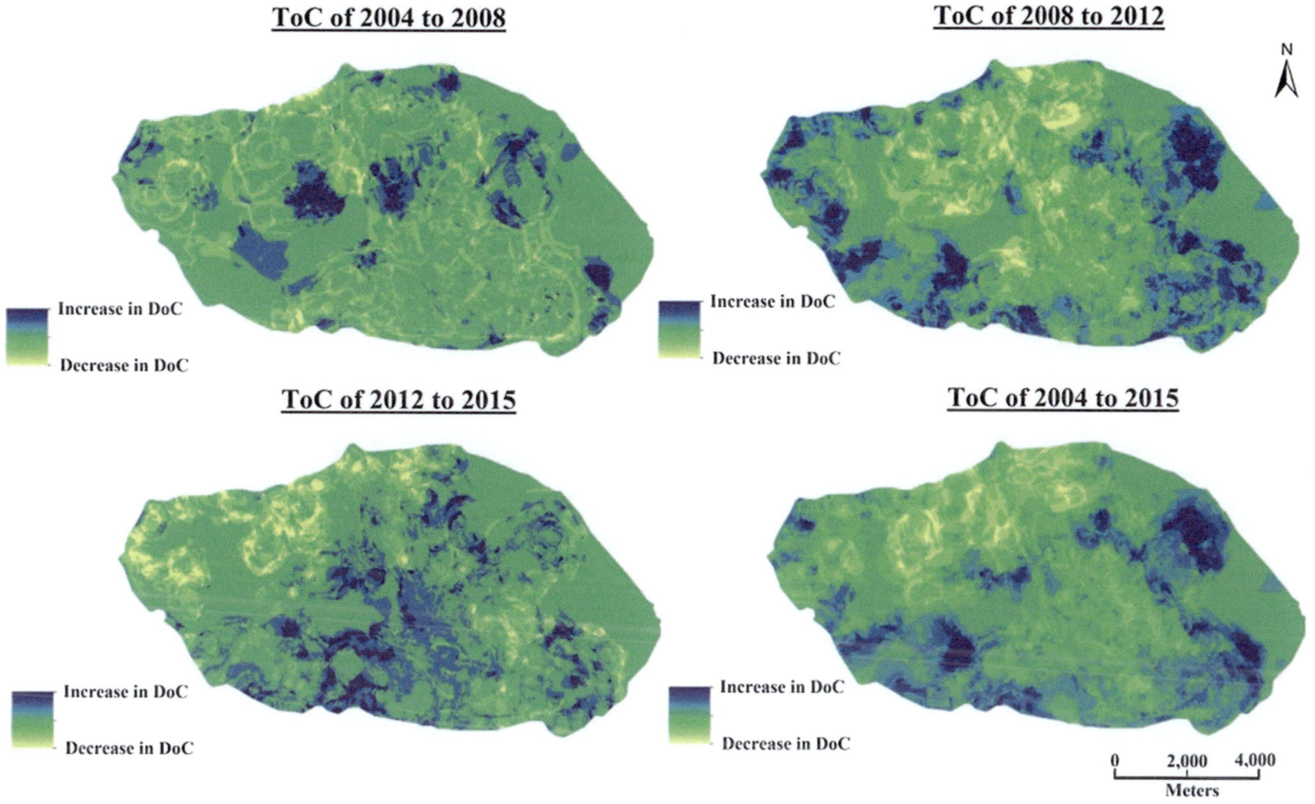

Fig. 5.31 Graphical presentation of ToC for four different time periods

5.6 Conclusion and Future Recommendations

For the analysis and modeling of a compact city, the evaluation of the existing condition of the study area based on city compactness is an essential task that should be conducted before any further analysis is performed. For this purpose, a comprehensive and standard compactness assessment was performed based on physical and functional perspectives. Physical compactness considering several urban metrics analysis and Shannon's entropy method deals with the physical composition and spatial configuration of the city. The analysis of urban indices highlighted the land use changes and assessed the compactness of the study area using several physical properties such as size, number of edges, number of patches (patch density), shape (irregularity, linearity), extension of patches, pattern (dispersion, interspersion, subdivision, and isolation), and diversity of various land use categories. The conversion of large agricultural and natural fields to small residential and commercial parcels affects the physical properties of surrounding neighborhoods and is consequently detected by urban metrics analyses. Shannon's entropy also assigned Kajang City with a low compactness level in all time periods and revealed the slight reduction at this level from 2004 to 2015. However, on the one hand, in the last period, a different development pattern was observed. On the other hand, functional compactness was performed by considering actual activities in the neighborhood as well as the land use planning and development pattern. This process was conducted based on three compactness indicators, namely, urban density, urban intensity, and land use diversity. These indicators evaluated the study area with 1 square meter cell size to extract highly accurate and precise information on the compactness condition. In general, the developed or saturated neighborhoods (zones) with high compactness level have low potential for growth and change. Thus, in the case of land consumption and eventually green environment preservation objective, focusing on the areas with high potential for new development and growth such as the eastern zones of this case study is better. In addition, unlike internal zones that are mainly dependent on the city characteristics, the neighborhood zones along the city borders are normally affected by the other side of the boundaries, which is either developed or natural environments. Thus, considering external factors from outside the borders and analyzing the study area in large perspectives are important. Furthermore, regarding zoning analysis, non-zonal basis provides more detailed and accurate results during the selected period than zonal basis.

Finally, the ToC of the study area was extracted, revealing a gradual growth of compactness from 2004 to 2015. ToC provided a clear perspective regarding city compactness and eventually sustainable urban development. In addition, ToC provided straightforward and valuable information for local planning authorities and decision makers to have better judgment about the existing situation and future trend of Kajang City growth based on urban sustainability.

In terms of limitations and future recommendation, this study selected Kajang City as a small region with a high potential of land use growth and changes because of its proximity to three metropolitan cities of Malaysia. Although we attempted to involve all the effective factors in the methodological process, neighborhood zones along the city boundaries are affected by external factors, which are either developed or natural environment. Thus, considering the external forces from outside the city boundary is another important issue that can improve the assessment and modeling of compact cities. This task can be performed by considering a buffer around the study area boundaries to consider their characteristics and their effects on the city border regions.

In addition, although network analysis was used to assess and extract strategic roads of the city, detailed information on traffic loads and vehicle miles traveled can increase the comprehensiveness of compactness assessment and provide informative vision regarding the negative effects of extreme compactness of the current and future situations.

Regarding natural and green environment preservation, this study mainly focused on the rural environment and the conservation of agricultural and forest areas. Thus, for future studies, we suggest concentrating on urban parks and green spaces within the built-up areas. In fact, identifying the basis for the creation of urban green corridors network with proper buffer zones (100, 200, and/or 500 m) is important in reducing car dependency for leisure accessibility. The implementation of this process positively affects the proposed compact land use modeling of this study in the allocation of areas for green spaces and residential purposes.

Finally, measuring urban compactness reveals the current situation of urban forms regarding density, diversity, and intensity. However, in addition to this knowledge, estimating the urban capacities is essential in serving the community or local neighborhood for high compactness. In fact, no standard threshold exists for various aspects of urban compactness. For instance, various departments and agencies proposed different population density values suitable for the urban environment. Thus, we suggest linking urban compactness and urban capacity assessment in various aspects to identify a baseline and threshold for a compact city to avoid the negative effects of overly compact neighborhoods.

References

Abdullahi, S., Pradhan, B., & Jebur, M. N. (2015a). GIS-based sustainable city compactness assessment using integration of MCDM, Bayes theorem and RADAR technology. *Geocarto International, 30*(4), 365–387.

Abdullahi, S., Pradhan, B., Mansor, S., & Shariff, A. R. M. (2015b). GIS-based modeling for the spatial measurement and evaluation of mixed land use development for a compact city. *GIScience and Remote Sensing, 52*(1), 18–39.

Adolphson, M. (2010). Kernel densities and mixed functionality in a multicentred urban region. *Environment and Planning. B, Planning and Design, 37*(3), 550.

Arifwidodo, S. D. (2012). Exploring the effect of compact development policy to urban quality of life in Bandung, Indonesia. *City, Culture and Society, 3*(4), 303–311.

Bhat, C. R., & Gossen, R. (2004). A mixed multinomial logit model analysis of weekend recreational episode type choice. *Transportation Research Part B: Methodological, 38*(9), 767–787.

Burton, E. (2002). Measuring urban compactness in UK towns and cities. *Environment and Planning B, 29*(2), 219–250.

Burton, E., Jenks, M., & Williams, K. (2003). *The compact city: A sustainable urban form?* Routledge.

Carruthers, J. I. (2002). The impacts of state growth management programmes: A comparative analysis. *Urban Studies, 39*(11), 1959–1982.

Carruthers, J. I., & Ulfarsson, G. F. (2008). Does smart growth matter to public finance? *Urban Studies, 45*(9), 1791–1823.

De Chiara, J. (1990). *Time-saver standards for building types*. New York: McGraw-Hill.

Hoppenbrouwer, E., & Louw, E. (2005). Mixed-use development: Theory and practice in Amsterdam's Eastern Docklands. *European Planning Studies, 13*(7), 967–983.

Huang, J., Lu, X., & Sellers, J. M. (2007). A global comparative analysis of urban form: Applying spatial metrics and remote sensing. *Landscape and Urban Planning, 82*(4), 184–197.

Li, X., Yang, Q., & Liu, X. (2008). Discovering and evaluating urban signatures for simulating compact development using cellular automata. *Landscape and Urban Planning, 86*(2), 177–186.

Li, X., & Yeh, A. G.-O. (2004). Analyzing spatial restructuring of land use patterns in a fast growing region using remote sensing and GIS. *Landscape and Urban Planning, 69*(4), 335–354.

Lin, J., & Yang, A. (2006). Does the compact-city paradigm foster sustainability? An empirical study in Taiwan. *Environment and Planning B: Planning and Design, 33*(3), 365.

Manaugh, K., & Kreider, T. (2013). What is mixed use? Presenting an interaction method for measuring land use mix. *Journal of Transport and Land Use, 6*(1), 63–72.

McGarigal, K., & Marks, B. J. (1995). Spatial pattern analysis program for quantifying landscape structure. *Gen. Tech. Rep. PNW-GTR-351.* US Department of Agriculture, Forest Service, Pacific Northwest Research Station.

Mubareka, S., Koomen, E., Estreguil, C., & Lavalle, C. (2011). Development of a composite index of urban compactness for land use modelling applications. *Landscape and Urban Planning, 103*(3), 303–317.

Musakwa, W., & Van Niekerk, A. (2013). Implications of land use change for the sustainability of urban areas: A case study of Stellenbosch, South Africa. *Cities, 32,* 143–156.

Ramachandra, T., Bharath, H., & Sowmyashree, M. (2013). Analysis of spatial patterns of urbanisation using geoinformatics and spatial metrics. *Theoretical and Empirical Researches in Urban Management, 8*(4), 5–24.

Regan, S. (2000). Towards an urban renaissance: The final report of the urban task force. *The Political Quarterly, 71*(1), 115–118.

Saaty, T. L. (1980). *The analytic hierarchy process: Planning, priority setting, resources allocation.*

Scoffham, E., & Vale, B. (1996). How compact is sustainable—How sustainable is compact. *The compact city: A sustainable urban form* (pp. 66–73).

Shannon, C. E. (2001). A mathematical theory of communication. *ACM SIGMOBILE Mobile Computing and Communications Review, 5*(1), 3–55.

Song, Y., & Knaap, G.-J. (2004). Measuring the effects of mixed land uses on housing values. *Regional Science and Urban Economics, 34*(6), 663–680.

Song, Y., Merlin, L., & Rodriguez, D. (2013). Comparing measures of urban land use mix. *Computers, Environment and Urban Systems, 42,* 1–13.

Thinh, N. X., Arlt, G., Heber, B., Hennersdorf, J., & Lehmann, I. (2002). Evaluation of urban land-use structures with a view to sustainable development. *Environmental Impact Assessment Review, 22*(5), 475–492.

Turskis, Z., Zavadskas, E. K., & Zagorskas, J. (2006). Sustainable city compactness evaluation on the basis of GIS and Bayes rule. *International Journal of Strategic Property Management, 10*(3), 185–207.

Van Eck, J. R., & Koomen, E. (2008). Characterising urban concentration and land-use diversity in simulations of future land use. *The Annals of Regional Science, 42*(1), 123–140.

Zagorskas, J., Burinskienė, M., Zavadskas, E., & Turskis, Z. (2007). Urbanistic assessment of city compactness on the basis of GIS applying the COPRAS method. *Ekologija, 53*(2), 55–63.

6 Assessing the Relationship Between City Compactness and Residential Land Use Growth

Saleh Abdullahi, Biswajeet Pradhan and Hossein Mojaddadi

6.1 Introduction

Land use change arises from the complex interaction of various environmental, physical, political, cultural, and other factors. Land use change models are generally based on the following four core principles (Lantman et al. 2011; Abdullahi and Pradhan 2015): historical basis (Kuijpers et al. 2007), suitability basis (Abdullahi et al. 2014), neighborhood basis (Li et al. 2008), and actor interaction basis (Matthews et al. 2007). Several models for land use change application are based on these core principles, such as Markov chain, economic-based systems, agent-based systems, statistical analysis, cellular automata (CA), and artificial neural network (ANN). The growth and changes of various urban land use categories affect the surrounding natural environment or social and economic issues. Thus, evaluating these consequences has become a topic of interest in this field, particularly the effects of rapid urbanization and built-up areas on the natural environment (Burnside et al. 2003; Li and Yeh 2004; Dadhich and Hanaoka 2011). These urban morphologies are necessary to achieve a sustainable urban environment. Urban characteristics, such as density, shape, and size, significantly affect energy consumption and cause various environmental issues (Alberti et al. 2007; Petsch et al. 2011; Beatley 2012). Therefore, a compact urban form is one of the most sustainable urban patterns (Livingstone and Authority 2003) because of its high urban density, central area revitalization, land use diversity, natural environment preservation, and promotion of public transportation facilities (Burton 2000; Burton et al. 2003; Chang and Sheppard 2013; Abdullahi et al. 2015b).

Although compact urban form is recognized as an attempt to achieve urban sustainability, few researchers have studied the reciprocal relationships of compact city indicators of the surrounding environment. An example is the relationship between city compactness and land use diversity or urban density, or the growth and loss of specific land use categories. Particularly, residential land use growth needs to be evaluated with respect to various aspects of urban environment in urban applications (Middel et al. 2011; Xu 2011). Thus, this chapter attempts to assess the correlation between residential land use growth and city compactness, that is, the way by which residential area growth increases or decreases the degree of city compactness. In the second phase, the performance and accuracy of two modeling approaches, machine learning land transformation model (LTM) and statistical-based weight-of-evidence (WoE), is examined in predicting the growth of residential land use with respect to compact urban pattern.

6.2 Data and Methodological Process

Rajang City in Malaysia is a tropical region that consists of an urban developed area and large portions of forest and agricultural fields (Fig. 6.1). Therefore, the effects of growth and changes in various land use types can be adequately observed. The proximity of this city to the major cities of Malaysia has resulted in recent urban sprawl developments, especially a significant growth of residential areas (Fig. 6.2). Thus, this study is conducted to provide a clearer view of these growths with respect to the city compactness of Rajang City.

The flowchart of the methodological process is shown in Fig. 6.3. The land use maps for 2008 and 2012 were the main data used to evaluate the residential growth. Several other data layers, such as road network, public transportation, and population, were utilized in addition to land use maps for the city compactness assessment. All these data were collected from the local planning authority of Rajang City.

B. Pradhan (ed.), *Spatial Modeling and Assessment of Urban Form*,
DOI 10.1007/978-3-319-54217-1_6

Fig. 6.1 Rajang city and Peninsular Malaysia

6.2.1 Residential Land Use Growth Versus City Compactness

Cross-tabulation technique was applied on the land use maps of both years to extract and analyze the growth of residential areas from 2008 to 2012. Cross-tabulation is a mathematical matrix that provides unbiased information concerning the entire area of interest to derive unbiased summary statistics (Pontius Jr and Millones 2011). This process revealed that residential land use significantly exhibited the most evident growth among the urban land uses. Thus, this study focused on evaluating the effect on and relationship of these growths with the compactness pattern of Rajang City.

The city compactness of Rajang City was evaluated using both available land use maps. Several compactness assessment studies, such as those by Thinh et al. (2002), Li and Yeh (2004), Mubareka et al. (2011), employ different approaches. However, some drawbacks of these studies are that they neglect the concentration on built-up areas and assessment of compactness on physical and cellular bases, especially that of large cell size (for instance 500 m × 500 m), and ignore the functional characteristics and real behaviors and activities of local residents. The land use modeling process at these resolutions and pixel sizes is incapable of identifying subtle information regarding urban patterns (Houet et al. 2010). Thus, the current study assessed the urban compactness of the study area based on three main indicators of the compact city: urban density, intensity, and land use diversity (Burton 2002).

Urban density was evaluated based on population, residential, built-up, and road densities, as discussed in detail by Abdullahi et al. (2015a, b). Land use diversity, as another main indicator of urban compactness (Gainza and Livert 2013; Gu et al. 2013; Song et al. 2013), has been evaluated using various approaches, such as those in the studies conducted by Burton (2002), Song and Knaap (2004), Van Eck and Koomen (2008), Manaugh and Kreider (2013), and Musakwa and Van Niekerk (2013). However, as explained by Abdullahi et al. (2015a, b), land use diversity is evaluated in the current study based on the proximity of main urban land use categories, such as residential, commercial, industrial, recreational, and community facilities. In this manner, not only the richness of mixed land use development can be assessed, but also the distribution pattern of various land use categories. This process highlights neighborhoods with high and low land use diversities. The urban intensity of Rajang City is evaluated based on distribution pattern and the quality and quantity of community facilities, such as educational, medical, and recreational, with respect to the characteristics and demands of local residents (Abdullahi et al. 2015a).

Thus, the compactness pattern of both land use maps (2008 and 2012) with fine spatial resolution (5 m × 5 m cell size) was evaluated based on these three compactness indicators. Next, the trends of growth and loss of compactness during this 4 year period were extracted. The same process of compactness assessment was applied to the 2015 land use map for and the master plan of Rajang City.

For the next phase, a similarity assessment analysis was performed to evaluate the relationship between the residential land use growth and city compactness pattern. This process was conducted using the relative operating characteristic-based area under the curve (AUC) method (Pontius and Schneider 2001; Pradhan 2011; Kolb et al.

Fig. 6.2 The growth of residential area in Rajang City. *Photo* taken in September 2014

2013; Chen et al. 2014a, b). This method is capable of comparing a map of actual change to the maps of modeled suitability for land cover change. Thus, the trend of the compactness pattern of the study area was utilized in this study as the suitability map. In this map, negative values indicate a decrease in the degree of compactness, whereas positive values denote an increase. Meanwhile, a residential growth map was used as the actual change map to assess its similarity and fitness to the degree of compactness map for each cell. This assessment ranges from zero to one, where one indicates a perfect match and fit.

6.2.2 Land Use Change Modeling Using Land Transformation Model (LTM)

The LTM integrates multi-layer perceptron ANN and geographic information system using socioeconomic and biological factors (Zurada 1992; Pijanowski et al. 2002, 2014). Multi-layer perceptron applies a supervised learning algorithm that computes a function between input–output pairs without previous knowledge of functional forms. LTM has been used numerous times to simulate land use change and urban growth in the literature (Pijanowski et al. 2005, 2014;

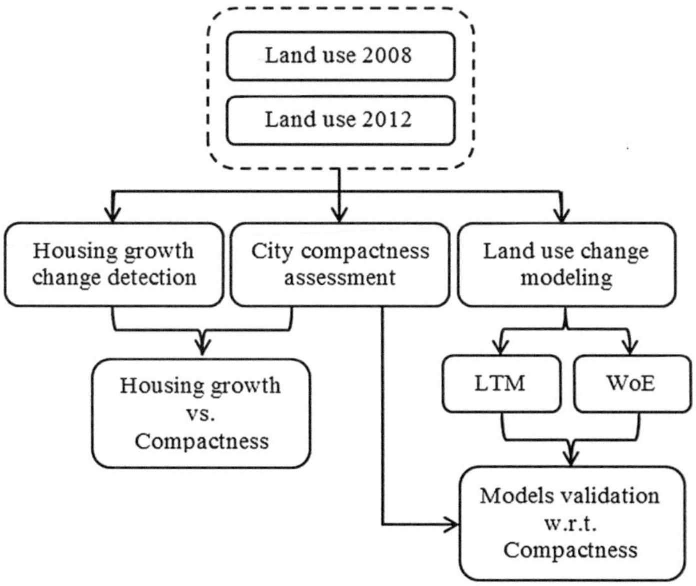

Fig. 6.3 The overall methodological flowchart

Tayyebi and Pijanowski 2014). This model uses knowledge from at least two historical land use maps to train the network. The current study applied this model to assign the location of residential land use types with the help of land use maps for 2008 and 2012. In fact, LTM investigates the previous trend of land use changes and creates functional relationships to project future land use change and growth. Both available maps were classified into two classes: one and zero as residential and nonresidential areas, respectively. Several urban-related parameters were used as input drivers. These parameters were applied to represent land use types, such as residential and industrial, and urban features, such as roads and water bodies. All these parameters were created in raster format as either binary or continuous variables. Euclidean distance analysis was implemented to distance-based variables. In this manner, their spatial extents were subsequently covered, and every cell was assigned a distance class. These parameters were used to define a set of transition rules to quantify the spatial effects of predictor cells on land use transitions (Pijanowski et al. 2002).

ANN model trains and tests all input data to create a network with suitable predictive capacity. The training process is implemented to adjust the weights for each node according to the learning algorithm, while testing is conducted to calculate the error rates (Pijanowski et al. 2002). Therefore, three main phases were included in the process: (i) designing the network and input from previous data, (ii) testing the neural network using the full dataset, and (iii) using the output information of the neural network to forecast residential growth. Stuttgart's Neural Network Simulator (SNNS) was used for the design, training, and prediction of the ANN (Zell et al. 1994). The neural network was designed to contain several numbers of inputs depending on the selected variables and an equal number of hidden layers and a single output layer as the final prediction. All input layers were normalized and converted to ASCII format.

The model was tested after producing the real change map that indicates the change in land uses from 2008 to 2012. This process was conducted by comparing the cells changed to residential during the given period with cells that have the highest probability to change based on the model. Then, the percent correct metric (PCM) was computed to assess the reliability of the projected maps. This process was implemented by considering the proportion of true positive areas and number of pixels that changed to other categories during the selected period. The model computes the PCM for every cycle, and the best results are selected to project the residential growth. The model normally predicts the same proportion of the area that actually changed according to historical trends. Nevertheless, other variables can be integrated into the model at this stage to produce different scenarios. This study considers 1.5 and 2 times the proportion of change (from 2008 to 2012) as two other scenarios of growth in addition to the same proportion of growth to determine the appearance of the region by using different transition rules (Table 6.1). This task increases the comprehensiveness of the process because the model is allowed to project the proportion of residential land use cells according to the change probability of the cells.

6.2.3 Land Use Change Modeling Using Weights-of-Evidence (WoE)

In addition to LTM as a machine learning approach, statistical-based method is also applied to the residential land use growth projection to evaluate both methods in terms of

Table 6.1 Different scenarios of both land use change modeling

Model	No.	Scenarios details	Remarks
LTM	1	Projected map with same amount of previous residential growth	Modeling based on historical trend
	2	Projected map with 1.5 times of previous residential growth	Decrease the limitation to evaluate the performance of the model
	3	Projected map with 2 times of previous residential growth	
WoE	1	Probability map with all available evidences	Modeling based on historical trend
	2	Probability map with same factors used for LTM model	To compare both models

accuracy and performance. Regression approach possesses higher explanatory power and outperforms ANN, particularly when functional correlations among dependent and independent parameters are known (Tayyebi et al. 2014). WoE is a data-driven approach based on Bayes rule in a log-linear form using prior and posterior probabilities. This approach is suitable with enough information to estimate the relative priority of evidential themes by statistical means (Bohman-Carter 1994). WoE assesses the degree to which evidence supports the hypothesis and the degree to which the evidence does not refute the hypothesis (Dempster 1967; Shafer 1976). This approach conducts an assessment of combined pieces of evidence from a variety of dependent variables. This method can estimate uncertainty and involve evidence from various data sources (Thiam 2005). WoE has been used in several studies, such as mineral and geological potential assessment and mapping (Chen et al. 2014a, b), cliff instability mapping (Zahiri et al. 2006), and natural disaster management (Tehrany et al. 2014; Youssef et al. 2015). This method has also been used in a few urban application studies, such as that of Almeida et al. (2003), Abdullahi et al. (2015a, b), and Abdullahi and Pradhan (2015). This approach is simple, time effective, and can be easily integrated with GIS (Dahal et al. 2008). Detailed descriptions of the theoretical concepts and mathematical formulation of this method were provided in the studies of Bonham-Carter (1994), Regmi et al. (2010), and Pradhan et al. (2010). The model computes the weight of each independent variable based on the occurrence and nonoccurrence of the dependent variable within the study area. In the current study, residential land use growth is the dependent parameter and the other selected effective factors are the independent parameters (Table 6.2).

The WoE based on the selected evidence can produce the probability map for the growth of residential areas. The probability value is computed based on the prior probability of occurrence and nonoccurrence of residential pixels in each class of evidence. The transitional probability was estimated based on the magnitude of transition in each class. The classification of the evidence was defined according to type. Proximity analysis was performed on the distance-based parameters. These distances were then divided into classes, which include their spatial extents. In the case of nominal parameters, such as soil and geological types, each type was considered as one class. The entire base layer of all pieces of evidence was converted into a grid cell to assess the growth of residential land use in their classes.

The spatial association of each residential pixel and each class of evidence was computed by subtracting the natural logarithm of occurrence and natural logarithm of nonoccurrence. A positive value indicates a high number of residential pixels, while a negative value indicates a low number of residential pixels growing in this class. Finally, the standardized value, which represents the significance of the spatial association and measures the relative certainty of the posterior probability, was estimated based on the variance and standard deviation of the contrast (Bohman-Carter 1994).

The WoE method can assess and optimize the factor data sets and select the most effective factors among all available data sets. This process is important, especially for this factor-based approach where the priority of independent variables with respect to dependent variable is important. This assessment was implemented by observing the relationships between the dependent variables with respect to each independent variable. Thus, the existence and changes (increase or decrease) of residential cells in relation to each class of parameters were examined in this case.

The WoE model was designed to produce two scenarios to evaluate its performance and conduct a proper comparison with the LTM output maps. In the first scenario, all available pieces of evidence were utilized to produce the residential growth probability map (Table 6.1). The second scenario was conducted with factors similar to those utilized in the LTM model. This scenario was used to make a one-to-one comparison between the WoE and LTM model output maps.

6.2.4 Land Use Change Model Validation with Respect to City Compactness

The performances of both LTM and WoE were evaluated with respect to the actual land use map for 2015 and city compactness. Similarity assessment was applied between the projected and actual land use maps for this process. The AUC approach was used to conduct these assessments

Table 6.2 WoE and LTM input parameters

No.	WoE Parameters	LTM Parameters
1	Residential growth from 2008 to 2012	Residential growth from 2008 to 2012
2	Proximity to residential	Proximity to residential
3	Proximity to commercial	Proximity to commercial
4	Proximity to industrial	Proximity to industrial
5	Proximity to main roads	Proximity to main roads
6	Proximity to public transportation	Proximity to public transportation
7	Proximity to community facilities	Proximity to community facilities
8	Proximity to recreational facilities	Proximity to recreational facilities
9	Proximity to infrastructure	Proximity to infrastructure
10	Proximity to agricultural fields	Proximity to agricultural fields
11	Proximity to river and water bodies	Proximity to river and water bodies
12	Proximity to restricted areas	Proximity to restricted areas
13	Proximity to flood zones	
14	Geological properties	
15	Soil properties	

(Pontius and Schneider 2001; Van Eck and Koomen 2008; Chen et al. 2014a, b). The results of both models for all selected scenarios were validated with the actual land use map for 2015. First, the relationship of the residential growth in all scenarios was investigated in terms of the city compactness of the reference maps to evaluate the compatibility of the models with compact development modeling. In the second phase, the similarity of the residential growth in all scenarios was assessed using the actual residential land use of the reference maps to evaluate the land use change modeling performance of both models.

6.3 Results and Discussion

Extracting the historical trend of residential land use changes from 2008 to 2012 was essential to evaluate the relationship between the growth of residential areas and city compactness (Fig. 6.4). The change detection based on cross-tabulation analysis indicated the seemingly encompassing effect of residential land use on almost all types of activities, but mostly through the open spaces and agricultural fields of about 102 ha and 345 ha, respectively. The growth of residential areas through the open spaces or redevelopment of

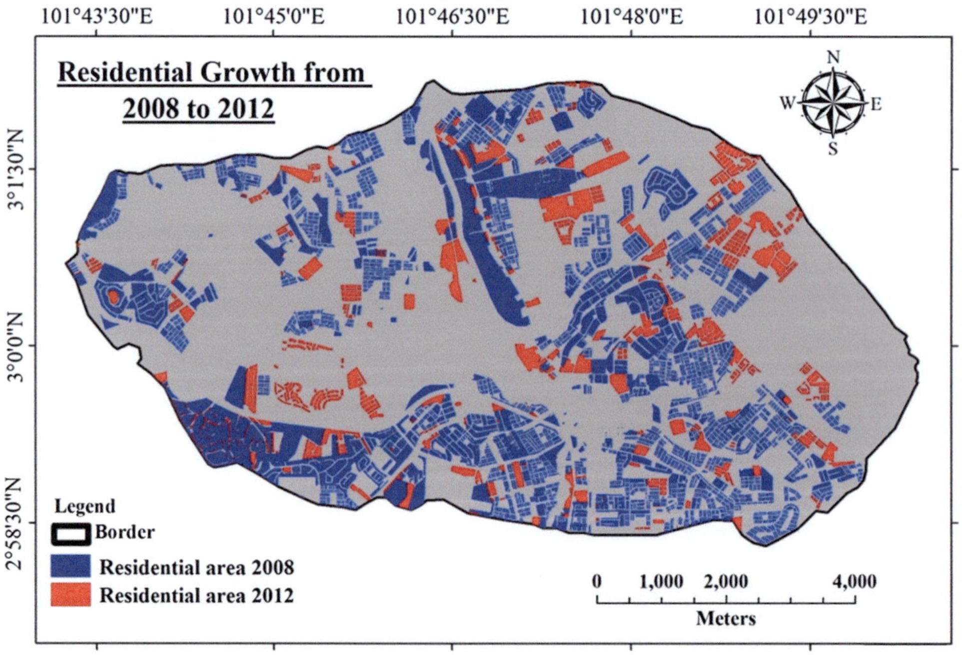

Fig. 6.4 Residential growth of Rajang City from 2008 to 2012

Table 6.3 Cross-tabulation between 2008 and 2012 land use maps

Land use	Res.	Infra.	Fac.	Open	Agri.	Indus.	Comm.	Growth (A)	Total (A–B)
Res.	**1277.6**	**2.2**	**13.5**	**102.8**	**344.5**	**20.6**	**20.2**	*503.9*	**366.8**
Infra.	**18.6**	115.2	1.2	18.0	7.2	3.6	1.2	*49.9*	**39.3**
Faci.	**19.3**	0.6	423.6	10.6	4.5	0.9	2.0	*38.0*	**8.0**
Open	**51.4**	3.4	8.6	251.2	20.1	5.8	6.0	*95.3*	**-213.8**
Agri.	**1.3**	0.2	0.4	95.3	489.9	8.7	0.4	*106.3*	**-347.5**
Indus.	**15.6**	0.9	0.0	38.0	59.5	491.7	6.0	*120.0*	**75.1**
Comm.	**30.8**	3.2	6.2	44.4	18.0	5.3	74.7	*107.9*	**72.1**
Loss (B)	*137.0*	*10.6*	*29.9*	*309.0*	*453.9*	*44.9*	*35.8*		

Res. Residential; *Infra.* Infrastructure; *Fac.* Facility; *Open* Open spaces; *Agri.* Agriculture; *Indus.* Industrial; *Comm.* Commercial
Bold letters indicate important/significant values

these areas to any other land use type is acceptable because of the brownfield redevelopment objectives. Nevertheless, the destruction of large amounts of agricultural fields and green environments because of residential growth is not acceptable and should be avoided as much as possible. Table 6.3 shows that commercial and industrial land uses also had significant growths through other land use and land covers.

A large proportion of agricultural fields, several brownfield sites, and a few commercial land use parcels along the main road in the north of the city were converted to residential land use during the 4 year period. This phenomenon indicates that accessibility and proximity to central business districts are important factors that control residential land use growth. In the next phase, this result was compared with the compactness trend in Rajang City during the same period.

6.3.1 City Compactness Assessment Versus Residential Growth

The compactness assessment process was performed based on urban density, intensity, and land use diversity for both available land use maps. Therefore, based on these three variables, single land use areas, such as agricultural fields and industrial areas, were indicated as less compact, whereas areas covered by a combination of several land use categories, such as residential, commercial, and facilities, were illustrated as more compact environments. In fact, such single land use neighborhoods usually have low population density with fewer community facilities. These are the main reasons why these areas were identified as less compact zones. However, mixed land use neighborhoods attract population, commercial services, and community facilities, which increase the degree of compactness of these areas. In Rajang City, the eastern zones were assigned as less compact and central zones because of agricultural fields and industrial buildings, while the southern regions were assigned as high compact areas because of the existence of a main public transportation station (Fig. 6.5). By investigating the relation of road network and compactness maps, we observed that the proximity to main roads increase the degree of compactness, which shows the importance of road network in city compactness.

Although insignificant changes were observed during the four-year period, these changes affect the degree of compactness of the city significantly. Figure 6.5 (c) depicts the changes in the degree of compactness from 2008 to 2012. This map is produced by subtracting the compactness map of 2008 from that of 2012. Thus, the dark brown color indicates areas with growth while yellow indicates areas with reduction in the degree of compactness. The western regions of Rajang City had compactness growth and the central and eastern regions lost compactness during the selected period. This result is caused by the proximity of the western parts of the Rajang City to the other city centers of the province and the existence of forest areas in the east of this city. Thus, insignificant development and changes can be expected in the eastern parts.

Next, the relationship between the trend of compactness maps and growth of residential areas was investigated. Although a random relationship between these two phenomena can be observed from the general statistical perspective (Fig. 6.6a), the curve of residential land use growth leaned toward positive degrees of compactness (Fig. 6.6b). This trend indicates that significant amounts of residential growth areas are located in neighborhoods where their compactness increased during the selected period. Thus, residential land use growth is effective in the growth of compactness of the local neighborhoods. In the next level, residential growth was compared with each of the three indicators of city compactness. The high residential land use growth substantially increased urban density, slightly increased land use diversity, but has no effect on urban intensity.

6.3.2 Land Use Change Modeling Evaluation

To address the next objective, an assessment of the performance of the two land use modeling approaches for residential growth was attempted with respect to city

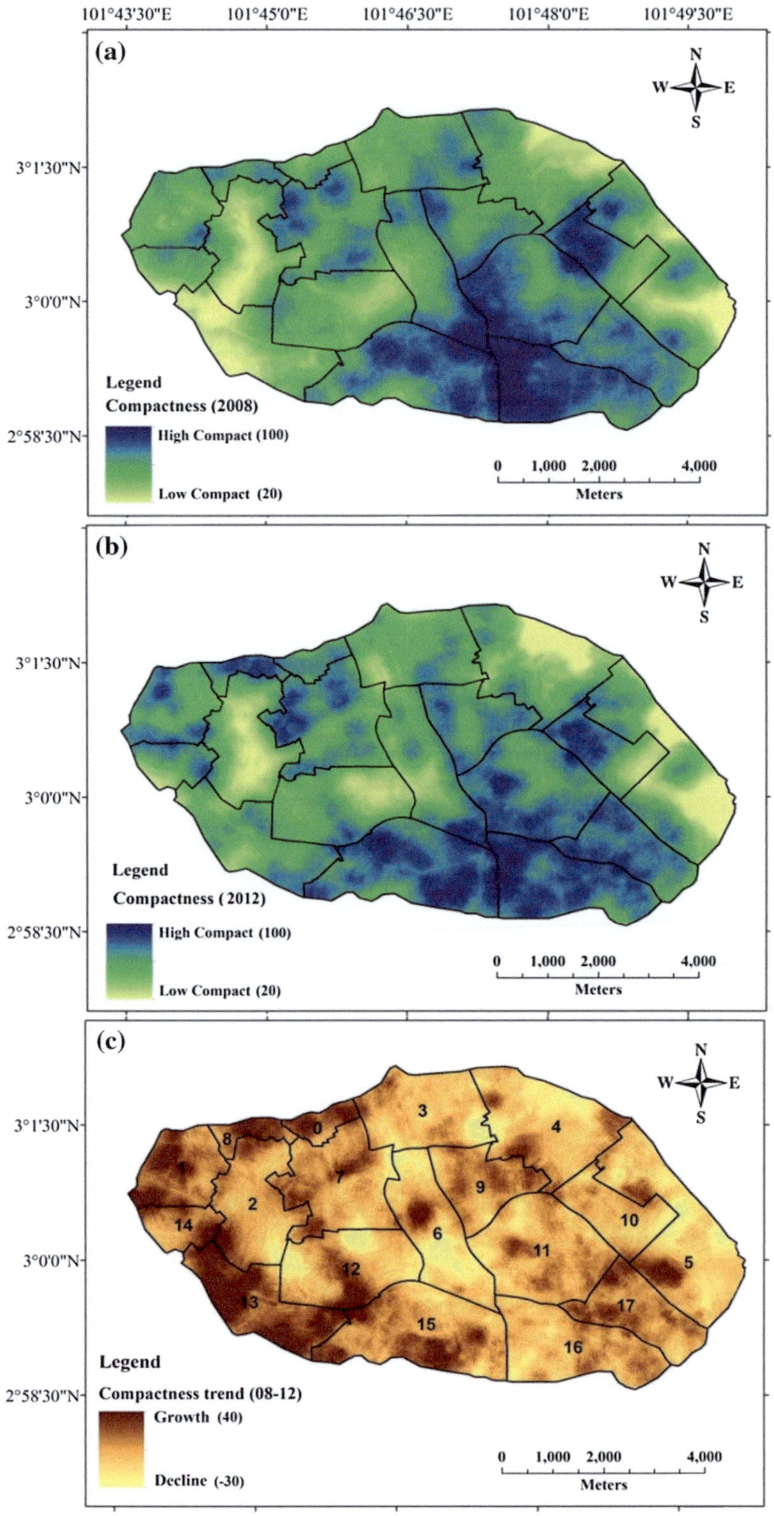

Fig. 6.5 City compactness assessment of; **a** land use map of 2008, **b** land use map of 2012; **c** the trend of the city compactness during the 4 year period

compactness. Figure 6.7 depicts the results of the three scenarios proposed by the LTM land use modeling.

Based on the assumption of less limitation for growth (in case of second and third scenarios), it can be seen that more agricultural areas are converted to residential in eastern regions compared to first scenario. Several large brownfield sites near the industrial areas and many small brownfield sites in the south were projected to change to residential areas, especially in the third scenario. More unsuitable areas are projected to change to residential use as the limitation of the model decreases. Similarly, the first scenario with more than 90% PCM value has higher accuracy than the second

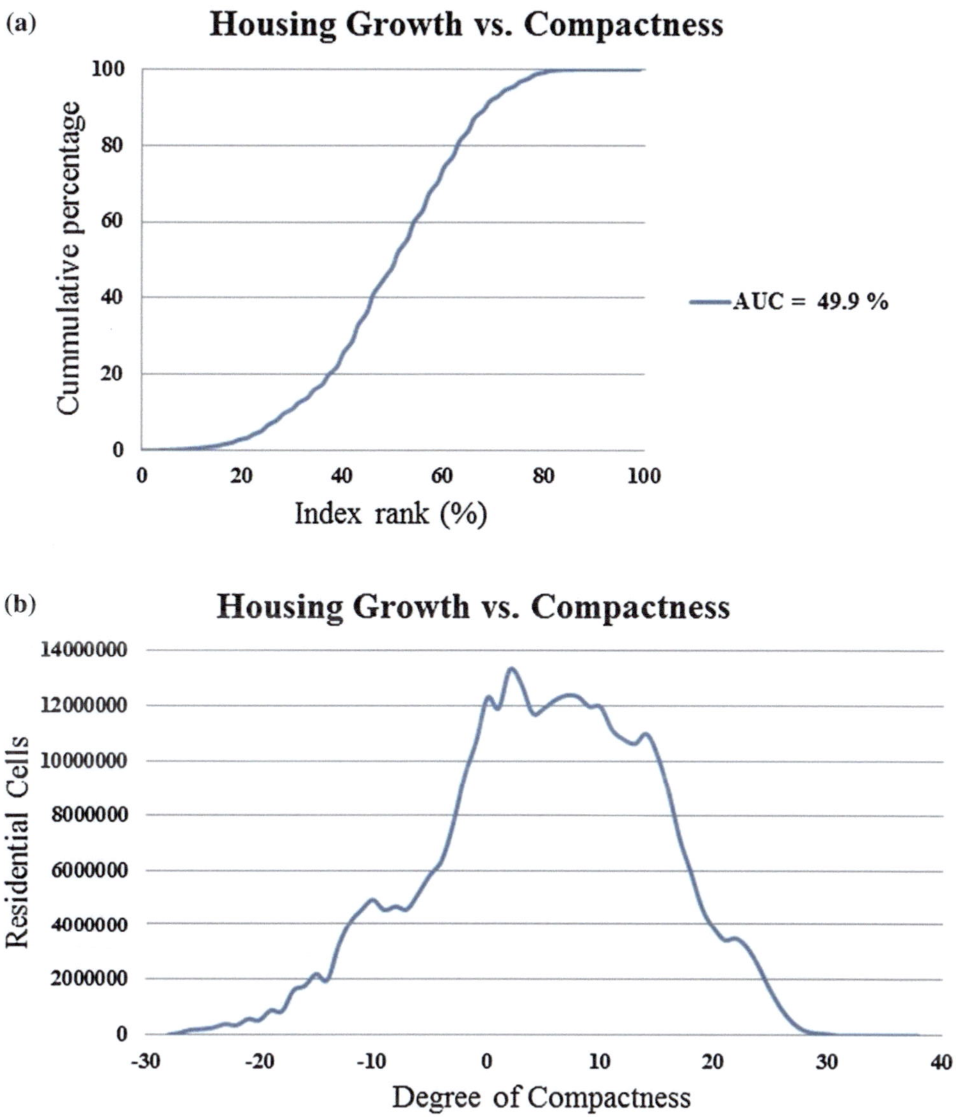

Fig. 6.6 **a** AUC graph of residential land use growth versus compactness trend of Rajang City. **b** Comparison graph of residential land use growth and compactness trend of Rajang City

and third scenarios according to the calculated PCM. The spatial comparison between these scenarios and the actual land use map for 2015 indicated that the similarity of residential maps with respect to compactness decreases slightly as the proportion of residential growth increases. Thus, more residential land uses grow in areas with less compactness, such as agricultural fields or industrial zones (Fig. 6.8a). By contrast, the similarity of residential growths with respect to the residential map for 2015 increases as the proportion of residential growth increases (Fig. 6.8b). Thus, the second and third scenarios have more similarity with the real changes despite the lower PCM values. Therefore, the third scenario generally performed better based on the residential growth perspective. However, more unsuitable areas (e.g., natural environment) are converted to residential areas, thereby causing unsustainable development, as the growth of the residential land use increases significantly.

Next, WoE was used as the statistical method based on the probabilistic concept to create probability of residential growth maps. The calculated contrast, W+ and W− values, was computed as weighting for each selected evidence (parameter) and overlaid to create a probability map for each scenario for this analysis. A few of these pieces evidence had high positive influences on residential growth, including proximity to main roads, public transportation, and recreational facilities (Table 6.4). By contrast, areas near agricultural lands and industrial buildings had an inverse effect on residential growth. Meanwhile, the middle class of proximity of some pieces of evidence, such as proximity to commercial areas, has the highest growth probability. This means that residential use is normally located neither very near nor extremely far from these land use/covers. Figure 6.9 illustrates the probability maps for both explained scenarios of the WoE method. In general, by the visual

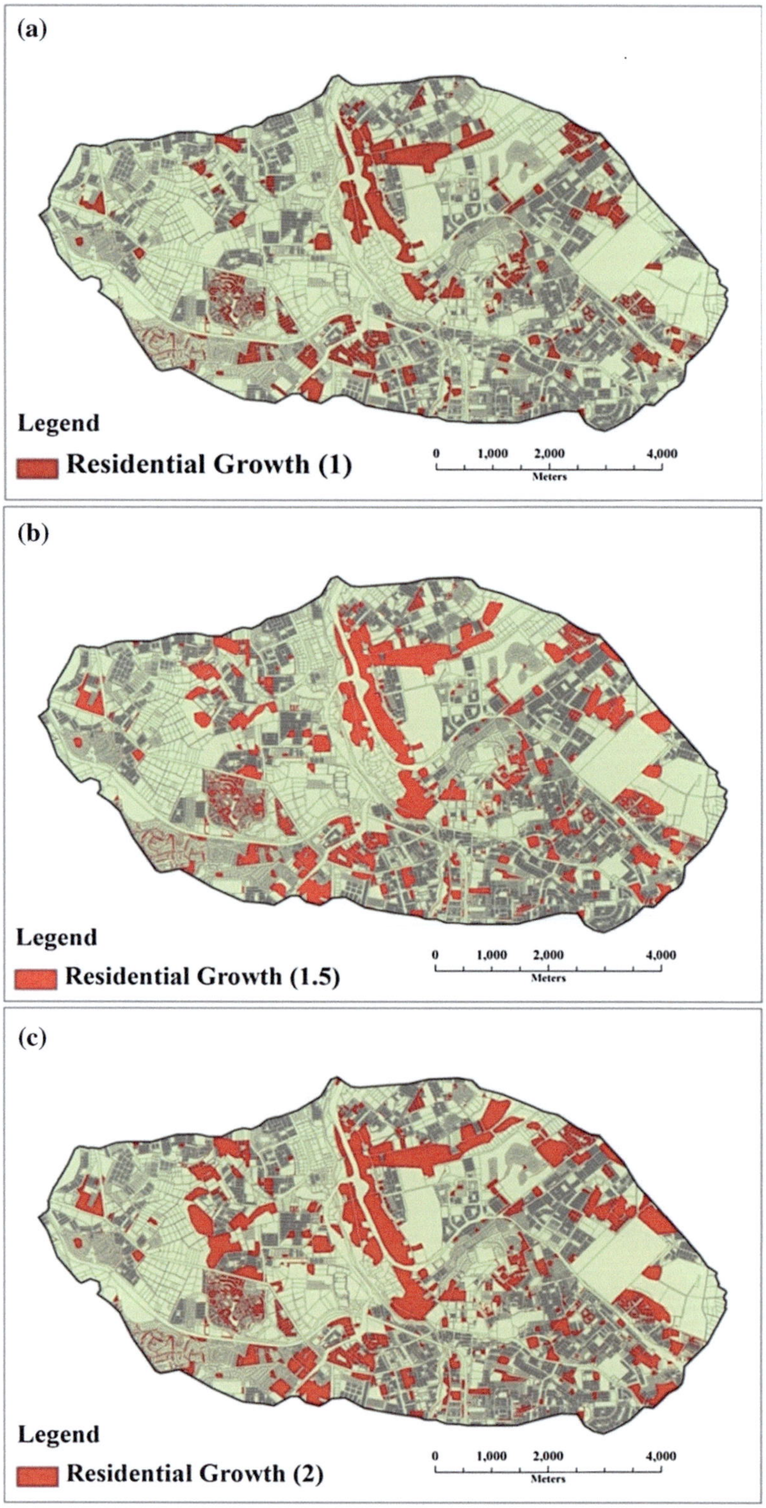

Fig. 6.7 Three LTM modeling scenarios of residential growth; **a** 1 scenario, **b** 2 scenario and **c** 3 scenario

comparison of Figs. 6.7 and 6.9, both models placed the new residential land uses in the same areas. For instance, no projected residential growth can be observed in the central west (industrial zones), east border (agricultural fields), and northeast areas. No significant differences in the visual interpretation can be observed between the two scenarios in terms of the location and distribution of residential growth. However, the legends of the maps illustrate the intensity of probability values, which show that a higher number of variables increase the intensity of the probability values. In

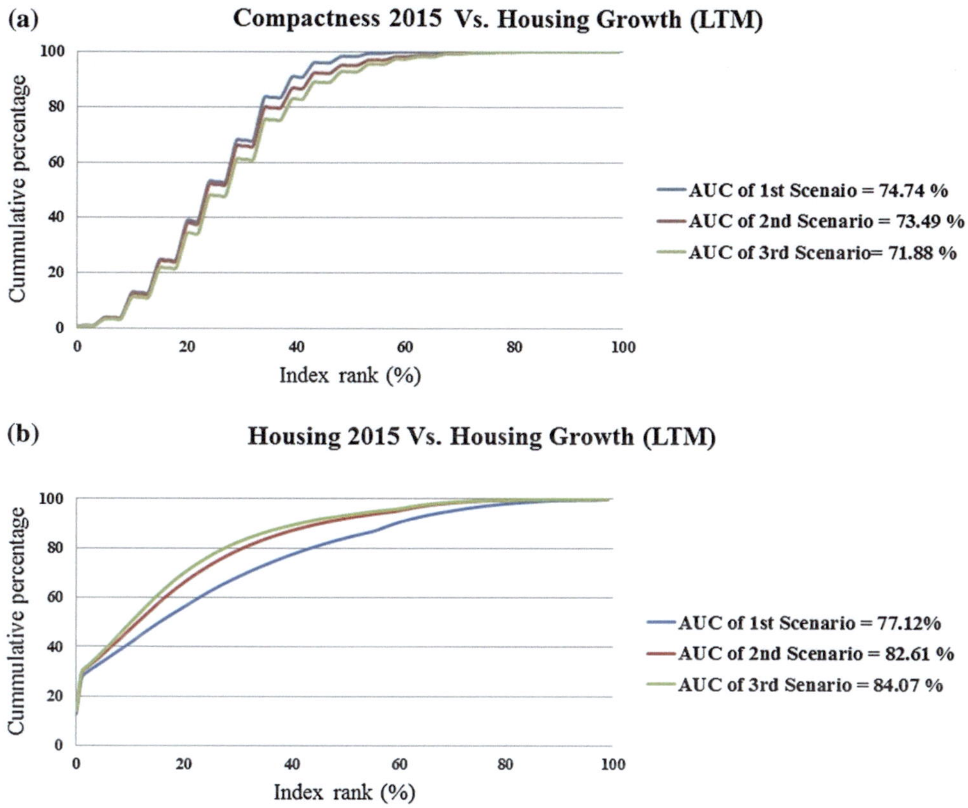

Fig. 6.8 LTM modeling scenarios evaluation with respect to year 2015 land use map; **a** 3 scenarios versus year 2015 compactness map, **b** 3 scenarios versus year 2015 residential land use map

Table 6.4 Summarized weights-of-evidence for residential land use growth

Evidence	Class	C/S(C)	Evidence	Class	C/S(C)
Proximity to residential-2012	Near	363.27	Proximity to community facilities-2012	Near	51.54
	Middle	−264.07		Middle	109.01
	Far	−156.13		Far	−177.41
Proximity to commercial-2012	Near	−11.66	Proximity to recreational facilities-2012	Near	129.38
	Middle	45.24		Middle	38.84
	Far	−35.10		Far	−185.97
Proximity to industrial-2012	Near	−232.07	Proximity to infrastructure-2012	Near	39.38
	Middle	99.20		Middle	62.97
	Far	105.06		Far	−109.34
Proximity to main roads-2012	Near	53.76	Proximity to agricultural fields-2012	Near	−132.02
	Middle	29.54		Middle	19.97
	Far	−86.90		Far	100.26
Proximity to public transportation-2012	Near	62.67	Proximity to river and water bodies-2012	Near	24.57
	Middle	−2.57		Middle	73.25
	Far	−62.48		Far	−103.09
Proximity to flood zones-2012	Near	4.38	Proximity to restricted areas-2012	Near	56.11
	Middle	70.43		Middle	42.42
	Far	−79.03		Far	−104.18
Soil properties-2012	2nT	76.03	Geological properties-2012	Acid	−24.40
	2Gn	4.35		Quartz	5.19
	5H(u)	42.78		Schist	−9.26
	5H(m)	17.06		Filit	16.16
	2Gnt	−106.87			
	5G	−78.52			

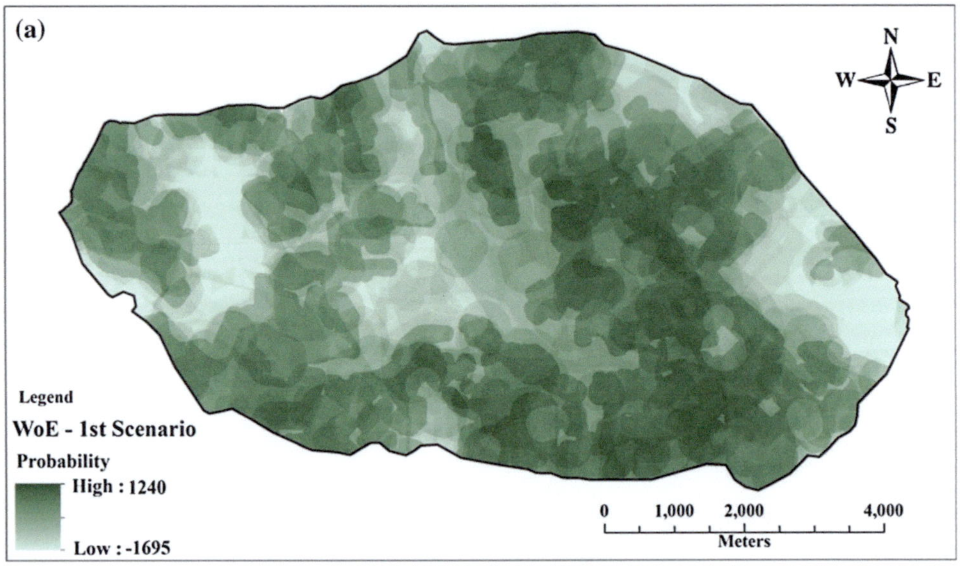

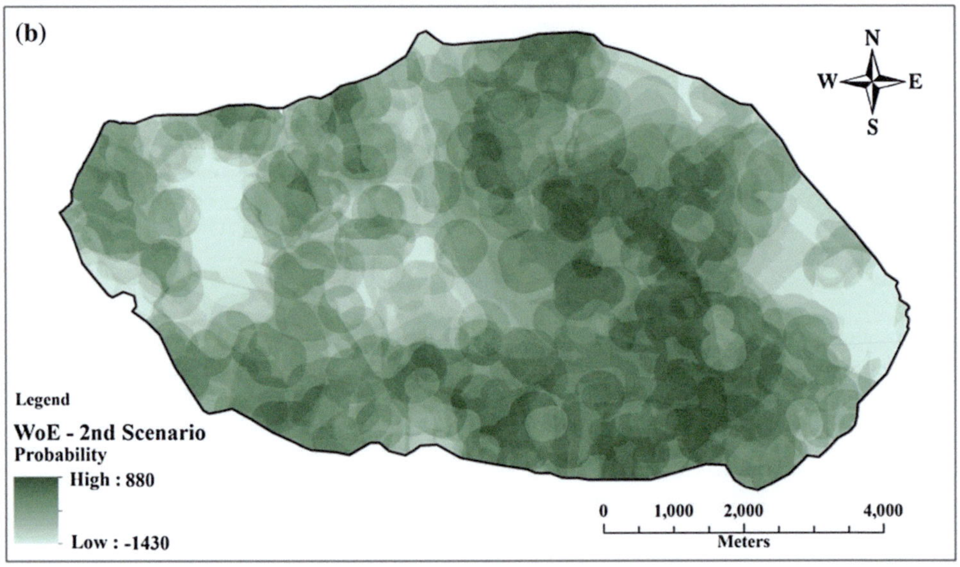

Fig. 6.9 Two WoE modeling scenarios of residential growth; **a** 1 scenario, **b** 2 scenario

addition, subtle information and differences, especially in relation to LTM modeling outputs can be extracted by quantitatively evaluating these scenarios.

Similar to the three scenarios of the LTM approach, the probability maps produced by WoE were also compared with the actual land use map for 2015. Figure 6.10 illustrates the relationship between the compactness and residential map for 2015 with respect to each scenario. According to this figure and Table 6.5, no significant difference can be observed between the scenarios and compactness map. However, the relationship between different scenarios of WoE and the compactness map is higher than that in the LTM scenarios. This result means that the process of land use change modeling in WoE has a higher relationship with city compactness and its indicators than in LTM. On the contrary, the WoE scenarios have lower similarities with the real change residential map than the LTM scenarios. This result indicates the better performance of the LTM technique for land use change modeling based on the parameters and assumptions of the present study.

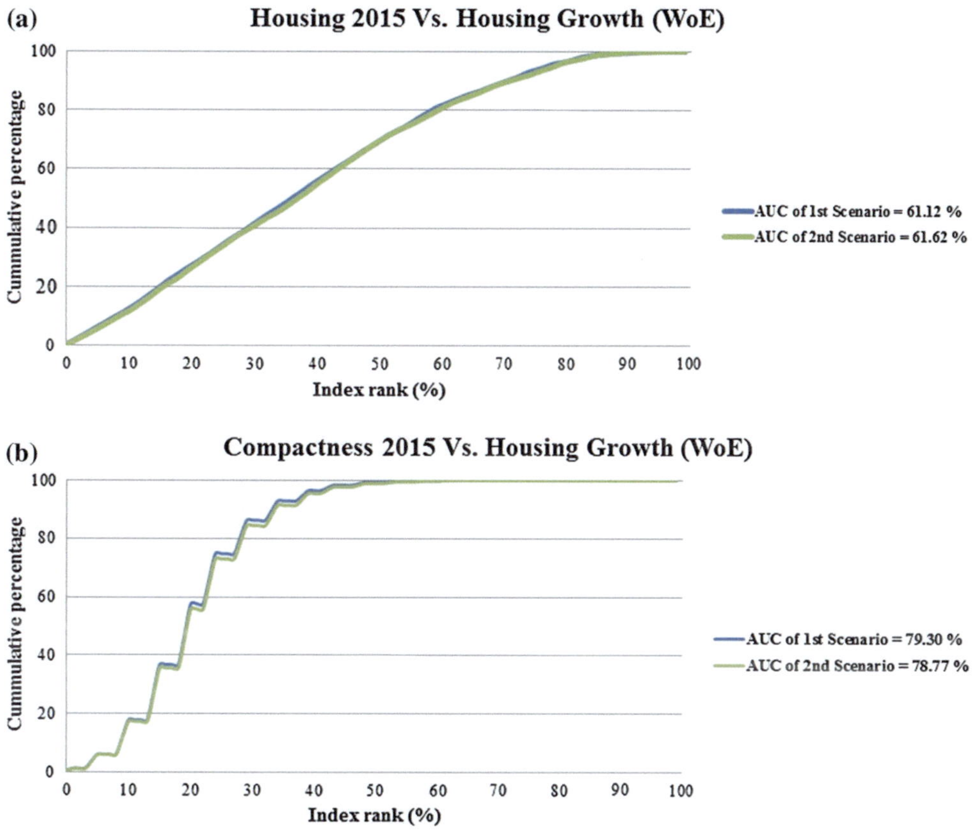

Fig. 6.10 WoE modeling scenarios evaluation with respect to year 2015 land use map; **a** 2 scenarios versus year 2015 compactness map, **b** 2 scenarios versus year 2015 residential land use map

Table 6.5 Overall results of both techniques with respect to year 2015 land use map

		AUC (%) similarity assessment with respect to year 2015 land use map	
Techniques	Scenarios	Compactness map	Residential land use map
LTM	1	74.74	77.12
	2	73.49	82.61
	3	71.88	84.07
WoE	1	79.30	61.12
	2	78.77	61.62

6.4 Conclusion

Compactness assessment generally examines the sustainability of urban development by utilizing specific indicators. The evaluation and relation of these indicators and other urban phenomena, such as growth and loss of various land use categories, are essential in performing this assessment. This study quantified the proportion and direction of residential growth to examine the trend of changes in relation to compact urban development and to identify how these changes can influence city compactness. According to these assessments, the reduction and/or growth in the degree of compactness during the selected period can be evaluated. This information is a valuable reference for local planning authorities to obtain an overall perspective of recent development patterns. The relationship between the growth and city compactness can be investigated by comparing this assessment with residential or any other land use growth. In this study, most of the newly developed residential areas were located in zones where their degrees of compactness grew. This result indicates the positive effect of residential growth on city compactness and vice versa. The effects of residential growth on city compactness can be sufficiently evaluated in detail by decomposing the city compactness concept into its indicators further.

Furthermore, this study also examined two land use change modeling techniques based on computer learning (LTM) and statistical concept (WoE) by projecting the future residential land use growth of Rajang City in Malaysia. LTM generally performs better and more accurately models residential growth than statistical-based WoE. However, WoE provides clearer and more informative results that were indirectly interpreted by LTM in the case of functional relationships among the dependent and independent variables. In addition, the number of parameters in the LTM depended on several factors, including type of neural network, number of hidden layers, and number of drivers and outputs. The results indicate that unsuitable areas are also converted to residential areas, which causes unsustainable development, by increasing the growth of residential cells and decreasing the limitations of LTM. In the compact development perspective, WoE provides more related and effective performance and results than LTM.

Acknowledgements We wish to thank the Selangor town and country planning department (JPBD) in Rajang City for providing various thematic and social information of Rajang area. In addition, we wish to thank the Ministry of Higher Education, Malaysia for financial supporting of this research.

References

Abdullahi, S., Mahmud, A. R. B., & Pradhan, B. (2014). Spatial modelling of site suitability assessment for hospitals using geographical information system-based multicriteria approach at Qazvin city, Iran. *Geocarto International, 29*(2), 164–184.

Abdullahi, S., & Pradhan, B. (2015). Sustainable brownfields land use change modeling using GIS-based weights-of-evidence approach. *Applied Spatial Analysis and Policy*, 1–18.

Abdullahi, S., Pradhan, B., & Jebur, M. N. (2015a). GIS-based sustainable city compactness assessment using integration of MCDM, Bayes theorem and RADAR technology. *Geocarto International, 30*(4), 365–387.

Abdullahi, S., Pradhan, B., Mansor, S., & Shariff, A. R. M. (2015b). GIS-based modeling for the spatial measurement and evaluation of mixed land use development for a compact city. *GIScience & Remote Sensing, 52*(1), 18–39.

Alberti, M., Booth, D., Hill, K., Coburn, B., Avolio, C., Coe, S., et al. (2007). The impact of urban patterns on aquatic ecosystems: an empirical analysis in Puget lowland sub-basins. *Landscape and Urban Planning, 80*(4), 345–361.

Beatley, T. (2012). *Green urbanism: Learning from European cities*. Island Press.

Bohman-Carter, G. (1994). *Geographic information systems for geoscientists*. Pergamon: Oxford.

Burnside, N. G., Smith, R. F., & Waite, S. (2003). Recent historical land use change on the South Downs, United Kingdom. *Environmental Conservation, 30*(01), 52–60.

Burton, E. (2000). The compact city: Just or just compact? A preliminary analysis. *Urban Studies, 37*(11), 1969–2006.

Burton, E. (2002). Measuring urban compactness in UK towns and cities. *Environment and Planning B, 29*(2), 219–250.

Burton, E., Jenks, M., & Williams, K. (2003). *The compact city: A sustainable urban form?* Routledge.

Chang, I.-C. C., & Sheppard, E. (2013). China's eco-cities as variegated 1 urban sustainability: Dongtan eco-city and Chongming eco-island. *Journal of Urban Technology, 20*(1), 57–75.

Chen, C., He, B., & Zeng, Z. (2014a). A method for mineral prospectivity mapping integrating C4. 5 decision tree, weights-of-evidence and m-branch smoothing techniques: a case study in the eastern Kunlun Mountains, China. *Earth Science Informatics, 7*(1), 13–24.

Chen, J., Chang, K.-T., Karacsonyi, D., & Zhang, X. (2014b). Comparing urban land expansion and its driving factors in Shenzhen and Dongguan, China. *Habitat International, 43*, 61–71.

Dadhich, P. N., & Hanaoka, S. (2011). Spatio-temporal urban growth modeling of Jaipur, India. *Journal of Urban Technology, 18*(3), 45–65.

Dahal, R. K., Hasegawa, S., Nonomura, A., Yamanaka, M., Masuda, T., & Nishino, K. (2008). GIS-based weights-of-evidence modelling of rainfall-induced landslides in small catchments for landslide susceptibility mapping. *Environmental Geology, 54*(2), 311–324.

de Almeida, C. M., Batty, M., Monteiro, A. M. V., Câmara, G., Soares-Filho, B. S., Cerqueira, G. C., et al. (2003). Stochastic cellular automata modeling of urban land use dynamics: Empirical development and estimation. *Computers, Environment and Urban Systems, 27*(5), 481–509.

Dempster, A. P. (1967). Upper and lower probabilities induced by a multivalued mapping. *The Annals of Mathematical Statistics*, 325–339.

Gainza, X., & Livert, F. (2013). Urban form and the environmental impact of commuting in a segregated city, Santiago de Chile. *Environment and Planning B: Planning and Design, 40*(3), 507–522.

Gu, Z., Sun, Q., & Wennersten, R. (2013). Impact of urban residences on energy consumption and carbon emissions: An investigation in Nanjing, China. *Sustainable Cities and Society, 7*, 52–61.

Houet, T., Loveland, T. R., Hubert-Moy, L., Gaucherel, C., Napton, D., Barnes, C. A., et al. (2010). Exploring subtle land use and land cover changes: A framework for future landscape studies. *Landscape Ecology, 25*(2), 249–266.

Kolb, M., Mas, J.-F., & Galicia, L. (2013). Evaluating drivers of land-use change and transition potential models in a complex landscape in Southern Mexico. *International Journal of Geographical Information Science, 27*(9), 1804–1827.

Kuijpers, M. A. J., Bouwman, A., & Natuurplanbureau, M.-E. (2007). *Nederland later: Tweede duurzaamheidsverkenning: Deel fysieke leefomgeving Nederland*. Milieu en Natuur Planbureau Bilthoven.

Lantman, J. V. S., Verburg, P. H., Bregt, A., & Geertman, S. (2011). Core principles and concepts in land-use modelling: A literature review. *Land-Use Modelling in Planning Practice* (pp. 35–57). Springer.

Li, X., Yang, Q., & Liu, X. (2008). Discovering and evaluating urban signatures for simulating compact development using cellular automata. *Landscape and Urban Planning, 86*(2), 177–186.

Li, X., & Yeh, A. G.-O. (2004). Analyzing spatial restructuring of land use patterns in a fast growing region using remote sensing and GIS. *Landscape and Urban Planning, 69*(4), 335–354.

Livingstone, K., & Authority, G. L. (2003). *Housing for a compact city*. Greater London authority.

Manaugh, K., & Kreider, T. (2013). What is mixed use? Presenting an interaction method for measuring land use mix. *Journal of Transport and Land Use, 6*(1), 63–72.

Matthews, R. B., Gilbert, N. G., Roach, A., Polhill, J. G., & Gotts, N. M. (2007). Agent-based land-use models: A review of applications. *Landscape Ecology, 22*(10), 1447–1459.

Middel, A., Brazel, A., Hagen, B., & Myint, S. (2011). Land cover modification scenarios and their effects on daytime heating in the inner core residential neighborhoods of Phoenix, Arizona. *Journal of Urban Technology, 18*(4), 61–79.

Mubareka, S., Koomen, E., Estreguil, C., & Lavalle, C. (2011). Development of a composite index of urban compactness for land use modelling applications. *Landscape and Urban Planning, 103* (3), 303–317.

Musakwa, W., & Van Niekerk, A. (2013). Implications of land use change for the sustainability of urban areas: A case study of Stellenbosch, South Africa. *Cities, 32*, 143–156.

Petsch, S., Guhathakurta, S., Heischbourg, L., Müller, K., & Hagen, H. (2011). Modeling, monitoring, and visualizing carbon footprints at the urban neighborhood scale. *Journal of Urban Technology, 18*(4), 81–96.

Pijanowski, B. C., Brown, D. G., Shellito, B. A., & Manik, G. A. (2002). Using neural networks and GIS to forecast land use changes: A land transformation model. *Computers, Environment and Urban Systems, 26*(6), 553–575.

Pijanowski, B. C., Pithadia, S., Shellito, B. A., & Alexandridis, K. (2005). Calibrating a neural network-based urban change model for two metropolitan areas of the Upper Midwest of the United States. *International Journal of Geographical Information Science, 19*(2), 197–215.

Pijanowski, B. C., Tayyebi, A., Doucette, J., Pekin, B. K., Braun, D., & Plourde, J. (2014). A big data urban growth simulation at a national scale: Configuring the GIS and neural network based land transformation model to run in a high performance computing (HPC) environment. *Environmental Modelling and Software, 51,* 250–268.

Pontius, R. G., Jr., & Millones, M. (2011). Death to Kappa: birth of quantity disagreement and allocation disagreement for accuracy assessment. *International Journal of Remote Sensing, 32*(15), 4407–4429.

Pontius, R. G., & Schneider, L. C. (2001). Land-cover change model validation by an ROC method for the Ipswich watershed, Massachusetts, USA. *Agriculture, Ecosystems & Environment, 85* (1), 239–248.

Pradhan, B. (2011). An assessment of the use of an advanced neural network model with five different training strategies for the preparation of landslide susceptibility maps. *Journal of Data Science, 9*(1), 65–81.

Pradhan, B., Oh, H.-J., & Buchroithner, M. (2010). Weights-of-evidence model applied to landslide susceptibility mapping in a tropical hilly area. *Geomatics, Natural Hazards and Risk, 1*(3), 199–223.

Regmi, N. R., Giardino, J. R., & Vitek, J. D. (2010). Modeling susceptibility to landslides using the weight of evidence approach: Western Colorado, USA. *Geomorphology, 115*(1), 172–187.

Shafer, G. (1976). *A mathematical theory of evidence* (Vol. 1). Princeton university press Princeton.

Song, Y., & Knaap, G.-J. (2004). Measuring the effects of mixed land uses on housing values. *Regional Science and Urban Economics, 34*(6), 663–680.

Song, Y., Merlin, L., & Rodriguez, D. (2013). Comparing measures of urban land use mix. *Computers, Environment and Urban Systems, 42,* 1–13.

Tayyebi, A., & Pijanowski, B. C. (2014). Modeling multiple land use changes using ANN, CART and MARS: Comparing tradeoffs in goodness of fit and explanatory power of data mining tools. *International Journal of Applied Earth Observation and Geoinformation, 28,* 102–116.

Tayyebi, A., Pijanowski, B. C., Linderman, M., & Gratton, C. (2014). Comparing three global parametric and local non-parametric models to simulate land use change in diverse areas of the world. *Environmental Modelling and Software, 59,* 202–221.

Tehrany, M. S., Pradhan, B., & Jebur, M. N. (2014). Flood susceptibility mapping using a novel ensemble weights-of-evidence and support vector machine models in GIS. *Journal of Hydrology, 512,* 332–343.

Thiam, A. K. (2005). An evidential reasoning approach to land degradation evaluation: Dempster-Shafer theory of evidence. *Transactions in GIS, 9*(4), 507–520.

Thinh, N. X., Arlt, G., Heber, B., Hennersdorf, J., & Lehmann, I. (2002). Evaluation of urban land-use structures with a view to sustainable development. *Environmental Impact Assessment Review, 22*(5), 475–492.

Van Eck, J. R., & Koomen, E. (2008). Characterising urban concentration and land-use diversity in simulations of future land use. *The Annals of Regional Science, 42*(1), 123–140.

Xu, Z. (2011). *Application of system dynamics model and GIS in sustainability assessment of urban residential development.* In Paper presented at the Proceedings of the 29th International Conference of the System Dynamics Society, Washington, DC, July.

Youssef, A. M., Pradhan, B., Pourghasemi, H. R., & Abdullahi, S. (2015). Landslide susceptibility assessment at Wadi Jawrah Basin, Jizan region, Saudi Arabia using two bivariate models in GIS. *Geosciences Journal, 19*(3), 449–469.

Zahiri, H., Palamara, D., Flentje, P., Brassington, G. M., & Baafi, E. (2006). A GIS-based weights-of-evidence model for mapping cliff instabilities associated with mine subsidence. *Environmental Geology, 51*(3), 377–386.

Zell, A., Mache, N., Hübner, R., Mamier, G., Vogt, M., Schmalzl, M., & Herrmann, K.-U. (1994). *SNNS* (stuttgart neural network simulator) *Neural Network Simulation Environments* (pp. 165–186). Springer.

Zurada, J. (1992). *Introduction to artificial neural systems, 1992*. St. Paul: Went Publishing Company.

7 Urban Expansion and Change Detection Analysis

Saleh Abdullahi and Biswajeet Pradhan

7.1 Introduction

The urbanization process is the key factor in urban growth and land use change. Unorganized and unexpected urban expansion causes poor and unplanned land use changes and consequently results in poor infrastructure and facility provision. Generally, sprawl development refers to the development in which built-up areas have higher growth than the population in a specific area (Barnes et al. 2001; Soffianian et al. 2010). Thus, controlling the growth of built-up areas through green and open spaces is essential to avoid sprawl development. In this regard, urban expansion and change detection analysis produce baseline information on the previous and existing trends of urban growth behavior in various urban applications, such as political and historical processes (Flint 2002), urban crime assessment (Craglia et al. 2001), land suitability assessment (Abdullahi et al. 2014), estimation of urban population (Zhang 2003), urban heat island research (Kaveckis and Bechtel 2014), merging urban ecology and socioeconomics (Zhang et al. 2006), land use/land cover evaluation (López et al. 2001), urban change analysis and the modeling of growth (Hu and Lo 2007; Al-shalabi et al. 2013; Abdullahi and Pradhan 2015).

Urban expansion is evaluated by determining the amount of built-up areas between two time instances (Bhatta 2009). In addition to the growth of built-up areas, evaluation of the amount of changes among various other land use categories also provides a clear understanding of the urban growth trend. Thus, the modeling of urban expansion and land use changes allows the quantification of these changes in urban environment.

Urban growth and expansion has a direct relationship with socioeconomic information and population figures. Growth in population induces the spread of the built-up area. Therefore, the population in an area is one of the important metrics in the urban sprawl process. A simple and acceptable measure to identify and quantify urban sprawl is the proportion of total built-up area to the total population (Sharma et al. 2012; Sandhya Kiran and Joshi 2013). The built-up area and population percentages should be computed by dividing the amount of built-up area and population in every district by the overall built-up area and population of the whole study area, respectively. The interchangeable relationship of urban growth and population can be evaluated by subtracting the population ratio from the built-up area ratio in each district. The results will fall in the range of −1 to 1, where 0 indicates moderate conditions. A positive value reflects higher built-up area consumption per capita, which, in turn, indicates better environment and extended urban services. A negative value indicates population crowding, which may result in serious negative effect at the social, economic, and urban levels.

Land absorption rate is another urban expansion and sprawl assessment method based on the relationship between the built-up area and population figures (Al-sharif et al. 2013). It measures the urban sprawl as a process. The land absorption rate technique is based on the evaluation of changes in the built-up area and the population data in the defined period. In addition to the population consideration in urban growth assessment, the observed expansion should be compared with the forecasted urban expansion to understand the divergence of urban growth. The divergence of urban growth in each zone and each temporal period can be easily identified by subtracting the calculated theoretical expected urban growth from the observed growth. Positive values confirm more growth than the expectations, whereas negative values indicate less growth. The level of variation can also be identified by the magnitude of differences. Higher deviations between the observed urban expansions in the selected zones reflect the freedom and independence of the urban expansion process, that is, high deviation means that the studied variable is independent from other similar types of variables. However, developing countries normally have no clear urban plan or wise estimations of urban expansions unlike developed countries.

In the urbanization process, expansion differs per region and direction because of the policy on urban driving factors and their spatial effects. Such factors include road network, population density, slope, and economics. The differences in

B. Pradhan (ed.), *Spatial Modeling and Assessment of Urban Form*,
DOI 10.1007/978-3-319-54217-1_7

expansions are referred to as the preference of urban growth. Urban expansion intensity index is another equation to quantitatively assess and analyze the differences in urban spatial expansions. This process can recognize the preference of urban growth in a certain period (Ren et al. 2013). It reflects the probable future direction and potentials of urban expansions and compares the speed or intensity of urban land use change in different periods. Another statistical urban growth assessment is Pearson's chi-square, which estimates the freedom or degree of variation for the observed urban growth over the expected urban growth (Almeida et al. 2005; Bhatta et al. 2010). Pearson's chi-square method can check the degree of freedom between pairs of variables selected to describe the same class of land cover change. The estimation of this index for the urbanization process is normally used as a supportive analysis tool to provide another perspective on urban sprawl. Pearson's chi-square method assesses the deviation of real urban expansion from planned or expected growth; a high deviation of urban growth is considered as a sign of urban sprawl occurrence. A higher degree of freedom implies the need for consistency in planning, managing, and controlling urban growth. A high degree of freedom in a zone is a warning about unbalanced growth within the zone over time, and high degrees of freedom in a period implies high interzone inconsistency in urban growth. However, a high degree of freedom cannot be considered as a clear sprawl but as a disparity in urban growth instead. The overall degree of freedom of the study area can also be calculated by the summation of the degrees of freedom across all periods or by summing up the degrees of freedom of all zones. If the lower limit of the chi-square becomes 0, the observed growth value is exactly equal to the expected growth value.

The following sections present the urban growth and land use change analysis of Kajang City (Malaysia) as a case study based on the historical trends of growth and changes with respect to the master plan of the city.

7.2 Implementation of Growth and Change Analysis

Land use change assessments provide a clear understanding of the built-up growth through various land uses and land cover categories. These assessments reveal the rates, amount, and directions of the growth. Thus, significant growth and/or loss of a specific land use type can be highlighted precisely. Cross-tabulation analysis is applied to each pair of available land use maps of the study area (2004, 2008, 2012, and 2015) to implement this analysis for Kajang City. This city is located 21 km away from Kuala Lumpur, the capital city of Malaysia (Fig. 7.1). The eastern parts of the city are mainly covered by agriculture and forest lands. Recent rapid urban developments have mushroomed in Kajang City because of its proximity to three main cities of Malaysia. Although many abandoned brownfields exist in the city, most of these new developments have been constructed at the outskirts of the agricultural and forest environments. Thus, this chapter attempts to assess the growth and changes of various land use categories with respect to each other. Next, these growths were evaluated and compared with the master plan of the study area, which was proposed by the local planning authority.

Cross tabulation is a mathematical matrix that provides unbiased information concerning the entire area of interest to derive unbiased summary statistics (Pontius Jr and Millones 2011). This matrix is also known as confusion, error, and contingency matrices and is a quantitative research method for analyzing the relationship and interrelation between two or more variables. For the current study, the matrix provided unbiased information to quantify the persistence and land use changes and growth between all land use maps. Figure 7.2 shows that four land use maps of Kajang City were collected from the local planning authority in temporal bases of years 2004, 2008, 2012, 2015, and the master plan (Figs. 7.3, 7.4, 7.5, 7.6, 7.7).

Generally, $n(n - 1)$ possible land use conversions are possible, where n is the number of land uses. Thus, 72 possible conversions are expected in the case of nine land use types. Land conversion is symmetric at the high spatial resolution used for this study, and changes in both ways can be observed. Thus, some of the 72 possible land use conversions are expected to be observable and a few to be negligible. For instance, the conversion of water bodies from and to other land use categories rarely occurs. By contrast, the growth of residential, commercial, and industrial areas through agricultural and green spaces is common all around the world.

All four shapefile land use maps were converted to raster format with 1 m pixel size to perform the change detection process. Next, all raster layers were converted to ASCII data and then imported to IDRISI software to conduct the cross-tabulation analysis. The input to this process for each run were two land use maps, for instance, 2004 and 2008 as older and newer input images, respectively. This process can produce four types of output: (1) cross-classification image, (2) full cross-tabulation table, (3) both cross-classification image and cross-tabulation table, and (4) image similarity/association data only. The third output type was selected for this study to obtain both cross-tabulation image and full cross-tabulation table. Therefore, in addition to the land use changes in two periods that can be observed visually, all the quantitative information of different land use changes can be extracted in a matrix format. In these matrices, the land use types of the older land use map are arranged in columns against the land use types of the newer land use map in rows.

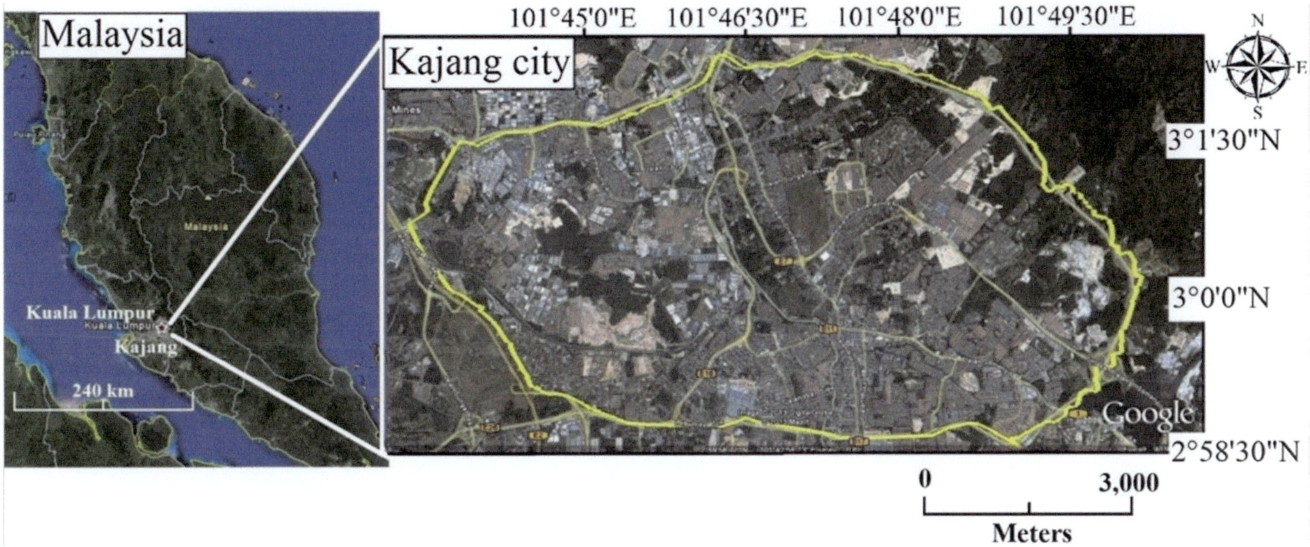

Fig. 7.1 The map of Malaysia and Kajang City

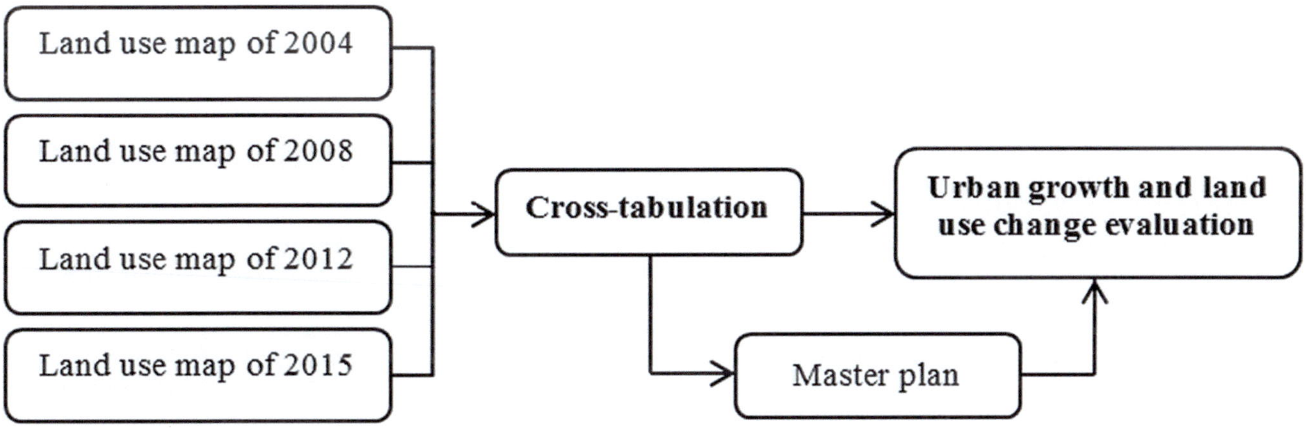

Fig. 7.2 Process flow of urban growth and change analysis

The extra column and row at the end of the matrix represent the total amount (area in m^2) of each land use type in each column and row. These values show the total change (growth and loss) of each land use category to and from other categories. The diagonal line of the matrix shows the amount of each land use type that remained unchanged during the given period.

Some statistical values, such as chi-square, degree of freedom, P-Level, Cramer's *V*, Kappa index of agreement (KIA), and overall Kappa, were calculated for each matrix (Kuzera and Pontius 2008). Chi-square test was constructed from a sum of squared errors or through the sample variance. This test examined independent variables for independency and determined whether splitting nodes generated a significant improvement. Degree of freedom indicates the number of independent ways by which the variables in the matrix can move without violating any constraint imposed on them. This value indicates the number of interactions among all land use types. P-Level is a function of the observed sample results that is used for testing the hypothesis. A threshold value was selected, which is called significance level, before the test was conducted. A P-level lower or equal to the threshold indicates inconsistency in the observed data. Cramer's *V* calculated the correlation in the matrix (Almeida et al. 2005). It was used as a posttest to determine the strengths of association after chi-square determined the significance. In fact, chi-square indicated the relationship among the land use types, but Cramer's *V* showed how important and significant they were. Cramer's *V* was in the range of 0 (minimal association among variables) to 1 (high association among variables). Two land use maps with significant difference and changes in land use types had a low Cramer's *V* value (close to 0) and two land use maps with insignificant changes had a high Cramer's *V* value (close to

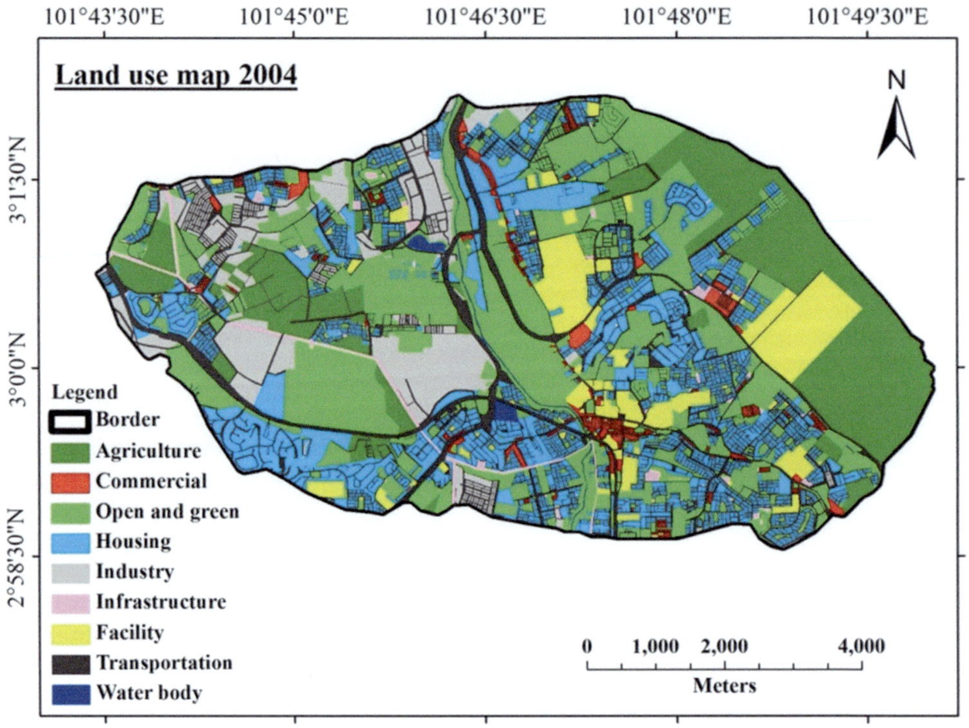

Fig. 7.3 Land use map of year 2004

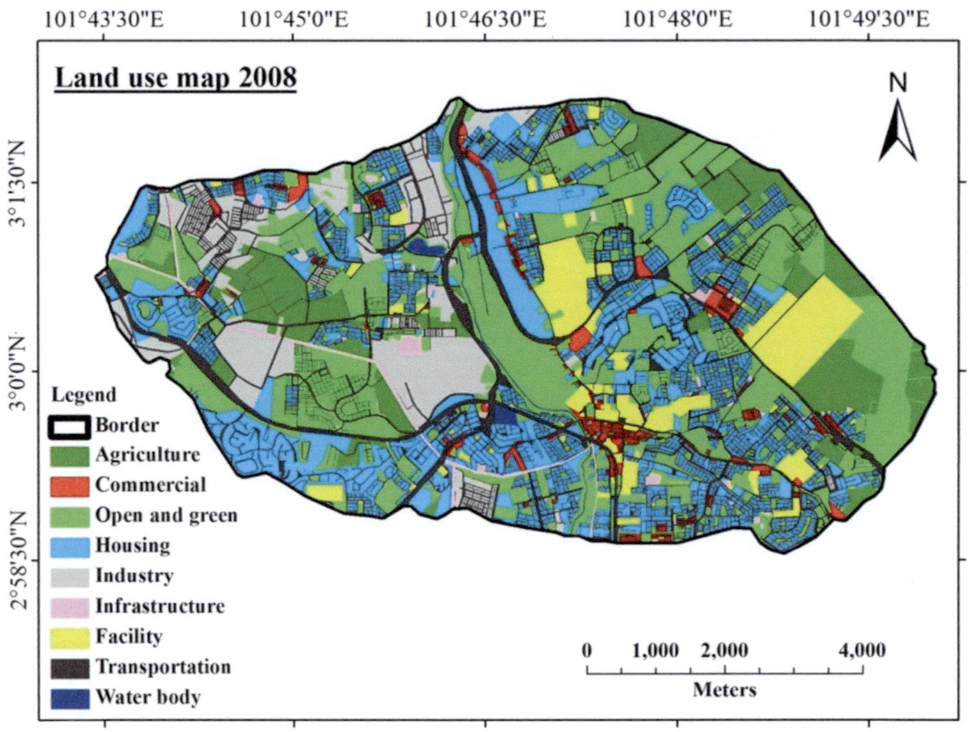

Fig. 7.4 Land use map of year 2008

1). This assessment provided valuable information regarding the growth and changes between two land use maps. The Kappa index of agreement was calculated for all land use categories twice, first using the older land use map as the reference map and second using the newer land use map. This measurement was originally developed for accuracy assessment of remotely sensed images. In the current study, the Kappa index of agreement was used to test whether the differences between two land use maps were due to chance or real (dis) agreement. In fact, this index evaluated the degree of agreement between the two land use maps on a per category basis. This assessment was also in the

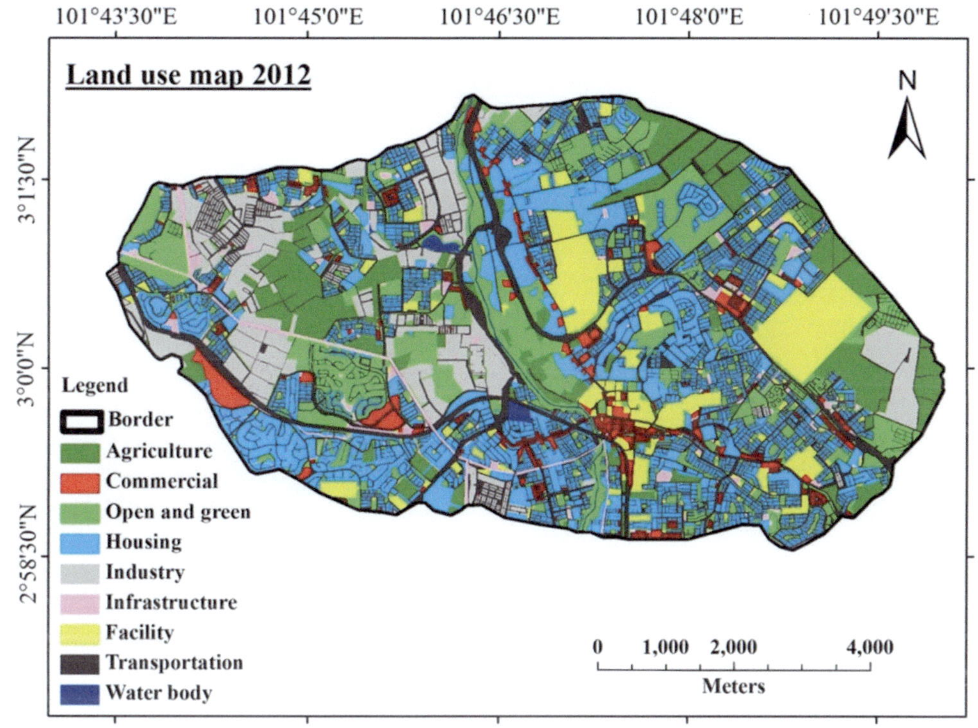

Fig. 7.5 Land use map of year 2012

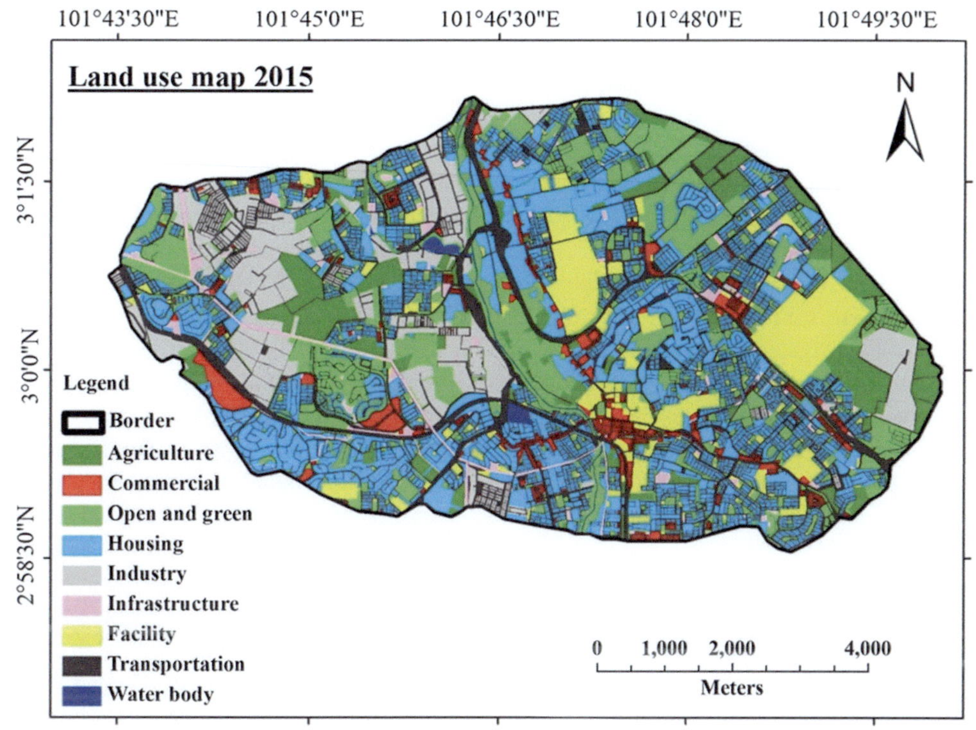

Fig. 7.6 Land use map of year 2015

range of 0 (total agreement from chance) to 1 (perfect true agreement). Finally, overall Kappa was calculated, which discriminated between errors of quantity and errors of location between two land use maps.

Cross tabulation in addition to matrix showed that areas of land use changes created another matrix known as proportional cross tabulation. This matrix determined the proportional conversion of each land use type, which was calculated from each area of change with respect to the total area of analysis.

The amount of built-up area for each temporal land use map was measured from this quantitative information. This

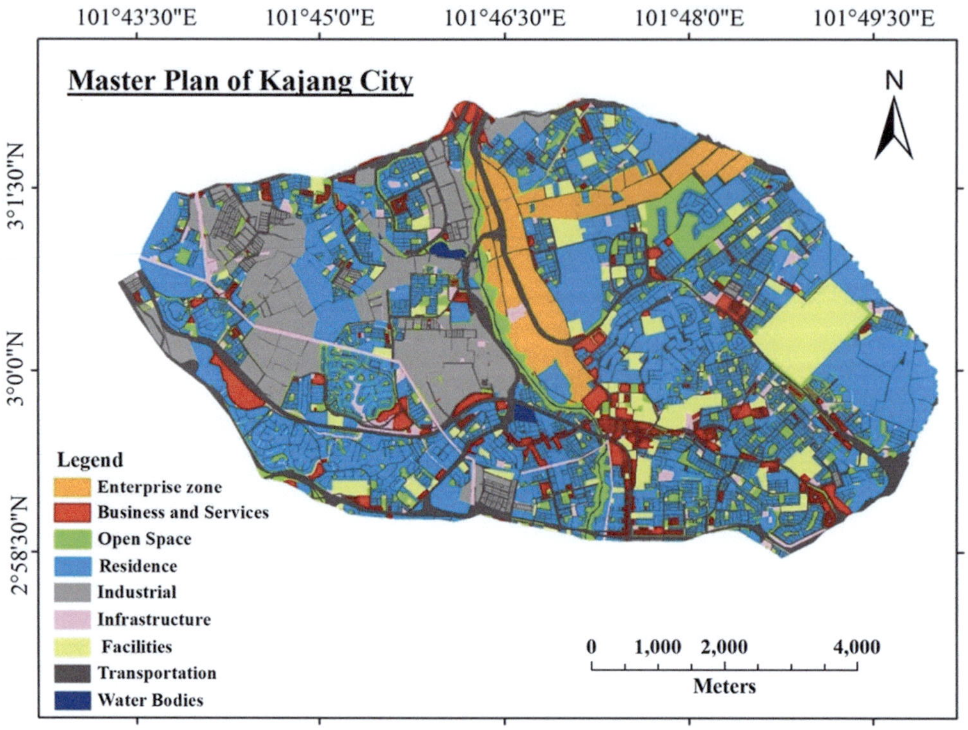

Fig. 7.7 Master plan of Kajang City

process for shapefile format was easy to conduct using the ArcGIS spatial analysis tool. However, the number of pixels for the output maps of IDRISI software, which are in raster format and which belong to built-up land use categories, was multiplied by the cell size, which was 1 m^2.

After the land use growth and change evaluation of all temporal land use maps (2004, 2008, 2012, and 2015), these development patterns were compared with the master plan of Kajang City to evaluate whether the current development patterns were following the master plan or not. Thus, this process investigated the amount, direction, and reason for the similarity and dissimilarity of the current development trend with respect to the master plan. The master plan layer consisted of one extra class for enterprise zones, which showed the areas in Kajang City that need to be revitalized or regenerated to improve the economy and livability of certain neighborhoods.

7.3 Results of Urban Growth and Change Analysis

This section demonstrates and discusses the results of urban growth analysis based on several spatiotemporal land use maps and the evaluation of various land use change patterns. As the first and an essential step in most urban planning applications, change detection was performed using the cross-tabulation approach to extract the land use change historical trend of the study area. This process was conducted for each pair of available land use maps. Cross tabulation provided two-dimensional tables to show land use changes of all cells of the study area for each pair of selected periods. In addition, cross tabulation determined the geographical distribution of changes throughout the study area. Cross tabulation was conducted among all available land use maps to produce a cross-classification image and cross-tabulation table for each pair of land use maps.

Tables 7.1, 7.2, 7.3 show the quantitative information of land use changes during selected periods in matrix format. These tables show the area for each land use type that was converted to another type. In these tables, the older land use maps (column) are cross-tabulated with the newer land use maps (row). The last rows and second to the last columns (sum) in all tables show the calculated sum of each row and column, which is the sum of growth and loss for the corresponding land use category. The last columns of all tables (total loss) are the summarized growth or loss for each land use category. These values were calculated by subtracting the value of growth from the loss. Thus, a positive value meant growth and a negative value indicated loss of the corresponding land use type during the selected period. These tables show that the land use categories are listed in number format (agriculture = 1, commercial = 2, open spaces = 3, housing = 4, industry = 5, infrastructure = 6, community facility = 7, road network = 8, and water body = 9).

The values in the given tables are the numbers of pixels of the study area. The given values in the tables are in m^2 unit because the process was conducted with 1 m^2 pixel size.

Table 7.1 Cross tabulation of land use map 2004 (column) against 2008 (row) (m^2)

	1	2	3	4	5	6	7	8	9	Sum (*A*)	Total (*A* − *B*)
1	4,479,839	0	572,389	278	41,868	34	356	5544	0	5,100,308	−3,720,252
2	13,341	976,420	114,310	48	32	3	0	3831	0	1,107,985	123,245
3	3,379,430	1402	10,313,110	493,569	6180	22,260	46,037	25,586	0	14,287,574	−155,521
4	321,811	113	1,984,856	10,580,336	220	9383	3895	25,738	0	12,926,352	1,766,064
5	16,328	0	240,742	8	4,589,970	36	0	839	0	4,847,923	197,057
6	28,157	0	153,938	18	77	864,658	5	1485	0	1,048,338	132,468
7	8280	1	92,510	23	0	0	3,867,271	511	0	3,968,596	40,518
8	566,748	6800	971,139	86,003	12,518	19,495	10,513	11,257,536	256	12,931,008	1,609,660
9	0	0	91	0	0	0	0	16	332,438	332,545	−149
Sum (*B*)	8,820,560	984,740	14,443,095	11,160,288	4,650,866	915,870	3,928,078	11,321,348	332,694		

Note Agriculture = 1, commercial = 2, open spaces = 3, housing = 4, industry = 5, infrastructure = 6, community facility = 7, road network = 8 and water body = 9. Positive value of last column means growth and negative value means loss of land use category

Table 7.2 Cross tabulation of land use map 2008 (column) against 2012 (row) (m^2)

	1	2	3	4	5	6	7	8	9	Sum (*A*)	Total (*A* − *B*)
1	3,165,026	3426	2,047,234	214,850	99,859	2505	7677	80,995	0	5,621,572	518,242
2	42,750	663,906	586,049	276,972	36,316	2563	59,799	75,740	0	1,744,095	636,103
3	444,946	51,677	5,581,021	750,289	598,365	118,098	59,904	346,298	753	7,951,351	−6,338,200
4	612,824	120,515	2,713,650	10,653,226	106,177	10,878	41,576	498,328	27	14,757,201	1,830,828
5	338,067	57,180	1,170,208	148,901	3,704,952	8759	0	177,248	0	5,605,315	757,389
6	67,988	9713	313,279	137,347	33,059	865,759	1981	99,510	0	1,528,636	480,298
7	6730	20,242	547,355	127,217	8680	3625	3,770,909	45,694	0	4,530,452	561,856
8	420,918	181,333	1,309,232	605,781	260,518	36,149	23,719	11,547,620	712	14,385,982	1,453,096
9	4007	0	21,523	11,790	0	1	3031	61,453	331,053	432,858	100,313
Sum (*B*)	5,103,330	1,107,992	14,289,551	12,926,373	4,847,926	1,048,338	3,968,596	12,932,886	332,545		

Table 7.3 Cross tabulation of land use map 2012 (column) against 2015 (row) (m^2)

	1	2	3	4	5	6	7	8	9	Sum (*A*)	Total (*A* − *B*)
1	4,210,981	0	0	0	0	0	0	0	0	4,210,981	−1,410,738
2	0	1,744,100	13,751	0	0	0	0	0	0	1,757,851	13,751
3	876,405	0	7,477,620	130	27,681	0	0	123	0	8,381,959	430,601
4	109,794	0	458,426	14,757,081	0	1	0	68	0	15,325,370	568,159
5	424,539	0	0	0	5,577,638	0	0	0	0	6,002,177	396,858
6	0	0	1173	0	0	1,528,636	0	1	0	1,529,810	1173
7	0	0	0	0	0	0	4,530,455	0	0	4,530,455	0
8	0	0	282	0	0	0	0	14,385,842	0	14,386,124	90
9	0	0	0	0	0	0	0	0	432,858	432,858	0
Sum (*B*)	5,621,719	1,744,100	7,951,358	14,757,211	5,605,319	1,528,637	4,530,455	14,386,034	432,858		

In general, the result of the cross-tabulation analysis indicated that residential, transportation, commercial, and industrial land use types had significant growths compared with other land use types. Furthermore, the growth of these three land use types mainly resulted in the reduction of agricultural fields and open spaces. In most of the periods, the seemingly encompassing effect of residential land use on almost all types of activities can be observed. However, this growth through open spaces and agricultural areas was more significant. Followed by residential land use, transportation and industrial were the two other land use categories that had substantial growth.

With the 1 m^2 spatial resolution of the change detection process, a symmetric and two-way land use conversion was expected. This condition means that although the land cover change of water bodies happens rarely, the conversion of water bodies to other land use types and vice versa can be observed in this process. Nevertheless, these insignificant changes are also measured and presented in terms of amount and frequency. Lower precision and accuracy can be achieved from this assessment with a coarser spatial resolution process.

Table 7.1 shows the cross tabulation between the land use maps for 2004 and 2008. Each column of this table shows that the pixels of the study area belong to one land use category that was converted to another category indicated by each row. The first and third columns, which are agricultural and open spaces, respectively, have higher values than the other columns, such as commercial (2) and residential (4). This condition means that the conversions of the open spaces and agricultural areas to other land uses were higher than those of the residential and commercial areas. Open spaces mainly consist of brownfields or abandoned lands, which are undeveloped because of several reasons. Thus, the changes of these lands to any other land use category with the aim of brownfield redevelopment are preferable.

The diagonal values of these tables show the number of pixels (the area) that remained unchanged during the corresponding period. The diagonal values are higher than the other values in all the cross-tabulation tables because the majority of land use types usually remain the same. This hypothesis can be observed clearly in Table 7.3, which shows that no significant change occurred in Kajang City during the period 2012–2015. Thus, most of the values in this specific table are zero and the maximum values are placed in diagonal cells, indicating that most of the areas remained unchanged. Unlike in this period (2012–2015), Kajang City encountered a large proportion of changes in the other periods. This can be seen from the higher values in the tables corresponding to the changes between two land use types. However, most of the land use types, such as residential, commercial, agricultural, and open spaces, have larger proportion of changes than water bodies, which have less change during selected periods. Water body with class value of nine in these tables usually has zero or very small value.

Table 7.4 shows the overall period of land use changes from the oldest to newest available land use map, that is, 2004–2015. This table presents the same and/or summarized information of previous tables. For instance, the residential area had total growths of 176.6, 183.0, and 56.8 ha from 2004 to 2008, from 2008 to 2012, and from 2012 to 2015, respectively. Thus, the total growth of the residential area from the oldest to the newest land maps (2004 to 2015) can be observed in Table 7.4, which is the sum of the growth rates in the three periods (416.4 ha).

Table 7.5 and Fig. 7.8 present the overall changes (growth and loss) of each land use type during the selected periods. Table 7.5 shows that agricultural fields were reduced by 372 ha from 2004 to 2008, and then increased by 52 ha from 2008 to 2012, and again reduced by 141 ha from 2012 to 2015. Similarly, open spaces decreased from 2004 to

Table 7.4 Cross tabulation of land use map 2004 (column) against 2015 (row) (m^2)

	1	2	3	4	5	6	7	8	9	Sum (*A*)	Total (*A* − *B*)
1	2,846,634	3426	1,108,332	99,939	90,956	2226	7,760	48,522	0	4,207,795	−4,612,765
2	46,234	583,320	752,665	253,339	25,412	1046	58,028	37,807	0	1,757,851	773,111
3	1,556,335	36,073	5,622,998	400,534	530,654	26,626	58,332	149,094	734	8,381,380	−6,061,715
4	1,414,198	119,108	3,589,923	9,594,798	102,577	7080	49,765	446,807	27	15,324,283	4,163,995
5	1,376,475	57,218	717,746	57,014	3,631,480	8735	0	153,508	0	6,002,176	1,351,310
6	179,193	3491	301,225	119,950	26,974	828,368	2036	68,572	0	1,529,809	613,939
7	66,456	14,214	560,914	118,694	8680	3552	3,714,563	43,382	0	4,530,455	602,377
8	1,324,412	167,890	1,769,707	502,110	234,133	38,235	34,563	10,311,994	944	14,383,988	3,062,640
9	4007	0	19,479	13,910	0	1	3031	61,440	330,989	432,857	100,163
Sum (*B*)	8,820,560	984,740	14,443,095	11,160,288	4,650,866	915,870	3,928,078	11,321,348	332,694		

2012. By contrast, other land use types grew in all selected periods with different amounts and fluctuations. A slight change can be observed in the case of water bodies. A growth of 10 ha, which accounted for the change of private water bodies to public and/or the consideration of the wetlands as water bodies, occurred only from 2008 to 2012.

The low values in the columns representing the period of 2012–2015 indicate the similarity and smaller amount of changes between these two land use maps. Finally, the last column provides the summarized amount of changes in the land use types from the first to the last period (2004–2015) as explained previously. Proportional matrices were also produced from this process in addition to the area analysis by cross tabulation, which determined the proportional conversion of each land use type (Tables 7.6, 7.7, 7.8, 7.9). The values of proportional cross tabulation were calculated based on the area change in the previous trend with respect to the total area of the analysis. Defining a rectangular boundary around the current study area was necessary to perform cross-tabulation analysis. Thus, the total study area in this process was about 80,548,936 m^2 rather than actual area of Kajang City of 60,000,000 m^2. Therefore, all the land use change values in Tables 7.1, 7.2, 7.3 and 7.4 were divided based on the value of the rectangular study area (80,548,936 m^2).

The geographical distribution of the land use changes throughout the study area, which was produced from the cross-tabulation process, is shown in Figs. 7.9, 7.10, 7.11, 7.12. Many land use changes were expected in these figures because of the nine land use categories (maximum of 72

Table 7.5 Summarized cross tabulation during selected time period (hectare)

Land use category		2004–2008	2008–2012	2012–2015	2004–2015
Agriculture	1	−372	52	−141	−461
Commercial	2	12	64	1	77
Open spaces	3	−15	−634	43	−606
Housing	4	177	183	57	417
Industry	5	20	76	40	135
Infrastructure	6	13	48	0	61
Facility	7	4	56	0	60
Transportation	8	161	145	0	306
Water body	9	0	10	0	10

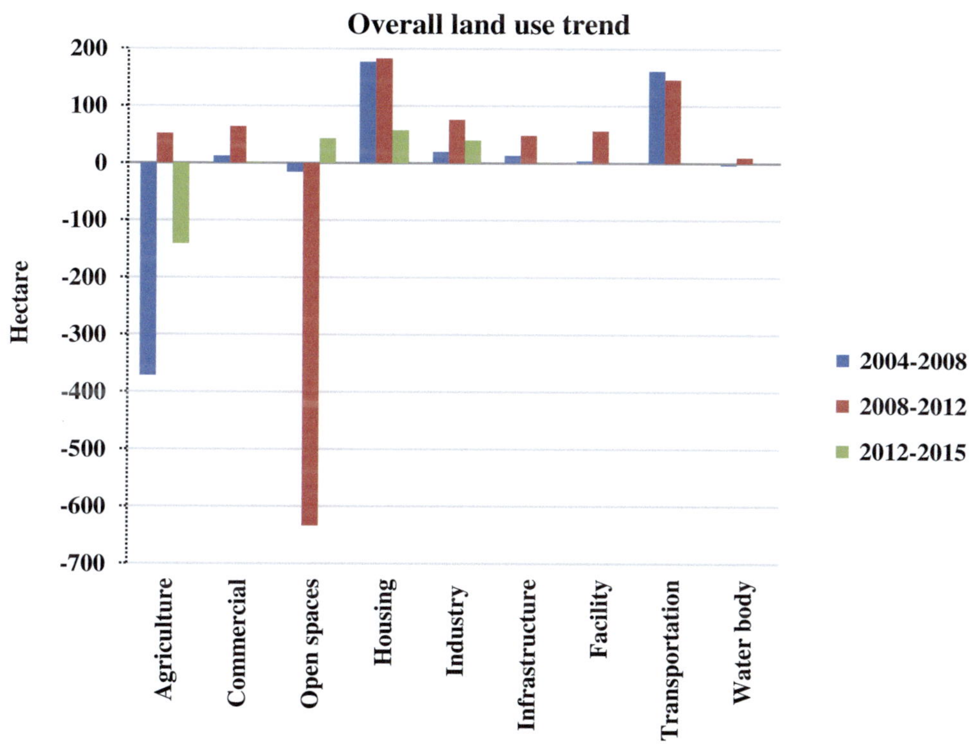

Fig. 7.8 Overall growth and loss of each land use type for selected period of times (ha)

changes). Showing all these changes was not possible, especially in the legend of the maps. In addition, only a few of these changes were significant in amount and value. Thus, unlike the cross-tabulation tables that present all the changes, the legends in these figures show only the main changes. Land use categories are listed in number format (agriculture = 1, commercial = 2, open spaces = 3, housing = 4, industry = 5, infrastructure = 6, community facility = 7, road network = 8, and water body = 9). Therefore, in Fig. 7.9, the legend can be explained as follows:

- 1–3: conversion of agriculture area to open spaces;
- 1–4: conversion of agriculture area to residential area;
- 3–2: conversion of open spaces to commercial use;
- 3–4: conversion of open spaces to residential area, and so on.

In addition to the quantitative information provided in tabular format, these figures provided a better view of the changes in land use types in the study area. As previously presented and explained, the east of Kajang City mainly consists of agricultural and forest lands. The eastern parts within the border of the city are also occupied by these land covers. Thus, a large proportion of the agricultural fields located in the eastern parts were converted to open spaces (1–3) as shown in Fig. 7.9 (land use changes from 2004 to 2008). These conversions mainly represent the land clearance process for redevelopment of new urban structures. However, these changes in some cases can be caused by agricultural phonological processes that appear similar to land use conversion from agriculture field to underused land. For instance, this assumption can be confirmed by checking three parcels in the northeast locations of Kajang City in both Figs. 7.9 and 7.10 (land use changes from 2004 to 2008 and from 2008 to 2012). These parcels were converted from agriculture to open space in the first map and from open space to agriculture in the second map. Some areas changed from agriculture to open spaces from 2004 to 2008, and then changed from open spaces to residential area from 2008 to 2012. Another main land use conversion during the period

Table 7.6 Proportional cross tabulation between 2004 and 2008

	1	2	3	4	5	6	7	8	9	Sum
1	0.0556	0	0.0071	0	0.0005	0	0	0.0001	0	0.0634
2	0.0002	0.0121	0.0014	0	0	0	0	0	0	0.0138
3	0.042	0	0.128	0.0061	0.0001	0.0003	0.0006	0.0003	0	0.1774
4	0.004	0	0.0246	0.1314	0	0.0001	0	0.0003	0	0.1605
5	0.0002	0	0.003	0	0.057	0	0	0	0	0.0602
6	0.0003	0	0.0019	0	0	0.0107	0	0	0	0.013
7	0.0001	0	0.0011	0	0	0	0.048	0	0	0.0493
8	0.007	0.0001	0.0121	0.0011	0.0002	0.0002	0.0001	0.1398	0	0.1606
9	0	0	0	0	0	0	0	0	0.0041	0.0041
Sum	0.1095	0.0122	0.1793	0.1386	0.0577	0.0114	0.0488	0.1406	0.0041	

Table 7.7 Proportional cross tabulation between 2008 and 2012

	1	2	3	4	5	6	7	8	9	Sum
1	0.0393	0	0.0254	0.0027	0.0012	0	0.0001	0.001	0	0.0698
2	0.0005	0.0082	0.0073	0.0034	0.0005	0	0.0007	0.0009	0	0.0217
3	0.0055	0.0006	0.0693	0.0093	0.0074	0.0015	0.0007	0.0043	0	0.0987
4	0.0076	0.0015	0.0337	0.1323	0.0013	0.0001	0.0005	0.0062	0	0.1832
5	0.0042	0.0007	0.0145	0.0018	0.046	0.0001	0	0.0022	0	0.0696
6	0.0008	0.0001	0.0039	0.0017	0.0004	0.0107	0	0.0012	0	0.019
7	0.0001	0.0003	0.0068	0.0016	0.0001	0	0.0468	0.0006	0	0.0562
8	0.0052	0.0023	0.0163	0.0075	0.0032	0.0004	0.0003	0.1434	0	0.1786
9	0	0	0.0003	0.0001	0	0	0	0.0008	0.0041	0.0054
Sum	0.0634	0.0138	0.1774	0.1605	0.0602	0.013	0.0493	0.1606	0.0041	

Table 7.8 Proportional cross tabulation between 2012 and 2015

	1	2	3	4	5	6	7	8	9	Sum
1	0.0523	0	0	0	0	0	0	0	0	0.0523
2	0	0.0217	0.0002	0	0	0	0	0	0	0.0218
3	0.0109	0	0.0928	0	0.0003	0	0	0	0	0.1041
4	0.0014	0	0.0057	0.1832	0	0	0	0	0	0.1903
5	0.0053	0	0	0	0.0692	0	0	0	0	0.0745
6	0	0	0	0	0	0.019	0	0	0	0.019
7	0	0	0	0	0	0	0.0562	0	0	0.0562
8	0	0	0	0	0	0	0	0.1786	0	0.1786
9	0	0	0	0	0	0	0	0	0.0054	0.0054
Sum	0.0698	0.0217	0.0987	0.1832	0.0696	0.019	0.0562	0.1786	0.0054	

Table 7.9 Proportional cross tabulation between 2004 and 2015

	1	2	3	4	5	6	7	8	9	Sum
1	0.0353	0	0.0138	0.0012	0.0011	0	0.0001	0.0006	0	0.0523
2	0.0006	0.0072	0.0093	0.0031	0.0003	0	0.0007	0.0005	0	0.0218
3	0.0193	0.0004	0.0698	0.005	0.0066	0.0003	0.0007	0.0019	0	0.1041
4	0.0176	0.0015	0.0446	0.1191	0.0013	0.0001	0.0006	0.0055	0	0.1903
5	0.0171	0.0007	0.0089	0.0007	0.0451	0.0001	0	0.0019	0	0.0745
6	0.0022	0	0.0037	0.0015	0.0003	0.0103	0	0.0009	0	0.019
7	0.0008	0.0002	0.007	0.0015	0.0001	0	0.0461	0.0005	0	0.0562
8	0.0164	0.0021	0.022	0.0062	0.0029	0.0005	0.0004	0.128	0	0.1786
9	0	0	0.0002	0.0002	0	0	0	0.0008	0.0041	0.0054
Sum	0.1095	0.0122	0.1793	0.1386	0.0577	0.0114	0.0488	0.1406	0.0041	

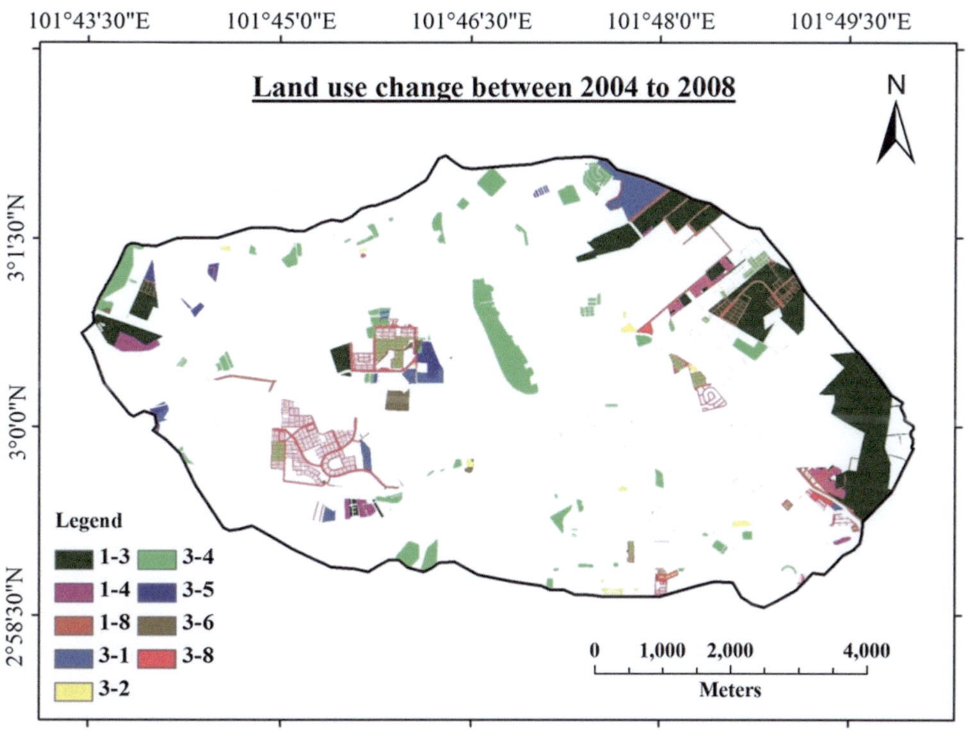

Fig. 7.9 Cross tabulation between years 2004 and 2008

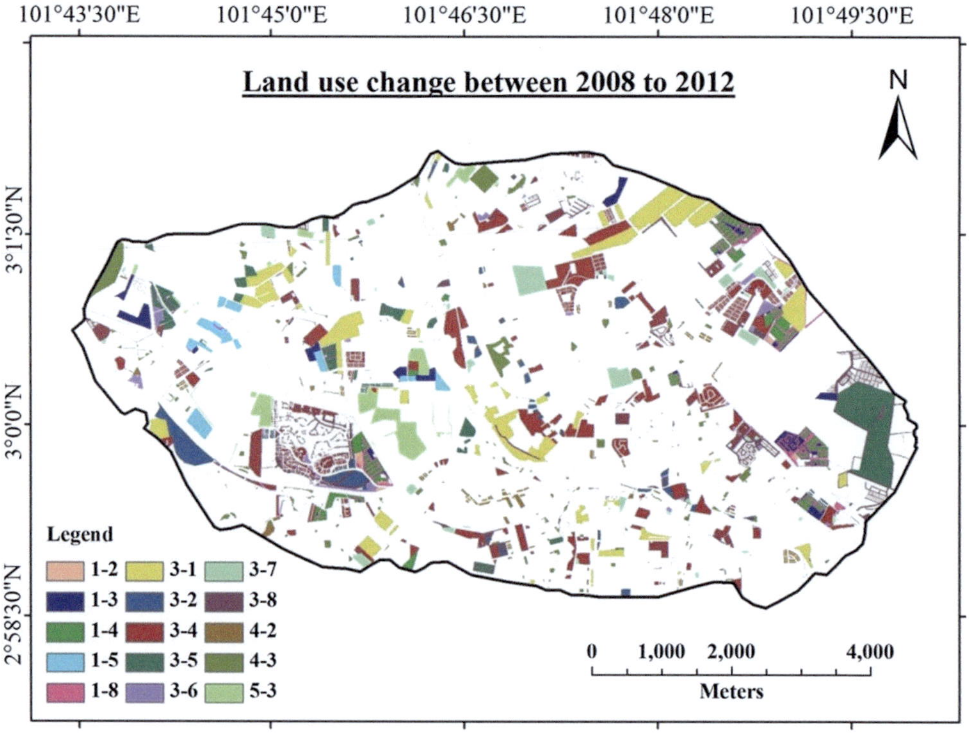

Fig. 7.10 Cross tabulation between years 2008 and 2012

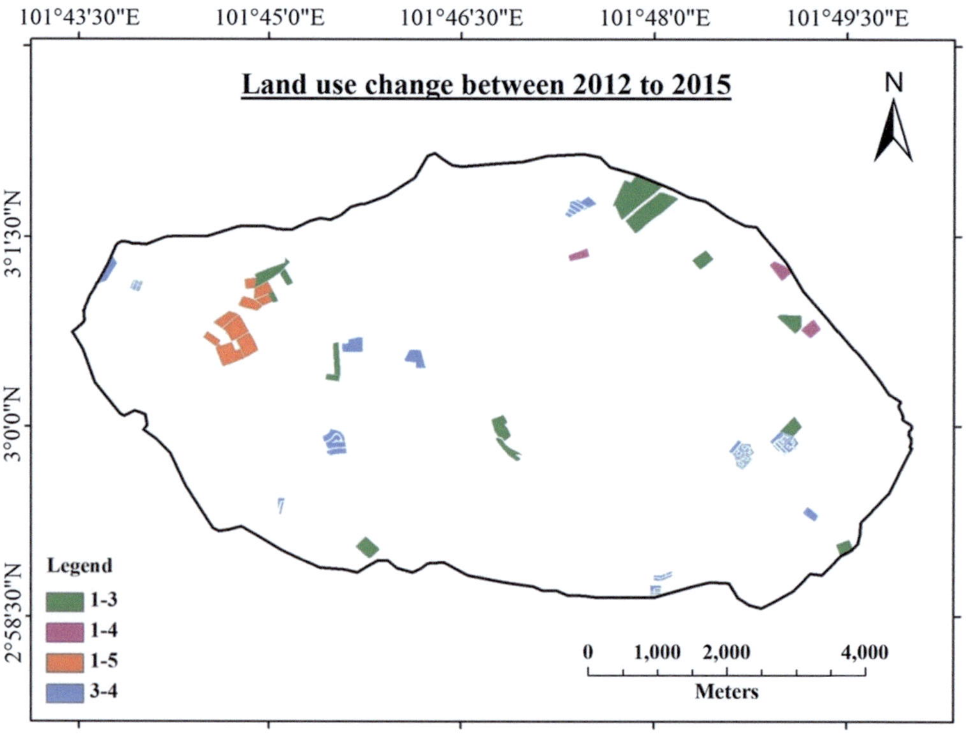

Fig. 7.11 Cross tabulation between years 2012 and 2015

from 2004 to 2008 is the change from open spaces to residential use (198.4 ha), which was distributed in the study area, especially in the central area and along the northern borders. Furthermore, the conversion of open spaces and agriculture fields to these land use types can be distinguished in all illustrated maps easily because of the texture of road networks. The quantitative outputs also showed that the road networks had a growth of more than 300 ha from 2004 to 2015 (Fig. 7.12).

The changes in other land uses to commercial and consequently the growth of the commercial area cannot be observed easily because the area of commercial land use

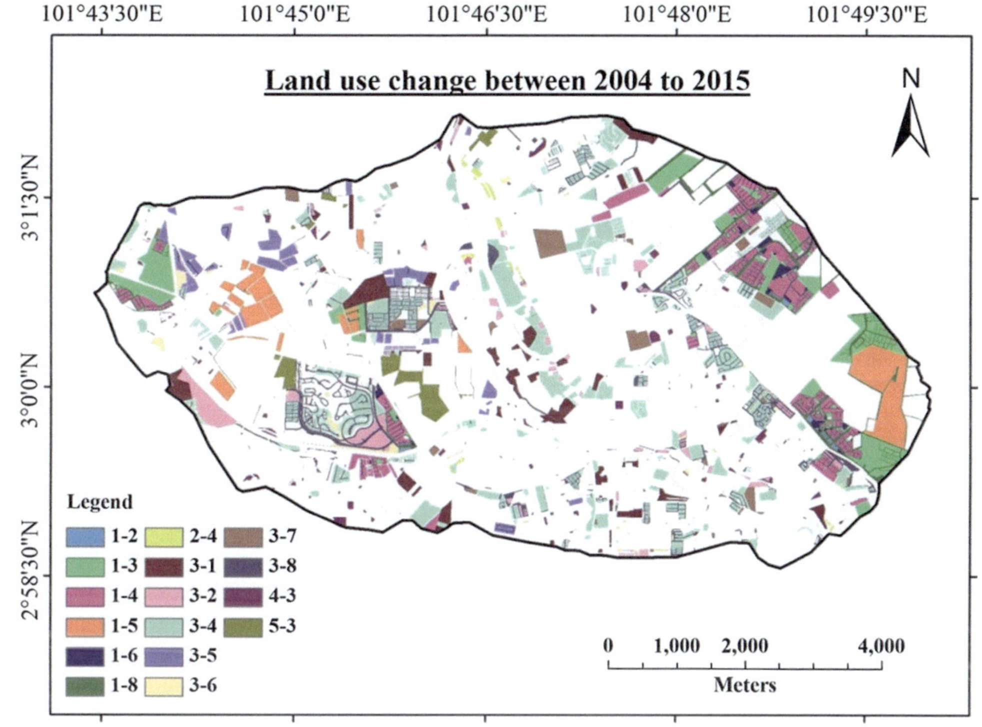

Fig. 7.12 Cross tabulation between years 2004 and 2015

types in Kajang City is small relative to other land use types. However, overlaying the road network of the study area and these land use change maps shows that commercial areas mainly grew along the major roads and public transportation.

These four cross-tabulation illustration maps show that the numbers of significant land use changes from 2008 to 2012 and from 2004 to 2015 are higher than others, as indicated in the figures. Figure 7.11 (land use change from 2012 to 2015) shows only four land use change types. These variations were dependent on the trend of changes during the selected period. A large proportion of land use changes, such as those from 2008 to 2012 or 2004 to 2015, caused a large number of land use changes, whereas similar land use maps (or a smaller proportion of changes), such as those in 2012 and 2015, caused a smaller number of land use change types. This assumption can also be evaluated from the calculation of the Kappa index. The overall Kappa value in the period from 2008 to 2012 was about 0.75 and the overall Kappa value from 2012 to 2015 was 0.97. Thus, the larger similarity between land use maps resulted in a higher Kappa value and fewer significant changes. By contrast, less similarity resulted in a lower Kappa value and a large amount of significant changes.

After examining the distribution of land use changes in the tabular and graphical presentations, the next task was to observe the relationships between each pair of land use maps. These assessments were conducted using degree of freedom, Cramer's *V*, KIA, and overall Kappa statistical approaches. Table 7.10 presents the results of these statistical tests to evaluate the cross-tabulation process.

These assessments generally revealed the dependency or similarity of each pair of land use maps. Previous results showed that among the three successive periods (2004–2008, 2008–2012, and 2012–2015), Kajang City had the highest amount of land use changes in 2008–2012. In addition, the land use maps for 2012 and 2015 had high similarity, indicating a very low amount of change during this period. Finally, large differences were expected between the land use maps for 2004 and 2015 because of the long period, indicating a large amount of land use change. These findings can be observed in Table 7.10 as well. The chi-square values in 2008–2012 and 2004–2015 were lower than those in the other periods. In addition, the Cramer's *V* and overall Kappa values of these two periods were lower than those of the other two. The 2012–2015 period had the highest chi-square, Cramer's *V*, and overall Kappa values because of their high similarity in land use pattern. All periods had constant degrees of freedom because of the same number of land use types and sampling assumptions. The KIA was calculated for each land use category using each pair of old and newer land use maps as reference and ranged from 0 to 1. Similar to the overall Kappa value, the KIA values of all the land use categories from 2012 to 2015 were 1 or near to 1. By contrast, the KIA values of land use types for 2004 to 2015 were from 0.2 to 0.99. The land use types with large changes during the

Table 7.10 Statistical assessment of cross-tabulation process

Variables	2004–2008	2008–2012	2012–2015	2004–2015
Degree of freedom	81	81	81	81
Cramer's *V*	0.8992	0.736	0.9713	0.7164
Overall Kappa	0.8594	0.7547	0.9711	0.7135

selected period had low KIA values, such as 0.2 for agricultural and 0.3 for open spaces, and land use types with very minor changes had high KIA values, such as 0.99 for water bodies. Thus, these statistical assessments provided an overall view of the land use change process and relationship or similarity among the various land use maps and even among the land use categories.

Table 7.11 presents the overall number of land use types in each year of the available data. This table provides the same information obtained by the cross-tabulation process from another perspective. By subtracting the area of agriculture in 2004 (882) from the area in 2015 (421), the value of −461 is achieved, which is provided in Table 7.5. The growth of the residential area (1533–1116) during the selected period (2004–2015) can also be calculated from this table.

The residential and transportation land uses had the most significant growth rates during the selected period, followed by the commercial and industrial land uses, based on the results of the change detection process for the available period to assess the historical trend in the study area. Considering that the growth of some land use types is dependent on the growth of others is important. For instance, community facilities are usually provided for the neighborhood based on the local demand. Thus, the growth of facilities and services are mainly dependent on the population and/or residential area of a community. Similarly, transportation and road network is developed where a proper residential, commercial, and/or build-up area exist in a neighborhood. In most cases, construction companies provide suitable road networks with their corresponding developments (whether commercial, industrial and/or residential development). Therefore, some land use types, such as community facility, infrastructure, and road network, are not developed independently. By contrast, residential, commercial, and industrial land uses are growing because of driving factors and parameters rather than the effects of other land use types. Industrial growth is mainly based on economic issues and government policies. Residential growth is not a new problem and is mainly due to population growth and migration of people from rural to urban areas (as discussed in detail in previous chapters). Commercial growth also depends on local demands, governmental policies, and economic perspectives.

Cross-tabulation provides suitable measurements, identification, and illustration of land use change to evaluate the recent growth and development of a study area. Thus, this assessment provided a clear understanding of the growth rate and direction of built-up areas through various land uses and land cover categories. Finally, this process revealed the significant growth and/or loss of each land use type to highlight the unsustainability of the current development pattern and to propose new alternative scenarios.

Finally, the land use growths in the selected years were compared with the master plan of the study area, as shown in Table 7.12. This table indicates that the agricultural area is completely removed from the Kajang City master plan, but proper green environment areas are proposed within the city. Enterprise zones are the areas that have potential to increase the livability of the city, such as central business district and/or transit-oriented development.

Most of the land use categories still have capacities to grow according to the master plan, but this area is still capable and has growth potential without clearing the

Table 7.11 Total area of each land use category in selected years (hectare)

Land use category		2004	2008	2012	2015
Agriculture	1	882	510	562	421
Commercial	2	98	111	174	176
Open spaces	3	1444	1429	795	838
Housing	4	1116	1293	1476	1533
Industry	5	465	485	561	600
Infrastructure	6	92	105	153	153
Facility	7	393	397	453	453
Transportation	8	1132	1293	1439	1439
Water body	9	33	33	43	43
Sum		5656	5656	5656	5656

Table 7.12 Comparison of land use maps with master plan of the Kajang City (hectare)

Land use	2004	2008	2012	2015	Master plan	
Agriculture	882	510	562	421	Enterprise zone	334
Commercial	98	111	174	176	Commercial	265
Open spaces	1444	1429	795	838	Green spaces	380
Housing	1116	1293	1476	1533	Housing	2208
Industry	465	485	561	600	Industry	672
Infrastructure	92	105	153	153	Infrastructure	186
Facility	393	397	453	453	Facility	418
Transportation	1132	1293	1439	1439	Transportation	1528
Water body	33	33	43	43	Water body	41
					Agriculture	0

agricultural fields. The agricultural fields can also be preserved by proposing a compact land use pattern with higher density.

7.4 Conclusion

This study successfully highlighted and discovered the spatiotemporal urban land use change patterns in the Kajang City area. In general, this process found significant land use growths and changes in Kajang City from 2004 to 2012. By contrast, in the last period (2012–2015), very insignificant growths and changes occurred in this region because of several reasons, such as shorter time step, development saturations, and/or planning policies. In addition, the results of this analysis indicated that residential, commercial, and industrial land use types have significant growths compared with other land use types. Furthermore, the growth of these three land use types mainly resulted in the reduction of agricultural fields and open spaces. In most of the periods, the seemingly encompassing effect of residential land use on almost all types of activities was observed. However, this growth through open spaces and agricultural areas was more significant. Each land use type had different behaviors, amounts, and directions of growth, which were evaluated with respect to various land use types and several external factors in fulfillment of the second objective. Nevertheless, a comprehensive quantitative assessment was performed for the first objective to deal with various changes among all land use categories. This assessment was performed through several cross-tabulation matrices for each pair of land use maps: 2004–2008, 2008–2012, 2012–2015, and 2004–2015. A similar finding from the cross tabulations of the first to the last period was the increase in the land consumption rate, which shows the growth of urban areas through natural environments. Thus, an alternative development pattern is required to minimize the quantity of urban land consumptions based on these assessments.

References

Abdullahi, S., Mahmud, A. R. b., & Pradhan, B. (2014). Spatial modelling of site suitability assessment for hospitals using geographical information system-based multicriteria approach at Qazvin city, Iran. *Geocarto International, 29*(2), 164–184.

Abdullahi, S., & Pradhan, B. (2015). Sustainable brownfields land use change modeling using GIS-Based weights-of-evidence approach. *Applied Spatial Analysis and Policy*, 1–18.

Al-shalabi, M., Billa, L., Pradhan, B., Mansor, S., & Al-Sharif, A. A. (2013). Modelling urban growth evolution and land-use changes using GIS based cellular automata and SLEUTH models: the case of Sana'a metropolitan city, Yemen. *Environmental Earth Sciences, 70*(1), 425–437.

Al-sharif, A. A., Pradhan, B., Shafri, H. Z. M., & Mansor, S. (2013). Spatio-temporal analysis of urban and population growths in Tripoli using remotely sensed data and GIS. *Indian Journal of Science and Technology, 6*(8), 5134–5142.

Almeida, C., Monteiro, A. M. V., Câmara, G., Soares-Filho, B. S., Cerqueira, G. C., Pennachin, C. L., et al. (2005). GIS and remote sensing as tools for the simulation of urban land-use change. *International Journal of Remote Sensing, 26*(4), 759–774.

Barnes, K. B., Morgan, J. M., III, Roberge, M. C., & Lowe, S. (2001). *Sprawl development: its patterns, consequences, and measurement* (pp. 1–24). Towson: Towson University. Available online at: http://chesapeake.towson.edu/landscape/urbansprawl/download/Sprawl_white_paper.pdf. Last accessed February 13, 2008.

Bhatta, B. (2009). Analysis of urban growth pattern using remote sensing and GIS: a case study of Kolkata, India. *International Journal of Remote Sensing, 30*(18), 4733–4746.

Bhatta, B., Saraswati, S., & Bandyopadhyay, D. (2010). Quantifying the degree-of-freedom, degree-of-sprawl, and degree-of-goodness of urban growth from remote sensing data. *Applied Geography, 30*(1), 96–111.

Craglia, M., Haining, R., & Signoretta, P. (2001). Modelling high-intensity crime areas in English cities. *Urban Studies, 38*(11), 1921–1941.

Flint, C. (2002). The theoretical and methodological utility of space and spatial statistics for historical studies: The Nazi Party in geographic context. *Historical Methods: A Journal of Quantitative and Interdisciplinary History, 35*(1), 32–42.

Hu, Z., & Lo, C. (2007). Modeling urban growth in Atlanta using logistic regression. *Computers, Environment and Urban Systems, 31*(6), 667–688.

Kaveckis, G., & Bechtel, B. (2014). *Land use based urban vulnerability to climate change assessment.* Paper presented at the

Environmental Engineering. Proceedings of the International Conference on Environmental Engineering. ICEE.

Kuzera, K., & Pontius, R. G. (2008). Importance of matrix construction for multiple-resolution categorical map comparison. *GIScience & Remote Sensing, 45*(3), 249–274.

López, E., Bocco, G., Mendoza, M., & Duhau, E. (2001). Predicting land-cover and land-use change in the urban fringe: A case in Morelia city, Mexico. *Landscape and Urban Planning, 55*(4), 271–285.

Pontius, R. G., Jr., & Millones, M. (2011). Death to Kappa: Birth of quantity disagreement and allocation disagreement for accuracy assessment. *International Journal of Remote Sensing, 32*(15), 4407–4429.

Ren, P., Gan, S., Yuan, X., Zong, H., & Xie, X. (2013). Spatial expansion and sprawl quantitative analysis of Mountain City built-up area. *Geo-Informatics in Resource Management and Sustainable Ecosystem* (pp. 166–176). Springer.

Sandhya Kiran, G., & Joshi, U. B. (2013). Estimation of variables explaining urbanization concomitant with land-use change: A spatial approach. *International Journal of Remote Sensing, 34*(3), 824–847.

Sharma, L., Pandey, P. C., & Nathawat, M. (2012). Assessment of land consumption rate with urban dynamics change using geospatial techniques. *Journal of Land Use Science, 7*(2), 135–148.

Soffianian, A., Nadoushan, M. A., Yaghmaei, L., & Falahatkar, S. (2010). Mapping and analyzing urban expansion using remotely sensed imagery in Isfahan, Iran. *World Applied Sciences Journal, 9*(12), 1370–1378.

Zhang, B. (2003). Application of remote sensing technology to population estimation. *Chinese Geographical Science, 13*(3), 267–271.

Zhang, Y., Yang, Z., & Li, W. (2006). Analyses of urban ecosystem based on information entropy. *Ecological Modelling, 197*(1), 1–12.

8 Spatial Land Use Change Modeling Techniques

Abubakr A.A. Al-sharif, Biswajeet Pradhan and Saleh Abdullahi

8.1 Introduction

A proper understanding of the reason, degree, direction, and consequences of urban growth and expansion is essential for most urban application projects. Urban growth and land use changes are the main reasons for environmental, social, and economic issues, such as hydrological problems, destruction of forests and agricultural fields, natural and wildlife disturbance, and global warming (Wang 2012). In this regard, understanding these change behaviors improve environmental sustainability through several actions, such as managing land use and land cover, overseeing rural development, and advancing land use change modeling and prediction (Veldkamp and Lambin 2001; Abdullahi et al. 2015). Thus, models and simulation techniques are required to deal with these issues effectively. The utilization of models in scientific research represents the natural behaviors and reactions in the real world (Liu 2008). However, the behaviors of real-world phenomena are very complex and multidimensional. Some simplifications and predefined assumptions are required to understand and investigate these processes. The proposed models should be comprehensive and applicable enough to support urban growth and create a better and clearer view of the function of this process. These models can be used as powerful tools to increase our mental capabilities and make more informed decisions regarding land use changes (Costanza and Ruth 1998).

In 1994, the importance of land use change process was understood through the core project of the International Geosphere–Biosphere Program (IGBP) and the International Human Dimensions Program (IHDP) (Verburg et al. 2006; Veldkamp 2009). This project was proposed to advance and improve the knowledge on human and biophysical dynamics of the land use change process and to propose strong models for future land use prediction (Wang 2012).

Historically, many models were applied to urban applications after the quantitative revolution in geographical science from the 1950s to the 1960s, as shown in Fig. 8.1 (Wrigley and Bennett 1981). In the beginning of the nineteenth century, Johann Heinrich von Thünen developed a simple land use and urban growth modeling theory, which explains how market processes affect and control the spatial distribution of land use changes and urban growth (Candau 2002; Parker 2015). In 1926, Burgess proposed the concentric zone theory, which presents a city as a series of concentric land use circular zones centered on the central business district (CBD) (King 1985). In 1933, Walter Christaller formulated the central place theory, which explains the size, number, spatial distribution, and hierarchical arrangement of cities (King 1985). This theory is also concerned with the distribution of retail and wholesale administrative land uses and community facilities. Sector theory is based on the idea that functional land use regions expand in wedge-shaped zones radiating outward from the CBD (Torrens 2000). Multiple nuclei theory is based on the simple fact that most large cities have various hubs that serve as centers of agglomerative growth instead of the simple CBD (Torrens 2000). Finally, bid-rent theory was also based on the theory of von Thünen and considers several other urban factors, such as transportation. It assumes that rents generally tend to decrease correspondingly when transportation costs increase with distance from the markets.

Discovering the behaviors of urban growth and land use changes historically was possible based on these simple principles. However, the dynamism and continuous growth characteristics of urban areas led to the complexity of modeling these environments. The land use change phenomenon within an urban area is also a complex process because it is the result of the interaction between various issues, such as environmental, physical, and political (Medley et al. 1995). This phenomenon is mainly dependent on the spatial location, scale, and current state of land use (Lambin et al. 2001; Lesschen et al. 2005). Understanding the reasons and rate of land use change is important because

B. Pradhan (ed.), *Spatial Modeling and Assessment of Urban Form*,
DOI 10.1007/978-3-319-54217-1_8

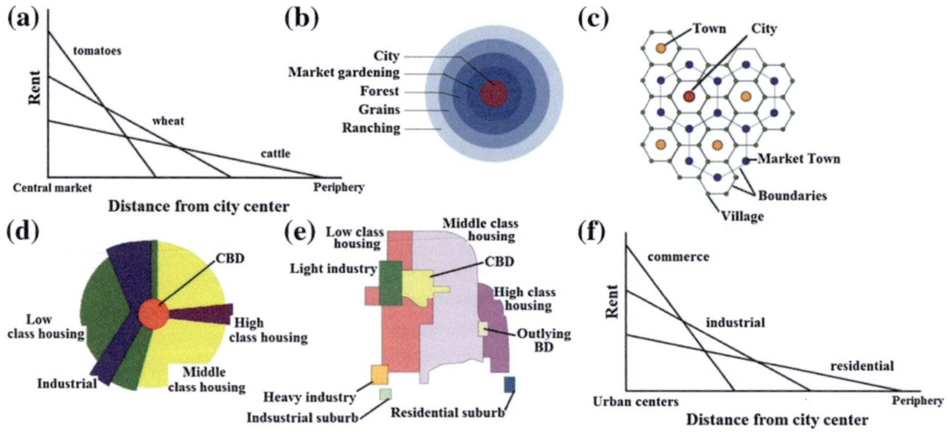

Fig. 8.1 Historical urban growth modeling: **a** Johann Heinrich von Thünen theory, **b** concentric zone theory, **c** central place theory, **d** sector theory, **e** multiple nuclei theory and **f** bid-rent theory (King 1985; Candau 2002)

of their significant effects on the surrounding natural environment, air and water quality, local temperature, and urban economy, as well as other social impacts (Burchell 1996; Pijanowski et al. 2002; Mellino and Ulgiati 2015). Thus, models are required to:

- investigate the various issues that cause land use change,
- project the effects of these changes on the environment and economy, and evaluate the effects of policies and scenarios on land use growth and development (Pijanowski et al. 2002).

In recent years, advanced models provided artificial environments and used various statistic-, factor-, and cellular-based concepts to conduct different analyses to understand and explain urban behaviors. In spite of all attempts, no clear superior approach toward modeling land use change and urban growth exists, and no methodology can answer all questions in these processes (Verburg et al. 2006). The existing models are able to analyze, simulate, and predict land use changes based on different concepts and theories. According to the literature, the following four core principles are the bases of all land use change simulation models (Koomen and Borsboom-van Beurden 2011);

- Historical bases,
- Suitability bases,
- Neighbourhood bases, and
- Actor interaction bases.

The logic behind historical bases is "the past is the key to the future." Therefore, background information can be helpful in predicting future land use changes, as demonstrated by Kuijpers-Linde et al. (2007). Suitability bases may consist of several factors (such as physical, social, and environmental) of a land parcel to evaluate the allocation for a specific purpose. For example, site suitability evaluation for a specific use, such as hospital, and school (Abdullahi et al. 2014). Therefore, the underlying premise is to achieve maximum profit and minimize the negative effects. Neighborhood bases deal with the neighborhood interaction of each cell that affects the transition of one land use to another. Numerous studies based on this cellular concept, which is implemented by cellular automata approach, exist (Wu 1998; Kocabas and Dragicevic 2007; Liu 2008). Actor interaction bases assume that land use change is the result of the interaction among several actors or agents. This core principle is one of the promising research tools for land use change modeling (Matthews et al. 2007).

Verburg et al. (2004a, b), Heistermann et al. (2006) and Koomen and Stillwell (2007) categorized land use change simulation models based on six main concepts:

- Markov chain,
- Economic based,
- Agent based modeling (ABM),
- Statistical analysis,
- Cellular automata (CA), and
- Artificial neural network (ANN).

Most of these categories have some factors in common, but the variety of approaches makes these studies difficult to compare. On a one-to-one basis, each concept and approach has its own merits and demerits. However, all these concepts are always based on the four core principles, as explained, and their main aim is to explain and translate reality into a model.

The Markov Chain concept was first proposed by Burnham (1973) and is based on the continuation of historical trend of development. This concept calculates the probability matrix of change of one land use type to another. The main disadvantage of this model is the lack of spatial bases of the results (Dadhich and Hanaoka 2011). Therefore, integration with other spatial-based methods is required eventually (Koomen and Borsboom-van Beurden 2011). Although the economy-based model is not exactly a concept, it is an important reason for land use changes based on the suitability principle of land. This concept is based on the von Thünen theory, which states that the land continues to be used to produce a commodity as long as its profit is higher than its transportation costs (Koomen and Borsboom-van Beurden 2011). A landowner usually seeks to maximize profits to change the land use type or to sell the land.

An agent-based model of land use change modeling consists of two main components: a map of the study area and a model with agents that represent human decisions (Parker et al. 2003). An agent is a representation of actors important in the process and can be either a single actor or a group of actors with their own preferences (Grimm et al. 2006). The preferences of these actors can be determined by expert knowledge or ANN. A variety of statistical computations can be derived from land use maps. For example, logistic regression, frequency ratio, and weights of evidence techniques can be used to analyze the probability of occurrence of dependent variable in each class of independent variables (Verburg et al. 2004b). The coefficients of each variable can be calculated from the historical land use changes and projected for future land use changes. A logit model using neighborhood interaction, historical change, and suitability factors is a good example of a statistical concept. Land use scanner (Hilferink and Rietveld 1999) has been used to achieve urban sustainability for the Netherlands as a logit model for land use policy and management (Beurden et al. 2007; Kuijpers et al. 2007).

The use of ANN in land use change modeling has increased recently because of the advancement in computing performance and availability of powerful and flexible software (Skapura 1996). ANN is very useful for land use change modeling because of its self-learning and pattern recognition abilities (Pijanowski et al. 2002). Another advantage of ANN is its ability to relate past and future land use changes and their linkage to suitability maps. In this manner, a model can train itself to corresponding land use maps of different years to recognize and reproduce the pattern of land use types (Mas et al. 2004; Pijanowski et al. 2005).

Therefore, appropriate knowledge of land use change models based on their concepts allows modelers to select the most appropriate approach for the study area under investigation. More explanations on these common approaches will be provided in the following sections.

8.2 Statistical-Based Approaches

Several types of statistical information can be extracted from urban or land use maps. Such information can be based on the four core principles mentioned (historical, suitability, neighborhood, and actor interaction) depending on the research objective. Several land use change models, which are based on the statistical relationship between different land use periods, exist to predict future changes. Among these models, conversion of land use and its effects (CLUE) (Veldkamp and Fresco 1996), CLUE-S (Verburg et al. 2004a), and GEOMOD (Pontius et al. 2001) are some well-known examples. Traditional models such as logistic regression (LR) and Markov chain, with their own merits and demerits, are widely used in urban applications (López et al. 2001; Harrell 2015; Malaitham et al. 2015). The Markov chain model provides two matrices, namely, transitional probability and transitional area matrices, which are proven stronger at the descriptive than at the predictive level (López et al. 2001).

One of the most common statistical analyses can be performed by the computer program FRAGSTATS by McGarigal and Marks (1995), which is able to process many indices related to urban statistics. In urban applications, land use change modeling, linear regression (Verburg et al. 2004b), probit regression, binomial logit and multinomial logit models (DeMaris 1992) are some techniques that are utilized to evaluate the statistical relations of land uses and consequently projects future changes (Lantman et al. 2011). For instance, LR analyzes the probability of occurrence, which is dependent on several variables, of a specific land use type (Verburg et al. 2004b), such as social and physical properties. Historical land use maps are used to compute coefficients for land use changes to predict future patterns. Wu and Yeh (1997) and Hu and Lo (2007) presented the ability of the LR model to interpret urban growth based on probability assessments of land use changes. A logit model can be based on neighborhood interaction, historical land use change, soil suitability, spatial location, or combinations of these. Binary and multinomial logit are based on the same mathematical concept but use prediction of single land use and different land use changes, respectively (DeMaris 1992; Liao 1994). Walsh et al. (2003) used a multinomial approach to analyze the spatial association of various land use categories.

The prediction of land use change requires a good understanding of the actual processes that drive the change (Riebsame et al. 1994). Although statistic-based models are easy to build, they lack a theoretical basis to understand and simulate the actual driving forces of land use change (Koomen and Stillwell 2007). Statistical models have the disadvantage of ignoring high spatial and biophysical variability of land cover types, as well as socioeconomic and institutional driving forces of change even though these are suited to simulate the possible changes over a short period of time (Serneels and Lambin 2001; Koomen and Stillwell 2007).

In general, two alternatives can be used to apply multiple classifications (Tayyebi and Pijanowski 2014). One approach is to run the model by several binary classifications that are solved using binary classifiers. In this approach, the probability of growth for each main land use type is evaluated separately by decomposing the model into several binary classifications. The change or growth of one class is normally evaluated with respect to all others for these types of classifications (One-Versus-All), or all possible mutual binary classifiers between n available classes can be considered (All-Versus-All). These processes are lengthy and difficult to analyze, especially for large numbers of binary classifiers. This concept can be extended for multiple land use change as well. Thus, numerous binary classifiers are used to solve multiple binary classification problems simultaneously (Tayyebi and Pijanowski 2014).

8.2.1 Frequency Ratio Model

One of the common concepts of statistical analysis is evaluating the frequency of occurrence of a phenomenon with respect to several driving factors (Lee and Pradhan 2006, 2007; Pradhan and Lee 2010; Naghibi et al. 2015). In this regard, frequency ratio (FR) is a univariate probability-based approach that is widely applied in natural hazard management studies, such as landslide modeling, flood modeling, and earthquakes, to produce susceptibility maps and identify and analyze hazard occurrence (Ozdemir and Altural 2013). FR is based on the observed relationship between one dependent variable (any phenomenon) and several independent variables (driving factors that affect the corresponding phenomenon). Thus, this process shows the correlation among these variables. In urban application, land use growth and change occurrence can be evaluated with respect to several urban-related factors, such as proximity and population density. This model is simple and its input, calculation process, and results are clear and understandable (Lee and Pradhan 2007; Pradhan and Lee 2010). FR provides a straightforward geospatial assessment technique to calculate the probability relationship between dependent and independent factors. In fact, FR is the ratio of the area where the phenomenon (land use change) occurred to the total area of interest and the ratio of the land use change occurrence probability to the nonoccurrence for the selected factors. Park et al. (2011) applied this model to simulate urban growth patterns and predict the future probable urban growth of the entire country. In addition, the overall urban extent for model calibration and validation processes was used instead of real net urban expansion.

8.2.2 Weights of Evidence Model

Statistic-based techniques are able to apply and integrate transitional rules in the evaluation process of land use change modeling application. In this regard, Weight of Evidence (WoE) is a well-known statistical method based on the Bayes theorem of conditional probability. This method is a global parametric approach that estimates probabilities from the linear regression of a dependent phenomenon against independent variables. Tayyebi et al. (2014) stated that regression models provide better explanatory power and outperform certain methods, such as ANNs, when the functional relationships between the dependent and independent variables are known.

In general, WoE evaluates the degree to which evidence supports the hypothesis (for instance land use change occurrence) and the degree to which the evidence does not refute the hypothesis (Dempster 1967; Shafer 1976). This model is applicable when enough information is available to evaluate the relative importance of evidential themes through statistical concepts (Bohman-Carter 1994). This method allows the identification of the influence of spatial determinants on the analyzed transitions. The WoE can combine spatial data from several disciplines and sources to explain and evaluate interaction, support the decision-making process, and create predictive models (Almeida et al. 2008).

In urban growth and land use change processes, this model can detect the reason for a certain land use change through several variables, which are known as pieces of evidence in this model. Weights are calculated by evaluating the relation between the land use change occurrence and values of the evidence. As discussed in the literature, WoE has been extensively used in a variety of applications, such as geological and mineral mapping (Gettings et al. 2004; He et al. 2010; Chen et al. 2014) and natural disaster management (Pradhan et al. 2010; Althuwaynee et al. 2012; Bui et al. 2012; Pourghasemi et al. 2013). However, this approach in urban applications and land use dynamic modeling has been utilized in few studies, such as those by Thapa and Murayama (2011), Teixeira et al. (2009), de Almeida et al. (2003) and Abdullahi and Pradhan (2015).

Checking the independency among pairs of selected variables is important in the WoE modeling process (Almeida et al. 2002). Cramer's coefficient (V) and joint information uncertainty (U) are two common methods for this evaluation. Obtaining values from an area cross-tabulation between pairs of maps of variables under analysis is required for both methods.

8.2.3 Evidential Belief Functions (Dempster–Shafer) Model

The Dempster–Shafer theory of evidence refers to the generalization of the Bayesian theorem of subjective probability. Proposed by Dempster in 1967 and developed by Shafer in 1976, this theory combines beliefs from several sources of evidence and the relative flexibility to accept uncertainty (Thiam 2005). The theory predicts how closely the evidence shows the certainty of a hypothesis rather than guessing at the possibilities that a hypothesis is correct (Pearl 1990). The Dempster–Shafer theory has been applied effectively in many applications using GIS.

In applying the EBF model in the urban expansion modeling process, a set of urban growth driving factors $C = (C_i, i = 1, 2, 3, \ldots, n)$ is assumed to comprise mutually exhaustive and exclusive factors C_i. C is named the frame of discrimination. The function m: $P(C) \rightarrow [0, 1]$ is a simple probability assignment, where $P(C)$ is the set of whole subsets of C, including the empty set and the C set itself. The m function can be considered as a mass function that satisfies $m(\Phi) = 0$ and $\sum_{AC} m(A) = 1$, where Φ is the empty set and A can be any subset of C. $m(A)$ estimates the level to which the evidence supports A and is a belief function Bel(A).

The theory demonstrates four basic EBF functions, namely, Dis (degree of disbelief), Bel (degree of belief), Pls (degree of plausibility), and Unc (degree of uncertainty). Dis represents the belief of the suggestion being untrue based on given evidence. Bel and Pls provide the upper and lower bounds, respectively, of the probability for the suggestion (Awasthi and Chauhan 2011). Unc means ignorance, that is, the difference between plausibility and the belief. Therefore, $1 - \text{Unc} - \text{Bel}$ or $\text{Dis} = 1 - \text{Pls}$, and always $\text{Dis} + \text{Bel} + \text{Unc} = 1$. For cases of C_{ij} with no urban expansion, that is, Bel = 0, Dis is reset to 0 even though $D \neq 0$ (Carranza et al. 2008).

8.2.4 Logistic Regression Model

Several researchers have employed various empirical and theoretical modeling techniques to model, simulate, and predict urban sprawl or growth and land-use changes. One of these techniques is an empirical estimation model called the LR model. According to the literature, LR in urban growth or sprawl modeling results in a good understanding of the urbanization process and provides a clear picture of the weight of independent variables and their respective functions (Hu and Lo 2007; Eyoh et al. 2012).

The LR model enables the integration of demographic and socioeconomic factors that are not available in many models. In addition, it considers spatial effects, autocorrelation, and heterogeneity (Devkota et al. 2013). However, the model requires caution regarding spatial autocorrelations that typically exist in spatially referenced data because such autocorrelations may violate the hypothesis of the LR model (Lin et al. 2011). Understanding and quantifying the interaction between the driving forces of land use/cover change in the LR models is a complex and difficult process, hence the need to overcome the misunderstanding and lack of information on driving forces. Certain drawbacks of the LR restrictions also need to be considered (Lin et al. 2011). This model is used to demonstrate and explain the relationship of a number of X independent variables to a dichotomous single dependent variable Y, which represents the occurrence or nonoccurrence of an event. LR empirically finds the relationship between the independent variables and the function of the probability of an event happening (Kleinbaum and Klein 2010).

The use of LR can yield the coefficients of independent variables (both continuous and categorical); the dependent variable is a binary categorical variable with a value of either 1 or 0 and can be computed using the well-known LR equation (Huang et al. 2009). The LR model is applied in urban expansion modeling and land use change analysis, as shown in the literature. It provides the probability of the existence or nonexistence of each type of land-use/cover in every location based on driving factors. LR is a powerful empirical method used when the outcome-dependent variable is dichotomous. Spatial urban expansion is the dependent variable represented in a raster binary map. A value of 1 on the produced probability map indicates the presence of urban growth, and a value of 0 indicates the absence of urban growth.

8.3 Agent-Based Models

The agent-based modeling (ABM) approach is based on the science of artificial intelligence and the object-oriented technique of modeling the interaction of the individual units of a system (Parker et al. 2003; Matthews et al. 2007). ABM consists of various and interrelated agents, which are the decision-making components identified with a series of rules or behaviors that allow the agents to acquire information, process the input, and effect changes in the external environment (Arsanjani et al. 2013; Pooyandeh and Marceau 2013). In land change modeling, agents can be land owners, farmers, collectives, migrants, and management agencies. In fact, anyone who makes decisions or take actions that cause change in land use patterns and processes (Brown and Geist 2006; Arsanjani et al. 2013) can be considered an agent. In urban areas, multi-agent systems are very good tools to represent movable entities, such as vehicles and people. Multi-agent systems have been used to model the relocation of households (Benenson 1998) and to simulate pedestrian movement in dense urban environments (Kerridge et al. 2001). In urban land change modeling using ABM, decisions often depend on the physical environment of the agent (the landscape), but may also depend on what other agents do (Parker et al. 2003). Compared with top-down models, such as statistical models, ABM has the following advantages (Brown and Geist 2006):

- ABM can deal with emergent phenomena as characteristics of complex systems, such as self-organization, chaos, and adaptation.
- ABM is flexible in terms of designing geospatial models, which means that ABM can use different levels of description and aggregation of single, aggregate, or subgroups of agents. Furthermore, ABM provides a structure for adjusting the complexity of the agents in terms of their behaviours.
- ABM describes a system naturally. It represents a natural behaviour in the description and simulation of agent behaviours and makes the models close to reality.

However, ABM also has disadvantages in several cases, such as land change modeling and urban growth modeling (Crooks et al. 2008).The disadvantages are as follows:

- Although ABM has been integrated with GIS, it does not have powerful tools to represent patterns of phenomena.
- A number of agents and their attributes interact with one another and with their environment. Thus, a single run of an ABM cannot provide any trustworthy information, and agent computing must be performed through multiple runs that systematically change initial conditions or parameters to assess the robustness of results.
- Critical issues regarding the validation and calibration of ABMs hinder the practice of such models.

8.4 Rule-Based Models

Rule-based models enable users to include explicit decision rules that direct their behavior (Brown et al. 2005; Klosterman and Pettit 2005). A well-known example of rule-based land-use model is the "What If?" model. "What if?" is a commercial, stand-alone GIS-based software package and a scenario-based and policy-oriented planning support system (PSS) used to conduct land suitability analysis, estimate future land-use demand, and allocate these projected demands to the most suitable locations (Klosterman 1999). "What If?" is capable of simulating future land-use patterns by balancing the supply of and demand for lands suitable for different uses at different locations. The flexible features of "What If?" allow the modeler to simulate the consequences of spatial decisions, which makes it a useful PSS (Geertman and Stillwell 2004; Koomen et al. 2008).

8.5 Artificial Neural Networks

ANN is a system that consists of several processing nodes that work in equivalent ways. These processing elements are defined by the network structure, connection strength, and processing performance at computing nodes. The development of an ANN model requires the description of a learning paradigm, a learning algorithm, and a network topology (Fig. 8.2). ANN is different from other commonly used analytical approaches because it does not depend on specific functional relationships, does not adopt any assumptions on data distribution properties, and does not require prior understanding of variable relationships. These properties make ANN models powerful modeling techniques for exploring complex nonlinear problems (Olden and Jackson 2002; Olden et al. 2004; Lakshminarayana and Rao 2010). ANN applications in geographical sciences include transport planning (Cheng et al. 2012; Kumar et al. 2013), spatial interpolation (Merwin et al. 2009), image classification (Arslan 2009; Sadeghi et al. 2013), transport and land use interaction (Rodrigue 1997), land cover classification (Foody 2002), land cover transformation (Pijanowski et al. 2002; Isik et al. 2013), and urban change detection (Tayyebi and Pijanowski 2014).

In urban development modeling, ANN was integrated with GIS to predict urban land use change, where GIS was used to develop the spatial urban driving factors (Pijanowski et al. 2002). The researchers followed four important steps: (1) design the network and input historical data, (2) use a suitable subset of inputs for network training, (3) use the full data set of the inputs for the neural network testing, and (4) finally, use the acquired neural network information to predict future changes. However, among the major limitations of ANN models are their static nature and black box, which limit the modeling of the urban expansion process.

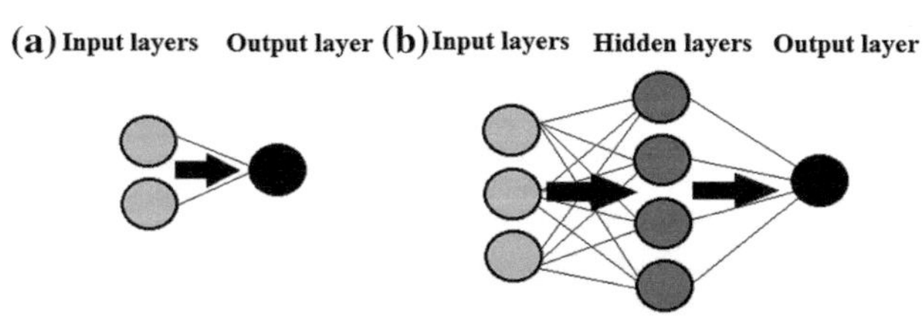

Fig. 8.2 Schematic illustration of two ANN models, a simple and complex version

The land transformation model (LTM), as an ANN-based model that combines multilayer perceptron (Zurada 1992) and GIS using socioeconomic and biophysical factors (Pijanowski et al. 2002, 2014). This model has been used worldwide to simulate land use changes (Pijanowski et al. 2005, 2014; Tayyebi et al. 2014). Multilayer perceptron uses a supervised learning algorithm that estimates a function between input–output pairs without the knowledge of the functional form (Tayyebi et al. 2014). This model utilizes information from at least two land use maps in different periods to train the network. This model determines the location of specific land use changes by using temporal land use maps. The actual land use changes are typically observed from the historical trend and used to establish functional relationships to extrapolate land use change probabilities for future prediction. Land use maps should be classified, with a value of 1 to selected land use and 0 to other categories. For input drivers, various spatial urban-related variables generated from a series of base layers should be created and stored within a GIS environment. These base layers represent land use categories, such as residential, commercial, agricultural, and facilities. These base layers can also represent features of the urban area, such as roads, water bodies, and rivers. Raster layers should be coded to represent these predictors as either binary or continuous variables. Input variables are developed to define a set of spatial transition rules that quantify the spatial effects of predictor cells on land use transitions (Pijanowski et al. 2002).

In an ANN model, all input data must be trained and tested to develop a network with proper predictive capacity. Training is performed to adjust the weights for each node according to the learning algorithm, whereas testing is conducted to calculate error rates (Pijanowski et al. 2002). Thus, the ANN process is usually conducted in three phases: (1) designing the network and inputs from historical data, (2) testing the neural network using the full dataset, and (3) using the output information of the neural network to forecast residential growth. Stuttgart's Neural Network Simulator (SNNS) can be used for the design, training, and prediction of the ANN (Zell et al. 1994). The neural network is designed to contain several numbers of inputs depending on the selected variables and an equal number of hidden layers and a single output layer as the final prediction.

8.6 Cellular Automata Model

Cellular automaton (CA) is one of the most common approaches in urban analysis because of its cellular concept (Lantman et al. 2011). This model is a dynamic discrete space and based on time systems. CA models are individual-based spatial models that are increasingly used to simulate and forecast the dynamics of natural and human-made environments. CA is an efficient bottom-up tool that provides an environment to examine the decision-making processes in complex urban spatial systems (O'Sullivan and Torrens 2001; Barredo et al. 2003). Tobler (1979) was the first person who introduced the use of CA in geographical aspects. This was further developed by Couclelis (1985), Batty and Xie (1994) and White and Engelen (1993). More recently Hagoort and te Utrecht (2006) presented a detailed overview of the history of CA, and Norte Pinto and Pais Antunes (2007) also described CA modeling in various urban applications.

The main logic behind CA modeling for land use changes is the current state of each cell and its interaction with neighborhood cells (Fig. 8.3). CA land use change modeling is also based on historical concept, that is, the trend of changes in the past has significant effects on future changes. For example, if a forest area is paved for the construction of a road to connect the regions on both sides of the forest, the conversion probability of the forest area to the urban area increases significantly. CA consists of four main elements: cell space, cell state, time steps, and transition rules (White and Engelen 1993). CA models represent space in a matrix format of regular cells with a state value that develops based on the transition rules applied at each time step of the simulation (Marceau and Moreno 2008; Gong et al. 2015).

Transition rules can be derived from expert knowledge or statistical analysis. Two types of CA modeling exist: unconstrained and constrained (Li and Yeh 2000). Unconstrained is the most "true" CA because it only uses decision rules to calculate land-use change. In constrained CA, the amount of land-use change per land use class is limited; the limit of a certain land-use class is either expert-based or calculated from historical land use (Koomen and Borsboom-van Beurden 2011). Dietzel and Clarke (2006) defined two more types of CA modeling in the case of land

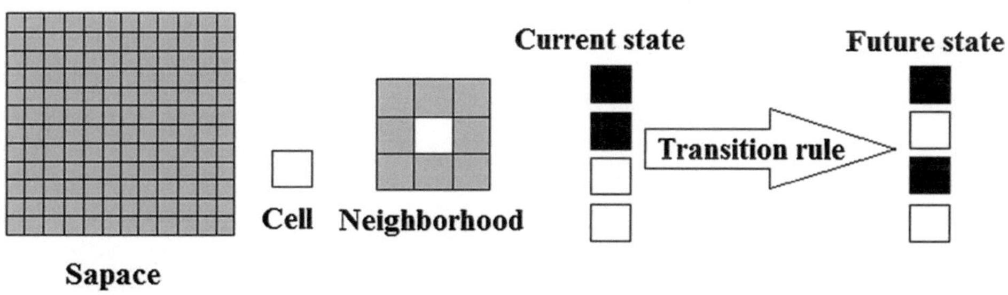

Fig. 8.3 The main components of cellular automata

use change processing. The first type deals with an urban system as a fundamental entity, urban and nonurban units (Ward et al. 2000; Wu 2002). The second type of CA disaggregates the urban areas into several categories of land uses within a city. The second model of CA is definitely more compatible with the research on land use types with fine resolution.

CA modeling has several advantages with respect to other modeling approaches. For instance, applying dynamic spatial variable during the iterative looping of CA models leads to better performance than general urban modeling techniques (Li and Yeh 2002). In addition, more factors or features can emerge during the simulation process, such as new aggregate centers (Wu 1998), fractal properties (White and Engelen 1993), and/or complex global pattern from local interactions (Batty and Xie 1994) that can be listed as capabilities of CA modeling. Therefore, the number of CA modeling studies in this field has been increasing because of these potential and capabilities.

Although CA modeling has many advantages, some drawbacks need to be considered. Modern CA models attempt to establish strong links with the decision-making process to achieve a more reliable modeling of land use changes (Koomen and Stillwell 2007). Furthermore, the determination of parameter values is another concern of CA modeling. Conventionally, CA models are used to simulate urban growth from rural to urban land use. CA processing becomes more complex when several land use types (residential, commercial, industrial, and so on) are included in the model (Batty et al. 1999). Another important issue is defining transition rules and model structures, which are generally application dependent (Li and Yeh 2002). Therefore, the calibration of CA models is usually required to ensure the performance accuracy of the models (Wu 1998; Li and Yeh 2002; Kocabas and Dragicevic 2007). However, calibration is more difficult in the case of multiple land use changes. Statistical approach, such as logistic regression, is one type of calibration method to obtain parameter values for urban simulation (Wu 1998). The integration of multi-criteria evaluation (MCE) with CA is another approach for CA calibration to define more behavior-oriented transition rules for land use change modeling (Wu 1998). In this manner, simulation interacts with decision makers to apply priority in the development process.

The integration of the Markov chain with CA enables transition cells to change their current status based on the suitability of changes derived from decision makers and the probability of changes calculated from Markov analysis, rather than the deterministic transition rules of conventional CA models (Koomen and Stillwell 2007; Kamusoko et al. 2009; Arsanjani et al. 2011; Gong et al. 2015). Al-sharif and Pradhan (2014) applied this integration approach to simulate urban land use changes and predict the spatial patterns in Tripoli's metropolitan areas. Markov chain was used to predict the land use change quantitatively and then CA was applied to simulate the dynamic spatial pattern of the changes explicitly. They assessed the performance of the CA–Markov integration approach and then used this model to compute the optimal transition rules and predict future land use changes. In another paper, Al-sharif and Pradhan (2015) proposed a hybrid model that integrated CA–Markov with the chi-squared automatic integration detection decision tree (CHAID-DT). In addition to the application of CA–Markov, CHAID-DT model was applied to investigate the contributions of urban factors, explore their interactions, and provide future urban probability maps. This integration significantly improved the capability of cellular-based urban modeling approaches. Wu (1998) presented a prototype of a simulation model integrating CA, analytical hierarchy process (AHP), and GIS. This integration model was written in the C programming language and built within ARC/INFO GIS. Finally, he stated that this combination has several benefits in terms of decision-making visualization, easier access to spatial information, and creating a more realistic definition of transition rules in CA. Al-shalabi et al. (2013) applied the SLEUTH (slope, land use, exclusion, urban extent, transportation, and hillshade) model to predict the shape and direction of spatial urban sprawl from 2004 to 2020 in Sana'a, Yemen. SLEUTH is a type of CA model that has been widely applied to urban growth modeling and studies in various parts of the world (Jantz et al. 2004; Leao et al. 2004; Zeug et al. 2006). Kocabas and Dragicevic (2007) developed a novel CA model within a GIS environment that consists of Bayesian network and influence diagram. Bayesian network is used to encode the drivers with the conditional probabilities computed from historical information. The influence diagram based the decision of land use conversion on utility theory. The proposed model was intended to simplify the definition of parameter values, transition rules, and model structure. Finally, they stated that the model is able to detect spatiotemporal drivers and generate various scenarios of land use change.

Almeida et al. (2005) also utilized a statistic-based WoE model that employs Bayesian conditional probabilities to compute the transitional probabilities for CA-based land use change modeling. Li and Yeh (2002) simulated the evolution of multiple land use changes based on the integration of CA and neural network. The neural network is used to calculate conversion probabilities for multiple land uses. The model involved the iterative looping of the neural network to simulate gradual land use conversion processes. Wang et al. (2011) stated that the methodologies for identifying the dominant factors that drive the landscape dynamics should be improved. Therefore, they evaluated the potential of rough set theory (RST) in factor selection for the calibration of the CA model. RST is a type of data mining and knowledge discovery method that can ascertain an optimal subset of features from an original dataset based on several factors. They selected a smaller set of factors from 18 original factors by using this approach to assess the conversion of forest and vegetation to built-up areas. An alternative way to control the spatial transition of each cell was conducted by using a higher level of constraints, such as the degree of land use change, through a regional level spatial interaction model (Koomen et al. 2008). Monitoring land use/cover dynamics (MOLAND) is an example of this type of CA modeling. In MOLAND, spatial dynamics are estimated by transition rules or weighting techniques that indicate the interaction among neighboring land use categories.

8.7 Decision Tree Models

Decision tree (DT) technique is a popular multidisciplinary data mining method used to extract many decision rules. DTs are frequently utilized in decision analysis to identify and support the ideal strategies to achieve a certain goal (Lee and Park 2013). DT is a method of hierarchical classification and logical deductive reasoning that is composed of decision rules that recursively split the inputs of independent variables (predictors). It uses these inputs to project the value of a dependent variable (target) (Cho and Kurup 2011; Pradhan 2013). DT uses conditional techniques to separate complex sets into a number of simpler sets based on the significance of the independent variables. This process generates simple and understandable solutions. Based on a set of independent attributes, the classification tree (DT) estimates the value of a discrete target variable with finite classes (Quinlan 1986). The attributes can be either discrete or continuous variables. The tree structure is recursive and begins with the entire set of training cases. At each stage, the ideal informative attribute is considered as the DT root, and data from the current training set are divided into subsets based on the values of the selected attributes. Given continuous attributes, DT branches are generated based on a selected threshold. Given discrete attributes, each possible attribute value typically produces a DT branch. The DT building algorithm is applied repeatedly to the subsets of the training cases in each branch, and the tree is completed when the stopping criteria are fulfilled. The end nodes are called leaves and are characterized by consistent class values (Quinlan 2014).

The constructed DT then determines a set of decision rules that can be employed to forecast a result based on a set of independent variables (Debeljak and Džeroski 2011). In the DT, the independent variables need not be related to the dependent variable (target) in advance because the structure of the DT model can determine and describe the structural patterns of data (Saito et al. 2009). Moreover, this technique can extract acceptable outcomes under imperfect conditions and reduce model construction time. The DT model also simplifies data conversion because it can deal directly with continuous variables (Quinlan 2014).

Figure 8.4 demonstrates the basic DT formation. A DT consists of three elements: node, condition, and production. The nodes are categorized into three types: chance, decision, and end. Figure 8.4 shows that (A) represents a decision node; (B), (C), and (D) denote chance nodes; and (R) indicates end nodes. End nodes (leaves) correspond to the estimations of a solution to the study case. The descending arrangement of the DT suggests that independent variables in the high order of the DT structure are more significant than the others (Saito et al. 2009).

DTs are advantageous over other numerically oriented methods, including ANNs, LR, genetic algorithms, and linear and nonlinear regressions (Kheir et al. 2010) because they are easily built and interpreted. Furthermore, DTs can automatically address the interactions among categorical (nominal) and continuous variables (Althuwaynee et al. 2014). They can identify the most important (decisive) variables, which are those closer to the top of the tree structure. These variables generate splits. Moreover, DTs do not require specific function forms to fit the modeling data, unlike other modeling methods (e.g., nonlinear regression) (Kheir et al. 2010). They also indicate the relative weights of independent variables (predictors) and describe training data input, whereas bivariate modeling approaches demonstrate only the relationship between the target variable and a single predictor variable. However, despite the advantages of DT, future trends remain difficult to predict.

DT differs from other statistical techniques in that it does not make any statistical assumptions. Moreover, it accommodates different data measurement scales and is computationally fast (Yeon et al. 2010). However, the DT model is limited by its susceptibility to noisy data (Zhao and Zhang

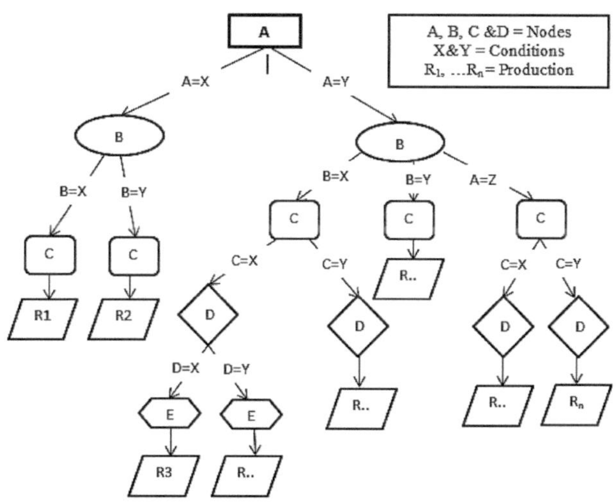

Fig. 8.4 Architecture of the decision tree model (Lee and Park 2013)

2008). DT has been coupled with GIS and has been used to analyze, classify, understand, and predict spatial data in various geospatial applications, such as the mapping of ground subsidence hazards (Lee and Park 2013), landslide susceptibility (Pradhan 2013; Althuwaynee et al. 2014), and environmental and ecological assessments (Zhang et al. 2012), as well as the prediction of heavy metal contamination (Kheir et al. 2010).

Many DT algorithms can be utilized to construct a DT model, such as C4.5 (Quinlan 2014), CART (Olshen and Stone 1984), ID3 (Quinlan 1986), QUEST (Loh and Shih 1997), and Decision-Tree-based Chi-squared Automatic Interaction Detection (CHAID) (Berry and Linoff 1997). The DT model must be calibrated to avoid model overtraining and confirm that the developed model does not fit only the training data considered and to achieve reliable modeling results with these algorithms. However, some researchers reported that the DT technique can be improved by combining the DT model with other models, such as those based on CA (Kim et al. 2006) and LR (Althuwaynee et al. 2014), in geospatial simulations.

8.8 Validation of Urban Modeling Techniques

Most land use change models are based on raster-based GIS, which predicts the future changes of each cell (Pontius and Schneider 2001). Thus, a method to validate the accuracy and reliability of these predictive models is necessary. These validation processes are very important and critical components of urban growth and change modeling (Pontius Jr and Chen 2006). Validation process in this specific field compares the output maps of the modeling approach with real changes to evaluate their similarity. Thus, such validation needs another real land use map for future years. The period between land use maps should be sufficient to compare the observed and simulated dynamics. Ideally, this duration should be as long as the period for which future scenario predictions are made (Verburg et al. 2004b). Validation also ensures that the structure of the models is properly built in terms of conceptual and operational aspects and accurately represents the real world (Henninger et al. 2010). The validation methods should make a clear distinction between the quantity of changes and quality of spatial allocation of the land use changes in the simulation model performance (Verburg et al. 2004a).

One of the most common validation techniques in this field is the quantitative-based method called relative operating characteristic (ROC) (Pontius and Schneider 2001; Nykänen et al. 2015). ROC is an excellent method to evaluate the reliability of the class change occurrence (i.e., urban expansion) by comparing a probability image that represents the likelihood of that class occurrence (i.e., the input map) with a Boolean image that indicates where that class actually exists in the real map. This technique has been used in many land use change modeling studies and is accepted as a reliable validation approach (Pontius and Schneider 2001; Hu and Lo 2007; Wang and Mountrakis 2011).

ROC offers a method of statistical analysis that answers one important question: "*How well is the category of interest concentrated at the locations with relatively high suitability for that category?*" The answer to this question allows the researcher to answer the general question "*How well do the maps agree in terms of the location of cells in a category?*" while not being forced to answer the question "*How well do the maps agree in terms of the quantity of cells in each category?*" Therefore, ROC is useful when the researcher aims to see how well the suitability map is produced by the model that represents the location of a specific class, but does not have an estimated quantity of the class (Pontius and Schneider 2001). This method is able to apply any model that projects a homogenous class or category in each grid cell. ROC has three main advantages, which are required for all kinds of land use change validations: (1) using measurements other than percent success assessment, (2) measuring its performance over a variety of scenarios of quantity of changes, and (3) presenting the validation with figures that show clearly how high similarity differs from low similarity (Pontius and Schneider 2001).

In addition to ROC, a frequently used validation method is the area under the ROC curve, which is commonly known as AUC (Pontius Jr and Parmentier 2014). AUC can range from 0 to 1, where a higher AUC value represents a stronger positive association. AUC is a unit less summary metric that synthesizes the relationships between the reference Boolean feature and several diagnoses by the index. However, AUC and other similar techniques have limitations, such as the inability to predict probability values, and the goodness of fit of the model does not provide information on the spatial distribution of model errors (Lobo et al. 2008; Pontius Jr and Parmentier 2014). Thus, these limitations should be considered when applying these validation methods of urban modeling.

Contingency table is another common method of land use model validation that is based on two-by-two comparison between projected and actual land use maps for each land use category. Moreover, Kappa index of agreement is commonly applied to analyze the accuracy of classification, but the confusion matrix is currently the main focus among the accuracy assessment methods (Foody 2002). Land use change modeling using the CA and Markov land change models show that, up to the present, the vast majority of such models have been validated with the aid of the Kappa index of agreement (Araya and Cabral 2010; Mitsova et al. 2011; Thapa and Murayama 2011). The Kappa statistic index assesses the validity and reliability of the projected maps in terms of quantity and location of the changes (Arsanjani et al. 2011). The Kappa index of agreement is a measure of proportional accuracy adjusted for chance agreement.

References

Abdullahi, S., Mahmud, A. R. B., & Pradhan, B. (2014). Spatial modelling of site suitability assessment for hospitals using geographical information system-based multicriteria approach at Qazvin city, Iran. *Geocarto International, 29*(2), 164–184.

Abdullahi, S., & Pradhan, B. (2015). Sustainable brownfields land use change modeling using GIS-based weights-of-evidence approach. *Applied Spatial Analysis and Policy*, 1–18.

Abdullahi, S., Pradhan, B., & Jebur, M. N. (2015). GIS-based sustainable city compactness assessment using integration of MCDM, Bayes theorem and RADAR technology. *Geocarto International, 30*(4), 365–387.

Al-shalabi, M., Billa, L., Pradhan, B., Mansor, S., & Al-Sharif, A. A. (2013). Modelling urban growth evolution and land-use changes using GIS based cellular automata and SLEUTH models: The case of Sana'a metropolitan city, Yemen. *Environmental Earth Sciences, 70*(1), 425–437.

Al-sharif, A. A., & Pradhan, B. (2014). Monitoring and predicting land use change in Tripoli Metropolitan City using an integrated Markov chain and cellular automata models in GIS. *Arabian Journal of Geosciences, 7*(10), 4291–4301.

Al-sharif, A. A., & Pradhan, B. (2015). A novel approach for predicting the spatial patterns of urban expansion by combining the chi-squared automatic integration detection decision tree, Markov chain and cellular automata models in GIS. *Geocarto International* (ahead-of-print), 1–24.

Almeida, C., Gleriani, J., Castejon, E. F., & Soares-Filho, B. (2008). Using neural networks and cellular automata for modelling intra-urban land-use dynamics. *International Journal of Geographical Information Science, 22*(9), 943–963.

Almeida, C., Monteiro, A., Câmara, G., Soares-Filho, B., Cerqueira, G., & Pennachin, C. (2002). *Modeling urban land use dynamics through Bayesian probabilistic methods in a cellular automaton environment.* Paper presented at the Proceedings of the 29th International Symposium on Remote Sensing of the Environment.

Almeida, C., Monteiro, A. M. V., Câmara, G., Soares-Filho, B. S., Cerqueira, G. C., Pennachin, C. L., et al. (2005). GIS and remote sensing as tools for the simulation of urban land-use change. *International Journal of Remote Sensing, 26*(4), 759–774.

Althuwaynee, O. F., Pradhan, B., & Lee, S. (2012). Application of an evidential belief function model in landslide susceptibility mapping. *Computers & Geosciences, 44,* 120–135.

Althuwaynee, O. F., Pradhan, B., Park, H.-J., & Lee, J. H. (2014). A novel ensemble decision tree-based CHi-squared automatic interaction detection (CHAID) and multivariate logistic regression models in landslide susceptibility mapping. *Landslides, 11*(6), 1063–1078.

Araya, Y. H., & Cabral, P. (2010). Analysis and modeling of urban land cover change in Setúbal and Sesimbra, Portugal. *Remote Sensing, 2*(6), 1549–1563.

Arsanjani, J. J., Helbich, M., & de Noronha Vaz, E. (2013). Spatiotemporal simulation of urban growth patterns using agent-based modeling: The case of Tehran. *Cities, 32,* 33–42.

Arsanjani, J. J., Kainz, W., & Mousivand, A. J. (2011). Tracking dynamic land-use change using spatially explicit Markov chain based on cellular automata: The case of Tehran. *International Journal of Image and Data Fusion, 2*(4), 329–345.

Arslan, O. (2009). A novel confidence estimation method for neural networks in multispectral image classification. *International Journal of Digital Earth, 2*(4), 343–358.

Awasthi, A., & Chauhan, S. S. (2011). Using AHP and Dempster–Shafer theory for evaluating sustainable transport solutions. *Environmental Modelling and Software, 26*(6), 787–796.

Barredo, J. I., Kasanko, M., McCormick, N., & Lavalle, C. (2003). Modelling dynamic spatial processes: Simulation of urban future scenarios through cellular automata. *Landscape and Urban Planning, 64*(3), 145–160.

Batty, M., & Xie, Y. (1994). From cells to cities. *Environment and Planning B, 21,* s31–s31.

Batty, M., Xie, Y., & Sun, Z. (1999). Modeling urban dynamics through GIS-based cellular automata. *Computers, Environment and Urban Systems, 23*(3), 205–233.

Benenson, I. (1998). Multi-agent simulations of residential dynamics in the city. *Computers, Environment and Urban Systems, 22*(1), 25–42.

Berry, M. J., & Linoff, G. (1997). *Data mining techniques: For marketing, sales, and customer support*. New Jersey: Wiley.

Beurden, J., Bakema, A., & Tijbosch, H. (2007). A land-use modelling system for environmental impact assessment. Recent applications of the LUMOS toolbox. In *Modelling land-use change* (Progress and Applications) (pp. 281–296). Dordrecht: Springer.

Bohman-Carter, G. (1994). Geographic information systems for geoscientists: Pergamon. Oxford.

Brown, D., & Geist, H. (2006). *The earth's changing land: An encyclopedia of land-use and land-cover change*. CT: Greenwood Publishing Group Westport.

Brown, D. G., Riolo, R., Robinson, D. T., North, M., & Rand, W. (2005). Spatial process and data models: Toward integration of agent-based models and GIS. *Journal of Geographical Systems, 7* (1), 25–47.

Bui, D. T., Pradhan, B., Lofman, O., Revhaug, I., & Dick, O. B. (2012). Spatial prediction of landslide hazards in Hoa Binh province (Vietnam): A comparative assessment of the efficacy of evidential belief functions and fuzzy logic models. *Catena, 96,* 28–40.

Burchell, R. W. (1996). *Economic and fiscal impacts of alternative land-use patterns.* Paper presented at the Center for Urban Policy Research, Rutgers University. Prepared for Michigan State Land Use Forum Conference.

Burnham, B. O. (1973). Markov intertemporal land use simulation model. *Southern Journal of Agricultural Economics, 5*(01), 253–258.

Candau, J. T. (2002). *Temporal calibration sensitivity of the SLEUTH urban growth model.* Master's thesis, University of California, Santa Barbara, CA.

Carranza, E. J. M., Van Ruitenbeek, F., Hecker, C., van der Meijde, M., & van der Meer, F. D. (2008). Knowledge-guided data-driven evidential belief modeling of mineral prospectivity in Cabo de Gata, SE Spain. *International Journal of Applied Earth Observation and Geoinformation, 10*(3), 374–387.

Chen, C., He, B., & Zeng, Z. (2014). A method for mineral prospectivity mapping integrating C4. 5 decision tree, weights-of-evidence and m-branch smoothing techniques: A case study in the eastern Kunlun Mountains, China. *Earth Science Informatics, 7*(1), 13–24.

Cheng, M.-Y., Su, C.-W., Tsai, M.-H., & Lin, K.-S. (2012). Data preprocessing for artificial neural network applications in prioritizing railroad projects—A practical experience in Taiwan. *Journal of Civil Engineering and Management, 18*(4), 483–494.

Cho, J. H., & Kurup, P. U. (2011). Decision tree approach for classification and dimensionality reduction of electronic nose data. *Sensors and Actuators B: Chemical, 160*(1), 542–548.

Costanza, R., & Ruth, M. (1998). Using dynamic modeling to scope environmental problems and build consensus. *Environmental Management, 22*(2), 183–195.

Couclelis, H. (1985). Cellular worlds: A framework for modeling micro-macro dynamics. *Environment and Planning A, 17*(5), 585–596.

Crooks, A., Castle, C., & Batty, M. (2008). Key challenges in agent-based modelling for geo-spatial simulation. *Computers, Environment and Urban Systems, 32*(6), 417–430.

Dadhich, P. N., & Hanaoka, S. (2011). Spatio-temporal urban growth modeling of Jaipur, India. *Journal of Urban Technology, 18*(3), 45–65.

de Almeida, C. M., Batty, M., Monteiro, A. M. V., Câmara, G., Soares-Filho, B. S., Cerqueira, G. C., et al. (2003). Stochastic cellular automata modeling of urban land use dynamics: Empirical development and estimation. *Computers, Environment and Urban Systems, 27*(5), 481–509.

Debeljak, M., & Džeroski, S. (2011). Decision trees in ecological modelling. *Modelling complex ecological dynamics* (pp. 197–209). Berlin: Springer.

DeMaris, A. (1992). *Logit modeling: Practical applications* (Vol. 86). Sage.

Dempster, A. P. (1967). Upper and lower probabilities induced by a multivalued mapping. *The Annals of Mathematical Statistics, 38*(2), 325–339.

Devkota, K. C., Regmi, A. D., Pourghasemi, H. R., Yoshida, K., Pradhan, B., Ryu, I. C., et al. (2013). Landslide susceptibility mapping using certainty factor, index of entropy and logistic regression models in GIS and their comparison at Mugling–Narayanghat road section in Nepal Himalaya. *Natural Hazards, 65* (1), 135–165.

Dietzel, C., & Clarke, K. (2006). The effect of disaggregating land use categories in cellular automata during model calibration and forecasting. *Computers, Environment and Urban Systems, 30*(1), 78–101.

Eyoh, A., Olayinka, D. N., Nwilo, P., Okwuashi, O., Isong, M., & Udoudo, D. (2012). Modelling and predicting future urban expansion of Lagos, Nigeria from remote sensing data using logistic regression and GIS. *International Journal of Applied Science and Technology, 2*(5), 116–124.

Foody, G. (2002). Hard and soft classifications by a neural network with a non-exhaustively defined set of classes. *International Journal of Remote Sensing, 23*(18), 3853–3864.

Geertman, S., & Stillwell, J. (2004). Planning support systems: An inventory of current practice. *Computers, Environment and Urban Systems, 28*(4), 291–310.

Gettings, M. E., Bultman, M. W., & Fisher, F. S. (2004). A complex systems model approach to quantified mineral resource appraisal. *Environmental Management, 33*(1), 87–98.

Gong, W., Yuan, L., Fan, W., & Stott, P. (2015). Analysis and simulation of land use spatial pattern in Harbin prefecture based on trajectories and cellular automata—Markov modelling. *International Journal of Applied Earth Observation and Geoinformation, 34,* 207–216.

Grimm, V., Berger, U., Bastiansen, F., Eliassen, S., Ginot, V., Giske, J., et al. (2006). A standard protocol for describing individual-based and agent-based models. *Ecological Modelling, 198*(1), 115–126.

Hagoort, M. J., & te Utrecht, R. (2006). *The neighbourhood rules: Land-use interactions, urban dynamics and cellular automata modelling*. Koninklijk Nederlands Aardrijkskundig Genootschap.

Harrell, F. (2015). *Regression modeling strategies: With applications to linear models, logistic and ordinal regression, and survival analysis*. Berlin: Springer.

He, B., Chen, C., & Liu, Y. (2010). Mineral potential mapping for Cu-Pb-Zn deposits in the East Kunlun Region, Qinghai Province, China, integrating multi-source geology spatial data sets and extended weights-of-evidence modeling. *GIScience & Remote Sensing, 47*(4), 514–540.

Heistermann, M., Müller, C., & Ronneberger, K. (2006). Land in sight?: Achievements, deficits and potentials of continental to global scale land-use modeling. *Agriculture, Ecosystems & Environment, 114*(2), 141–158.

Henninger, H. B., Reese, S. P., Anderson, A. E., & Weiss, J. A. (2010). Validation of computational models in biomechanics. *Proceedings of the Institution of Mechanical Engineers. Part H, Journal of Engineering in Medicine, 224*(7), 801–812.

Hilferink, M., & Rietveld, P. (1999). Land use scanner: An integrated GIS based model for long term projections of land use in urban and rural areas. *Journal of Geographical Systems, 1*(2), 155–177.

Hu, Z., & Lo, C. (2007). Modeling urban growth in Atlanta using logistic regression. *Computers, Environment and Urban Systems, 31*(6), 667–688.

Huang, B., Zhang, L., & Wu, B. (2009). Spatiotemporal analysis of rural–urban land conversion. *International Journal of Geographical Information Science, 23*(3), 379–398.

Isik, S., Kalin, L., Schoonover, J. E., Srivastava, P., & Lockaby, B. G. (2013). Modeling effects of changing land use/cover on daily streamflow: An artificial neural network and curve number based hybrid approach. *Journal of Hydrology, 485,* 103–112.

Jantz, C. A., Goetz, S. J., & Shelley, M. K. (2004). Using the SLEUTH urban growth model to simulate the impacts of future policy scenarios on urban land use in the Baltimore-Washington metropolitan area. *Environment and Planning B, 31*(2), 251–272.

Kamusoko, C., Aniya, M., Adi, B., & Manjoro, M. (2009). Rural sustainability under threat in Zimbabwe—Simulation of future land use/cover changes in the Bindura district based on the Markov-cellular automata model. *Applied Geography, 29*(3), 435–447.

Kerridge, J., Hine, J., & Wigan, M. (2001). Agent-based modelling of pedestrian movements: The questions that need to be asked and answered. *Environment and Planning B: Planning and Design, 28* (3), 327–341.

Kheir, R. B., Greve, M. H., Abdallah, C., & Dalgaard, T. (2010). Spatial soil zinc content distribution from terrain parameters: A GIS-based decision-tree model in Lebanon. *Environmental Pollution, 158*(2), 520–528.

Kim, J., Kang, Y., Hong, S., & Park, S. (2006). *Extraction of spatial rules using a decision tree method: A case study in urban growth modeling*. Paper presented at the Knowledge-Based Intelligent Information and Engineering Systems.

King, L. J. (1985). Central place theory. Regional Research Institute, West Virginia University Book Chapters.

Kleinbaum, D. G., & Klein, M. (2010). *Logistic regression: A self-learning text*. Springer Science & Business Media.

Klosterman, R. E. (1999). The what if? Collaborative planning support system. *Environment and Planning B: Planning and Design, 26*(3), 393–408.

Klosterman, R. E., & Pettit, C. J. (2005). An update on planning support systems. *Environment and Planning B: Planning and Design, 32*(4), 477–484.

Kocabas, V., & Dragicevic, S. (2007). Enhancing a GIS cellular automata model of land use change: Bayesian networks, influence diagrams and causality. *Transactions in GIS, 11*(5), 681–702.

Koomen, E., & Borsboom-van Beurden, J. (2011). *Land-use modelling in planning practice* (Vol. 101). Springer Science & Business Media.

Koomen, E., Rietveld, P., & de Nijs, T. (2008). Modelling land-use change for spatial planning support. *The Annals of Regional Science, 42*(1), 1–10.

Koomen, E., & Stillwell, J. (2007). Modelling land-use change. Theories and methods. *Modelling land-use change: Progress and applications* (pp. 1–24). Berlin: Springer.

Kuijpers, M. A. J., Bouwman, A., & Natuurplanbureau, M. -E. (2007). *Nederland later: tweede duurzaamheidsverkenning: deel fysieke leefomgeving Nederland.* Milieu en Natuur Planbureau Bilthoven.

Kumar, K., Parida, M., & Katiyar, V. (2013). Short term traffic flow prediction for a non urban highway using artificial neural network. *Procedia-Social and Behavioral Sciences, 104,* 755–764.

Lakshminarayana, B., & Rao, K. G. (2010). Artificial neural networks in spectral-spatial landuse classification. *ISH Journal of Hydraulic Engineering, 16*(supl), 64–73.

Lambin, E. F., Turner, B. L., Geist, H. J., Agbola, S. B., Angelsen, A., Bruce, J. W., et al. (2001). The causes of land-use and land-cover change: Moving beyond the myths. *Global Environmental Change, 11*(4), 261–269.

Lantman, J. v. S., Verburg, P. H., Bregt, A., & Geertman, S. (2011). Core principles and concepts in land-use modelling: A literature review. *Land-use modelling in planning practice* (pp. 35–57). Berlin: Springer.

Leao, S., Bishop, I., & Evans, D. (2004). Spatial—Temporal model for demand and allocation of waste landfills in growing urban regions. *Computers, Environment and Urban Systems, 28*(4), 353–385.

Lee, S., & Park, I. (2013). Application of decision tree model for the ground subsidence hazard mapping near abandoned underground coal mines. *Journal of Environmental Management, 127,* 166–176.

Lee, S., & Pradhan, B. (2006). Probabilistic landslide hazards and risk mapping on Penang Island, Malaysia. *Journal of Earth System Science, 115*(6), 661–672.

Lee, S., & Pradhan, B. (2007). Landslide hazard mapping at Selangor, Malaysia using frequency ratio and logistic regression models. *Landslides, 4*(1), 33–41.

Lesschen, J. P., Verburg, P. H., & Staal, S. J. (2005). *Statistical methods for analysing the spatial dimension of changes in land use and farming systems.* Citeseer.

Li, X., & Yeh, A. G.-O. (2000). Modelling sustainable urban development by the integration of constrained cellular automata and GIS. *International Journal of Geographical Information Science, 14*(2), 131–152.

Li, X., & Yeh, A. G.-O. (2002). Neural-network-based cellular automata for simulating multiple land use changes using GIS. *International Journal of Geographical Information Science, 16*(4), 323–343.

Liao, T. F. (1994). *Interpreting probability models: Logit, probit, and other generalized linear models.* Sage.

Lin, Y.-P., Chu, H.-J., Wu, C.-F., & Verburg, P. H. (2011). Predictive ability of logistic regression, auto-logistic regression and neural network models in empirical land-use change modeling—A case study. *International Journal of Geographical Information Science, 25*(1), 65–87.

Liu, Y. (2008). *Modelling urban development with geographical information systems and cellular automata.* CRC Press.

Lobo, J. M., Jiménez-Valverde, A., & Real, R. (2008). AUC: A misleading measure of the performance of predictive distribution models. *Global Ecology and Biogeography, 17*(2), 145–151.

Loh, W. -Y., & Shih, Y. -S. (1997). Split selection methods for classification trees. *Statistica Sinica*, 815–840.

López, E., Bocco, G., Mendoza, M., & Duhau, E. (2001). Predicting land-cover and land-use change in the urban fringe: A case in Morelia city, Mexico. *Landscape and Urban Planning, 55*(4), 271–285.

Malaitham, S., Vichiensan, V., Fukuda, A., & Wasuntarasook, V. (2015). *Determinants of land use change using geographically weighted logistic regression.* Paper presented at the Transportation Research Board 94th Annual Meeting.

Marceau, D., & Moreno, N. (2008). An object-based cellular automata model to mitigate scale dependency. *Object-based image analysis* (pp. 43–73). Berlin: Springer.

Mas, J.-F., Puig, H., Palacio, J. L., & Sosa-López, A. (2004). Modelling deforestation using GIS and artificial neural networks. *Environmental Modelling and Software, 19*(5), 461–471.

Matthews, R. B., Gilbert, N. G., Roach, A., Polhill, J. G., & Gotts, N. M. (2007). Agent-based land-use models: A review of applications. *Landscape Ecology, 22*(10), 1447–1459.

McGarigal, K., & Marks, B. J. (1995). Spatial pattern analysis program for quantifying landscape structure. Gen. Tech. Rep. PNW-GTR-351. US Department of Agriculture, Forest Service, Pacific Northwest Research Station.

Medley, K. E., Okey, B. W., Barrett, G. W., Lucas, M. F., & Renwick, W. H. (1995). Landscape change with agricultural intensification in a rural watershed, southwestern Ohio, USA. *Landscape Ecology, 10* (3), 161–176.

Mellino, S., & Ulgiati, S. (2015). Monitoring regional land use and land cover changes in support of an environmentally sound resource management. *Sustainable development, knowledge society and smart future manufacturing technologies* (pp. 309–321). Berlin: Springer.

Merwin, D., Cromley, R., & Civco, D. (2009). A neural network-based method for solving "nested hierarchy" areal interpolation problems. *Cartography and Geographic Information Science, 36*(4), 347–365.

Mitsova, D., Shuster, W., & Wang, X. (2011). A cellular automata model of land cover change to integrate urban growth with open space conservation. *Landscape and Urban Planning, 99*(2), 141–153.

Naghibi, S. A., Pourghasemi, H. R., Pourtaghi, Z. S., & Rezaei, A. (2015). Groundwater qanat potential mapping using frequency ratio and Shannon's entropy models in the Moghan watershed, Iran. *Earth Science Informatics, 8*(1), 171–186.

Norte Pinto, N., & Pais Antunes, A. (2007). Cellular automata and urban studies: A literature survey. *ACE: Arquitectura, Ciudad y Entorno, núm. 3, Febrero 2007.*

Nykänen, V., Lahti, I., Niiranen, T., & Korhonen, K. (2015). Receiver operating characteristics (ROC) as validation tool for prospectivity models—A magmatic Ni–Cu case study from the Central Lapland Greenstone Belt, Northern Finland. *Ore Geology Reviews, 71,* 853–860.

O'Sullivan, D., & Torrens, P. M. (2001). Cellular models of urban systems. *Theory and practical issues on cellular automata* (pp. 108–116). Berlin: Springer.

Olden, J. D., & Jackson, D. A. (2002). Illuminating the "black box": A randomization approach for understanding variable contributions in artificial neural networks. *Ecological Modelling, 154*(1), 135–150.

Olden, J. D., Joy, M. K., & Death, R. G. (2004). An accurate comparison of methods for quantifying variable importance in artificial neural networks using simulated data. *Ecological Modelling, 178*(3), 389–397.

Olshen, L., & Stone, C. J. (1984). Classification and regression trees. *Wadsworth International Group, 93*(99), 101.

Ozdemir, A., & Altural, T. (2013). A comparative study of frequency ratio, weights of evidence and logistic regression methods for landslide susceptibility mapping: Sultan Mountains, SW Turkey. *Journal of Asian Earth Sciences, 64,* 180–197.

Park, S., Jeon, S., Kim, S., & Choi, C. (2011). Prediction and comparison of urban growth by land suitability index mapping

using GIS and RS in South Korea. *Landscape and Urban Planning, 99*(2), 104–114.
Parker, D. C., Manson, S. M., Janssen, M. A., Hoffmann, M. J., & Deadman, P. (2003). Multi-agent systems for the simulation of land-use and land-cover change: A review. *Annals of the Association of American Geographers, 93*(2), 314–337.
Parker, S. (2015). *Urban theory and the urban experience: Encountering the city*. Routledge.
Pearl, J. (1990). Reasoning under uncertainty. *Annual Review of Computer Science, 4*(1), 37–72.
Pijanowski, B. C., Brown, D. G., Shellito, B. A., & Manik, G. A. (2002). Using neural networks and GIS to forecast land use changes: A land transformation model. *Computers, Environment and Urban Systems, 26*(6), 553–575.
Pijanowski, B. C., Pithadia, S., Shellito, B. A., & Alexandridis, K. (2005). Calibrating a neural network-based urban change model for two metropolitan areas of the Upper Midwest of the United States. *International Journal of Geographical Information Science, 19*(2), 197–215.
Pijanowski, B. C., Tayyebi, A., Doucette, J., Pekin, B. K., Braun, D., & Plourde, J. (2014). A big data urban growth simulation at a national scale: Configuring the GIS and neural network based land transformation model to run in a high performance computing (HPC) environment. *Environmental Modelling and Software, 51*, 250–268.
Pontius, R., Jr., & Chen, H. (2006). *GEOMOD modeling. Idrisi 15: The Andes Edition*.
Pontius, R. G., Cornell, J. D., & Hall, C. A. (2001). Modeling the spatial pattern of land-use change with GEOMOD2: Application and validation for Costa Rica. *Agriculture, Ecosystems & Environment, 85*(1), 191–203.
Pontius, R. G., Jr., & Parmentier, B. (2014). Recommendations for using the relative operating characteristic (ROC). *Landscape Ecology, 29*(3), 367–382.
Pontius, R. G., & Schneider, L. C. (2001). Land-cover change model validation by an ROC method for the Ipswich watershed, Massachusetts, USA. *Agriculture, Ecosystems & Environment, 85* (1), 239–248.
Pooyandeh, M., & Marceau, D. J. (2013). A spatial web/agent-based model to support stakeholders' negotiation regarding land development. *Journal of Environmental Management, 129*, 309–323.
Pourghasemi, H., Pradhan, B., Gokceoglu, C., & Moezzi, K. D. (2013). A comparative assessment of prediction capabilities of Dempster–Shafer and weights-of-evidence models in landslide susceptibility mapping using GIS. *Geomatics, Natural Hazards and Risk, 4*(2), 93–118.
Pradhan, B. (2013). A comparative study on the predictive ability of the decision tree, support vector machine and neuro-fuzzy models in landslide susceptibility mapping using GIS. *Computers & Geosciences, 51*, 350–365.
Pradhan, B., & Lee, S. (2010). Landslide susceptibility assessment and factor effect analysis: Backpropagation artificial neural networks and their comparison with frequency ratio and bivariate logistic regression modelling. *Environmental Modelling and Software, 25* (6), 747–759.
Pradhan, B., Oh, H.-J., & Buchroithner, M. (2010). Weights-of-evidence model applied to landslide susceptibility mapping in a tropical hilly area. *Geomatics, Natural Hazards and Risk, 1*(3), 199–223.
Quinlan, J. R. (1986). Induction of decision trees. *Machine Learning, 1* (1), 81–106.
Quinlan, J. R. (2014). *C4. 5: Programs for machine learning*. Elsevier.
Riebsame, W. E., Meyer, W. B., & Turner, B. L., II. (1994). Modeling land use and cover as part of global environmental change. *Climatic Change, 28*(1–2), 45–64.
Rodrigue, J.-P. (1997). Parallel modelling and neural networks: An overview for transportation/land use systems. *Transportation Research Part C: Emerging Technologies, 5*(5), 259–271.
Sadeghi, V., Ebadi, H., & Ahmadi, F. F. (2013). A new model for automatic normalization of multitemporal satellite images using artificial neural network and mathematical methods. *Applied Mathematical Modelling, 37*(9), 6437–6445.
Saito, H., Nakayama, D., & Matsuyama, H. (2009). Comparison of landslide susceptibility based on a decision-tree model and actual landslide occurrence: The Akaishi Mountains, Japan. *Geomorphology, 109*(3), 108–121.
Serneels, S., & Lambin, E. F. (2001). Proximate causes of land-use change in Narok District, Kenya: A spatial statistical model. *Agriculture, Ecosystems & Environment, 85*(1), 65–81.
Shafer, G. (1976). *A mathematical theory of evidence* (Vol. 1). Princeton University Press Princeton.
Skapura, D. M. (1996). *Building neural networks*. Addison-Wesley Professional.
Tayyebi, A., & Pijanowski, B. C. (2014). Modeling multiple land use changes using ANN, CART and MARS: Comparing tradeoffs in goodness of fit and explanatory power of data mining tools. *International Journal of Applied Earth Observation and Geoinformation, 28*, 102–116.
Tayyebi, A., Pijanowski, B. C., Linderman, M., & Gratton, C. (2014). Comparing three global parametric and local non-parametric models to simulate land use change in diverse areas of the world. *Environmental Modelling and Software, 59*, 202–221.
Teixeira, A. M. G., Soares-Filho, B. S., Freitas, S. R., & Metzger, J. P. (2009). Modeling landscape dynamics in an Atlantic Rainforest region: Implications for conservation. *Forest Ecology and Management, 257*(4), 1219–1230.
Thapa, R. B., & Murayama, Y. (2011). Urban growth modeling of Kathmandu metropolitan region, Nepal. *Computers, Environment and Urban Systems, 35*(1), 25–34.
Thiam, A. K. (2005). An evidential reasoning approach to land degradation evaluation: Dempster–Shafer theory of evidence. *Transactions in GIS, 9*(4), 507–520.
Tobler, W. (1979). Cellular geography. *Philosophy in geography* (pp. 379–386). Berlin: Springer.
Torrens, P. M. (2000). How cellular models of urban systems work (1. Theory).
Veldkamp, A. (2009). Investigating land dynamics: Future research perspectives. *Journal of Land Use Science, 4*(1–2), 5–14.
Veldkamp, A., & Fresco, L. (1996). CLUE: A conceptual model to study the conversion of land use and its effects. *Ecological Modelling, 85*(2), 253–270.
Veldkamp, A., & Lambin, E. F. (2001). Predicting land-use change. *Agriculture, Ecosystems & Environment, 85*(1), 1–6.
Verburg, P. H., Kok, K., Pontius, R. G., Jr., & Veldkamp, A. (2006). Modeling land-use and land-cover change. *Land-use and land-cover change* (pp. 117–135). Berlin: Springer.
Verburg, P. H., Schot, P. P., Dijst, M. J., & Veldkamp, A. (2004a). Land use change modelling: Current practice and research priorities. *GeoJournal, 61*(4), 309–324.
Verburg, P. H., Van Eck, J. R., de Nijs, T. C., Dijst, M. J., & Schot, P. (2004b). Determinants of land-use change patterns in the Netherlands. *Environment and Planning B, 31*(1), 125–150.
Walsh, S., Soranno, P., & Rutledge, D. (2003). Lakes, wetlands, and streams as predictors of land use/cover distribution. *Environmental Management, 31*(2), 0198–0214.
Wang, F. (2012). *A cellular automata model to simulate land-use changes at fine spatial resolution*. Doctor of Philosophy University of Calgary, Calgary (20365).
Wang, F., Hasbani, J.-G., Wang, X., & Marceau, D. J. (2011). Identifying dominant factors for the calibration of a land-use

cellular automata model using Rough Set Theory. *Computers, Environment and Urban Systems, 35*(2), 116–125.
Wang, J., & Mountrakis, G. (2011). Developing a multi-network urbanization model: A case study of urban growth in Denver, Colorado. *International Journal of Geographical Information Science, 25*(2), 229–253.
Ward, D. P., Murray, A. T., & Phinn, S. R. (2000). A stochastically constrained cellular model of urban growth. *Computers, Environment and Urban Systems, 24*(6), 539–558.
White, R., & Engelen, G. (1993). Cellular automata and fractal urban form: A cellular modelling approach to the evolution of urban land-use patterns. *Environment and Planning A, 25*(8), 1175–1199.
Wrigley, N., & Bennett, R. J. (1981). *Quantitative geography: A British view*. Routledge.
Wu, F. (1998). SimLand: A prototype to simulate land conversion through the integrated GIS and CA with AHP-derived transition rules. *International Journal of Geographical Information Science, 12*(1), 63–82.
Wu, F. (2002). Calibration of stochastic cellular automata: The application to rural-urban land conversions. *International Journal of Geographical Information Science, 16*(8), 795–818.
Wu, F., & Yeh, A. G.-O. (1997). Changing spatial distribution and determinants of land development in Chinese cities in the transition from a centrally planned economy to a socialist market economy: A case study of Guangzhou. *Urban Studies, 34*(11), 1851–1879.
Yeon, Y.-K., Han, J.-G., & Ryu, K. H. (2010). Landslide susceptibility mapping in Injae, Korea, using a decision tree. *Engineering Geology, 116*(3), 274–283.
Zell, A., Mache, N., Hübner, R., Mamier, G., Vogt, M., Schmalzl, M., et al. (1994). SNNS (stuttgart neural network simulator). *Neural network simulation environments* (pp. 165–186). Berlin: Springer.
Zeug, G., Eckert, S., Steiner, U., Kukuk, T., & Ehrlich, D. (2006). Monitoring urban growth and its impact on the environment: The case of Sana'a, Yemen. *Proceedings Digital Earth Summit on Geoinformatics, 2008,* 206–211.
Zhang, X., Jiang, H., Jin, J., Xu, X., & Zhang, Q. (2012). Analysis of acid rain patterns in northeastern China using a decision tree method. *Atmospheric Environment, 46,* 590–596.
Zhao, Y., & Zhang, Y. (2008). Comparison of decision tree methods for finding active objects. *Advances in Space Research, 41*(12), 1955–1959.
Zurada, J. M. (1992). *Introduction to artificial neural systems*. West St. Paul.

Spatial Urban Modeling and Prediction (Tripoli Metropolis Case Study)

9

Abubakr A.A. Al-sharif, Biswajeet Pradhan and Saleh Abdullahi

9.1 Introduction

In this chapter, the simulation process of urban growth in the Tripoli metropolis (Libya) case study will be presented and explained to understand its patterns and the role of each urban driving force behind the urbanization process. In the simulation process, the frequency ratio (FR) model was first applied based on the real urban expansion rather than on the entire urbanized area to present the role of classes within each urban factor and reflect actual urban expansion tendency. Second, the evidential belief function (Dempster–Shafer) model (EBF) was applied to provide further information by generating four maps representing belief, disbelief, uncertainty, and plausibility of predicted future urban growth. Third, the logistic regression (LR) model was applied to assess the overall effect of each urban driving factor, and subsequently combined with a simple growth ratio equation to present probable future scenarios. Fourth, the classic CA–Markov chain (MC) model was used to predict explicit future urban land use in Tripoli in 2020 and 2025. Finally, a novel hybrid model of CHAID–CA–Markov was proposed based on the advantages and shortcomings of the above mentioned models, and employed to model, explain, and predict explicit urban growth in 2020 and 2025. The general methodology flowchart of these processes is illustrated in Fig. 9.1.

9.2 Tripoli Metropolis from the 2nd to 3rd Generation Urban Plan

Libya lies along Africa's Mediterranean coast and stretches deep into the Saharan region. Although most parts of Libya consist of rocky plains and sandy seas, a narrow band of fertile lowlands stretches across the vast country's northern edge. Nearly three-fourths of Libya's population is located within urban areas that occupy only 1.5% of Libya's land area on the coast. Among these urban areas is the capital city, Tripoli. The Tripoli Agglomeration has the largest concentration of population and economic activities, not only in the Tripoli region, but also in the entire country. The area plays a highly important role in the socioeconomic development of Libya (UPA 2009). The study area is located along the Mediterranean coast in the northwestern part of Libya, between longitudes 12° 54′ 04″ E and 13° 26′ 38″ E and latitudes 32° 36′ 18″ N and 32° 54′ 17″ N. It occupies a total land area of approximately 1143.73 km^2. The Tripoli metropolitan area includes the districts of Tripoli Center, Hey Alandlus, Tajoura, Janzur, Kaser Ben Ghashir, Alswani, Ain Zara, Abuslim, and SuqAjumma (Fig. 9.2).

The traditional planning approach applied in the 2nd Generation Planning Project from 1980 to 2000 was a top-down process involving little direct input from the affected people and institutions. This time, advances in techniques and thinking in the international context transformed the urban planning process. The call for democratic planning processes played a huge role in enabling the public to influence important decisions as part of the planning process. In Libya, the number of actors in planning increased, and integrating the sector planning evolved. Large-scale developers, such as the Housing and Infrastructure Corporation, Ministry of Utilities, Roads and Bridges Authority, and Railway Authority, played a huge role in influencing the process of development, particularly through project-based planning.

Preparing the 3rd Generation Tripoli Agglomeration Plan involved a number of key processes. The strategic framework of the Agglomeration Plans was founded on the guidelines and recommendations presented in the Tripoli Agglomeration Report 2009 by UPA. The Agglomeration Plan also considered the Regional Development Plan for the Tripoli Region and the Sub-regional Plan for the Tripoli Sub-region. During the planning process, the problems and difficulties of implementing the 2nd Generation Agglomeration Plan was evaluated, which provided valuable information and data for the planning work. A more detailed analysis of contemporary conditions conducted in the planning process was based on statistical and geographical

B. Pradhan (ed.), *Spatial Modeling and Assessment of Urban Form*,
DOI 10.1007/978-3-319-54217-1_9

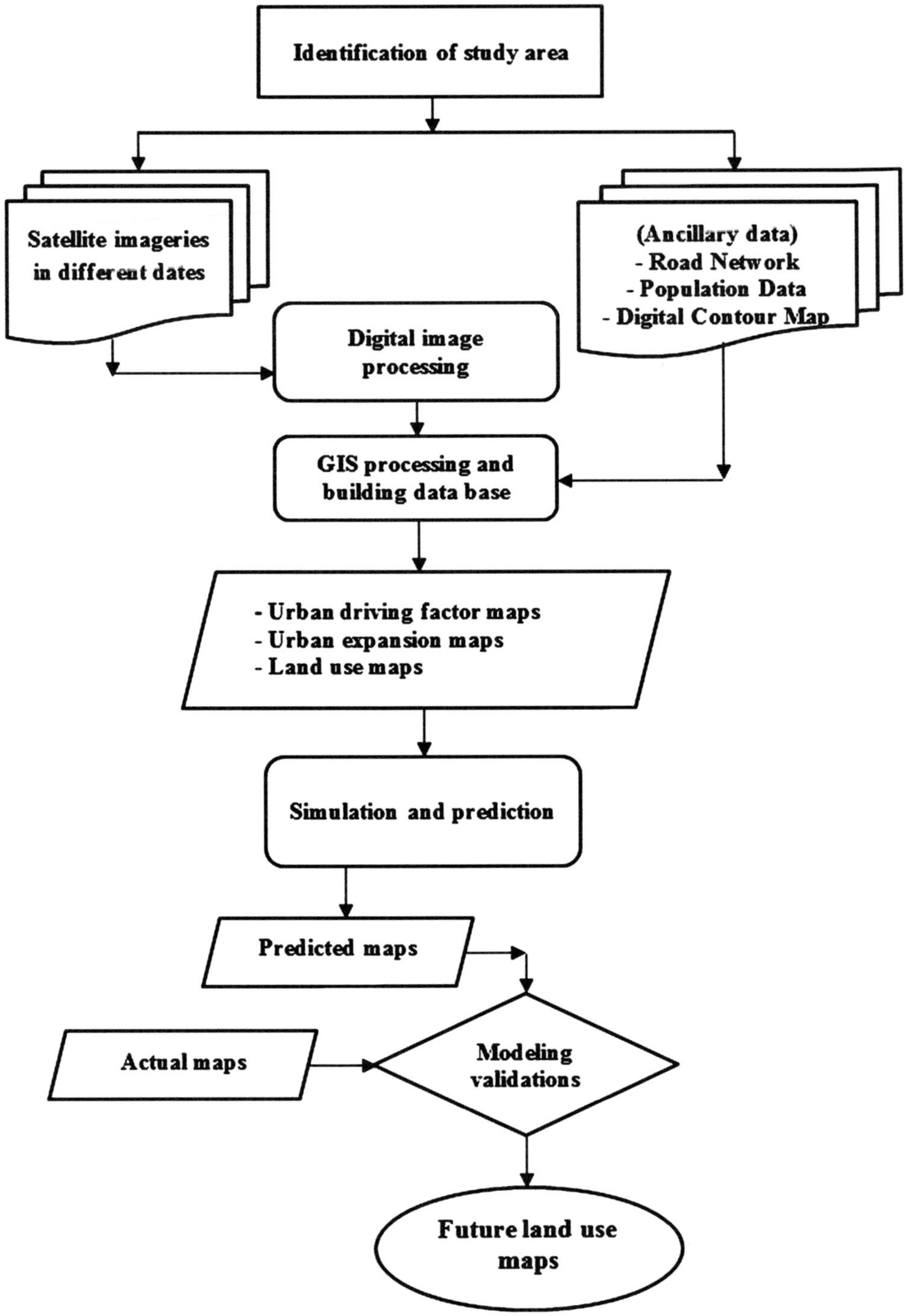

Fig. 9.1 Overall methodology flowchart

information. These data described the size and geographical distribution of the population, dwellings, social and technical services, land characteristics, and land use. Existing conditions and trends, including the development possibilities up to year 2025, were presented and addressed in the 3rd Generation Plan. As an integral part of the planning process, consultations were held with representatives from central and local authorities together with major stakeholders.

With housing as the main requirement in the Tripoli metropolis, the 2nd Generation Plan remains a valuable document for the housing issue. The Plan presented the housing stock picture for 1980 and broke down the housing stock figures. The total housing stock amounted to 106,000 dwellings, and the housing shortage was estimated to be 50,000 dwellings. Evaluating the situation and suggestions similar to those in the 2nd Generation Agglomeration Plan

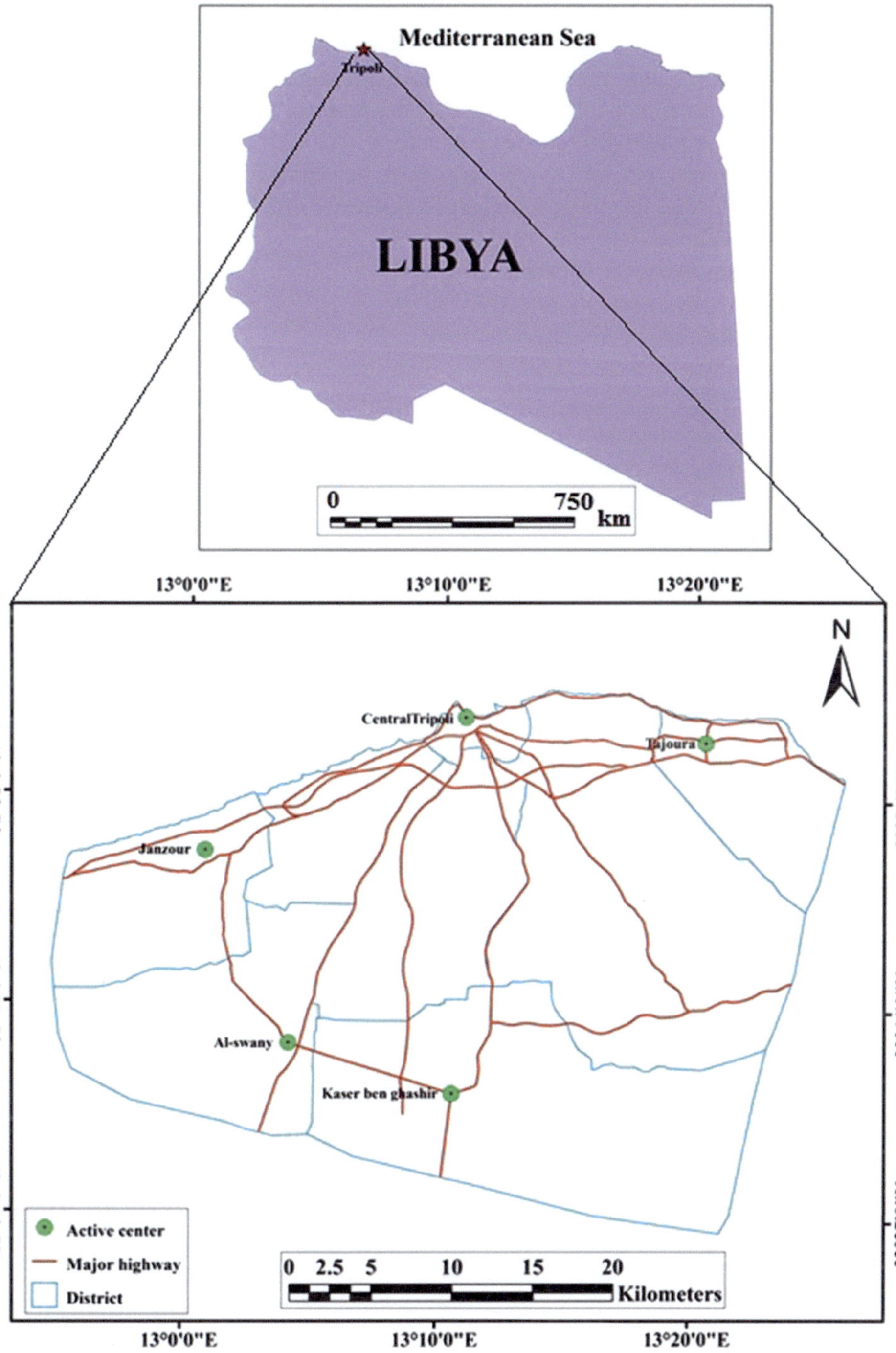

Fig. 9.2 The location map of Tripoli metropolitan (Libya)

largely depends on the country's state of economic growth. Hence, formulating programs within the framework of the total economic development is essential, and these programs should be realistic with regard to the financial ability of the State. The proposed housing investment exceeded the resources of the country and would have necessitated an output of 316,000 dwellings in the Agglomeration during the planning period, an increase of over 200,000 from 1980. However, no explanations or suggestions of policies or programs were forwarded to realize these changes. Moreover, the plans failed to consider the anticipated bottlenecks of implementing this huge number of housing projects, such as the shortage of skilled labor and building materials. These issues have restricted construction activities and led to price jumps. Thus, the 1980–2000 plans remained incomplete by year 2006. The government attempted to solve this problem by issuing laws stating that "anybody using or renting a house becomes the owner of that house." The newly implemented policy also prevented any person from obtaining more than one property. Nevertheless, the problem of severe housing shortage still requires solutions.

In the urban transportation system and road network context, the Tripoli Agglomeration is the largest generator of both passenger and goods traffic in the country. The greatest traffic flow is along the street networks of Tripoli, reaching 4000 vehicles in both directions in 1980. The main settlements in the Agglomeration are connected with paved roads. No construction of the proposed international roads has been done, namely, the new motorway (expected to replace the Coastal Road as an international road on some sections). However, parts of the 3rd Ring Road were recently designed and are undergoing construction.

The proposals for the road network in the 2nd Generation Plan have been implemented to a high degree. The traffic flow is approximately three times higher today than it was 25 years ago. The currently observed significant change is the more severe congestion during peak hours, because the traffic volume has increased beyond the increase in actual capacity of the road network.

In 1980, the Tripoli Agglomeration had low-quality infrastructure. For example, the municipal water network was in poor technical condition. This problem was due to the extremely small pipe diameters, unsatisfactory technical and sanitary conditions of water installations, and shallow pipelines or pipelines with unsatisfactory conduct. The worsening situation was attributed to illegal connections. Potable water was in many cases disinfected by gaseous chlorine.

Another example is the sewage systems, which were more unusual and only existed in the city of Tripoli. Most of the sewage was collected in septic tanks or in cesspools. Sewage treatment plants are in operation in Tripoli. However, numerous problems are connected with these treatment plants, most of which are working at reduced capacity. In other words, the municipal water system does exist for towns and cities within the Master Plan areas. The concept of replacing the ground water with treated effluent for agricultural production never materialized.

9.3 Input Data and Preprocessing

The data used for this process are shown in Table 9.1. The ENVI and ARC/INFO GIS software packages were used for image processing, generating classified land cover/land use maps, and spatial analysis and map preparation.

Resampling process was implemented to match the high-resolution images with low-resolution images. In this process, pixel sizes of images were unaltered to avoid changing the precision of the classification process with the various radiometric spectral and spatial resolutions. Next, the classification process was applied to separate built-up (impervious surfaces), non-built-up (agriculture), and restricted or excluded areas. Then the classified images were resampled to the same spatial resolution (30 m × 30 m), with each map containing 1,816,750 cells. Selecting the pixel size was intended to avoid the decrease in spatial details of the images. Therefore, resampling was conducted after the image classification.

For modeling input, thematic raster maps of all variables were prepared and calculated in the Arc-Info GIS environment and then presented in raster maps with a grid cell size of 30 m × 30 m (Fig. 9.3). The independent input data are as follows:

- Distance to active economy centers,
- Distance to CBD,
- Easting Coordinate,
- Northing Coordinate,
- Slope,
- Restricted areas,
- Distance to nearest urbanized area,
- Population density,
- Distance to educational area,
- Urban area,
- Distance to roads, and
- Distance to coast line

All the prepared data were converted to ASCII and IDRISI formats for further use in analysis and simulation using the IBM SPSS Statistics 20, IDRISI Selva, and FRAGSTATS software.

Table 9.1 Utilized data for urban growth modeling process

Data	Detail
Landsat image 1984	30 m resolution
Landsat image 1996	30 m resolution
Spot 5 image 2002	2.5 m resolution
Spot 5 image 2010	5 m resolution
Roads network	Shape file
Population data census	–
Digital contour map	5 m interval

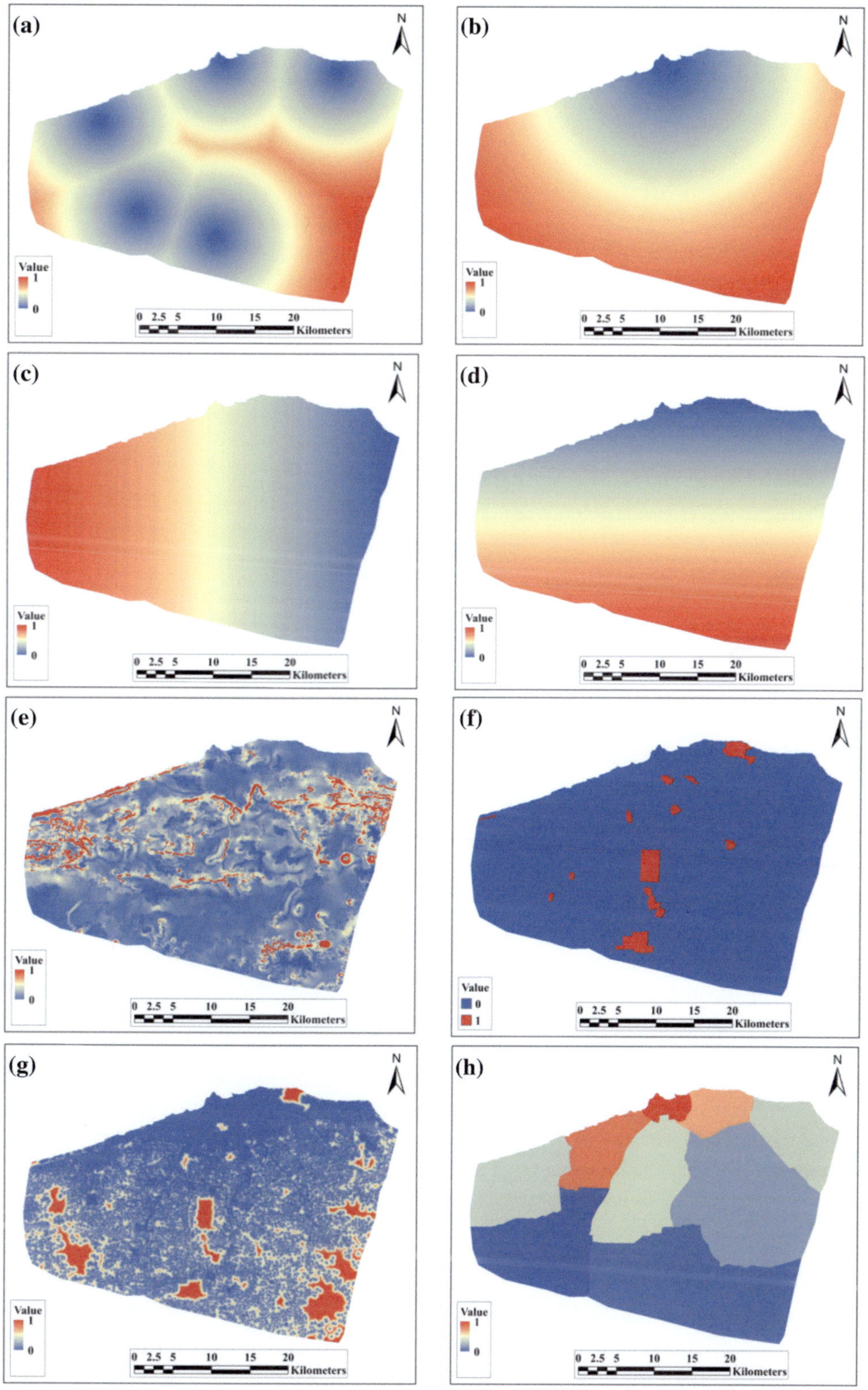

Fig. 9.3 Thematic raster maps of independent variables: **a** Distance to active economy centers, **b** Distance to CBD, **c** Easting coordinate, **d** Northing coordinate, **e** Slope, **f** Restricted areas, **g** Distance to nearest urbanized area, **h** Population density, **i** Distance to educational area, **j** Urban area, **k** Distance to roads, **l** Distance to coastal area

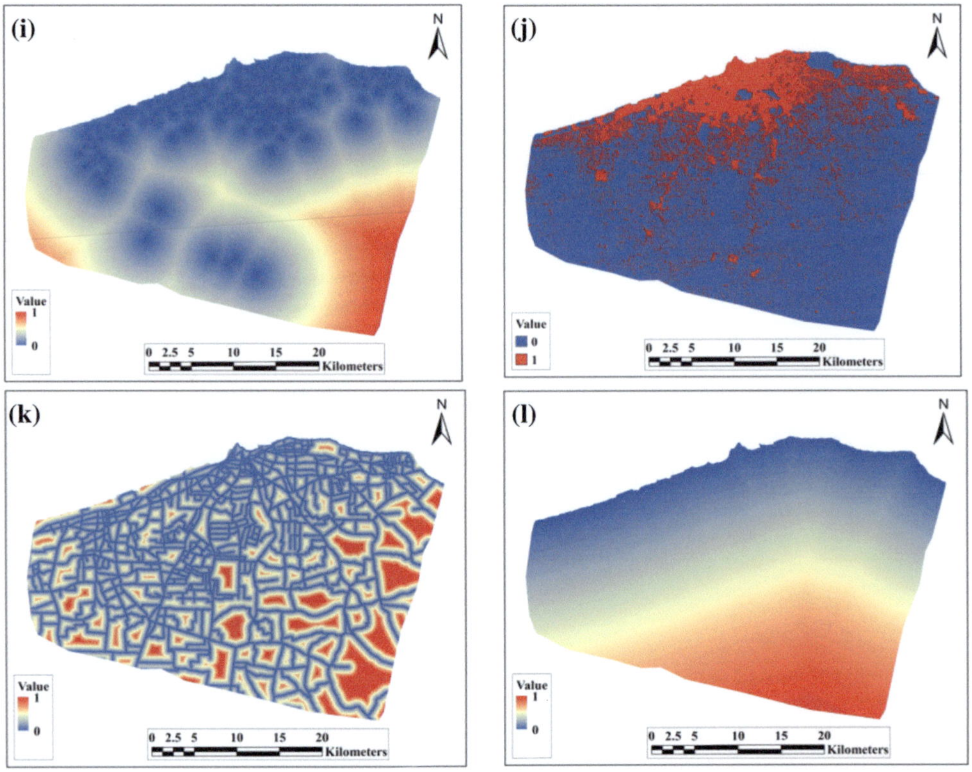

Fig. 9.3 (continued)

9.4 Urban Expansion Modeling and Prediction

This section presents different modeling methods to simulate and predict spatiotemporal urban growth and sprawl patterns. The techniques were used to identify and assess the role of each urban driving factor in the urban expansion process. The performance of each model was assessed and analyzed to identify its shortcomings and advantages and thus achieve optimum simulation. A new statistical method for urban growth modeling (i.e., EBF) was also presented. Furthermore, a novel hybrid model was developed to fulfill one of the research objectives. The proposed model considers the advantages and disadvantages of other applied models and analyzes the interactions of urban factors as well as their contributions to urban sprawl to explain and predict future trends of urban sprawl in the study area.

9.4.1 Frequency Ratio Model

The FR model was used to analyze the spatial effect of the location of urban growth in each class of the classified urban expansion driving factors (i.e., the FR model was based on the recorded associations among allocations of urban expansions and classified classes of driving factors). In this process, real urban growth was used for simulation and validation to reflect real urban expansion behaviors and their dynamics in metropolitan areas; such approach is another novelty of this study. The FR of urban growth is the ratio of the probability of urbanization occurrence to the probability of a non-occurrence for the given attributes. To produce future urban growth probability maps, the FR model was applied using the GIS technique, which spatially represents the information. FR was computed for all classes of each urban driving factor. The FR distribution maps were then summed up to derive the urban growth probability map via Eqs. 9.1 and 9.2.

FR values lower than 1 indicate a low relationship with urban growth, whereas values larger than 1 mean a high correlation with urban expansion. The value of 1 reflects an average condition:

$$FR_{ij} = \frac{N(L \cap C_{ij})/N(L)}{N(C_{ij})/N(C)}, \tag{9.1}$$

where $N(L)$ is the total number of urban growth pixels, $N(C)$ is the total number of pixels in the entire study area, C_{ij} is the jth class attribute of the urban growth driving factors, $C = (C_i, i = 1, 2, 3, \ldots\ldots\ldots n), N(C_{ij})$ is the total number of pixels in class C_{ij}, and $N(L \cap C_{ij})$ is the quantity of urban growth pixels in C_{ij}.

$$\text{Urban Growth Probability Map} = \sum \text{FR}, \tag{9.2}$$

where FR is the rating of the range of each urban factor.

9.4.2 Evidential Belief Function (Dempster–Shafer) Model

Dempster–Shafer theory of evidence refers to the generalization of the Bayesian theorem of subjective probability. Proposed by Dempster in 1967 and developed by Shafer in 1976, this theory combines the beliefs from several sources of evidence and the relative flexibility to accept uncertainty (Thiam 2005). The theory predicts how closely the evidence shows the certainty of a hypothesis rather than predicting the possible accuracy of a hypothesis (Pearl 1990). Dempster–Shafer theory has been applied effectively using GIS in many applications.

In applying the EBF model in the urban expansion modeling process, a set of urban growth driving factors $C = (C\_i,\ i = 1,2,3,\ldots\ldots\ldots n)$ are assumed to comprise mutually exhaustive and exclusive factors C_i. C is named the frame of discrimination. A simple probability assignment is the function m: $P(C) \rightarrow [0,1]$, where $P(C)$ is the set of whole subsets of C, including the empty set and the C set itself. The m function can be considered a mass function that satisfies $m(\Phi) = 0$ and $\sum_{AC} m(A) = 1$, where Φ is the empty set, and A can be any subset of C. $m(A)$ estimates the level to which the evidence supports A and is a belief function $Bel(A)$.

The theory demonstrated four basic EBF functions, namely, Dis (degree of disbelief), *Bel* (degree of belief), Pls (degree of plausibility), and Unc (degree of uncertainty). Dis represents the belief of false suggestion based on the given evidence. *Bel* and Pls provide the upper and lower bounds respectively of the probability for the suggestion (Awasthi and Chauhan 2011). Unc means ignorance, that is, the difference between plausibility and belief. 1—Unc—*Bel* or Dis = 1—Pls, and always Dis + *Bel* + Unc = 1. For cases of Cij with no urban expansion demonstrating that $Bel = 0$, Dis is reset to 0 even though $D \neq 0$ (Carranza et al. 2008).

Overlaying the urban growth map (L) on every thematic urban driving factor map determined the quantity of pixels with urban growth and those without for each factor class. Supposing $N(L)$ is the total number of urban growth pixels and $N(C)$ is the total number of pixels in the entire study area, C_{ij} is the jth class attribute of the urban growth driving factors $C = (C_i, i = 1, 2, 3, \ldots\ldots\ldots n)$, $N(C_{ij})$ is the total number of pixels in class C_{ij}, and $N(L \cap C_{ij})$ is the quantity of urban growth pixels in C_{ij}. The data-driven estimation of EBF can be obtained by

$$Bel(C_{ij}) = \frac{W_{C_{ij(\text{urban growth})}}}{\sum_{j=1}^{n} W_{C_{ij(\text{urban rowth})}}} \tag{9.3}$$

where

$$W_{C_{ij\,(\text{urban growth})}} = \frac{N(L \cap C_{ij})/N(L)}{[N(C_{ij}) - N(L \cap C_{ij})]/[N(C) - N(L)]} \tag{9.4}$$

$$Dis(C_{ij}) = \frac{W_{C_{ij\,(\text{non-urban growth})}}}{\sum_{j=1}^{n} W_{C_{ij(\text{non-urban growth})}}} \tag{9.5}$$

where

$$W_{C_{ij(\text{non urban growth})}} = \frac{[N(C_{ij}) - N(L \cap C_{ij})]/N(L)}{[N(C) - N(L) - N(C_{ij}) + N(L \cap C_{ij})]/[N(C) - N(L)]}. \tag{9.6}$$

The numerator in Eq. 9.4 is the percentage of urban growth pixels occurring in urban factor class C_{ij}. The numerator in Eq. 9.6 is the percentage of urban growth pixels not happening in factor class C_{ij}. The denominator in Eq. 9.4 is the percentage of nonurban growth pixels in factor class C_{ij}. The denominator in Eq. 9.6 is the percentage of nonurban growth pixels in other attributes outside factor class C_{ij}. Parameter W_{Cij} (urban growth) is the weight of C_{ij} supporting the belief that urban growth is more present than absent, whereas parameter W_{Cij} (nonurban growth) is the weight of C_{ij} supporting the belief that urban growth is more nonexistent than existent. When the EBF functions are computed for each urban growth factor, Dempster's rule of combination was applied to obtain the integrated EBF (Dempster 1967). The formulas for merging the two urban driving factors C_1 and C_2 are as follows (Carranza et al. 2005):

$$Bel_{C_1C_2} = \frac{Bel_{C_1}Bel_{C_2} + Bel_{C_1}Unc_{C_2} + Bel_{C_2}Unc_{C_1}}{1 - Bel_{C_1}Dis_{C_2} - Dis_{C_1}Bel_{C_2}} \tag{9.7}$$

$$Dis_{C_1C_2} = \frac{Dis_{C_1}Dis_{C_2} + Dis_{C_1}Unc_{C_2} + Dis_{C_2}Unc_{C_1}}{1 - Bel_{C_1}Dis_{C_2} - Dis_{C_1}Bel_{C_2}} \tag{9.8}$$

$$Unc_{C_1C_2} = \frac{Unc_{C_1}Unc_{C_2}}{1 - Bel_{C_1}Dis_{C_2} - Dis_{C_1}Bel_{C_2}}. \tag{9.9}$$

Thereafter, the remaining urban expansion factors were sequentially integrated using Eqs. 9.7–9.9. Finally, the integrated *Bel* was calculated by Eq. 9.7 and summed up to obtain the future urban expansion probability map of the study area. Disbelief, uncertainty, and plausibility maps of urban expansion were also produced.

9.4.3 Logistic Regression Model

The LR model was used to integrate the demographic and socioeconomic factors of the study area into the urban growth modeling process. This model demonstrates and explains the relationship of a number of Xs independent variables to a dichotomous single-dependent variable Y, which represents the occurrence or non-occurrence of an event. LR empirically finds the relationship between the independent variables and the function of the probability of an event happening (Kleinbaum and Klein 2010). The use of LR can yield the coefficients of independent variables (both continuous and categorical); the dependent variable is a binary categorical variable with a value of either 1 or 0 and can be computed using the well-known LR equation (Huang et al. 2009). This method provides the probability of the existence or nonexistence of each type of land use/cover in every location based on driving factors. LR is a powerful empirical method used when the outcome-dependent variable is dichotomous. Spatial urban expansion is the dependent variable represented in a raster binary map. A value of 1 on the produced probability map indicates the presence of urban growth, and a value of 0 indicates the absence of urban growth. The probability of urbanization for each cell in the raster map was produced based on the following LR equation:

$$f(z) = \frac{1}{1 + e^{-z}}$$

$$P(Y = 1|X_1, X_2, \ldots\ldots, X_k) = 1/(1 + e^{-(\alpha + \sum \beta_i X_i)}), \quad (9.10)$$

where X_i is an independent variable representing a driving factor of the urbanization process, which can be continuous or categorical by nature; α is the coefficient of the model formula; $P(Y = 1|X_1, X_2, \ldots\ldots, X_k)$ is the probability of the dependent variable Y being 1 given $(X_1, X_2, \ldots\ldots, X_k)$, that is, the probability of a cell being changed to a built-up area (urbanized); and β_i is the coefficient of variable X_i. The regression coefficient β_i reflects the function of independent explanatory variables. A negative sign indicates that the variable tends to decrease the possibility of change, and a positive sign indicates the opposite effect. The choice of variables conforms to previous urban modeling and simulation studies. These variables reflect socioeconomic factors, biophysical conditions, and spatial effects (Hu and Lo 2007; Eyoh et al. 2012).

Figure 9.3 and Table 9.2 indicate the independent variables used in this study. Figures 9.4, 9.5 and 9.6 show the dependent variable Y, which represents the urban growth in 1984–2002, 1996–2002, and 2002–2010. In the modeling process, data from 1984 to 2002 were initially used for model calibration and, later, to verify the spatial autocorrelation of regression results. Validation was conducted using the actual growth map of 2010, while prediction of future patterns used data from 2010.

9.4.4 Markov Chain Model

The MC method is a stochastic process system for predicting one status being changed to another known status (Muller and Middleton 1994). This model is frequently applied in modeling and simulation, specifically, the changes, dimensions, and tendencies of urban land use (López et al. 2001; Jianping et al. 2005; Sang et al. 2011). The Markovian stochastic process is one in which the state of a system at the second time can be predicted by the state of the system at the first time, given the matrix of transition probabilities from each cover class to every other cover class over a specified time.

The MC method summarizes and analyzes the change in urban land use and produces probabilities of transition areas that can be employed to predict and discover possible

Table 9.2 List of variables included in the modeling

Variable	Description	Type of variable
Dependent (Y)	0—no urban expansion; 1—urban expansion	Dichotomous
Independent (X_1)	Distance to main active economy centers	Continuous
Independent (X_2)	Distance to CBD	Continuous
Independent (X_3)	Easting coordinate	Continuous
Independent (X_4)	Northing coordinate	Continuous
Independent (X_5)	Slope (%)	Continuous
Independent (X_6)	1—restricted area; 0—non-restricted area	Design
Independent (X_7)	Distance to nearest urbanized area	Continuous
Independent (X_8)	Population density	Continuous
Independent (X_9)	Distance to educational area	Continuous
Independent (X_{10})	Distance to roads	Continuous
Independent (X_{11})	1—urbanized area; 0—nonurbanized area	Design
Independent (X_{12})	Distance to coastal areas	Continuous

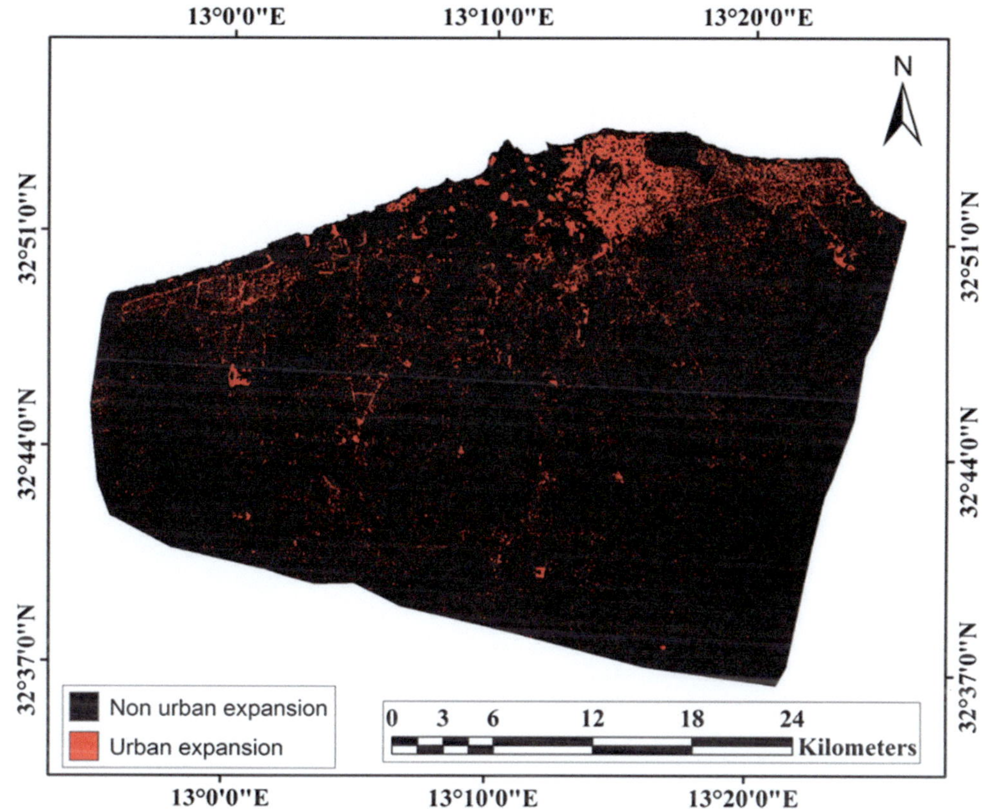

Fig. 9.4 Urban expansion from 1984 to 2002

Fig. 9.5 Urban expansion from 1996 to 2002

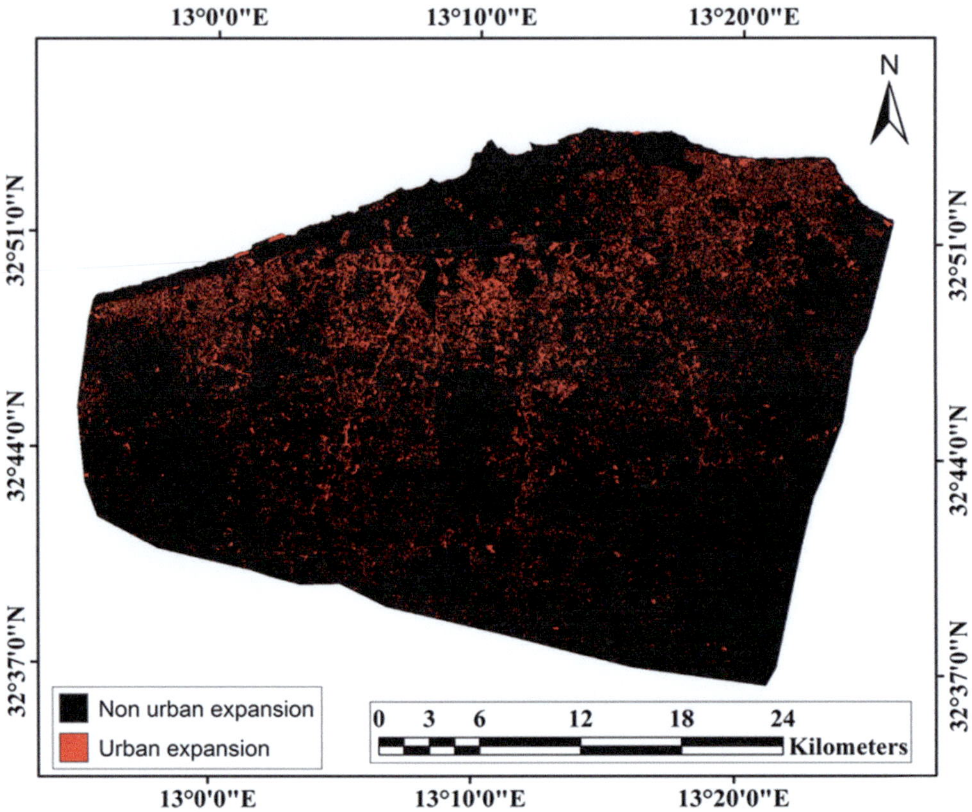

Fig. 9.6 Urban expansion from 2002 to 2010

situations of future urban land use changes and urban expansion patterns. This method is powerful and effective, with the capacity to predict the quantity of urban land use changes (Yang et al. 2012). Nevertheless, the model cannot simulate and model the changes in spatial distributions. The projection of future land use change quantity can be computed based on a conditional probability formula using the following equation:

$$S(t+1) = P_{ij} \times S(t), \tag{9.11}$$

where $S(t)$ is the state of the system at time t, $S(t+1)$ is the state of the system at time $(t + 1)$, P_{ij} is the matrix of transition probability in a state: $P_{ij} = \begin{bmatrix} P_{11} & P_{12} & P_{1n} \\ P_{21} & P_{22} & P_{2n} \\ P_{n1} & P_{n2} & P_{nn} \end{bmatrix}$, and $(0 \le P_{ij} < 1$ and $\sum_{j=1}^{N} P_{ij} = 1, (i,j = 1,2,\ldots..n))$.

9.4.5 Cellular Automata Model

Generally, CA models aim to simulate real natural regulations. Land use change modeling using the CA technique is a preferred method because it provides explicit spatial modeling results based on defined transition rules (White and Engelen 1993). Moreover, CA model types are suitable for representing, analyzing, and forecasting geographic processes owing to the relationships in a raster grid (Clarke and Gaydos 1998; Mitsova et al. 2011).

A CA-based model can represent nonlinear, spatial, and stochastic processes; model and control complex spatially distributed processes; and provide clear insights into local behaviors and global patterns of land use/cover change. Spatial and temporal complexities of land use change can also be well represented and simulated using a suitable transition rule in the CA model. However, the most important concern in the CA model is defining appropriate transition rules based on training data that control the model (Al-shalabi et al. 2013).

The model is affected by neighborhood type, neighborhood size, and cell size parameters. Accordingly, these parameters should be considered to obtain optimum simulation results. The CA is a practical tool in urban system simulations because population change and land use change can be presented together. Cells of the cellular lattice can likewise be aggregated efficiently with economic and transportation data. Thus, urban areas can be effectively simulated using the proper neighborhoods of cells on the cellular grid. Theories on the urbanization process can also be examined on the basis of used spatial models (Mitsova et al. 2011). Time and space are considered discrete units in the CA model, and space is considered a regular grid (lattice) in two dimensions. The main aspect of the CA model is the local interactions reflecting the dynamics of the urban system evolution (White and Engelen 1993; Wang 2012). CA models can simulate stochastic, nonlinear, and spatial process. Several studies have illustrated that the model can potentially exhibit an understandable complex spatiotemporal process of land use change and urban systems and its patterns (Batty et al. 1999; Wang 2012).

The major components of CA models include the cells, cell neighborhoods, and transition rules, such as the cell being the fundamental element of the automation system and the cell being organized in a lattice. The transition rule defining the state of each cell for the coming time step depends on the current state of that cell and its surrounding neighborhood cells. Thereafter, a land use change suitability map is required, and the dynamics should be defined into the system. The basic expression of the CA model can be expressed as follows:

$$S(t, t+1) = f(S(t), N), \quad (9.12)$$

where S is the state of a discrete cell, t is the time instant, $t + 1$ is the coming future time instant, N is the cellular field, and f is the transition rule of cellular states in local space.

9.4.6 CA–Markov Model

The reliability of land use change modeling methods can be improved by combining two or more simulation techniques to integrate the advantages of each model (Qiu and Chen 2008; Yang et al. 2012). The CA–MC model was recently used to simulate dynamic spatial phenomena and predict future land use change. The integrated model combines the MC of land use change quantity prediction and the dynamic explicit spatial simulation of the CA model. Thus, it can translate the results of the MC model using a CA function for spatially explicit outcomes required for urban planning and design (Guan et al. 2011).

The CA–MC model can be useful for the spatial modeling of land use change. Consequently, incorporating GIS and land use/cover maps derived from RS data with the CA–MC model can effectively model and simulate spatial and temporal land use change (Kamusoko et al. 2009). Moreover, the model provides reliable land use change simulation results and overcomes the lack of socioeconomic, statistical, and historical data. In the CA–MC modeling process, the temporal changes of land use classes are directed in the MC process based on produced transition matrices, whereas the spatial changes are controlled by transition potential maps, neighborhood configuration, and local transition rule during the CA modeling process.

9.4.7 Chi-Squared Automatic Integration Detection Decision Tree Model

Chi-squared automatic integration detection DT (CHAID-DT) algorithm is employed to analyze the urban expansion process, understand the interactions of urban driving factors, and predict future urban expansion probability maps. The CHAID-DT algorithm, which was proposed four decades ago, allows multiple splits of nodes (Kass 1980). The model depends on the chi-square test of association analysis (Althuwaynee et al. 2014). CHAID-DT is built by repetitively splitting variables into two or more child nodes (multi-way split) and starting with the entire dataset of variables. The chi-square test examines independent variables for independency and determines whether splitting nodes generate a significant improvement. The CHAID-DT algorithm involves splitting, merging, and stopping; dependent and independent variables fields can be categorical or continuous, and nodes can be split into two or more nodes at each level. In the case of categorical data, Eq. 9.13 of the Pearson chi-square is applicable as follows:

$$X^2 = \sum_{j=1}^{J} \sum_{i=1}^{I} \frac{(n_{ij} - m_{ij})^2}{m_{ij}}, \quad (9.13)$$

where

$$n_{ij} = \sum_{n \in D} \mathrm{fn} I(x_n = i \cap y_n = j),$$

where n_{ij} is the observed cell frequency, and m_{ij} is the estimated expected cell frequency for $(x_n = i, y_n = j)$ following the independence model. The corresponding p-value is given by p = Pr $(x_d^e > x^2)$ (Baker and Cousins 1984).

In the case of continuous data, Eq. 9.14 for the F test is employed. For ordinal data, a likelihood ratio test is considered (Miner et al. 2009).

$$F = \frac{\sum_{i=1}^{I} \sum_{n \in \mathrm{D}} \mathrm{w}_n \mathrm{f}_n I(x_n = i)(y_i - y)^2 / (I - 1)}{\sum_{i=1}^{I} \sum_{n \in D} \mathrm{w}_n \mathrm{f}_n I\left(\mathrm{x}_n = i\right)(y_n - y)^2 / (N_f - 1)}. \quad (9.14)$$

In addition, $p = \Pr(F(I-1, N_f - 1) > F)$,
where $y_i = \frac{\sum_{n \in D} w_n f_n y_n I(x_n = i)}{\sum_{n \in D} w_n f_n I(x_n = i)}$, $y = \frac{\sum_{n \in D} w_n f_n y_n I}{\sum_{n \in D} w_n f_n}$, $\mathrm{N_f} = \sum_{\mathrm{n \in D}} \mathrm{f_f}$,
and F (I-1, N_f-1) is a random variable following F-distribution with degrees of freedom I-1 and N_f-1.

The CHAID-DT model chooses the urban driving factor (predictor) that has the strongest relationship with urban expansion (dependent variable). The classes of every urban driving factor are then merged because of the absence of significant differences among the said factors and the target variable (urban expansion). The nodes in the CHAID-DT correspond to independent variables; every leaf is allocated to one class (e.g., urban growth/nonurban growth) representing the most frequent class value. The leaves also hold the probabilities demonstrating the probability of a target class occurrence (e.g., occurrence of urban growth). The urban driving factors with significant influence on urban expansion are used in the processing, whereas the insignificant factors are dropped by the program. The CHAID-DT algorithm was applied in a SPSS V.20 platform. Merging and splitting categories ranged between 0 and 1. In this work, the value of 0.3 was used to split and merge parameters.

A simple tree with few nodes was initially created to easily understand and analyze the interactions of urban factors. The CHAID-DT was limited to 82 nodes, including 64 terminal nodes (leaves), as shown in the Results section. However, a larger number of nodes performed better in the prediction process. The created tree used for prediction has a total of 57,024 nodes, including 38,249 terminal nodes (leaves).

In this work, the 1984 data of urban driving factors and 1984–2002 urban growth data were used to build and calibrate the CHAID-DT model. Thereafter, the created model was updated with urban driving factors in 2002; the urban expansion in 2002–2010 was used to check the model's validity. Finally, the model was updated once again with urban driving factors in 2010 to predict the future urban transition map. The urban factor of the northing coordinate was excluded from the inputs because of the high spatial correlation with the urban factor of distance to the coastal area. Furthermore, factors of restricted areas, urban extents, and easting coordinate factors were dropped by the model.

9.4.8 Development of a Novel Hybrid Model (CHAID–CA–Markov)

The hybrid model developed in this study is based on the integration of three models, namely, CHAID-DT model, MC analysis, and CA model. The proposed model considers the potentials and shortcomings of each model to present a robust modeling approach stronger than each single model. The suggested hybrid model is shown in Fig. 9.7.

The first step in the hybrid model is analyzing the urban driving factors in the study area and assessing the roles and influences of each factor on the others. Resulting from this first step, the model drops the unimportant urban factors and generates the future probability map of urban expansions and sprawl. The second step is validating the future probability map of the urbanization process produced by the CHAID-DT model using the ROC method. The third step is calculating the future demand of urban lands and estimating the quantities of land use change by analyzing urban land use maps in 1984 and 2002 in the MC model. The fourth step is using the CA model to distribute the estimated quantity of urban sprawl on the probability surface produced by the CHAID-DT model to obtain explicit future urban land use map. The final step is validating the produced explicit map using the Kappa statistics index to ensure optimum performance of the model.

9.5 Urban Growth Modeling Validation

The relative operating characteristic (ROC) technique is used to evaluate the performance of the LR, FR, EBF, and DT models and assess the resulting probability maps employed in this study. It measures the relationship between expected and real spatial changes by calculating the percentage of false positives (1-specificity) and true positives (sensitivity) for a range of thresholds and relating the values to one another in a chart. The ROC computes the area under the curve, which varies from 0.5 to 1. A value of 0.5 indicates a random assignment of the probabilities, that is, the expected agreement is due to chance; whereas a value of 1 indicates a perfect probability assignment, that is, an ideal spatial agreement exists between the real urban expansion and the predicted urban probability maps (Pontius and Schneider 2001; Wang and Mountrakis 2011). Model validation was conducted by comparing the probability image maps of future urban expansion produced from the LR, FR, EBF, and DT models alongside the real urban expansion in 2002–2010 to confirm the model's capability. The ROC curve is based on several two-by-two contingency tables sequentially based on the comparisons between actual and predicted probability images. Table 9.3 shows the contingency table form, where

- A is the amount of true positive cells, i.e., cells predicted as urban expansion and in agreement with the actual image;
- B is the amount of false positive cells, i.e., cells predicted as urban expansion but in disagreement with the actual image;
- C is the amount of false negative cells, i.e., cells predicted as nonurban expansion but in disagreement with the actual image; and
- D is the amount of true negative cells, i.e., cells predicted as nonurban expansion and in agreement with the actual image.

From every contingency table, a single data point (x, y) is created, where X and Y are the rates of false positives and true positives, respectively.

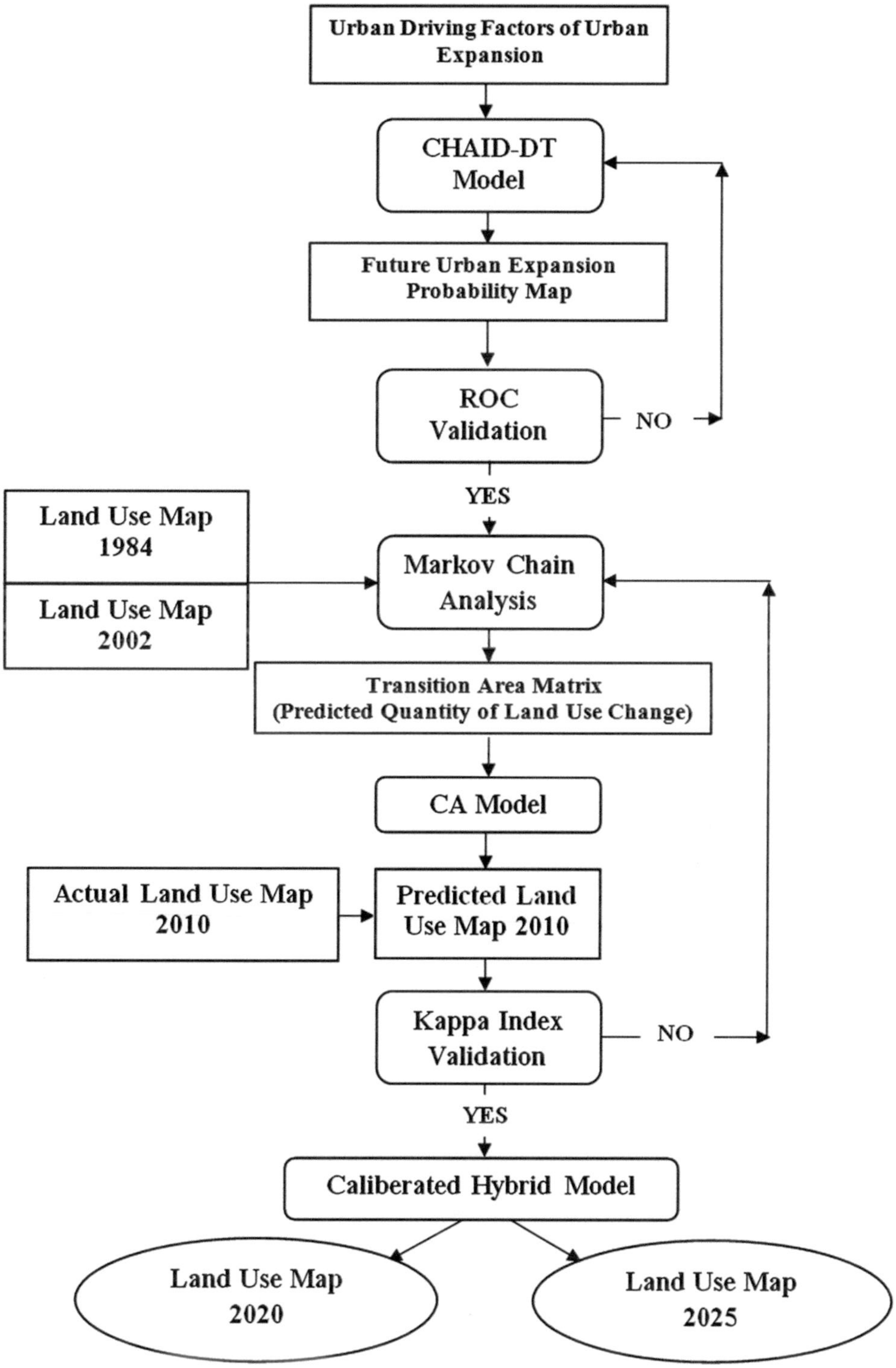

Fig. 9.7 Flowchart of the proposed hybrid model

$$(\text{true positive }\%) = \mathrm{A}/(\mathrm{A}+\mathrm{C}) \quad (9.15)$$

$$(\text{false positive }\%) = \mathrm{B}/(\mathrm{B}+\mathrm{D}). \quad (9.16)$$

The data points are joined to form the ROC curve from which the ROC value is computed (Fig. 9.8). The ROC value is the area under the curve created by the plotted points.

Apart from ROC, the projected land use maps for 2010 were compared with the actual maps using Kappa index statistic to check the validity in terms of quantity and location, as well as validate both the developed hybrid model and the produced explicit urban land use maps. Kappa index of agreement is a measure of proportional accuracy adjusted for the chance agreement (Pontius et al. 2004; Arsanjani et al. 2011).

Table 9.3 Two-by-two contingency table showing the number of grid cells in an actual map versus a predicted map

		Actual map		Total
		Urban expansion (1)	Nonurban expansion (0)	
Predicted probability map	Urban expansion (1)	A	B	A + B
	Nonurban expansion (0)	C	D	C + D
Total		A + C	B + D	A + B + C + D

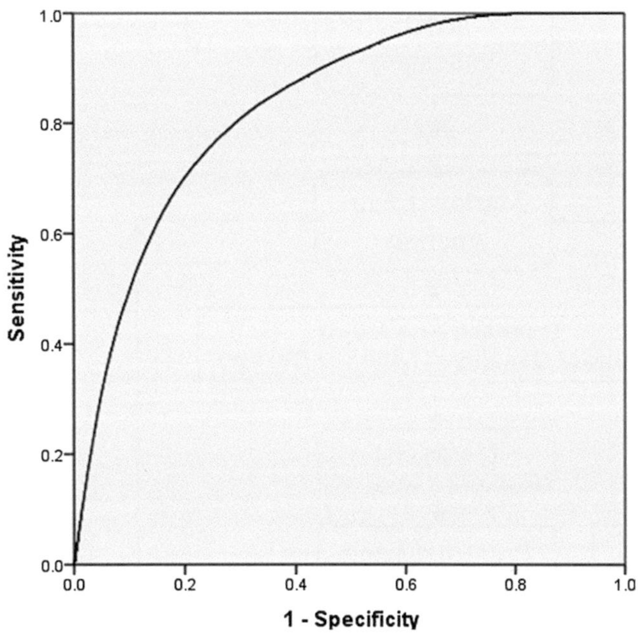

Fig. 9.8 ROC curve

The statistics mainly divided the agreement and disagreement of two comparison data into the following measures:

- Kappa for no information = K no
- Kappa for location = K location
- Kappa standard = K standard

The algorithm of statistics was conducted through the VALIDATE module in the Idrisi Selva platform environment.

9.6 Results and Discussion of Urban Growth Modeling

This section demonstrates the result and extensively discusses the findings from the applied GIS modeling techniques and the suggested hybrid model. The chapter also highlights and discusses the result of model validation, the accuracy of all predicted maps, and the advantages and shortcomings of the used models. It deliberates on the novel hybrid model performance and compares its results to other models.

9.6.1 Application of EBF and FR Models

The optimum set of urban driving factors for modeling was determined for both models of bivariate statistical techniques. The included urban independent variables were slope, distance to active economy centers, distance to CBD, distance to roads, distance to the nearest urbanized area, distance to the coastal area, and distance to the educational area. Table 9.4 illustrates the behaviors of urban expansions, the variations in EBF and FR values during different periods, and the relationship of such variations with urbanization factors. The recorded values of EBF (*bel*) and FR in Table 9.4 show the suitability level of urban expansion and the role of each class within every urban driving factor. The slope in the study area ranges from 0° to 17°. In this study, the slopes were classified into seven quantile classes. The values of EBF (*bel*) ranged from 0.114 to 0.164 in 1996–2002 and from 0.093 to 0.195 in 2002–2010. In general, the EBF (*bel*) values in each class of slope factor do not have high differences among them; thus, the role of the slope factor in the urban expansion of the study area is low. Furthermore, the outcomes of the FR model ranged from 0.793 to 1.131 in 1996–2002 and from 0.668 to 1.307 in 2002–2010. These results indicate that the slope factor insignificantly affected the urban growth process in Tripoli, especially when the low slope domains were saturated with urban areas. Consequently, the urban area extended to higher slope domains. Thus, the slopes of all domains in the study area allowed urban expansion with little variations.

The influence of the factor *distance to active economic centers* can clearly be assessed from the EBF (*bel*) values. The highest growth occurred in the distance of 0–630 m from active economic centers in 1996–2002 with an EBF (*bel*) value of 0.066, whereas the largest EBF (*bel*) was

Table 9.4 Values of EBF (*bel.*) and FR models for urban expansion driving factors

Urban factor	Class	EBF (*Bel*) (1996–2002)	EBF (*Bel*) (2002–2010)	FR (1996–2002)	FR (2002–2010)
Slope (°)	0–0.41	0.114	0.093	0.793	0.668
	0.41–1.04	0.129	0.095	0.897	0.679
	1.04–1.93	0.135	0.124	0.937	0.872
	1.93–3.18	0.143	0.145	0.993	1.005
	3.18–4.90	0.164	0.172	1.131	1.169
	4.90–9.39	0.153	0.195	1.061	1.307
	9.34–17.60	0.162	0.176	1.116	1.194
Distance to active economic centers (m)	0–629	0.066	0.072	1.643	1.732
	629–1258	0.059	0.075	1.479	1.800
	1258–1887	0.045	0.058	1.137	1.458
	1887–2516	0.048	0.052	1.201	1.330
	2516–3145	0.055	0.051	1.386	1.309
	3145–3775	0.050	0.042	1.259	1.103
	3775–4404	0.041	0.040	1.031	1.051
	4404–5033	0.042	0.037	1.071	0.975
	5033–5662	0.047	0.036	1.175	0.944
	5662–6291	0.049	0.035	1.227	0.932
	6291–6920	0.049	0.040	1.227	1.043
	6920–7549	0.043	0.039	1.079	1.025
	7549–8178	0.045	0.040	1.124	1.055
	8178–8807	0.044	0.047	1.100	1.200
	8807–9436	0.040	0.049	1.017	1.255
	9436–10,066	0.033	0.046	0.843	1.193
	10,066–10,695	0.037	0.047	0.928	1.200
	10,695–11,324	0.040	0.043	1.003	1.126
	11,324–11,953	0.035	0.040	0.887	1.048
	11,953–12,582	0.022	0.021	0.573	0.572
	12,582–13,211	0.019	0.017	0.498	0.484
	13,211–13,840	0.017	0.016	0.446	0.455
	13,840–14,469	0.016	0.014	0.401	0.406
	14,469–15,098	0.014	0.009	0.359	0.267
	15,098–15,727	0.013	0.011	0.341	0.307
	15,727–16,357	0.012	0.009	0.304	0.271
	16,357–16,986	0.005	0.005	0.126	0.132
	16,986–17,615	0.006	0.003	0.163	0.085
	17,615–18,244	0.005	0.002	0.135	0.068
	18,244–18,873	0.003	0.001	0.085	0.043
	18,873–19,502	0.000	0.001	0.000	0.024

(continued)

Table 9.4 (continued)

Urban factor	Class	EBF (*Bel*) (1996–2002)	EBF (*Bel*) (2002–2010)	FR (1996–2002)	FR (2002–2010)
Distance to CBD (m)	0–1144	0.005	0.001	0.318	0.038
	1144–2288	0.001	0.000	0.057	0.039
	2288–3432	0.013	0.001	1.201	0.045
	3432–4576	0.011	0.007	0.879	0.522
	4576–5719	0.013	0.008	1.033	0.559
	5719–6863	0.014	0.015	1.063	1.062
	6863–8007	0.021	0.025	1.601	1.715
	8007–9151	0.032	0.035	2.457	2.320
	9151–10295	0.022	0.037	1.573	2.259
	10295–11439	0.023	0.043	1.462	2.336
	11439–12583	0.033	0.044	1.893	2.108
	12583–13727	0.028	0.046	1.431	1.934
	13727–14871	0.032	0.050	1.369	1.752
	14871–16014	0.027	0.041	1.033	1.356
	16014–17158	0.036	0.050	1.173	1.376
	17158–18302	0.042	0.054	1.153	1.250
	18302–19446	0.039	0.058	0.942	1.164
	19446–20590	0.048	0.049	1.055	0.899
	20590–21734	0.051	0.053	0.933	0.808
	21734–22878	0.070	0.066	0.959	0.746
	22878–24022	0.071	0.057	0.867	0.585
	24022–25166	0.074	0.053	0.788	0.482
	25487–26595	0.070	0.056	0.581	0.393
	26595–27704	0.065	0.040	0.501	0.267
	27704–28812	0.060	0.054	0.499	0.381
	28812–29920	0.045	0.027	0.478	0.258
	29920–31028	0.028	0.015	0.370	0.184
	31028–32136	0.013	0.009	0.240	0.147
	32136–33244	0.007	0.005	0.241	0.159
	33244–34352	0.004	0.002	0.119	0.046
	34352–35460	0.000	0.001	0.000	0.009
Distance to roads (m)	0–93	0.152	0.147	1.551	1.404
	93–186	0.119	0.150	1.222	1.428
	186–279	0.110	0.120	1.130	1.182
	279–372	0.100	0.104	1.029	1.037
	372–465	0.089	0.078	0.920	0.802
	465–557	0.070	0.080	0.731	0.817
	557–650	0.056	0.062	0.584	0.651
	650–743	0.045	0.045	0.476	0.475
	743–836	0.046	0.034	0.482	0.369
	836–929	0.043	0.036	0.448	0.385
	929–1022	0.033	0.023	0.344	0.250
	1022–1115	0.023	0.024	0.238	0.265

(continued)

Table 9.4 (continued)

Urban factor	Class	EBF (*Bel*) (1996–2002)	EBF (*Bel*) (2002–2010)	FR (1996–2002)	FR (2002–2010)
	1115–1208	0.023	0.019	0.240	0.203
	1208–1301	0.021	0.016	0.219	0.174
	1301–1394	0.017	0.015	0.182	0.170
	1394–1486	0.019	0.012	0.204	0.128
	1486–1579	0.014	0.008	0.152	0.093
	1579–1672	0.012	0.007	0.128	0.077
	1672–1765	0.006	0.004	0.060	0.050
	1765–1858	0.001	0.009	0.009	0.102
	1858–1951	0.003	0.003	0.031	0.039
	1951–2044	0.000	0.001	0.000	0.007
	2044–2137	0.000	0.001	0.000	0.012
	2137–2230	0.000	0.001	0.000	0.010
	2230–2323	0.000	0.000	0.000	0.000
	2323–2415	0.000	0.000	0.000	0.000
	2415–2508	0.000	0.000	0.000	0.000
	2,508–2,601	0.000	0.000	0.000	0.000
	2601–2694	0.000	0.000	0.000	0.000
	2694–2787	0.000	0.000	0.000	0.000
	2787–2880	0.000	0.000	0.000	0.000
Distance to urban areas (m)	0–93	0.259	0.371	2.070	1.736
	93–185	0.142	0.316	1.165	1.519
	185–278	0.106	0.120	0.874	0.641
	278–370	0.083	0.063	0.691	0.346
	370–463	0.068	0.036	0.566	0.200
	463–555	0.059	0.018	0.493	0.102
	555–648	0.046	0.012	0.386	0.068
	648–740	0.035	0.009	0.293	0.051
	740–833	0.033	0.007	0.279	0.042
	833–926	0.031	0.011	0.263	0.060
	926–1018	0.024	0.007	0.200	0.038
	1018–1111	0.024	0.004	0.202	0.025
	1111–1203	0.017	0.007	0.142	0.039
	1203–1296	0.016	0.004	0.135	0.021
	1296–1388	0.010	0.003	0.088	0.017
	1388–1481	0.011	0.001	0.089	0.004
	1481–1574	0.009	0.001	0.074	0.007
	1574–1666	0.008	0.001	0.067	0.005
	1666–1759	0.004	0.001	0.035	0.006
	1759–1851	0.007	0.004	0.060	0.025
	1851–1944	0.004	0.002	0.037	0.009
	1944–2036	0.002	0.002	0.015	0.010
	2036–2129	0.000	0.000	0.000	0.000
	2129–2221	0.000	0.000	0.000	0.000

(continued)

Table 9.4 (continued)

Urban factor	Class	EBF (*Bel*) (1996–2002)	EBF (*Bel*) (2002–2010)	FR (1996–2002)	FR (2002–2010)
	2221–2314	0.000	0.000	0.000	0.000
	2314–2407	0.000	0.000	0.000	0.000
	2407–2499	0.000	0.000	0.000	0.000
	2499–2592	0.000	0.000	0.000	0.000
	2592–2684	0.000	0.000	0.000	0.000
	2684–2777	0.000	0.000	0.000	0.000
	2777–2869	0.000	0.000	0.000	0.000
Distance to educational areas (m)	0–523	0.083	0.099	1.515	1.592
	523–1045	0.084	0.119	1.535	1.855
	1045–1568	0.069	0.084	1.274	1.386
	1568–2090	0.067	0.076	1.228	1.276
	2090–2613	0.068	0.070	1.255	1.186
	2613–3135	0.066	0.061	1.213	1.050
	3135–3658	0.058	0.055	1.073	0.951
	3658–4180	0.049	0.050	0.903	0.881
	4180–4703	0.047	0.045	0.874	0.797
	4703–5225	0.047	0.044	0.871	0.769
	5225–5748	0.030	0.033	0.571	0.597
	5748–6270	0.025	0.029	0.469	0.524
	6270–6793	0.034	0.027	0.643	0.500
	6793–7316	0.027	0.026	0.504	0.472
	7316–7838	0.033	0.024	0.617	0.434
	7838–8361	0.032	0.021	0.599	0.386
	8361–8883	0.022	0.015	0.424	0.278
	8883–9406	0.023	0.016	0.426	0.297
	9406–9928	0.017	0.016	0.324	0.302
	9928–10,451	0.020	0.011	0.384	0.212
	10,451–10,973	0.020	0.013	0.374	0.241
	10,973–11,496	0.018	0.020	0.334	0.365
	11,496–12,018	0.017	0.014	0.318	0.263
	12,018–12,541	0.011	0.010	0.202	0.194
	12,541–13,064	0.012	0.006	0.221	0.115
	13,064–13,586	0.003	0.002	0.056	0.037
	13,586–14,109	0.005	0.003	0.097	0.048
	14,109–14,631	0.004	0.003	0.080	0.049
	14,631–15,154	0.007	0.003	0.133	0.055
	15,154–15,676	0.001	0.002	0.024	0.031
	15,676–16,199	0.002	0.003	0.047	0.050

(continued)

Table 9.4 (continued)

Urban factor	Class	EBF (*Bel*) (1996–2002)	EBF (*Bel*) (2002–2010)	FR (1996–2002)	FR (2002–2010)
Distance to coast (m)	0–852	0.021	0.044	4.173	2.916
	852–1825	0.045	0.069	3.397	2.681
	1825–2799	0.058	0.074	3.171	2.591
	2799–3773	0.059	0.062	2.369	2.026
	3773–4746	0.053	0.058	1.933	1.730
	4746–5842	0.059	0.067	1.592	1.599
	5842–6815	0.051	0.071	0.972	1.271
	6815–7789	0.054	0.074	0.601	1.074
	7789–8762	0.054	0.062	0.482	0.901
	8762–9858	0.049	0.051	0.539	0.842
	9858–10,831	0.042	0.043	0.483	0.736
	10,831–11,805	0.031	0.034	0.373	0.585
	11,805–12,778	0.026	0.029	0.298	0.487
	12,778–13,874	0.027	0.030	0.207	0.445
	13,874–14,847	0.027	0.027	0.212	0.420
	14,847–15,821	0.029	0.024	0.233	0.400
	15,821–16,794	0.027	0.022	0.177	0.350
	16,794–17,890	0.031	0.019	0.209	0.333
	17,890–18,863	0.023	0.016	0.151	0.259
	18,863–19,837	0.029	0.017	0.167	0.283
	19,837–20,810	0.032	0.020	0.268	0.375
	20,810–21,784	0.022	0.013	0.272	0.293
	21,784–22,758	0.022	0.009	0.134	0.179
	22,758–23,731	0.022	0.011	0.235	0.261
	23,731–24,705	0.024	0.013	0.169	0.240
	24,705–25,678	0.016	0.009	0.096	0.156
	25,678–26,652	0.014	0.008	0.079	0.139
	26,652–27,625	0.012	0.009	0.069	0.143
	27,625–28,599	0.016	0.006	0.085	0.117
	28,599–29,573	0.016	0.006	0.115	0.130
	29,573–31,033	0.010	0.005	0.057	0.084

recorded in the distance range of 630–1260 m in 2002–2010 with an EBF (*bel*) value of 0.075. However, the distance range of 0–630 m increased the FR value from 1.643 in the first period to 1.732 in the second period. This increase means that this distance range (0–630 m) will be denser with the built-up area. Accordingly, the suitability level of higher urbanization expanded to the next class domain, that is, 630–1260 m with an FR value of 1.800. The urban factor *distance to CBD* showed the highest EBF (*bel*) and FR values at a distance range of 8000–9150 m in both periods. However, the decrease in EBF (*bel*) and FR values in 2002–2010 when compared with the values obtained for 1996–2002 refers to the decrease in urban growth probability in that class with time increase (i.e., this class range is increasingly becoming compact). Distance ranges of less than 8000 m experienced dramatic decreases in urban growth possibility in 2002–2010 (i.e., these areas became extremely dense). Simultaneously, the distance range of 8000–19,500 m indicates a remarkable increase in FR and EBF (*bel*) values, suggesting that increased urban growth suitability may indicate uncontrolled growth.

The modeling results of the two applied bivariate models show that urban growth increased gradually as time progressed and with the increase in distance from roads. Earlier

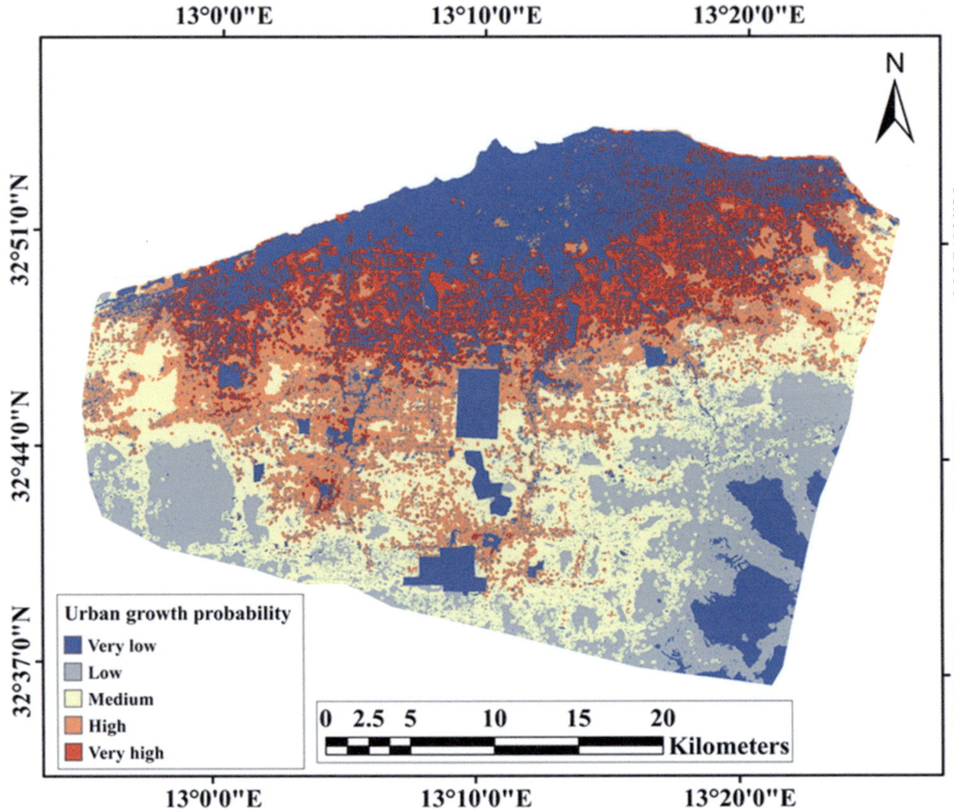

Fig. 9.9 Predicted urban growth probability map using the FR model

urban expansions tended to occur near roads, whereas urban areas expanded away from roads as time progressed. The general trend of EBF (*bel*) and FR values for the urban factor *distance to built-up areas* demonstrates that the probability of urban expansion increases with the decrease in distance to urbanized areas in all periods. The factor *distance to educational areas* showed behavior nearly similar to that of the *distance to roads* factor. In the case of compact urban classes nearer educational areas, growth will occur in the next domains. However, the most probable class of urban growth is in the distance range of 500–1000 m from educational areas. The last considered urban factor, *distance to coast*, indicated that the highest values of EBF (*bel*) were recorded in the fourth and sixth classes in 1996–2000, whereas the highest EBF (*bel*) values were in the third and eighth classes in 2002–2010. These results reflect different behaviors of spatiotemporal urban expansion.

However, the FR model showed a different urban growth behavior. The highest FR values were recorded in both periods in the first class (i.e., in distances below 852 m). A significant finding is that the FR value decreased from 4.173 in the first period to 2.916 in the second period. Therefore, future urban expansion will happen in the next classes rather than in the first class.

Finally, urban growth probability maps were produced from the applied methodological process. Thereafter, the restricted areas and last built-up area extents were excluded to produce the future maps (Figs. 9.9 and 9.10). As an advantage of the EBF model, the plausibility, disbelief, and uncertainty maps of urban expansion were produced to provide additional information regarding the probable occurrences of urban expansions in the study area, as shown in Figs. 9.11, 9.12 and 9.13.

9.6.2 Application of the LR Model

Table 9.5 lists the modeling results of the urban sprawl behavior. The LR model checks the multi-collinearity, which verifies the correlation of independent variables. The modeling results demonstrate the tolerance and variance inflation factor (VIF) that examines the multi-collinearity. The tolerance value ranges from 0.141 to 0.956, and the VIF (1/tolerance) varies from 1.046 to 7.069. A widely used standard provides that VIF should not exceed 10. In this study, the VIF value of the urban factor distance to coast exceeded 10, thereby compelling the model to drop the factor. The results shown in Table 9.5 reflect the highly effective model as per the multi-collinearity assessment (Menard 2012).

The model illustrates that urban growth was affected by the main active economic centers. This finding reflects the

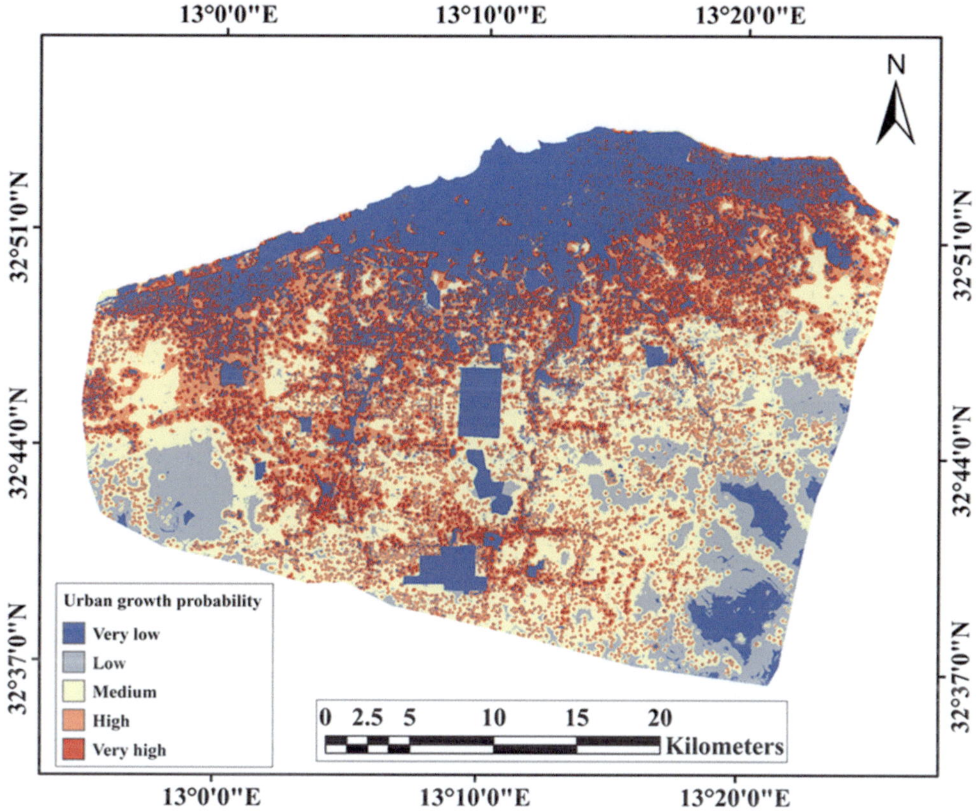

Fig. 9.10 Predicted urban growth probability map (*bel.* map) using the EBF model

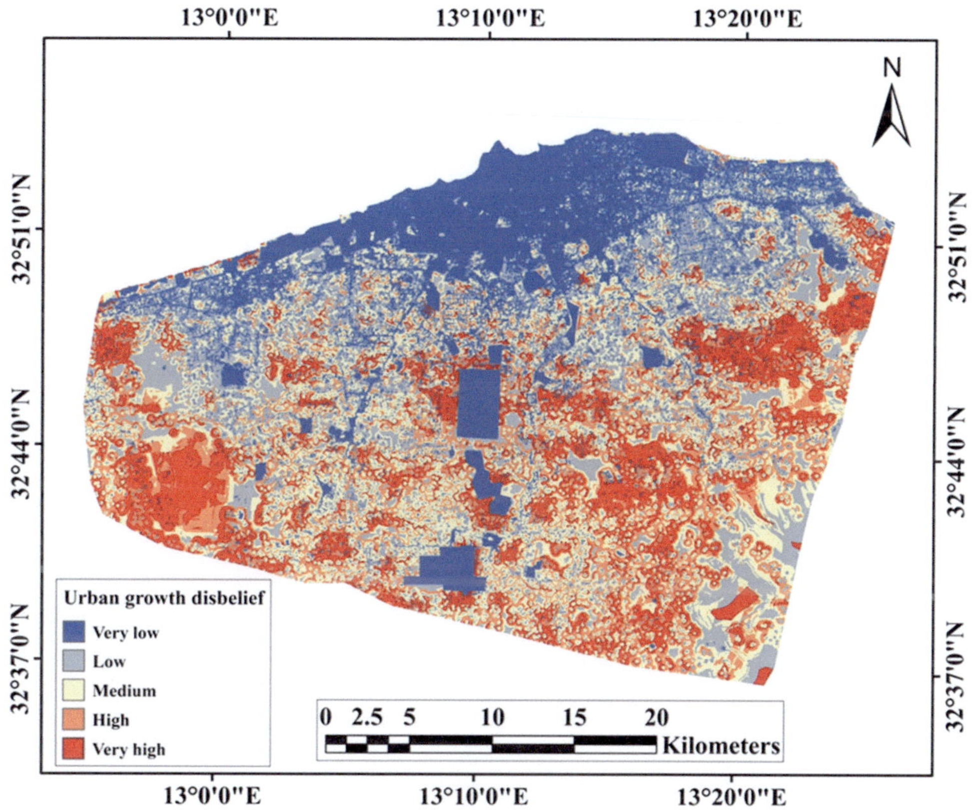

Fig. 9.11 Disbelief map of urban expansion

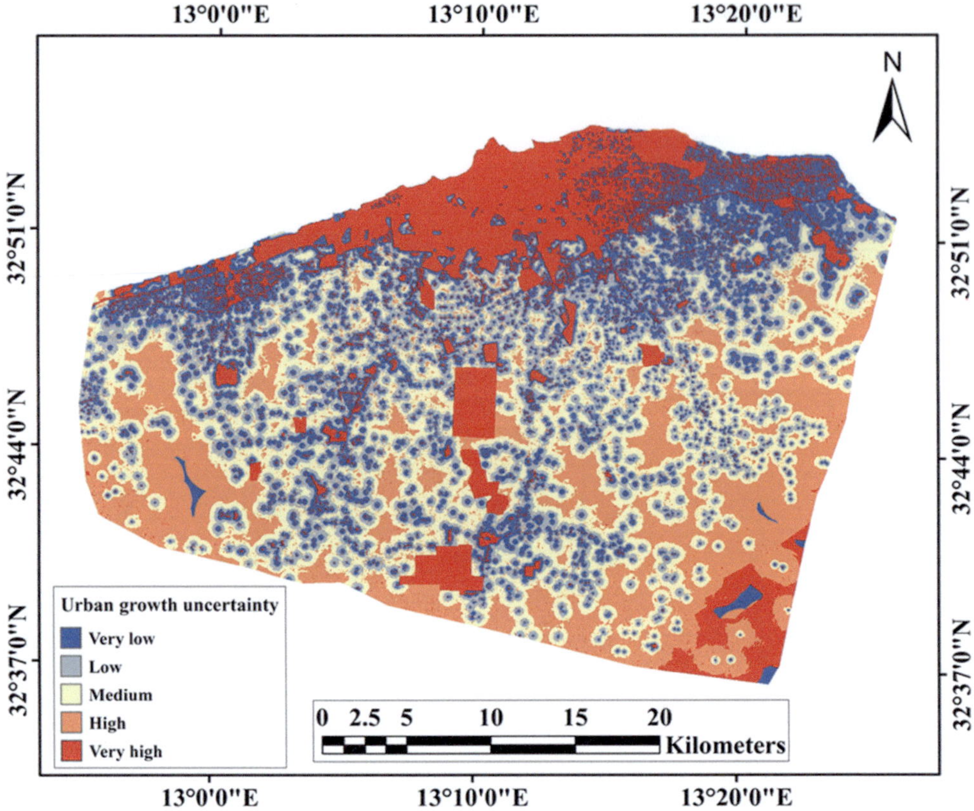

Fig. 9.12 Uncertainty map of urban expansion

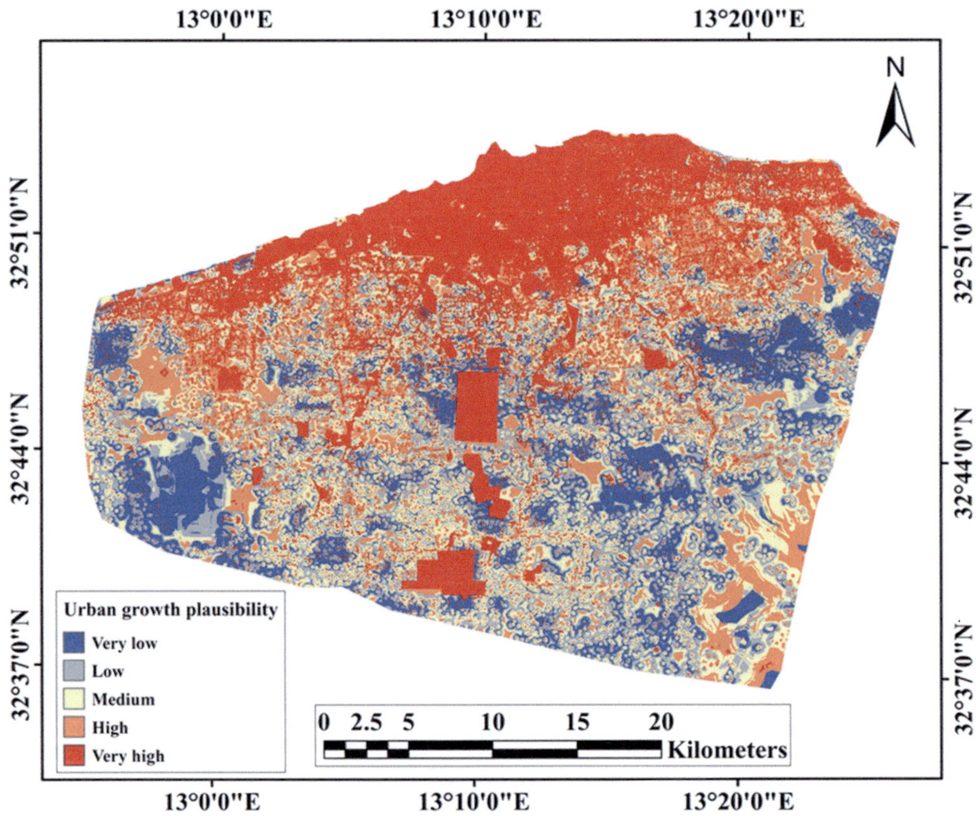

Fig. 9.13 Plausibility map of urban expansion

Table 9.5 Estimated coefficients of the implemented logistic regression model

Variable	Coefficient (β_i)	Odds ratio	Tolerance	VIF
Independent (X_1)	−1.563	0.210	0.230	4.351
Independent (X_2)	−1.889	0.151	0.180	5.563
Independent (X_3)	−0.169	0.844	0.760	1.315
Independent (X_4)	−1.983	0.138	0.269	3.719
Independent (X_5)	−3.004	0.050	0.925	1.082
Independent (X_6)	−19.188	0.000	0.956	1.046
Independent (X_7)	−4.576	0.010	0.399	2.505
Independent (X_8)	1.555	4.737	0.495	2.019
Independent (X_9)	−1.386	0.250	0.141	7.069
Independent (X_{10})	−4.305	0.014	0.785	1.274
Independent (X_{11})	−21.415	0.000	0.730	1.370
Constant	0.850	–	–	–

polycentric aspect of the study area. The distance to the main active economic center (X_1) has a coefficient that is equal to −1.563 and an odds ratio of 0.210. This result indicates that the odds of the urbanization process in the area nearer the active economic center are 1/0.210, that is, 4.76 times as large as that of an area 1 km further away from the main active economic centers. The model results also show that the distance to CBD (X_2) has an odds ratio equal to 0.151, which means the probability of urban development of the area closer to the CBD is equal to 6.622 times the odds of the area further away by 1 km from the CBD. The easting coordinate variable (X_3) has a coefficient of −0.169 and an odds ratio of 0.844, demonstrating that the urban sprawl is dissimilar in the east and west directions because urban expansion to the east direction is more probable. The northing coordinate variable (X_4) has a coefficient of −1.983 and an odds ratio of 0.138 or 1/7.246. The odds ratio indicates that increasing the distance by 1 km in the south direction decreases the urban expansion odds by 7.246. The probability of urban expansion in a lower slope area is greater than that of the urbanization in an area with a higher slope degree. This finding is a reasonable effect of the slope variable (X_5). The restricted area variable (X_6) and the urbanized area variable (X_{11}) have the highest, albeit negative, coefficients. The odds ratio of the two variables is 0, which means that the probability of urban development in those areas is nearly 0. These areas are either restricted or controlled against the urbanization process or have been previously urbanized. The urbanization process and urban sprawl tend to occur in areas nearer the urbanized clusters. The variable distance to the nearest urbanized area (X_7) has a coefficient of −4.576, which means that increasing the distance from the urban area decreases the probability of urbanization. The odds ratio is equal to 0.010, which shows that the probability of urban expansion in an area near an urbanized location will be 100 times greater than the probability of urban expansion in an area further away by 1 km. Urban expansion tends to occur in an area having an increment in population density (X_8). The odds ratio result is 4.737, which is more than 1, that is, the probability of urban expansion in some areas will increase 4.737 times with the unit increment in population density. The significance and function of distance to the educational areas variable (X_9) are shown by the modeling output with an odds ratio value of 0.250 or 1/4. Thus, the probability of urban development in an area nearer the educational area is estimated to be four times the probability of urbanization in an area further away by 1 km. For the variable distance to roads (X_{10}), the model demonstrates that roads significantly affect urban development, and that the odds ratio of distance to roads variable (X_{10}) is 0.014 or 1/71.428. This odds ratio reflects the strong influence of roads on urban spatial patterns, because roads cause the strip and ribbon urban expansion patterns.

9.6.3 Urban Expansion Probability Map Produced by the LR Model

Producing an urbanization probability map required using and applying the coefficients of the LR model in Eq. 9.10. Temporal dynamics was considered for improved model performance. The thematic raster maps of the independent variables distance to nearest urbanized area (X_7), population density (X_8), distance to educational area (X_9), distance to roads (X_{10}), and urbanized area (X_{11}) were updated using 2010 data, whereas the other independent variables remained unchanged. Figure 9.14 illustrates the predicted urban expansion probability map. The dark red color indicates a higher probability of urban expansion, whereas the dark blue color specifies the lowest probability of urbanization. The trend of future urban expansion process patterns is based on the future probability map (Fig. 9.14). Urban development will probably occur near roads and existing urbanized areas, particularly those areas associated with population growth.

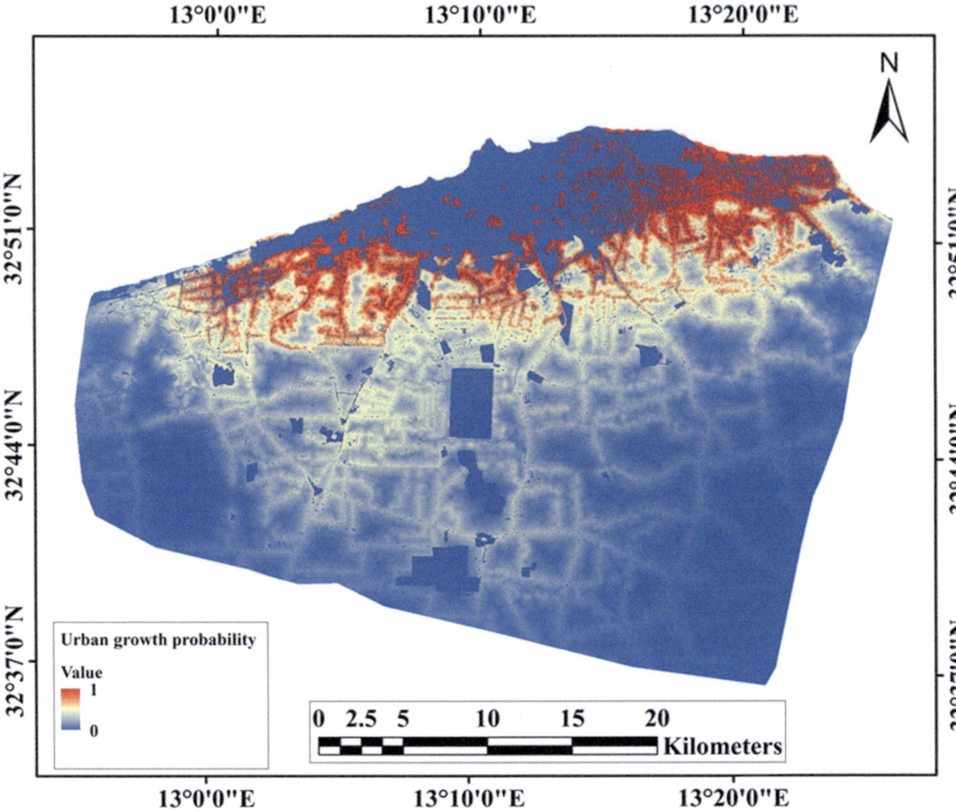

Fig. 9.14 Urbanization probability map of Tripoli in the future using the LR model

9.6.4 Urban Spatial Patterns in the Future Based on the LR Model

Identifying the location of the expected urban expansion is possible based on the urbanization probability map produced by the LR model. Thus, generating several maps can demonstrate future urban distribution patterns based on the current demand for land for urban use. The equation below is used to estimate future urbanization patterns (Campbell et al. 2008). This equation requires determining the size of the existing urbanized area, the anticipated future population, and the growth ratio, which is equal to the ratio of urbanized land to population growth:

$$A_{\text{future}} = A_{\text{existing}}\left(+ R\frac{\left(P_{\text{future}} - P_{\text{existing}}\right)}{P_{\text{existing}}}\right), \qquad (9.17)$$

where A_{future} is the future area of urbanized land; A_{existing} is the existing area of urbanized land; R is the growth ratio, which is equal to the ratio of change in urbanized land/ratio of population growth; P_{future} is the expected population in the future; and P_{existing} is the existing population.

The population and its growth rate data were obtained from the General Authority of Information, Libya. The population growth rate was 1.41% per year. The ratio of the urbanized area growth was 8.57% per year, calculated based on situations in 2002 and 2010. Thus, the growth ratio was approximately 6.

Forecasting the future population requires assuming an insignificant change in the population growth rate (Campbell et al. 2008). Various scenarios were also considered to predict the area of urbanized land for 2020 and 2025. The first scenario used the calculated growth ratio of 6, the second scenario assumed a decrease of 5 in the growth ratio, and the third scenario assumed an increase of 7 in the growth ratio. The rationale behind the different scenarios is the instability of economic, social, and political conditions. Urban planning policies were also unclear. Thus, different scenarios were expected to provide different perspectives to manage unexpected and uncontrolled urban growth. Table 9.6 summarizes the size of the predicted urbanized area (km^2) in the future.

Future spatial patterns were determined by allocating the estimated size of the urban area to the urban probability map. The increase in urbanized land was calculated by comparing the estimated area to the 2010 base year map. Subsequently, the increased urbanized area was converted to a number of cells. The number of predicted urbanized cells was allocated to the predicted probability map, starting from the highest probability cells to the lowest until the total area was equal to the estimated future area. The generated future urban spatial patterns are presented in Figs. 9.15 and 9.16.

Table 9.6 Predicted demand of urban land use in the future (km^2)

Year		2020	2025
Growth ratio	5	471	568
	6	509	626
	7	548	685

9.6.5 Land Use Change and Transition Probability Matrices Using the MC Model

The produced land use maps of 1984, 2002, 1996, and 2010 (Fig. 9.17a–d) were used to calculate the urban expansions occurring in the same time periods. The results of land use change analysis presented in Table 9.7 illustrate the land use change history and amount of change in each land use class. Studying these outcomes leads to knowledge regarding land use change rhythm, land use change behaviors, and change speed, all of which are extremely helpful to urban planners and decision-makers. However, in the time period 1984–1996, approximately 3.86 km^2 of agricultural areas were changed annually to built-up areas. Furthermore, the annual loss of agricultural areas increased from 3.86 to 5.18 km^2 in 2002. Unfortunately, the annual loss rate of fertile lands jumped dramatically and reached 13.97 km^2 annually in 2002–2010, and the total amount of increase in urban areas in 1984–2010 was over 189 km^2. This increase is considered extremely high in such a short time. However, this remarkable rate of land use change (agriculture area to built-up areas) raises various questions regarding urban growth patterns, driving forces of urban growth process, urban policies, and sustainable environmental regulations implemented in the study area. The transition probability matrices were calculated using MC analysis, as shown in Table 9.8. These transition probability matrices indicate the future probable percentages of land use change in 1984–1996, 1996–2002, and 2002–2010. However, as shown in Table 9.8, the probability of future change in agriculture lands to urban areas in 1984–1996 is 17%; this probability of change increased reasonably to 18.6% in 2002. In the last decade, the possibility of change to urban area jumped remarkably from 18.6 to 28.13%. However, the future change of built-up areas to agricultural areas had a probability of 18.68%, which was still lower than 28.13%. The large difference between these probabilities reflects the alarming dramatic decrease in fertile lands in Tripoli. Rapid urban growth and large consumption of urban lands were also observed along the study area's history. The analysis of results and a review of the classified maps further show that the metropolis of Tripoli is facing rapid urban sprawl instead of a normal urban growth; such phenomenon requires further analysis and simulation.

9.6.6 Future Spatial Land Use Change Prediction Using the CA–MC Model

The MC model was applied using 2002 and 2010 maps as inputs to quantify future land use changes, and the matrix of transition areas was estimated as output. The calculated transition area matrix records the total area expected to change from agricultural to urban land according to the time units. The quantities of land use change in the past and future in Tripoli are presented in Table 9.7.

The CA–MC model was applied to perform valid prediction of future land use patterns and calculate land use state in 2010. Various iteration numbers (i.e., number of optimum iterations) were used in the land use change map prediction to achieve the best performance of the used model. For model calibration, the projected land use map in 2010 was compared with the actual map using Kappa index statistic to check the validity in terms of quantity and location. Figure 9.18 shows the variation of the Kappa index with the change in iteration numbers. This figure shows that the optimum performance of the model was achieved at four iterations while predicting land use in 2010 with the Kappa standard index of 0.8584, Kappa location index of 0.886, and Kappa no index of 0.881. The model likewise predicted land use in 2002 and achieved simulation success with the Kappa standard index of 0.8502, Kappa location index of 0.9614, and Kappa no index of 0.88.

These validation results demonstrate very good agreement between the actual and projected maps. Through validation, the optimal transition rules for the model can be calculated using determined iteration numbers that can be used to predict land use maps in 2020 and 2025. The future land use maps in 2020 and 2025 were simulated based on the successful modeling of land use in 2010. The transition potential maps and transition area matrices of 2002–2010 future patterns of land use were predicted through the 2010 land use as base map, as shown in Fig. 9.17e, f. The classic CA–MC model simulations predicted that agricultural lands in the Tripoli metropolitan area will decrease from 829.26 to 647.49 km^2 in 2020 and to 606.99 km^2 in 2025 (Table 9.7). Unfortunately, this change in farm lands will be due to uncontrolled urban expansion. The simulated future scenarios of land use change show the growing pressure of rapid urban growth associated with important socioeconomic and environmental implications.

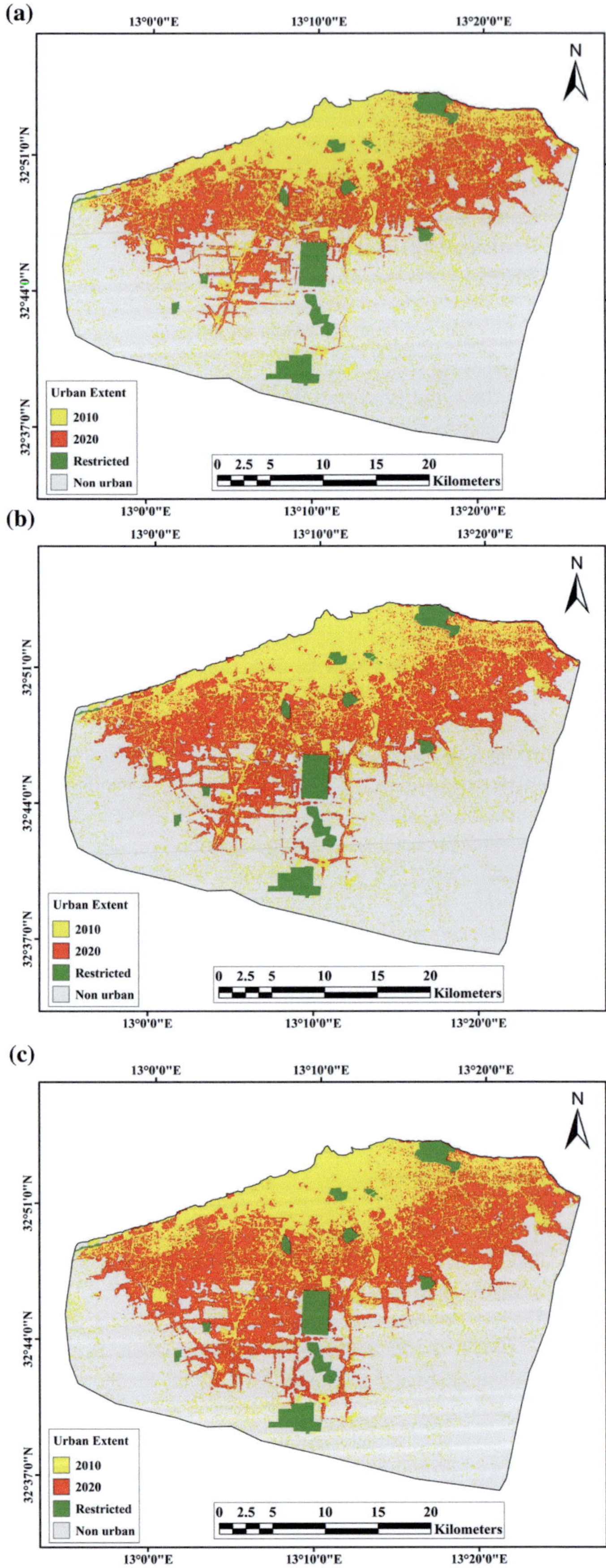

Fig. 9.15 Predicted urban patterns in 2020 with various growth ratios: **a** at 1:5, **b** at 1:6, and **c** at 1:7

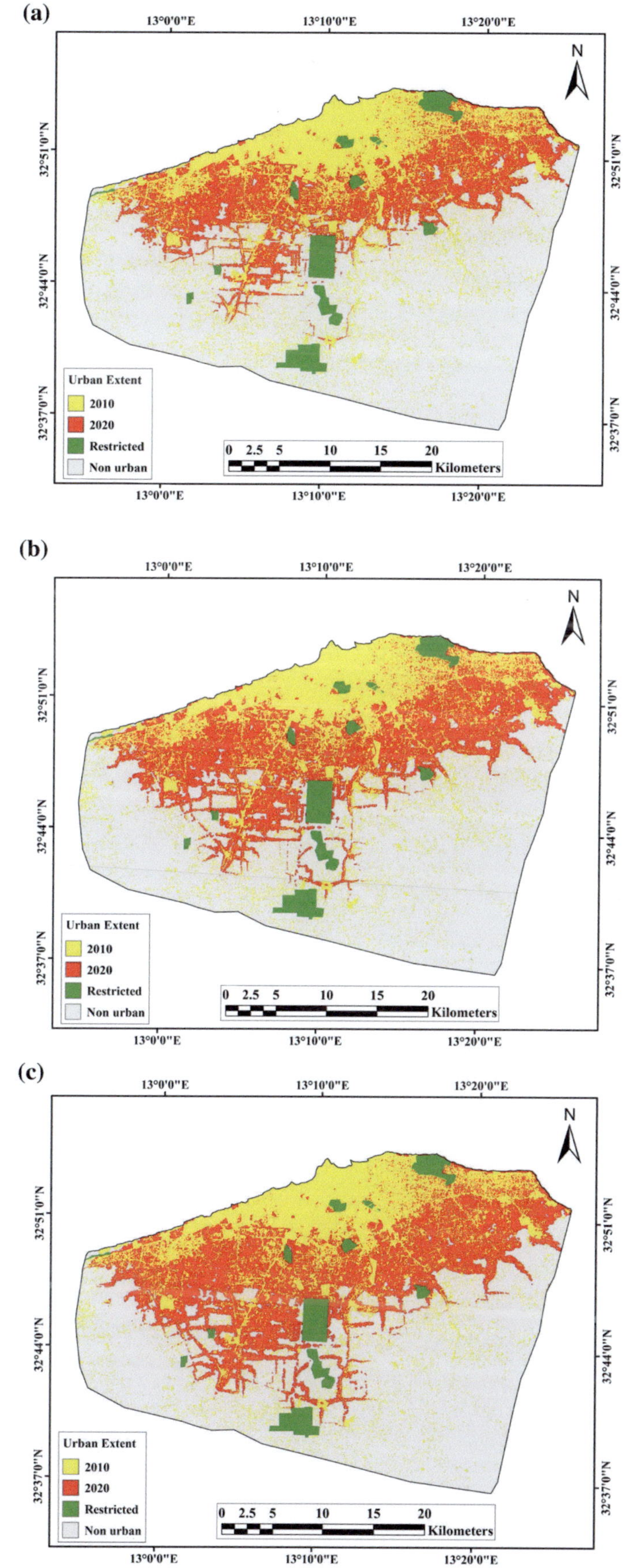

Fig. 9.16 Predicted urban patterns in 2025 with different growth ratio scenarios: **a** at 1:5, **b** at 1:6, and **c** at 1:7

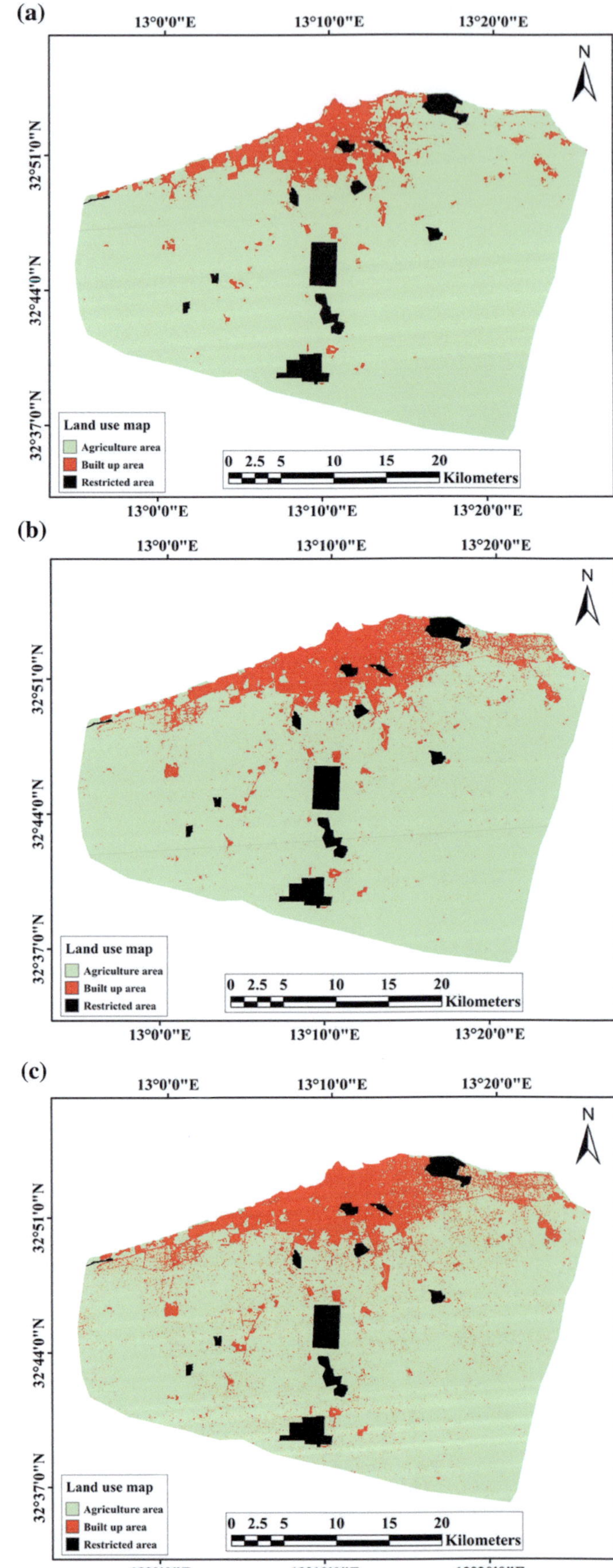

Fig. 9.17 Predicted land use maps in different years using the CA–Markov chain model: **a** 1984, **b** 1996, **c** 2002, **d** 2010, **e** 2020, and **f** 2025

Fig. 9.17 (continued)

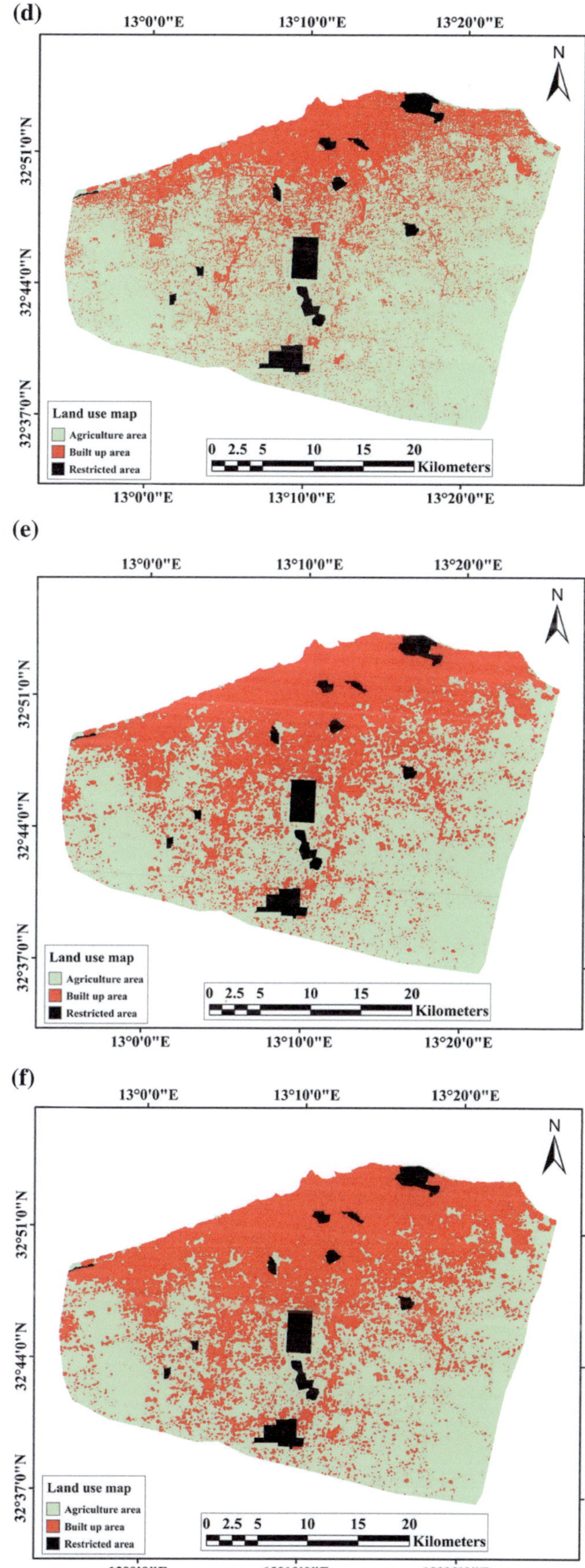

Table 9.7 Amount of land use changes (km^2)

	Agriculture area	Built-up area	Restricted area
1984	1018.47	86.40	38.89
1996	972.10	132.77	38.89
2002	941.05	163.82	38.89
2010	829.26	275.61	38.89
2020 (Predicted)	647.46	457.93	38.89
2025 (Predicted)	606.96	498.47	38.89
Annual change (1984–2002)	−4.30	4.30	0
Annual change (2002–2010)	−13.97	13.97	0
Annual change (2010–2020)	−18.18	18.18	0
Annual change (2020–2025)	−8.10	8.10	0
Total change up to 2010	−189.21	189.21	0
Total change up to 2025	−411.51	411.51	0

Table 9.8 Transition probability matrices of different time periods: 1984–1996, 1996–2002, and 2002–2010

		Agriculture area	Built-up area	Restricted area
1984–1996	Agriculture area	0.8296	0.1704	0.0000
	Built-up area	0.1582	0.8418	0.0000
	Restricted area	0.0750	0.0750	0.8500
1996–2002	Agriculture area	0.8140	0.1860	0.0000
	Built-up area	0.0750	0.8500	0.0750
	Restricted area	0.0750	0.0750	0.8500
2002–2010	Agriculture area	0.7187	0.2813	0.0000
	Built-up area	0.1868	0.8132	0.0000
	Restricted area	0.0750	0.0750	0.8500

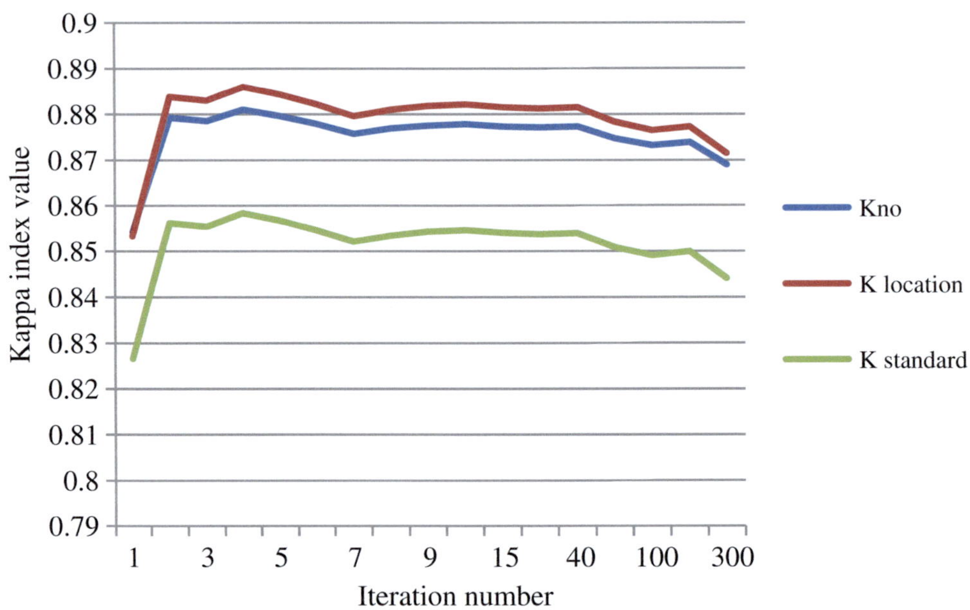

Fig. 9.18 Kappa index values versus number of iterations (for CA–Markov chain model)

9.6.7 Driving Factors of Urban Expansion and Their Interactions

The CHAID-DT model demonstrated that the factor distance to coast line is the most important in determining urban expansion in the study area, followed by distance to CBD, population density in each district, distance to roads, distance to built-up area, slope, and distance to active economic centers. Distance to educational areas had the least effect on the urbanization of Tripoli. The model classified the urban driving factors into various classes based on the history of urban expansion behavior and demonstrated how urban factors interact to determine the probabilities of future urban growth. The percentages in blue represent the tree leaves, that is, the probabilities of urban expansion. Distance to coastal area is at the top of the tree because it is the most effective urban driving force and is classified into 10 classes. Each of the 10 classes represents a different likelihood of urban expansion occurrence.

As per the CHAID tree results illustrated below, the probability of urban growth is approximately 0.84 when the distance to coastal area is less than or equal to 0.071. This probability changes as a result of the interactions between distance to CBD and distance to coast line. In this case, distances to CBD were classified as follows: less than or equal to 0.225, 0.225–0.375, 0.375–0.442, 0.442–0.502, 0.502–0.620, and greater than 0.620. The model results demonstrated that if the distance to coastal area is less than or equal to 0.071, the probabilities of urbanization occurrence in these six classes are 92, 90.8, 88.6, 87, 78, and 63%, respectively.

Similarly, other scenarios of interaction between urban factors can be described in different levels, in a presentation appropriate for urban planners and decision-makers. Nonetheless, numerous situations regarding urban growth in the study area are identified by examining the tree flowchart in the next page.

The third branch of CHAID-DT indicates that the probability of urban growth in the entire study area generally decreases when distance to coastal areas increases. However, the effects of population density and distance to active economic centers on this branch revealed that when distance to coast line ranged from 0.120 to 0.159 and population density is less than or equal to 0.156, growth probability increases when distance to active economic centers is less than or equal to 0.217. Nonetheless, distance to active economic centers became negligible when the population density was greater than 0.217. The possibility of urban expansion increases significantly rather than decreases, that is, expansion occurs at a distance from the economic centers. This situation indicates that the increase in population induced urban sprawl and unplanned urban development.

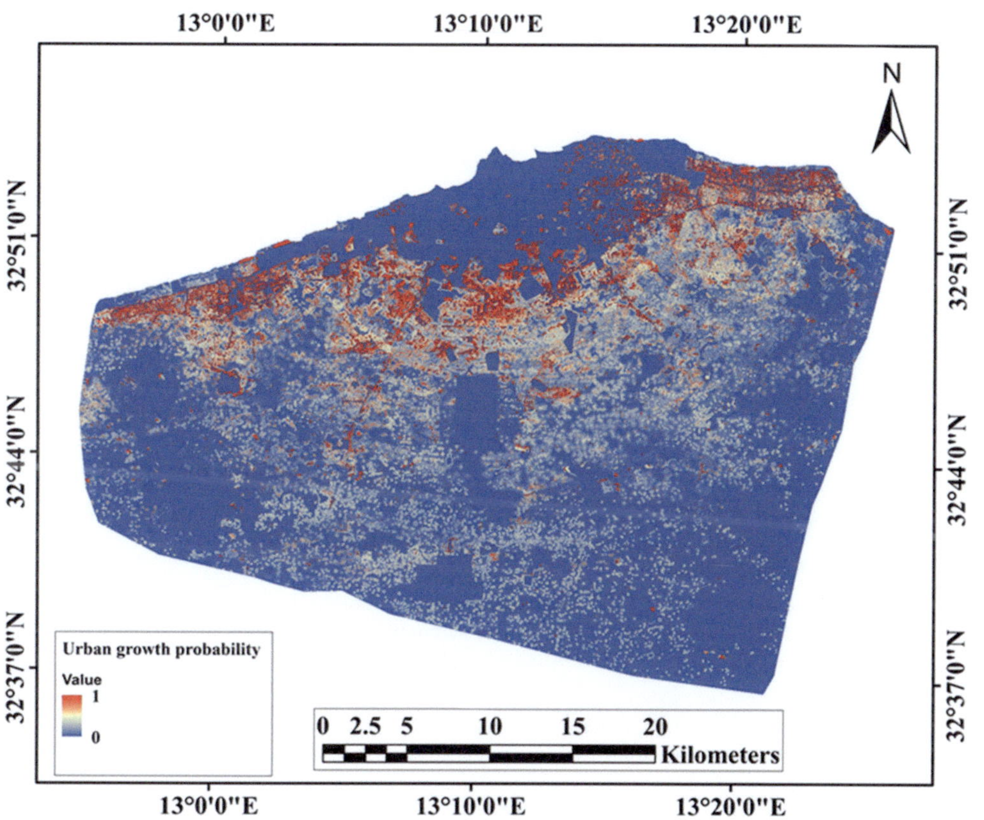

Fig. 9.19 Predicted urban expansion probability map using the CHAID-DT model

Urbanization probability is highest when distance to active economic centers exceeds 0.241 and when distance to coastal line varies from 0.159 to 0.200. This outcome suggests that expansion occurs at the fringes of urban areas and denotes possible urban sprawl. The influences of distances to educational areas and CBD are combined when the factor distance to coast ranged from 0.200 to 0.251. In this case, the probability of growth was highest when the distance to educational areas was over 0.221 and when distance to CBD extended from 0.442 to 0.674.

In the sixth branch of the tree, the effect of the slope factor was low. The possibility of urban expansion increased with the increases in distance to CBD and population growth (i.e., unplanned growth). This result suggests that a wise urban policy must be urgently developed for the study area to manage the urbanization process and accommodate the population increase in these regions. However, the last four branches of the tree model illustrate the declining probability of urban growth, although urban expansion may occur at nearby roads and built-up areas. Figure 9.19 shows the predicted probability map of urban expansion using the CHAID-DT model, and the analysis results are illustrated as follows:

0) Urban expansion

- 1) Distance to coastal area ≤ 0.071 [84%]
  - Distance to CBD≤ 0.225 [92%]
  - Distance to CBD > 0.225 and ≤ 0.375 [90.8%]
  - Distance to CBD > 0.375 and ≤ 0.442 [88.6%]
  - Distance to CBD > 0.442 and < 0.502 [87%]
  - Distance to CBD > 0.502 and ≤ 0.620 [78%]
  - Distance to CBD > 0.620 [63%]
- 2) Distance to coastal area > 0.071 and ≤ 0.120 [85%]
  - Population density ≤ 0.156 [85%]
    - Distance to roads ≤ 0.147 [81%]
      - Distance to built up area ≤ 0.028 [83%]
      - Distance to built up area > 0.028 [79%]
    - Distance to roads > 0.147 [72%]
  - Population density > 0.156 and ≤ 0.157 [72%]
  - Population density > 0.157 and ≤ 0.334 [92%]
    - Distance to CBD ≤ 0.225 [93%]
    - Distance to CBD > 0.225 [91%]
  - Population density > 0.334 [92%]
    - Distance to CBD≤ 0.304 [95%]
    - Distance to CBD> 0.304 [90%]
- 3) Distance to coastal area > 0.120 and ≤ 0.159 [81%]
  - Population density ≤ 0.156 [77%]
    - Distance to active economic centers ≤ 0.217 [83%]
    - Distance to active economic centers > 0.217 [67%]
  - Population density > 0.156 and ≤ 0.157 [71%]
  - Population density > 0.157 and ≤ 0.194 [85%]
  - Population density > 0.194 and ≤ 0.334 [84%]
  - Population density > 0.334[91%]
- 4) Distance to coastal area > 0.159 and ≤ 0.200[75%]
  - Distance to active economic centers ≤ 0.217 [71%]
  - Distance to active economic centers > 0.217 and ≤ 0.324 [70%]
    - Population density ≤ 0.157 [65%]
    - Population density > 0.157 [74%]
  - Distance to active economic centers > 0.324 and ≤ 0.372 [75%]
  - Distance to active economic centers > 0.372 and ≤ 0.421 [77%]
  - Distance to active economic centers > 0.421 and ≤ 0.478 [83%]
  - Distance to active economic centers > 0.478 [81%]

5) Distance to coastal area > 0.200 and ≤ 0.251 [67%]
  Distance to CBD ≤ 0.225 [68%]
  Distance to CBD > 0.225 and ≤ 0.304 [70%]
  Distance to CBD > 0.304 and ≤ 0.375 [55%]
  Distance to CBD > 0.375 and ≤0.442 [62%]
  Distance to CBD > 0.442 and ≤ 0.674 [76%]
    Distance to educational areas ≤ 0.221 [65%]
    Distance to educational areas > 0.221 [82%]
  Distance to CBD > 0.674 [47%]
6) Distance to coastal area > 0.251 and ≤ 0.309 [56%]
  Distance to CBD ≤0.304 [58%]
    Slope ≤ 0.021 [65%]
    Slope > 0.021 [52%]
  Distance to CBD > 0.304 and ≤ 0.442 [43%]
  Distance to CBD > 0.442 and ≤ 0.502 [68%]
  Distance to CBD > 0.502 and ≤ 0.562 [77%]
  Distance to CBD > 0.562 [54%]
    Population density ≤ 0.001 [30%]
    Population density > 0.001 [63%]

7) Distance to coastal area > 0.309 and ≤ 0.382[40%]
  Distance to educational areas ≤ 0.165 [29%]
  Distance to educational areas > 0.165 and ≤ 0.221 [40%]
  Distance to educational areas > 0.221 and ≤ 0.288 [38%]
  Distance to educational areas > 0.288 and ≤ 0.382 [48%]
    Population density ≤ 0.089 [57%]
    Population density > 0.089 [35%]
  Distance to educational areas > 0.382 [46%]
    Distance to roads ≤ 0.147 [50%]
    Distance to roads > 0.147 [42%]

8) Distance to coastal area > 0.382 and ≤ 0.469 [27%]
  Distance to CBD ≤ 0.442 [17%]
    Distance to active economic centers ≤ 0.544 [14%]
    Distance to active economic centers > 0.544 [42%]
  Distance to CBD > 0.442 and ≤ 0.502 [36%]
  Distance to CBD > 0.502 and ≤ 0.562 [16%]
  Distance to CBD > 0.562 and ≤ 0.620 [29%]
  Distance to CBD > 0.620 and ≤ 0.674 [31%]
  Distance to CBD > 0.674 and ≤0.748 [32%]
  Distance to CBD > 0.748 [33%]

9) Distance to coastal area > 0.469 and ≤ 0.582 [23%]
  Distance to active economic centers ≤0.142[1%]
  Distance to active economic centers > 0.142 and ≤ 0.217[7%]
  Distance to active economic centers > 0.217 and ≤ 0.273 [11%]
  Distance to active economic centers > 0.273 and ≤0.544 [21%]
    Distance to built up area ≤ 0.071 [23%]
    Distance to built up area > 0.071 [19%]
  Distance to active economic centers > 0.544 [35%]
    Distance to CBD ≤ 0.620 [27%]
    Distance to CBD > 0.620 [42%]

10) Distance to coastal area > 0.582 [40%]
  Distance to active economic centers ≤ 0.273 [6%]
  Distance to active economic centers > 0.273 and ≤ 0.421 [12%]
  Distance to active economic centers > 0.421 and ≤ 0.544 [15%]
  Distance to active economic centers > 0.544 and ≤ 0.653 [18%]
  Distance to active economic centers > 0.653 [32%]
    Distance to roads ≤ 0.313 [30%]
    Distance to roads > 0.313 [33%]

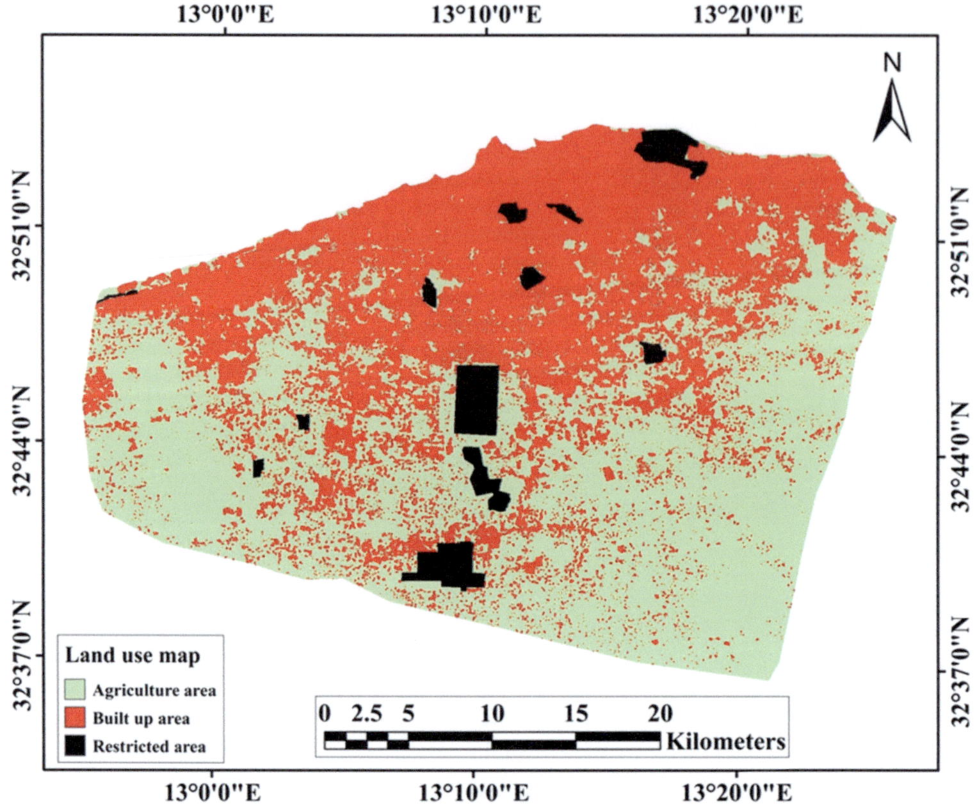

Fig. 9.20 Predicted urban land use in 2020 using the hybrid model

9.6.8 Prediction of Urban Spatial Patterns Using the Hybrid Model

The explicit location of urban development can be determined based on the urban expansion probability map produced by the CHAID-DT model (Fig. 9.19). This figure exhibits the predicted map of future urban transitions as the first input of the proposed model. The dark blue area indicates a low probability of urban growth, whereas the dark red area corresponds to a high probability of urban growth. However, the produced map presents only probable areas of urban expansions rather than exact points of expansion. Furthermore, the MC model cannot estimate change location (i.e., it is not spatially explicit). Accordingly, the amount of future urban expansion as calculated by the MC model is considered as another hybrid model input. However, the CA model can spatially allocate the predicted quantity of land use change over the predicted probability map. Therefore, Figs. 9.20 and 9.21 depict the predicted maps of explicit urban growth in 2020 and 2025 on basis of the integration of the three models.

9.7 Hybrid Model Validation and Future Land Use Change Prediction

In this study, the predicted urban expansion probability map was validated by comparing it with known real urban expansions using the ROC technique, which is considered as dependable in land use/cover change modeling studies. ROC measures the relationship among real and expected changes. In the ROC curve, model sensitivity (true positive) is plotted against 1-specificity (false positive). High sensitivity means large amounts of correct predictions, whereas high specificity means large amounts of false positives. The predicted urban growth probability maps produced by the models were compared against the net real urban growth in 2002–2010. The validation results using the ROC method showed accuracy levels of 84.4, 83.2, and 86% for the FR, EBF, and LR models, respectively. The proposed CHAID-DT model validation results indicated 94.9% prediction accuracy (Fig. 9.22) and reflected very acceptable reliability and good performance of the used model in such spatial applications.

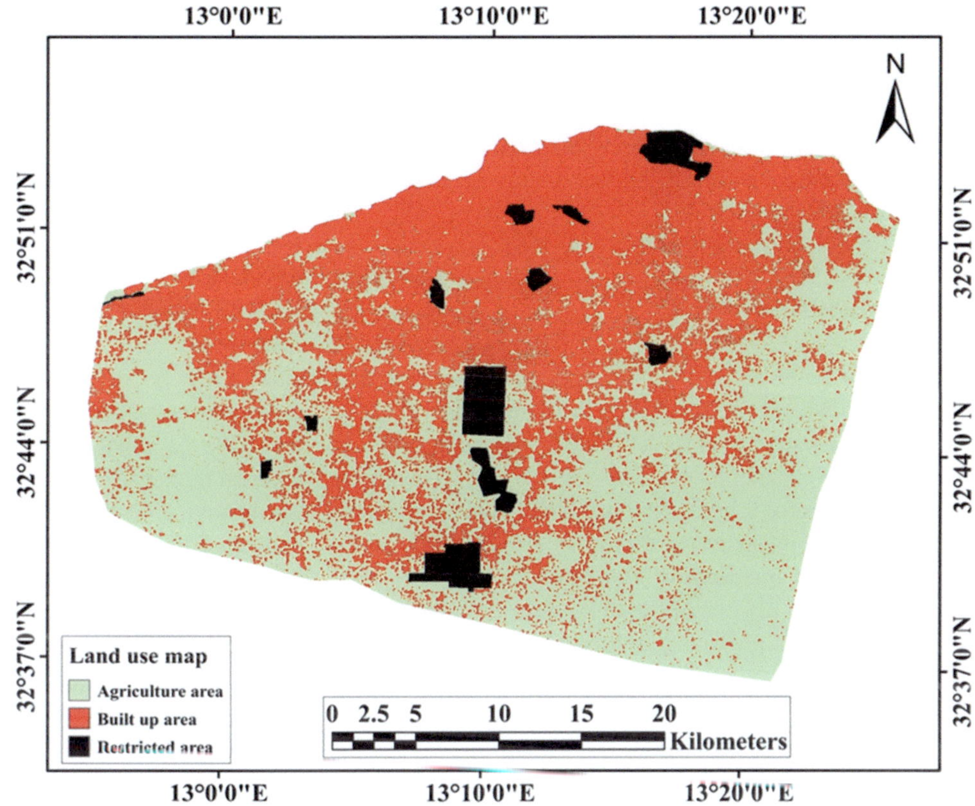

Fig. 9.21 Predicted urban land use in 2025 using the hybrid model

For the proposed hybrid model validation, the projected explicit urban land use map must be compared with the actual map. In this study, the well-known Kappa statistic index of agreement was used to check the hybrid model validity in terms of quantity and location. The developed hybrid model was initially applied to predict urban land use change in 2010. The urban land use change map was predicted using various iteration numbers (i.e., to obtain an optimum iteration number and ensure model reliability) to achieve the best performance of the proposed model. The best performance of the hybrid model in predicting urban land use in 2010 was achieved at 600 iterations. The

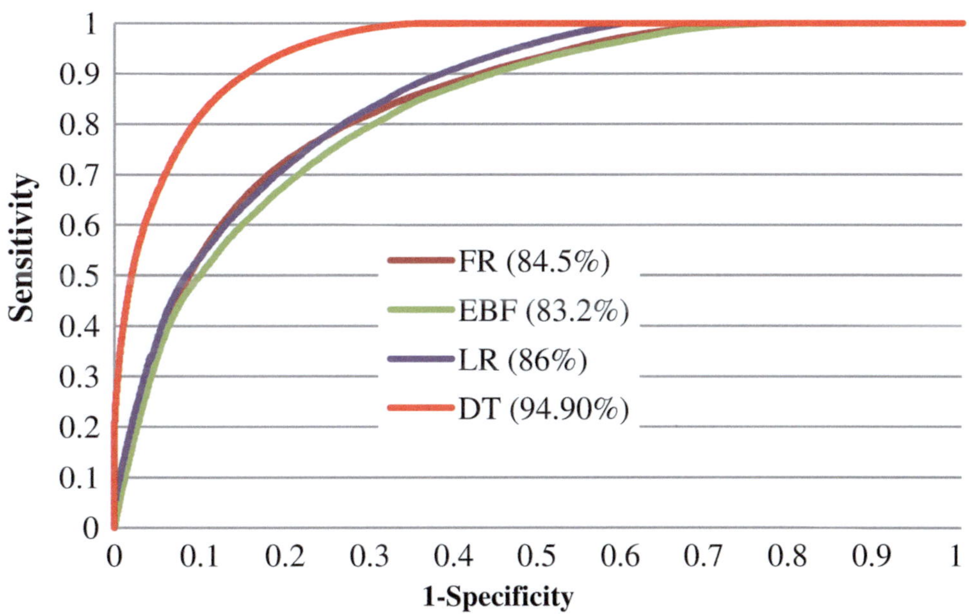

Fig. 9.22 ROC curve of models validation

achieved Kappa standard index was 0.8941, Kappa location index was 0.9227, and Kappa no index was 0.9110.

These validation results demonstrate very good agreement between the actual and the projected map. The developed model in this research achieved an accuracy rate higher than that recorded before for the same study area using the classic CA–MC model.

9.8 Conclusion

This chapter generally focused on the spatial patterns and extents of urban growth change in the Tripoli metropolitan area from 1984 to 2010 to assist in directing prospect urban plans and urbanization policies for Tripoli. The most important issues for urban planners include measuring urban expansions and determining the urban requirements to be accomplished in preparation for future urban demands, instead of waiting to see whether or not urban expansion will occur. The models presented and used in this study can be employed to guide the identification and measurement of the change likely to happen if the tendency of urban history persists. The analysis resulted in many figures to help understand and assess urban sprawl and growth in the Tripoli metropolis. Results further confirmed that the proposed models, remotely sensed data, and GIS are significantly practical for identifying urban growth/sprawl patterns and their general trends in the future.

Specifically, the importance levels and contribution of the urban driving forces of the urban development process were investigated at different levels. The interactions of these factors that design and lead the spatiotemporal patterns of urban sprawl were analyzed and clarified. The role of the urban driving factor was assessed using two applied bivariate models (FR and EBF). The modeling results demonstrate that both applied models can be used to analyze the urban expansion process and urban driving forces. Moreover, each individually analyzed urban causative factor provided superior understanding regarding the role of each single class within the considered factor. The models are simple but powerful in assessing the relationship between urban growth occurrences and their spatial factors. However, FR and EBF could not include the effects of other urban spatial factors within the same analysis time and could not consider or assess the interactions of urban factors.

The functions of spatial urban driving factors and their overall level of significance in the urban expansion process in the capital of Libya were analyzed using the LR model. Results revealed the quantitative relationships between urban sprawl and causative factors, as well as distinguished the effectiveness of variables and their functions in urban expansion. The LR model collectively assessed all urban factors and implicitly considered the effect of urban variables on one another. Such model found the best relation among the factors to explain the urbanization process. However, LR merely provided an overall assessment for each urban development variable; the model could not explain how the urban variables interacted. These details are extremely critical in understanding the urban situation and supporting urban planners.

The CHAID-DT model demonstrated the overall importance levels for all factors driving urban expansions in the study area. Each urban variable was also categorized into classes based on urban expansion occurrence in the study area. The classified urban driving factors provided additional detailed descriptions regarding the studied urban system and the behavior history. The CHAID-DT model successfully explained in detail how the urban factors interacted to produce future probabilities of urban growth. The CHAID-DT model also explicitly presented different situations leading to urban growth occurrence and showed how urban factors affect one another. The applied CHAID-DT model is highly advantageous in such studies because it overcomes the shortcomings of statistical methods, such as FR, EBF, and LR models. The advantages of the CHAID-DT model allow the identification of further effective conditions on the urban development process.

The urbanization process was modeled by several models, and future probable trends of urban developments in the studied area were predicted and presented in high-accuracy maps. The FR model resulted in a more accurate urban probability map based on the ROC validation technique. The EBF model favorably provided four maps, each offering additional spatial information and assessments regarding the urban expansion process. The multivariate LR model was employed to show the expected location of the possible future urban expansion. This model displayed greater accuracy than the bivariate models.

The main limitations of the abovementioned models include temporal determination of change and change quantification within an acceptable limit. The CA–MC model combining the CA and MC models effectively simulated and estimated the land use changes in the Tripoli metropolis. One advantage of the applied CA–MC model is that the model requires limited data to simulate and predict any future land use change explicitly (i.e., minimum of two land use maps in different dates). However, the model's disadvantage lies in its inability to analyze and include urban land use change driving factors, such as biophysical and socioeconomic factors, which are extremely important in managing, guiding, and controlling current urban situations as well as in predicting future trends.

The proposed modeling approach overcomes the restraints of the abovementioned models. The suggested model is capable of including, analyzing, and discovering various urban variables causing urban expansion and sprawl, such as socioeconomic and environmental variables. The hybrid model also estimates the quantity and place of urban growth. The two validation steps of the hybrid model ensure the accuracy of provided results. Furthermore, the validation results of the hybrid model demonstrated better performance than those of the other employed models in this research.

Finally, this study presented an exhaustive assessment of the urban development status in the studied area. The obtained results can be used by the national decision-maker and planner to recognize the past, present, and future of urban expansion in order to prepare, plan, and gear up for future demands.

References

Al-shalabi, M., Billa, L., Pradhan, B., Mansor, S., & Al-Sharif, A. A. (2013). Modelling urban growth evolution and land-use changes using GIS based cellular automata and SLEUTH models: The case of Sana'a metropolitan city, Yemen. *Environmental Earth Sciences, 70*(1), 425–437.

Althuwaynee, O. F., Pradhan, B., Park, H.-J., & Lee, J. H. (2014). A novel ensemble decision tree-based CHi-squared Automatic Interaction Detection (CHAID) and multivariate logistic regression models in landslide susceptibility mapping. *Landslides, 11*(6), 1063–1078.

Arsanjani, J. J., Kainz, W., & Mousivand, A. J. (2011). Tracking dynamic land-use change using spatially explicit Markov Chain based on cellular automata: The case of Tehran. *International Journal of Image and Data Fusion, 2*(4), 329–345.

Awasthi, A., & Chauhan, S. S. (2011). Using AHP and Dempster-Shafer theory for evaluating sustainable transport solutions. *Environmental Modelling and Software, 26*(6), 787–796.

Baker, S., & Cousins, R. D. (1984). Clarification of the use of chi-square and likelihood functions in fits to histograms. *Nuclear Instruments and Methods in Physics Research, 221*(2), 437–442.

Batty, M., Xie, Y., & Sun, Z. (1999). Modeling urban dynamics through GIS-based cellular automata. *Computers, Environment and Urban Systems, 23*(3), 205–233.

Campbell, C. E., Allen, J. & Shou Lu, K. (2008). Modelling Growth and Predicting Future Developed Land in the Upstate of South Carolina. http://tigerprints.clemson.edu/cgi/viewcontent.cgi?article=1038&context=scwrc.

Carranza, E. J. M., Van Ruitenbeek, F., Hecker, C., van der Meijde, M., & van der Meer, F. D. (2008). Knowledge-guided data-driven evidential belief modeling of mineral prospectivity in Cabo de Gata, SE Spain. *International Journal of Applied Earth Observation and Geoinformation, 10*(3), 374–387.

Carranza, E., Woldai, T., & Chikambwe, E. (2005). Application of data-driven evidential belief functions to prospectivity mapping for aquamarine-bearing pegmatites, Lundazi district, Zambia. *Natural Resources Research, 14*(1), 47–63.

Clarke, K. C., & Gaydos, L. J. (1998). Loose-coupling a cellular automaton model and GIS: long-term urban growth prediction for San Francisco and Washington/Baltimore. *International Journal of Geographical Information Science, 12*(7), 699–714.

Dempster, A. P. (1967). Upper and lower probabilities induced by a multivalued mapping. *The Annals of Mathematical Statistics*, 325–339.

Eyoh, A., Olayinka, D. N., Nwilo, P., Okwuashi, O., Isong, M., & Udoudo, D. (2012). Modelling and predicting future urban expansion of lagos, nigeria from remote sensing data using logistic regression and GIS. *International Journal of Applied Science and Technology, 2*(5), 116–124.

Guan, D., Li, H., Inohae, T., Su, W., Nagaie, T., & Hokao, K. (2011). Modeling urban land use change by the integration of cellular automaton and Markov model. *Ecological Modelling, 222*(20), 3761–3772.

Hu, Z., & Lo, C. (2007). Modeling urban growth in Atlanta using logistic regression. *Computers, Environment and Urban Systems, 31*(6), 667–688.

Huang, B., Zhang, L., & Wu, B. (2009). Spatiotemporal analysis of rural–urban land conversion. *International Journal of Geographical Information Science, 23*(3), 379–398.

Jianping, L., Bai, Z., & Feng, G. (2005). RS-and-GIS-supported forecast of grassland degradation in southwest Songnen Plain by Markov model. *Geo-spatial Information Science, 8*(2), 104–109.

Kamusoko, C., Aniya, M., Adi, B., & Manjoro, M. (2009). Rural sustainability under threat in Zimbabwe–simulation of future land use/cover changes in the Bindura district based on the Markov-cellular automata model. *Applied Geography, 29*(3), 435–447.

Kass, G. V. (1980). An exploratory technique for investigating large quantities of categorical data. *Applied Statistics*, 119–127.

Kleinbaum, D. G. & Klein, M. (2010). *Logistic regression: A self-learning text*. Berlin: Springer Science & Business Media.

López, E., Bocco, G., Mendoza, M., & Duhau, E. (2001). Predicting land-cover and land-use change in the urban fringe: a case in Morelia city, Mexico. *Landscape and Urban Planning, 55*(4), 271–285.

Menard, S. (2012). Six approaches to calculating standardized logistic regression coefficients. *The American Statistician, 58*(3), 218–223.

Miner, G., Nisbet, R. & Elder IV, J. (2009). *Handbook of statistical analysis and data mining applications*. Cambridge: Academic Press.

Mitsova, D., Shuster, W., & Wang, X. (2011). A cellular automata model of land cover change to integrate urban growth with open space conservation. *Landscape and Urban Planning, 99*(2), 141–153.

Muller, M. R., & Middleton, J. (1994). A Markov model of land-use change dynamics in the Niagara Region, Ontario, Canada. *Landscape Ecology, 9*(2), 151–157.

Pearl, J. (1990). Reasoning under uncertainty. *Annual Review of Computer Science, 4*(1), 37–72.

Pontius, R. G., Huffaker, D., & Denman, K. (2004). Useful techniques of validation for spatially explicit land-change models. *Ecological Modelling, 179*(4), 445–461.

Pontius, R. G., & Schneider, L. C. (2001). Land-cover change model validation by an ROC method for the Ipswich watershed, Massachusetts, USA. *Agriculture, Ecosystems & Environment, 85*(1), 239–248.

Qiu, B., & Chen, C. (2008). Land use change simulation model based on MCDM and CA and its application. *Acta Geographica Sinica-Chinese Edition-, 63*(2), 165.

Sang, L., Zhang, C., Yang, J., Zhu, D., & Yun, W. (2011). Simulation of land use spatial pattern of towns and villages based on CA–Markov model. *Mathematical and Computer Modelling, 54*(3), 938–943.

Thiam, A. K. (2005). An evidential reasoning approach to land degradation evaluation: Dempster-Shafer theory of evidence. *Transactions in GIS, 9*(4), 507–520.

UPA. (2009). Tripoli agglomeration plan. *3rd generation planning project.* Libya: Libyan government.

Wang, F. (2012). *A cellular automata model to simulate land-use changes at fine spatial resolution.* Doctor of Philosophy University of Calgary, Calgary (20365).

Wang, J., & Mountrakis, G. (2011). Developing a multi-network urbanization model: a case study of urban growth in Denver, Colorado. *International Journal of Geographical Information Science, 25*(2), 229–253.

White, R., & Engelen, G. (1993). Cellular automata and fractal urban form: a cellular modelling approach to the evolution of urban land-use patterns. *Environment and Planning A, 25*(8), 1175–1199.

Yang, X., Zheng, X.-Q., & Lv, L.-N. (2012). A spatiotemporal model of land use change based on ant colony optimization, Markov chain and cellular automata. *Ecological Modelling, 233,* 11–19.

Compact City Modeling (Case Study of Kajang City, Malaysia)

10

Saleh Abdullahi and Biswajeet Pradhan

10.1 Introduction

This chapter presents the methodological processes of land-use change modeling to simulate and predict future spatiotemporal urban growth and land-use changes. These processes were conducted to identify and assess the various aspects of land-use change modeling, especially regarding statistical (factor analysis) and cellular-based concepts. A hybrid land-use modeling approach based on applied modeling techniques was also developed to create a comprehensive projection of the future development pattern in two scenarios. The first scenario (business-as-usual scenario) is based on several urban-related factors and interaction among various land-use categories through a historical trend of land-use change and growth. In this scenario, weights-of-evidence (WoE) and MC were used to evaluate and create growth probability maps of various land-use types in Kajang City. Next, the results were integrated to the CA model to facilitate the application of contiguity filters and project future land-use maps based on the neighborhood concept. In the second scenario (compact land-use scenario), the proposed land-use modeling approach and evaluation of degree of compactness (DoC) and trend of compactness (ToC) were considered in proposing and implementing a compact land-use scenario using the city intensification process. The performance of each integrated modeling technique was validated during the analysis to confirm their accuracy and propose an optimum simulation approach. The proposed model considers the advantages and disadvantages of the existing models and analyzes the interactions of urban factors as well as their interaction among various land-use categories. Kajang City is selected as the case study because its proximity to the three main cities of Malaysia has resulted in rapid urbanization and sprawl developments in recent years. Furthermore, the availability of a large proportion of natural environments in this region presented an adequate observation of the effects of urban growth. The analyses and modeling approaches used in this study can be employed to guide the identification and measurements of the changes and growth likely to happen in urban areas. The output maps and results can likewise be helpful for town planning in order to design compact and eventually sustainable urban areas.

10.2 Kajang City

Kajang is a city in the eastern part of Selangor Province and the southwestern region of Peninsular Malaysia. This city is located 21 km away from Kuala Lumpur, the capital city of Malaysia (Fig. 10.1) and covers a 60 km^2 area with a population of 300,000 as of 2010. The current population of Kajang has grown rapidly in the past few years.

The eastern part of this region is mainly occupied by agricultural and forest lands. Agricultural land has a high proportion of land-use categories in this region. The central and border parts of the city are mainly occupied by community facilities and residential buildings. However, commercial buildings, such as shopping malls, have higher growth in the city center than in other categories. Industrial areas are mainly located in the central west of the city.

The proximity of this region to the three main cities of Malaysia (Kuala Lumpur, Putrajaya, and Seremben) has increased the urbanization rate, especially of rural developments. This proximity also means that Kajang City is well connected to major highways and expressways. As of 2004, a few townships have been developed in the periphery of Kajang, such as Taman Prima Saujana, Sungai Chua, and Taman Kajang Perdana. Lately, several high-end developments have mushroomed in Kajang, such as Twin Palms, Sri Banyan, Country Heights, Jade Hills, and Prima Paramount. The center of Kajang is the bustling Old Town, where all the roads meet. Most of the colonial era buildings were constructed around the 1920s–1930s. The architecture of these

B. Pradhan (ed.), *Spatial Modeling and Assessment of Urban Form*,
DOI 10.1007/978-3-319-54217-1_10

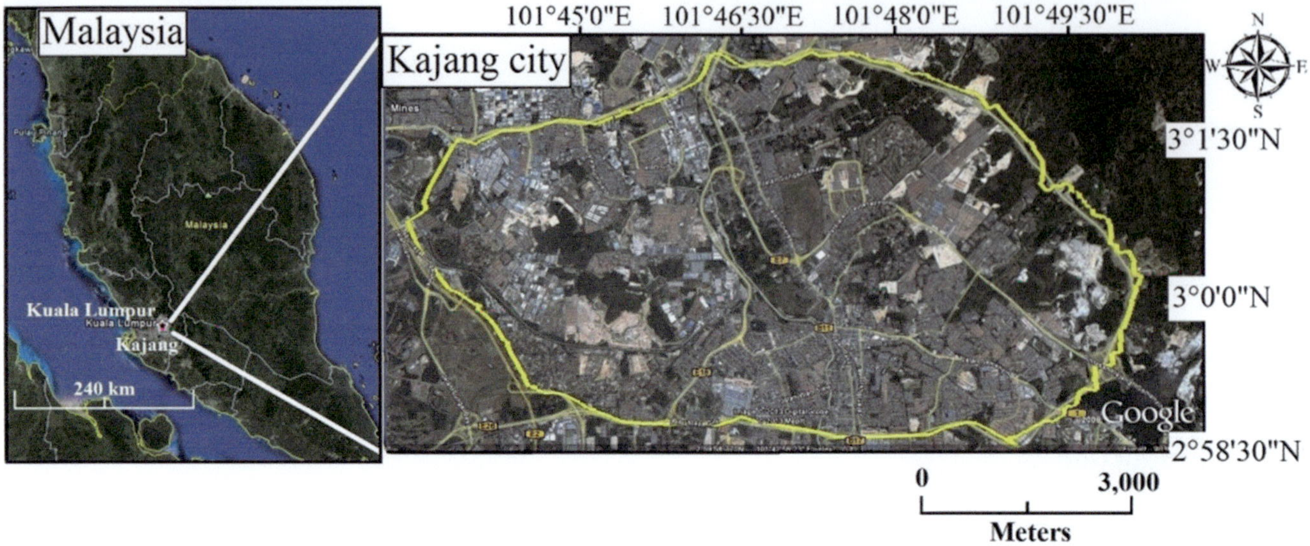

Fig. 10.1 The location of the study area (Kajang City)

shop houses combines traditional Malay, Chinese, and European designs. The ground floor is used mostly for commercial activities and the upper floor is for the family living quarters.

Although many available abandoned plots and brownfield (BF) sites are located within the municipality, improper planning and development force most of the current growth and developments to occur at the outskirts and go through rural environments. This phenomenon consequently destroys valuable forest and agricultural fields. An increasing proportion of BFs and unorganized agricultural lands in the central parts is another consequence of these developments.

The Kajang local authority has several development strategies related to various aspects of the city, as presented in Table 10.1. Planning the fast developing region of Kajang City provides a good opportunity for urban planners and managers to incorporate the ideas of urban sustainability. Moreover, this study can evaluate urban compactness as an input for further compact development analysis and modeling to enhance the sustainability of the region. The city's location near the border of the urban developed lands (east part of Selangor Province) also consists of numerous built-up areas and a large proportion of forest and agricultural fields. Accordingly, the effects of growth and changes of various land-use categories can be adequately observed, particularly on the natural environments. Finally, this study seeks to provide the local planning authority with information regarding the degradation of the natural environment and the possible solutions toward compact urban development.

Apart from these general development strategies, the local planning authority of Kajang (Jabatan Perancangan Bandar Dan Desa Negeri Selangor, JPBD) has specifically proposed other strategies to increase city compactness. These strategies consist of several aspects, such as mixed land-use development, building design, housing design, sense of place, public transportation, neighborhood, and the promotion of walking, cycling, and green environment (Table 10.2).

These strategies required several analyses and processing tools, such as site suitability analysis, readiness analysis, evaluation of land development potential, accessibility analysis, and network analysis, all of which can be carried out via GIS mapping and processing tools. Regardless of those related to planning and design (e.g., in housing and building design, creating and promoting walking and cycling environment), other strategies related to general urban sustainability (e.g., preservation of natural and green environment, less car dependency resulting in less carbon emission, promotion of public transportation) are involved in the processing of this section.

10.3 Data Used

The data utilized in this study were collected from different sources (Table 10.3). Conventional urban data collection is generally expensive and time consuming. In recent decades, the coupling of GIS and remote sensing has been widely applied in urban application, especially in data collection and processing. The capability to deal with several

Table 10.1 General development strategies of Kajang City to achieve more sustainable environment (Hassan et al. 2013)

Aspects	Development strategies
Land use and development	– Decrease the working, living and business uses in the central areas – Increase the linkages within the city – Properly distribution of the community facilities and services in the city – Consolidate the development and enhance the green environment of various parts of the city
Landscape and biodiversity	– Reserved forest conservation – Replant forest at abandoned fields – Increase development of recreation gardens in the town and homes
Security and safety	– Ensure a safe and healthy living environment that can cater for the need for all groups of people including the disabled, the disadvantaged and the aged – Ensure sufficient and well distributed police stations, police posts, and neighborhood watch centers – Place the closed-circuit television (CCTV) at the high rate crime activities areas – Increase the level of awareness of the local residence to help other in case of difficult situations – Build overhead bridge for pedestrians
Renewable energy	– Build the planted forest at vacant land – Protect the forest at Hulu Langat District and Hulu Semenyih District area – Every resident are encourage to do landscape at yard of their house – Turn off the light if no one in room and when not needed – Use compact fluorescent light bulbs to replace the lamp because these bulbs can produce same amount of light but using quarter of electricity only
Integrating transport and development	– Ensure the accessibility to train stations which is the main mode of transportation of the Kuala Lumpur – Providing traffic guards in front of the school during peak hours for safety of children – Provision of proper bus stations in the city to increase the level of services
Development accessibility	– Redirecting the movement pattern of using the private transport to public transport – Improving the quality and extending the public transportation services – Improving the traffic management system by implementing the smart systems – Reduce and increase the parking locations and parking charges – Improve accessibility for disabled people
Urban design principles	– Preservation of major corridors that relatives to the buildings – Implementation of the suitable landscape, soft scape street furniture and signage – Increase the quality of street lighting to create safeness – Support flexible development for mixed land use including entertainment, offices, commercials, educational, etc. – Installation of fountains, sculptures, water gardens and special lighting that can be contribute to the Kajang life and make experience more meaningful

Table 10.2 City compactness strategies of Kajang City

Aspect	Criteria
Mixed land use development	– Encourage residents to live within the workplace – Concentration of activities in line with the centralized public transport networks (TOD) – Implemented a mixed development area and development of potential/in-fill site – Building design: variety of activities/functions in one building creating an effective vertical mixed land use
Advantages of public transportation	– Create a wide range of public transport modes – User-friendly public transport system (appropriate age group)
Housing design	– Various types of residential design according to location and needs – Residential types are developed to suit the compact city – Residence district, integrated with transport convenience lay
Sense of place	– Safe and active open space – Characterized commercial development to local community activities
Cycling and walking neighborhood	– Building design incorporates pedestrian-friendly features – Accessibility of public transport nodes for pedestrian/cycling – Safe network of pedestrian/cycling, and uninterrupted between the neighborhood and the city center
Environment preservation	– The green area is maintained – Create green corridor and blue part of the redevelopment potential

Table 10.3 Utilized data in this study

Raw data	Scale/format
1. Land use map 2004 2. Land use map 2008 3. Land use map 2012 4. Land use map 2015 5. Master plan 6. Road network	1:5000
7. POI (points of interest) 8. Public transportation (bus, train, taxi)	Point data
9. Soil map	1:100,000
10. Geological maps	1:63,360
11. River and flood maps	Polygon data
12. Population map 2000 13. Population map 2010	

geospatial analyses, such as hydrological, interpolation, neighborhood, density, zonal, surface, and so on, is one of the main advantages of GIS, especially when faced with urban spatial issues.

All the data collected from the local planning authority of Kajang City (JPBD) were in shapefile layer format. Therefore, preparing and managing these data layers were conducted using ArcCatalog 10.0 software.

(1) **Land use map**: The detailed distribution map of various land uses and land covers is the most essential input for urban application projects. Land cover generally refers to the physical cover of the earth surface, such as soil, vegetation, water, and man-made. Land use refers to the activities performed on or the utilization purpose of a specific land, such as recreation, residence, agriculture, and so forth. Land-use and land cover maps can be used to extract the development trends on the landscape. Therefore, these layers provide fundamental information for evaluation, analysis, modeling, and predictions of the natural and man-made behaviors of the earth surface. Four land-use maps of Kajang City were in temporal basis for the years 2004, 2008, 2012, and 2015. The master plan of Kajang City was also collected and utilized to ensure that the performance of the proposed model is compatible with local policy and decisions. The land-use maps of Kajang consist of nine categories, namely, residential, commercial, industrial, community facilities, infrastructure, agriculture, green and open spaces, transportation, and water bodies. The master plan layer consists of one extra class, named enterprise zones, which shows the area in Kajang City requiring revitalization or regeneration to improve the economy and livability of the neighborhood (Fig. 10.2).

(2) **Road network**: Road network is an important variable for all types of urban applications, especially urban planning and development. All the urban land uses (residential, commercial, recreation, institutional, etc.) are connected to one another through various links by road or street networks. In addition, most of the community facilities, public transportation nodes (train, bus, and taxi stations), public attractions, commercial buildings, and institutional and governmental offices are located on the main roads. Therefore, living close to the main roads encourages local residents to use public transportation, walk, and cycle instead of using private vehicles. The road network map of Kajang City included several layers, namely, highways, streets, and dead-end alleys. Accordingly, network analysis was performed to extract strategic roads linking the main and populated city centers (Fig. 10.3).

(3) **Public attraction points**: Next to working places, public attraction points or points of interest (POIs), such as mega malls, markets, and places of worship, are generally the second most important destinations for community residences. Therefore, considering these places in the analysis and modeling of an urban area is extremely important. Specifically, proximity to these locations and/or proper distribution of these land uses within the municipality has several advantages regarding sustainable environment aspects. Information about this layer for Kajang City was obtained from recently developed specialized plans of dislocation of these places.

(4) **Public transportation facility**: Public transportation facility is a shared movement facility of local passengers that is available for the general public. This facility is one of the essential necessities of urban areas and normally consists of several modes, such as taxi, bus, and train (Domencich and McFadden 1975). Proper planning and designing of the transportation network in a community have several advantages. Such argument is supported by the urban sustainability perspective transit-oriented development (TOD) as one the most common concepts to achieve more sustainable neighborhoods (Kang 2012). TOD refers to the high mixed land-use area with available proper public transportation modes and stations. Hence, the distribution of various land-use categories of urban areas in the planning and development stage is extremely important because living close to public transportation nodes is one of the main solutions to induce residents to use these facilities for daily commuting. The public transportation facilities of Kajang City consist of taxis, buses, and trains (KTM and MRT). KTM (Keretapi

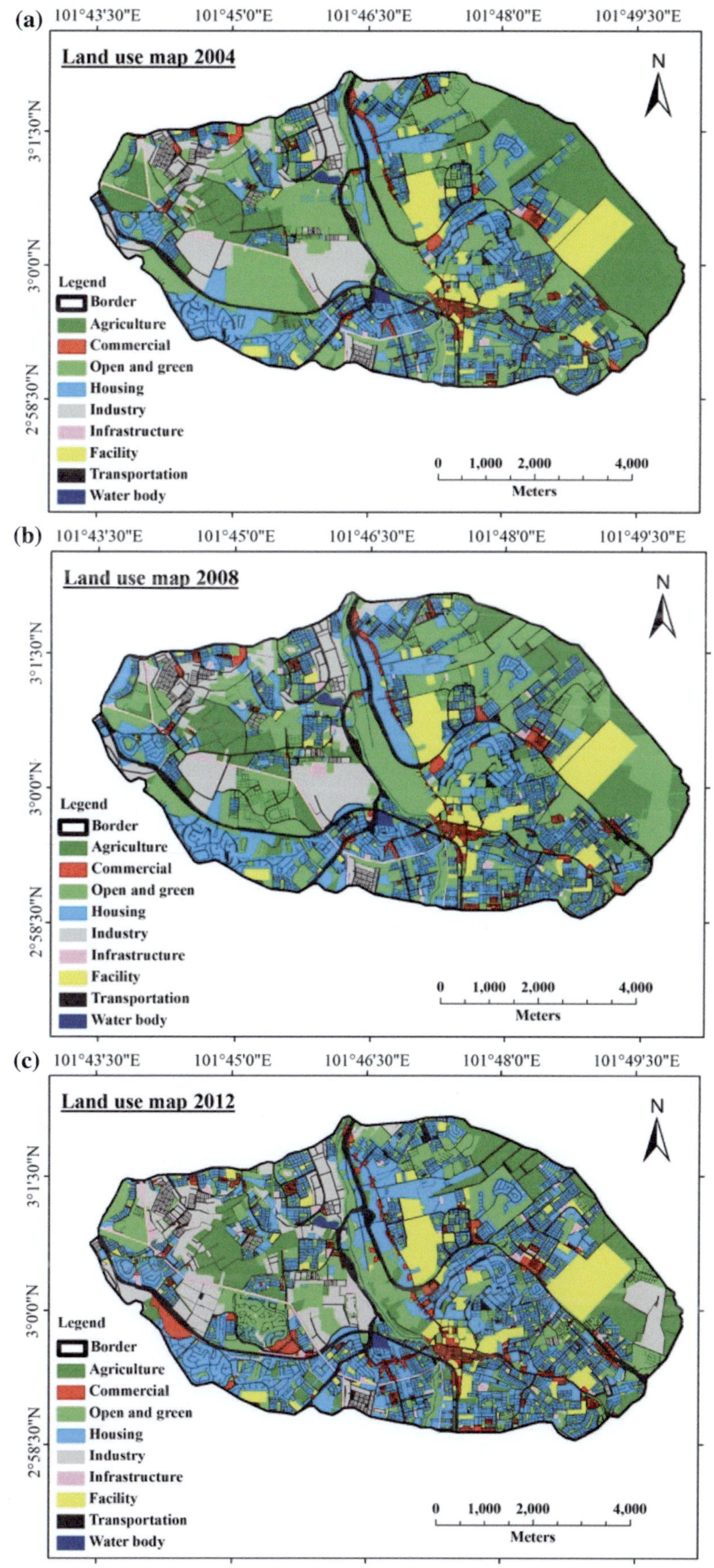

Fig. 10.2 Land use map of Kajang City; 2004, 2008, 2012, 2015 and Master plan of Kajang City

Fig. 10.2 (continued)

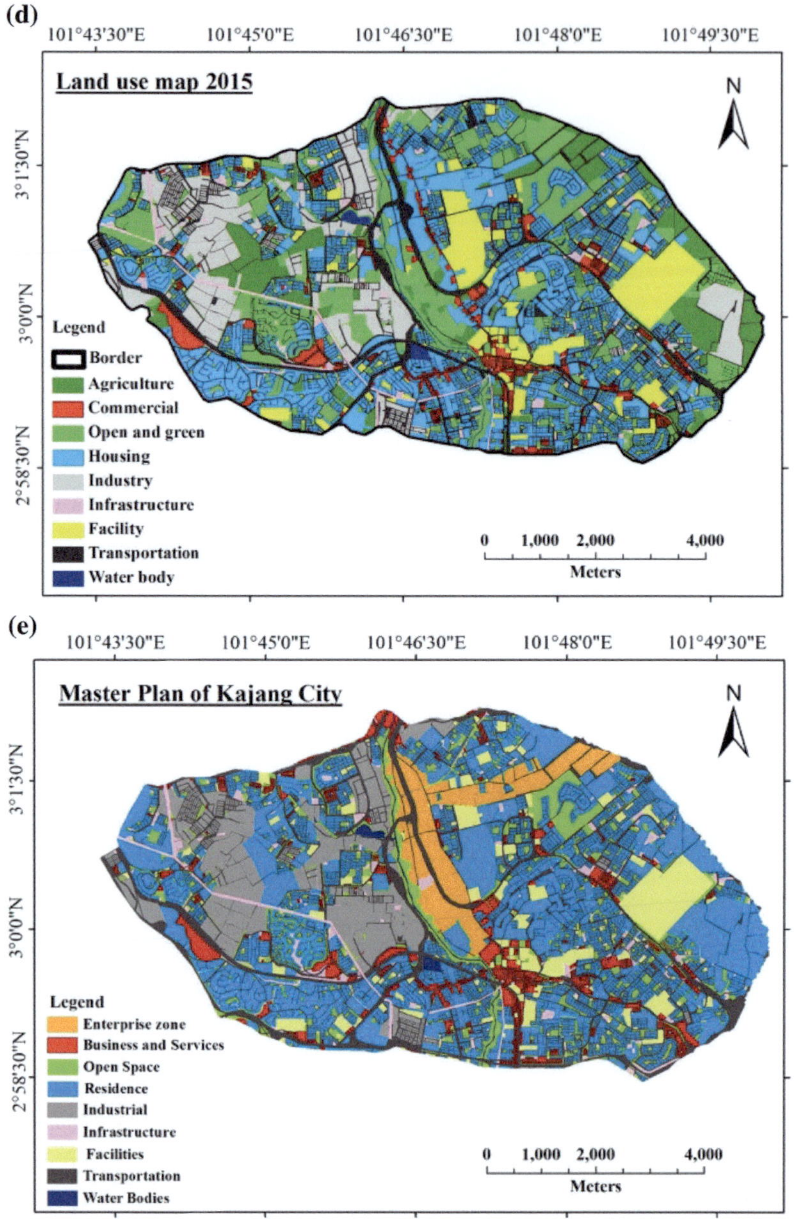

Tanah Melayu) commuter is one of the main transportation train systems in Malaysia, with one station in Kajang City (south-central), while the proposed MRT (Mass Rapid Transit) has several stations in central parts of the city. The public transportation facilities of Kajang are well developed in the central and southeastern parts of the city. In the western part, limited development can be observed (Fig. 10.4).

(5) **Hazardous map (Flood zones)**: Urban planning and development is a complex and long-term project. Therefore, the presence of natural hazards is the most important factor requiring consideration in the analysis. In Kajang City, the main risks mostly arise from flood zones located in the city center (Fig. 10.5). On the basis of the local council report, these zones should be buffered according to the severity of the hazard.

(6) **Population map**: Population analysis is one of the fundamental information required in urban studies. High and low population (or population density) is the main properties that characterize the sprawling development and/or social sustainability of a neighborhood. Numerous studies on the relation of population density and urban sustainability have been conducted, as explained in the Literature Review section. The population data collected for this study include detailed information (e.g., age, gender, ethnicity, and religion) regarding the local residents of Kajang City (JPBD-Department of Statistics). The population

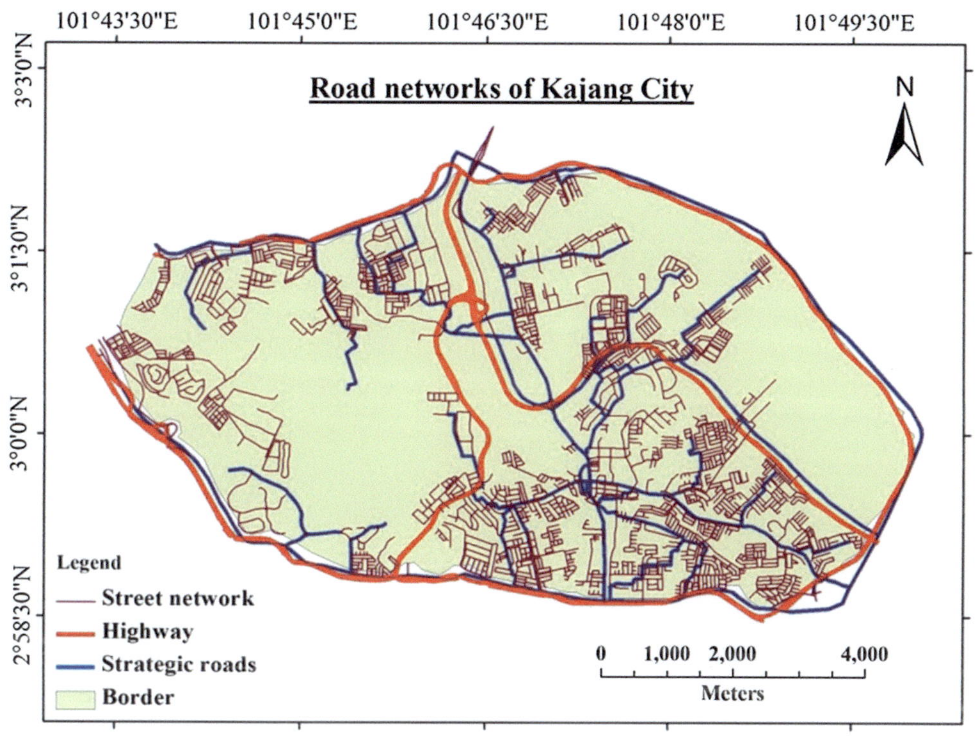

Fig. 10.3 Road networks of Kajang City

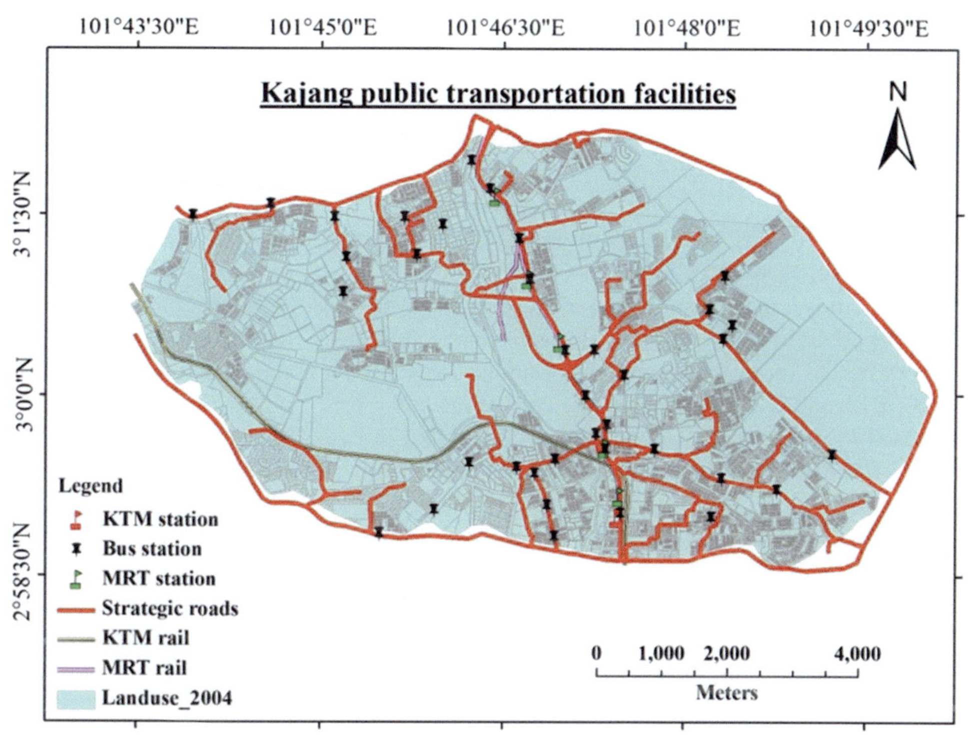

Fig. 10.4 Public transportation facility and main road network of Kajang City

density map was derived from the population map based on persons per hectare for each census block.

(7) **Soil and geological maps**: The physical characteristics of the site to measure the stability of land surface are extremely important for urban projects. Thus, soil and geological maps of Kajang City were included in the analysis (collected from the Department of Drainage and Irrigation and Department of Geoscience and Mineral Resources respectively). Various soil types are presented in the study area. The soil can be classified

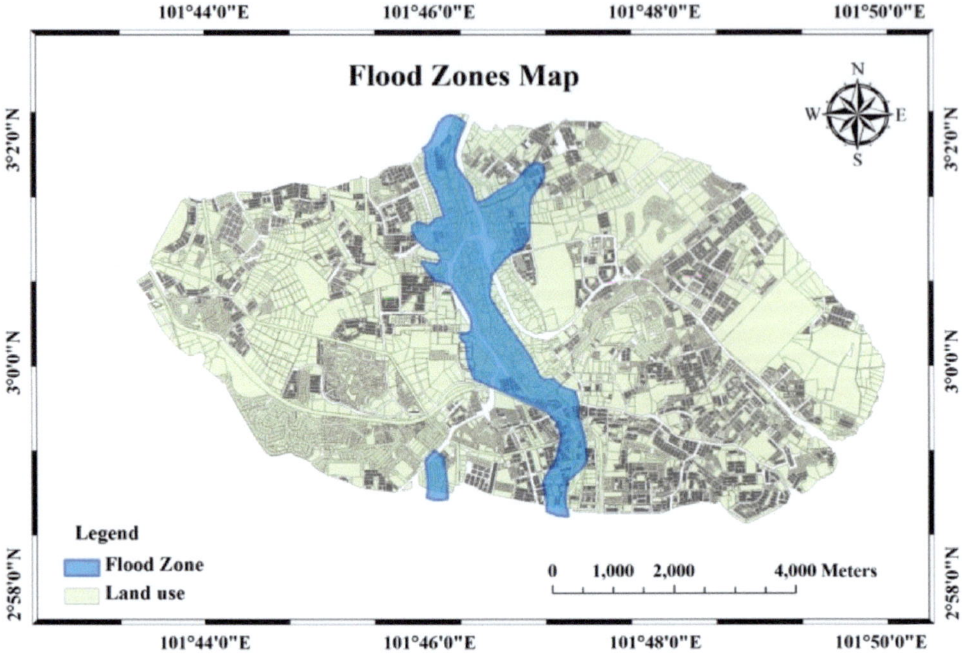

Fig. 10.5 Flood zone map of the Kajang City

into two main categories and three subcategories based on the homogeneity in terms of chemical composition and soil materials. According to the geological map, nearly all of the Kajang area (98%) is covered by acid, a non-differentiated granitoid, and various rocks, such as schist, slate, limestone, conglomerates, chert, and sandstone (Fig. 10.6).

After several digitized layers from the local planning authority (JPBD) were collected, a GIS database was prepared to retrieve, manage, analyze, and display the available data properly. Urban growth and changes of the study area were evaluated for available temporal land-use maps (2004, 2008, 2012, and 2015). Next, land-use change modeling was proposed to project future urban patterns (for 2026) in two scenarios (business as usual and compact land-use form).

10.4 Land Use Change Modeling and Prediction

10.4.1 Frequency Ratio Model

This study utilized the FR model to analyze the spatial effect of various factors on land-use changes. This model recorded the association among allocations of land-use change occurrence and classified the format of factors. This analysis can produce probability maps of land-use change occurrence and provide useful information for future trend of changes.

At the initial stage, all available data were used to derive several related factors affecting land-use change occurrence. The selection of important factors was conducted through a comprehensive literature review on land-use change and urban growth modeling studies (Table 10.4). Apart from these urban-related factors, evaluated city compactness indicators, such as urban density, intensity, and land-use diversity, were also included in the list of factors.

After selecting the driving factors, classifying them based on standard classification schemes available in the ArcGIS software is important. Accordingly, FR was used to evaluate the frequency of land-use change occurrence in each class of selected factors. Each factor was classified into the appropriate range or type. Proximity analysis was applied to distance-based factors, such as proximity to train and bus stations, proximity to community facilities, and so on. To generalize and standardize the analysis, these distances were divided into three classes, namely, (i) near, (ii) middle, and (iii) far, which cover their spatial extent and with every cell in a distance class. However, the classification of distances can be additionally complicated according to the research objective. For ordinal format factors, such as urban densities and land-use diversity (evaluated from previous sections), three classes were also considered as "high," "moderate," and "low" densities or diversity. In the case of nominal factors, such as soil and geology type, each type of these factors was used as one class. The entire layer of selected factors was converted to a raster-based format to assess the land-use growth and changes in their classes, thereby

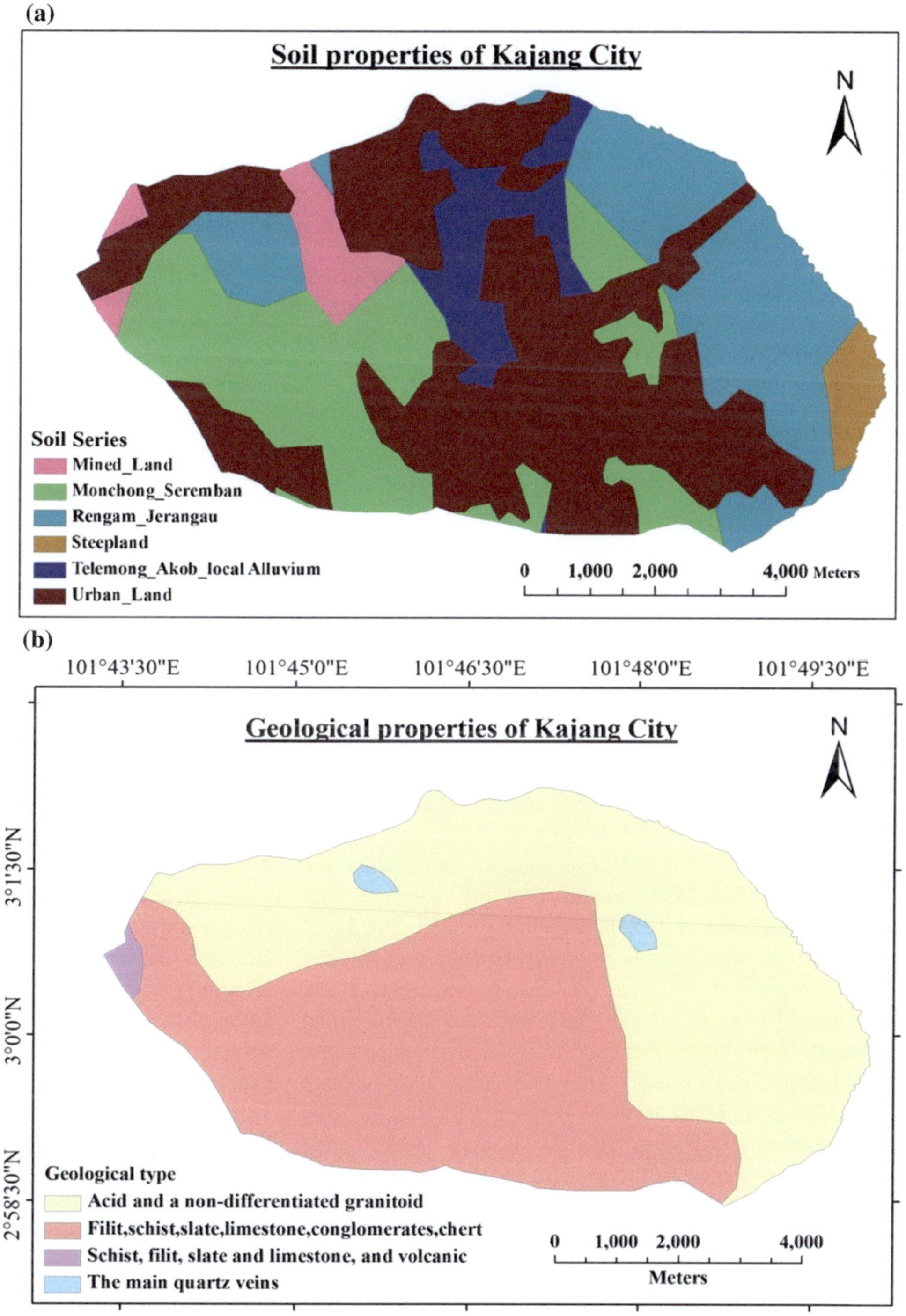

Fig. 10.6 **a** Soil and **b** geological properties of Kajang City

revealing the direct or indirect relation of land-use distribution patterns with respect to selected factors, especially the proximity and ordinal-based factors. The frequency of occurrence was calculated using the ratio of the area of each land-use type to the total area of each factor, with a value of 1 as the average value. A value higher than 1 denoted positive correlation, whereas a value less than 1 indicated negative correlation (Pradhan and Lee 2010; Abdullahi and Pradhan 2015).

10.4.2 Weights-of-Evidence Model

The WoE model combined urban-related and physical factors with the derived variables from compactness assessment. WoE was used to evaluate the existence of each land-use type in relation to some selected variables, and then create a probability of growth map for that corresponding land use. The evaluation was based on the weighting process estimated from the measured association between land-use

Table 10.4 Available raw data and derived factors

Raw data	Details	Derived factors
Land use maps	Various land use types	– Proximity to various land use types – Land use diversity – Built-up density – Residential density
POI map	Location of public attraction points	– Proximity to various points of interests
Road network	Highways, roads, streets, etc.	– Proximity to strategic roads – Road density
Population map	Age, gender, ethnic etc.	Population density map
Public transportation maps	Train, bus	– Proximity to train and bus stations
Soil map	Soil properties	– Soil categories
Geological map	Geological properties	– Geological categories
Flood map	Location of flood zones	– Proximity to flood zones

changes and values on the selected variables maps. The WoE can also be used to select the most important factors among all factors having direct effects on land-use change occurrence.

The FR process is actually the initial step of WoE. Hence, the same set of urban-related and city compactness factors $C = (C_i, i = 1, 2, 3, \ldots, n)$ were utilized to produce probability of growth maps depicting the integrated influence of proximity and other factors. The selected factors (commonly known as evidences in the WoE model) comprised biophysical and human variables that spatially analyze the location of the existing land-use changes and predict the future trend. The WoE model critically assumes that the selected evidences are spatially independent. An independency test was conducted using Cramer's coefficient (V) among pairs of evidences (Bohman-Carter 1994). If the estimated V value of all pairs is lower than the empirical threshold, then the evidences are spatially independent. Considering that the present study does not aim to evaluate the evidences affecting land-use changes and/or extract the most effective evidences, all the selected factors were included as evidence in the process to achieve the results based on all involved factors, regardless of their priority. However, as mentioned before, one of the advantages of the FR model is to reveal the behavior of land-use changes with respect to each class of evidences. Accordingly, estimating the direct or indirect influence and/or lack of influence behavior of the evidences becomes possible.

The list of evidences was divided into two categories, namely, constant evidences and non-constant evidences. Constant evidences do not change during the selected period and thereby do not depend on the year of each land-use map (2004, 2008, 2012, and 2015); examples of such evidences are proximity to water bodies, soil, and geological properties, and flood zone. In contrast, non-constant evidences change during the selected period and thereby depend on land-use maps; examples of these changing indications are proximity to residential, commercial, and industrial zones, among others; land uses, road density, and proximity, residential density, building density, and land-use diversity (Table 10.5).

The quantity of pixels of the corresponding land-use type in each class of evidences was observed and determined through overlaying the land-use maps representing each

Table 10.5 Constant and non-constant evidences

No.	Constant evidences	No.	Non-constant evidences
1	Proximity to public transportation stations	1	Proximity to residential
2	Proximity to water bodies	2	Proximity to commercial
3	Proximity to prison and cemetery	3	Proximity to industrial
4	Proximity to flood zones	4	Proximity to main road network
5	Geological type	5	Proximity to facilities
6	Soil type	6	Proximity to recreation
		7	Proximity to infrastructure
		8	Proximity to agriculture
		9	Urban density
		10	Land use diversity

land-use type, on every produced layer of the selected factors (evidences). For these determined number of pixels *N(L)* containing the occurrence of a specific land-use type (residential) and the total number of pixels of the study area (Kajang City), *N(C)*, the prior probability of the residential occurrence in general is expressed by the following:

$$P(L) = \frac{N(L)}{N(C)} \quad (10.1)$$

When the involved evidences, $C = (C_i, i = 1, 2, 3, \ldots, n)$, are considered, if the number of pixels of residential land use in a specific evidence is $N(L \cap C)$, then the probability of residential growth can be expressed by the conditional probabilities (Bohman-Carter 1994);

$$P(L|C) = \frac{P(L \cap C)}{P(C)} = P(L)\frac{P(C|L)}{P(C)} \quad (10.2)$$

Apart from calculating the occurrence of land use in evidence class C_{ij}, WoE evaluates the nonoccurrence of land use in the same class of evidence. Then, the natural logarithm of both values (occurrence and nonoccurrence) was calculated. These values are the weights to support the occurrence and nonoccurrence of land-use type with respect to each class of evidences. The subtraction value of these weights represents the spatial association of each land-use pixel and each class of evidences. The variance of both weights and standard deviation are calculated. Finally, the standardized value representing the significance of the spatial association and measuring the relative certainty of the posterior probability was computed.

The estimated probability value for each land-use category with respect to evidences should be presented in a raster-based map format. Hence, the classified factor layers were reclassified with the new values obtained from the WoE model. Finally, by integrating and summing up all reclassified evidences for one specific land-use type (residential), the probability of growth of that corresponding land use can be visually presented. However, the obtained map is a continuous layer with several classes, thereby requiring a proper and clear illustration of the output to apply an appropriate classification scheme.

10.4.3 Markov Chain Model

The main difference between the MC model and the utilized statistical models (FR and WoE) is that the input of the MC method only includes two successive land-use maps (e.g., 2004 and 2008) with several separated land-use classes. Thus, no other factors are involved in the processing. Integrating factor-based and cellular-based techniques to produce probability of growth maps was one of the advantages of this study to propose a strong land-use change modeling approach. MC produces two transition matrices, namely, transitional probability matrix and transitional area matrix. The transitional probability matrix estimates the probability of changes of each land-use type to other type. The transitional area matrix estimates the number of pixels expected to change from one land-use type to another over the next time period. In both matrices, the rows represent the earlier land-use maps, and the columns represent the later land-use maps.

The model analyzed land-use changes from earlier (2004) and later (2008) land-use maps, summarized the results in two matrices, and predicted and discovered the future changes and growth. Apart from transitional matrices, the Markov model produces a set of conditional probability images, which illustrate the probability that each land-use category would be found at each place in the future trend. However, the model cannot spatially simulate and model the changes, and its results lack spatial dependency.

10.4.4 Cellular Automata Model

One of the main advantages of the CA model is its spatial dependency. Hence, CA was integrated with the MC model to overcome the lack of spatial dependency. Land-use growth and change modeling by CA analysis provides explicit spatial outputs based on predefined transitional rules. The cellular basis of CA models provides the ability to represent, analyze, and project geographic systems. CA is suitable for this study because of its capability to represent spatial and stochastic processes, model and control complex spatially distributed urban activities, and provide a clear understanding of the behavior of land-use patterns. Unlike the statistical factor-based analyses (FR and WoE), which control the model by considering the driving forces (evidences), the CA model is affected by surrounding neighborhood properties, size, and state. These parameters were considered during the analysis to obtain optimum projection outputs. Given the temporal basis of the available data and the objective of the study, CA was also used to represent and simulate the spatial temporal complexities of land-use changes because of its ability. Time and space are discrete units in CA, and space is considered a regular grid in two dimensions. Local interactions within the 2-D space of the CA system illustrate the dynamics in the landscape pattern. During land-use change modeling, CA can involve population, economic, and transportation data, which are extremely important for this study.

However, the main concern in CA modeling lies in defining the transitional rule that controls the behavior of the system, which in the current study is to control the land-use change behavior. The transitional rules defining the state of each cell for the next time period depend on the current state of the cell, surrounding environment, and some external suitability maps that require integration into the system. The general expression for CA can be shown as follows (Li and Yeh 2000);

$$S^{t+1} = f(S^t, N), \tag{10.3}$$

where S is the state of a cell, t is the earlier time instant, $t + 1$ is the later time instant (future time), N is the cellular field, and f is the transition rule of cellular states in local space. However, apart from neighborhood interaction, the evolution of urban areas depends on a series of complex factors based on local, regional, and global scales. The neighborhood interaction cannot solely deal with urban structure and environmental problems. Some effective factors and constraints require utilization and integration into CA modeling to control the simulation and improve modeling accuracy. In fact, a CA model that does not consider external factors and constraints is known as a state-based CA. In this standard and/or conventional CA, the state is used as the main attribute to describe the development pattern. In this model, the neighborhood-developed cells increase the probability of development of the central cell. At the initial stage of land-use change modeling, this state-based model was run on two land-use maps to observe the land-use growth modeling based only on cellular aspects.

In the next stages, to conduct additional sophisticated modeling, the concept of development probability and development suitability were included and integrated into the CA. This model considers a relation between the states with higher suitability or higher probability of growth (Li and Yeh 2000);

$$S^{t+1}\{x, y\} = f\left(P_s^t\{x, y\}\right) \tag{10.4}$$

$$P_s^t\{x, y\} = f\left(DS_s^t\{x, y\}\right) \tag{10.5}$$

where $S\{x, y\}$ is the current state of the cell at the location $\{x, y\}$, $P_s\{x, y\}$ is the probability of transition to the state S at the same location, $DS_s\{x, y\}$ is the suitability of conversion to state S, and f is the transition function.

As mentioned earlier, in the cellular-based phase of the study, CA was integrated into the MC to combine their advantages and overcome their shortcomings. This integration approach combines the quantitative aspects of land-use change occurrence of the MC model with the spatial dynamic simulation aspect of the CA model. Accordingly, the quantitative information achieved from the Markov model can be translated by CA spatial dynamic capabilities required to analyze and predict urban growth and land-use changes. Merging these two models is a common integration approach in urban applications to simulate dynamic land-use change behavior and project future trends. Applying this integration within the GIS environment provides a strong and effective modeling approach to simulate spatial and temporal land-use changes. However, the importance of statistical concept and factor analysis cannot be ignored. Thus, in this study, these effective concepts using the WoE model were included in the analytical process.

In CA modeling, a 5×5 contiguity filter $\begin{bmatrix} 0 & 0 & 1 & 0 & 0 \\ 0 & 1 & 1 & 1 & 0 \\ 1 & 1 & 1 & 1 & 1 \\ 0 & 1 & 1 & 1 & 0 \\ 0 & 0 & 1 & 0 & 0 \end{bmatrix}$, were used to develop a spatially explicit contiguity weighting factor to change the cell's current land-use type on the basis of its neighborhood interactions, including the neighborhood effects in the modeling. This process will ensure that the lack of spatial distribution of the MC model is addressed, and that land-use change occurs based on neighborhood effects and is not entirely random.

10.4.5 Development of a Compact City Land-Use Modeling Approach

After the Kajang City compactness (DoC of each pixel) and its trend of compactness (ToC, as explained in the Urban Compactness Assessment chapter) were evaluated, the obtained results were integrated into the proposed hybrid land-use change modeling to calibrate the approach and produce a more compact land-use pattern.

City intensification is one of the main approaches to increase city compactness. Similar to mixed land-use development, city intensification has several advantages with respect to environmental, economic, and social sustainability. This process can be carried out in various scales, from urban infill development to the creation of entirely new developments. The current study focuses on the first process and attempts to improve and intensify the existing pattern through brownfield redevelopment (BR). BF sites are abandoned or underused properties that require redevelopment or reuse because of the real or suspected presence of substances, pollutants, or contaminants (Collins 2002; Oliver et al. 2005). Proper BR planning can have implications for all the three approaches of urban intensification. For example, a new residential building for a specific BF site will increase building and residential density and may increase population density by receiving a significant amount of population. Any type of community facility proposed in abandoned lands can improve the urban intensity of the local neighborhood. Finally, any land-use categories other than the existing surrounding categories can increase land-use diversity of the local neighborhood and eventually intensify the corresponding city.

A BR within urban areas can be serviced by existing infrastructures and utilities, whereas rural development requires essential provision of these systems. However, a BR requires a comprehensive effort to resolve and negotiate among several stakeholders with different interests (Gross 2010). Such complex issues have resulted in the continuing abandonment of most BFs. Several processes are available to conduct BR, such as risk assessment, policy analysis, optimization of remediation, remediation cost assessment, urban planning and site prioritization, and so forth (Schädler et al. 2011). Considering that the current study mainly deals with land-use analysis and modeling, the BR process was also conducted through an approach based on the land-use change modeling technique. The flowchart of the proposed compact land-use change modeling is shown in Fig. 10.7.

Land-use maps of 2004 and 2008 were utilized to create the probability of growth map for the main land-use types using the WoE model. These probability maps were validated by an actual land-use map of year 2012. The same process was implemented to create probability of growth maps using land-use maps of 2008 and 2012. Similarly, these maps were validated using the actual land-use map of 2015 (due to the lack of land-use map of 2016). After validating all probability maps, these results were integrated into the CA model to facilitate the application of a contiguity filter and consequently obtain the growth projection for 2012 and 2016. The process of validation was performed one more time for projected land-use maps of 2012 and 2016 to compare with actual land-use maps of 2012 and 2015 respectively.

After confirming the performance accuracy and reliability of the model, this process was conducted on land-use maps of 2004 and 2015 to project the land-use map for 2026. This modeling process can be called a "business-as-usual" scenario because it is based on the historical trend and current development pattern. However, achieving compact urban development requires proposing and conducting an alternative scenario, namely, a "compact land-use" modeling

Fig. 10.7 Detail flowchart of the proposed land use change modeling process. *Blue* input data, *orange* modeling process, *green* output maps, *red* final products

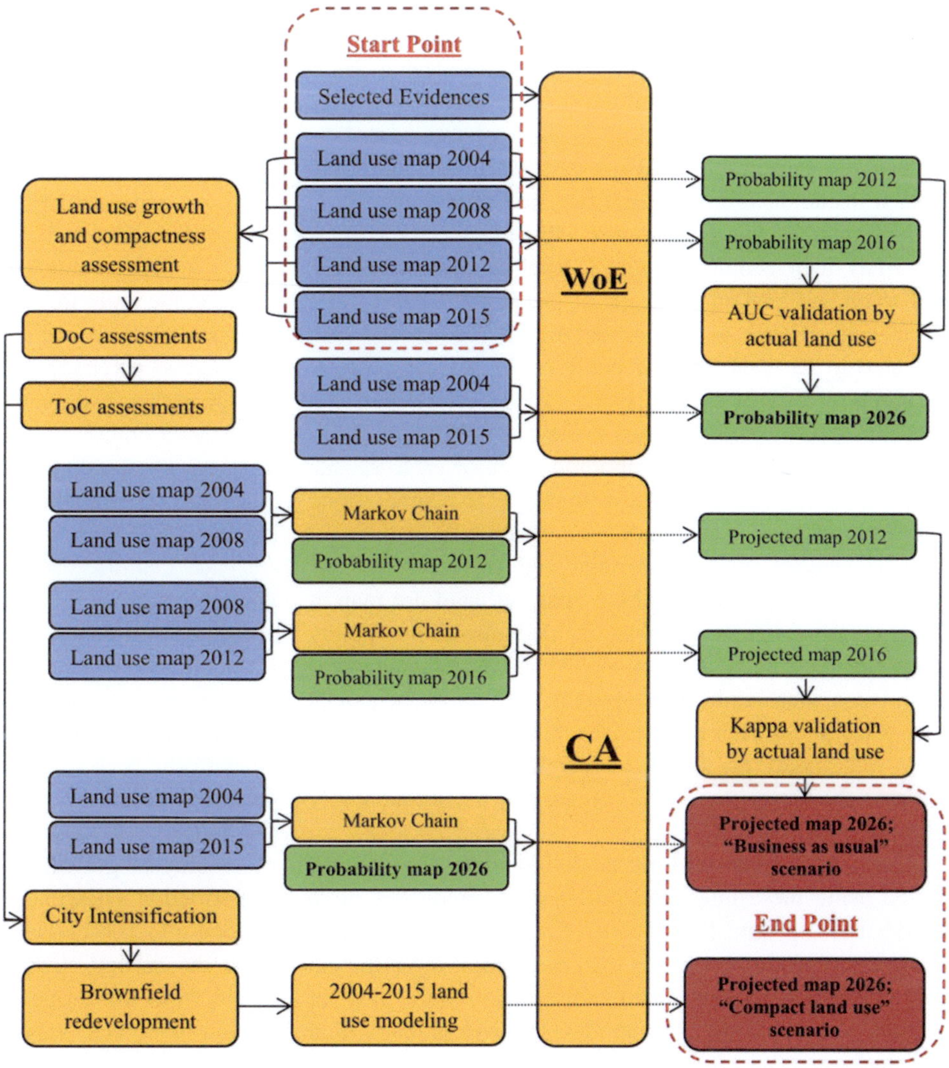

scenario. In this scenario, the results of compact city assessments (DoC and ToC) were integrated into the land-use modeling processes (Abdullahi et al. 2015a, b). The aim was to increase the DoC of those areas with low compact land-use pattern. One of the most feasible and cost-effective tasks in this process is to redevelop the existing BF sites (Abdullahi and Pradhan 2015). Hence, the BR process was also coupled with the land-use modeling process to achieve a more compact and sustainable land-use map.

The BR process was started by extracting the existing BF sites of land-use map of 2015. All open spaces, such as the buffer zone around rivers and highways; recreational fields, such as playgrounds; and natural landscapes, were excluded from the analysis. Evaluating the BF sites required the utilization of site indicators and criteria presented by Thomas (2002), such as existing and previous land use/cover, physical properties (soil and geological characteristics), neighborhood characteristics (neighborhood land-use types, availability of community facilities, recreational facilities, commercial and service buildings, etc.), proximity to transportations, and air and water quality. The list of siting guidelines and metrics applied for BF assessments has a greater number of items. However, some parameters (e.g., air and water quality) are useless to consider in case studies, such as the current study, as the BF sites in this study are located inside the urban areas; hence, no differences exist in air and water quality. Furthermore, other parameters, such as proximity to telecommunications, level of contaminations, and so on, are beyond the scope of this study. Hence, those parameters related to the objectives of this study were included in the analysis. For this process, gaining a wider perspective regarding the existing BF sites required the production of a proximity map to BF sites using the Euclidean distance tool. The neighborhood area of each BF site was investigated and evaluated with respect to neighborhood land-use categories, availability of community and recreational facilities, and commercial and service buildings, through overlaying this proximity map and land-use map of 2015. Apart from this evaluation, the growth probability maps created from the WoE modeling and master plan of the study area were utilized to propose the most beneficial land-use type for each BF site. Hence, the proposed land-use type was based on the local neighborhood demand to increase the compactness and, eventually, the sustainability of local environments.

Cellular automata land-use modeling approach was calibrated using these integrated processes to produce a more compact urban form. The modified CA model can produce a much better alternative rather than the current development pattern because of less land consumption, provision of required facilities, and concentration of development within the city borders.

10.5 Validation of Land-Use Change Modeling Process

In urban applications, especially in the field of land-use growth and change modeling, knowing the prediction accuracy and reliability of the model is essential. In this regard, quantitatively evaluating the degree of similarity between model outputs and reality provides a good performance assessment. In recent literature (Pontius and Schneider 2001; Van Eck and Koomen 2008; Chen et al. 2014), relative operating characteristic (ROC), error map and contingency matrix, and the Kappa statistic index are the common terminologies used to validate urban growth and change modeling processes.

The present study validates the performance of proposed land-use change modeling process in two stages. At the first stage, the area under the ROC curve (AUC) was utilized to evaluate the probability of growth maps created from the WoE model with actual land-use maps. In the second stage, a contingency table was calculated to evaluate the projected maps created from integrating CA and WoE.

The ROC validation technique measures the relationship between the projected and actual spatial changes by computing the percentage of false positive and true positive for a range of thresholds and relating the values to one another in a chart. The AUC is calculated as the area under the ROC curve and ranges from 0.5 to 1 (Pradhan et al. 2010). A value near 0.5 indicates a random relationship between input maps, while a value near 1 indicates a high relationship between the input maps, which is an ideal spatial agreement between modeled and actual land-use maps. In this process, the ordered pixel values according to the modeling process (which in this case is a probability value) were classified into 100 classes and set on an x-axis. The calculated index values were set in descending order on a y-axis. Hence, for this study, AUC validation was conducted by comparing the probability of growth maps produced from the WoE model for each land-use category. Probability maps created from land-use maps of 2004 and 2008 for year 2012 were compared with the actual land-use map of 2012, and probability maps created from land-use maps of 2008 and 2012 for year 2016 were compared with the actual land-use map of 2015. This process evaluates how well the evidences can be used to perform probability analysis. Accordingly, the model performance accuracy was evaluated, and the capability of the model was confirmed. To run AUC, the calculated probability of growth values of all pixels in the study area was sorted in a descending order. These ordered values were divided into 100 classes with accumulated 1% intervals.

In the second stage, the projected land-use maps produced from the proposed land-use change modeling integration approach was evaluated with respect to reality (reference map) by calculating the contingency table and

Table 10.6 Two-by-two contingency table showing the proportion of pixels in actual and modelled maps

		Reality		Total
		Change	Non-change	
Model	Change	True positive (TP)	False positive (FP)	TP + FP
	Non-change	False negative (FN)	True negative (TN)	FN + TN
	Total	TP + FN	FP + TN	TP + FP + FN + TN

True positive (TP) is the amount of pixels modeled to change and be changed in reality
True negative (TN) is the amount of pixels modeled to stay unchanged (remain as they are) and not be changed in reality
False positive (FP) is the amount of pixels modeled to change but not be changed in reality; and
False negative (FN) is the amount of pixels modeled to stay unchanged but be changed in reality

illustration. The contingency table is based on a two-by-two comparison between projected and actual land-use maps for each land-use category. This table summarizes the results for cases where each pixel is a homogenous land-use category (Table 10.6).

From every contingency table, a single data point can be created, where *X* and *Y* are the rate of false positive and true positive respectively:

- True positive % = TP/(TP + FN)
- False positive % = FP/(FP + TN)

However, this presentation of the contingency table is for coarse spatial scale analysis using the binary modeling process, where growth/non-growth or change/non-change of pixels is the main consideration. In the current study, given the high spatial resolution of data, the modeling process went through a more detailed analysis by considering several land-use category changes. Hence, the contingency table was expanded to evaluate all changes among every land-use type. Furthermore, final accuracy of the model was evaluated by the proportion of correct, which can be calculated by TP/(TP + FP + FN + TN) or the number of correctly modeled pixels divided by the total number of pixels (Pontius et al. 2001).

Apart from the contingency table, Kappa statistic index was calculated to assess the validity and reliability of the projected maps in terms of quantity and location of the changes. Kappa index of agreement is a measure of proportional accuracy adjusted for chance agreement (Arsanjani et al. 2011).

10.6 Results and Discussion of Compact Land-Use Modeling for Kajang City

10.6.1 Growth Probability Maps Using Weights-of-Evidence

This section presents and discusses the results of land-use change modeling and simulation of Kajang City on the basis of the previous land-use patterns. In brief, the process was started by selecting and evaluating several effective factors related to urban growth analysis. The FR method as an initial stage of the WoE model was used to conduct this process. FR evaluated the spatial effects of these selected factors on land-use change occurrence. Next, WoE evaluated the nonoccurrence of land-use changes with respect to the selected factors. Hence, using both analyses, WoE produces the probability of growth maps for selected land-use categories. The analysis was focused on the growth of three main land-use categories for each time period. Tables 10.7 and 10.8 respectively present an example of FR and WoE processes to show the details regarding the calculations of probability values.

The examples evaluate the growth of commercial land-use type with respect to proximity to road network for the time period of 2004–2008. The value of *C* was calculated by subtracting *W*+ (natural logarithm of occurrence) and *W*− (natural logarithm of nonoccurrence). This value represents the spatial association of each land-use pixel and each class of factors. A positive value represents a higher number of specific land-use pixels occurring in this class. In contrast, a negative value represents a lesser number of land-use pixels

Table 10.7 Frequency ratio of occurrence and nonoccurrence of commercial use with respect to road networks

Factor	Class	Deposit occurrences (+)					Non-deposit occurrences (−)				
		No. cell	% of cell	No. deposit	% of deposit	FR	No. cell	% of cell	No. deposit	% of deposit	FR
Proximity to road network	Near	18,480,314	**32.69**	767,133	**66.78**	**2.04**	38,057,361	67.31	381,530	33.22	0.49
	Middle	19,211,873	**33.98**	315,325	**27.45**	**0.81**	37,325,802	66.02	833,338	72.55	1.10
	Far	18,845,488	**33.33**	66,205	**5.76**	**0.17**	37,692,187	66.67	1,082,458	94.24	1.41
	Sum	56,537,675	100	1,148,663	100		113,075,350				

Bold letters indicate the important factors

Table 10.8 WoE calculation for commercial land use growth with respect to road networks

Factor	Class	FR (occurrence)	FR (non-occurrence)	*W*+	*W*−	*C*	*S*2(*W*+)	*S*2(*W*−)	*S*(*C*)	*C*/*S*(*C*)
Proximity to road network	**Near**	**2.04**	0.49	0.71	−0.71	1.42	0.00000136	0.000003	0.002	**709.66**
	Middle	**0.81**	1.10	−0.21	0.09	−0.31	0.00000322	0.000001	0.002	**−145.82**
	Far	**0.17**	1.41	−1.75	0.35	−2.10	0.00001516	0.000001	0.004	**−523.48**

Bold letters indicate the important factors

occurring in this class. *S*2(*W*+) and *S*2(*W*−) are variances of *W*+ and *W*− respectively, and *S*(*C*) is the standard deviation of the contrast. Finally, *C*/*S*(*C*) is the standardized value of *C*, which represents the significance of the spatial association and measures the relative certainty of the posterior probability. The probability value of land-use growth for every cell of the study area is calculated by considering the prior probability of occurrence and nonoccurrence of land-use type in each class of selected factors.

Tables 10.9, 10.10 and 10.11 present the results of FR and WoE for residential, commercial, and industrial land-uses, respectively, for each time period. The majority of the factors can be regarded as distance-based factors, and the FR and WoE values are calculated for each class of these

Table 10.9 Frequency ratio and weights-of-evidence calculation results for residential growth

Residential land use growth		2004–2008		2008–2012		2012–2015	
Factor	Class	FR	*C*/*S* (*C*) (WoE)	FR	*C*/*S* (*C*) (WoE)	FR	*C*/*S* (*C*) (WoE)
Proximity to housing	Near	2.57	**2700.92**	2.40	**2433.19**	3.55	**2900.11**
	Middle	0.37	**−1478.47**	0.44	**−1411.81**	0.04	**−1835.02**
	Far	0.12	**−1710.81**	0.29	**−1648.14**	0.06	**−1760.94**
Proximity to commercial	Near	1.33	**683.28**	1.18	**403.05**	1.08	**174.68**
	Middle	1.10	**217.41**	1.09	**215.69**	1.14	**321.30**
	Far	0.59	**−944.23**	0.73	**−647.96**	0.78	**−517.00**
Proximity to industrial	Near	0.73	**−599.47**	0.66	**−795.27**	0.52	**−1148.43**
	Middle	1.38	**815.17**	1.30	**673.47**	1.20	**451.20**
	Far	0.88	**−277.58**	1.02	**53.47**	1.27	**596.85**
Proximity to road network	Near	1.08	**172.23**	0.94	**−128.66**	0.83	**−393.30**
	Middle	1.27	**600.20**	1.20	**464.27**	1.14	**311.82**
	Far	0.64	**−817.31**	0.85	**−360.80**	1.03	**62.89**
Proximity to facility	Near	1.35	**736.57**	1.21	**467.06**	1.21	**457.57**
	Middle	1.11	**253.61**	1.11	**249.71**	1.22	**498.85**
	Far	0.55	**−1044.79**	0.69	**−752.38**	0.58	**−1028.37**
Proximity to recreation	Near	1.04	**81.93**	1.06	**142.60**	1.30	**665.63**
	Middle	0.88	**−271.43**	1.01	**32.38**	1.00	**8.91**
	Far	1.08	**185.17**	0.93	**−176.19**	0.70	**−729.82**
Proximity to infrastructure	Near	1.18	**398.13**	1.07	**150.46**	1.14	**322.03**
	Middle	1.12	**260.72**	1.06	**129.30**	1.13	**291.22**
	Far	0.70	**−682.92**	0.88	**−283.07**	0.73	**−642.91**
Proximity to agriculture	Near	0.56	**−996.04**	0.68	**−760.60**	0.74	**−616.72**
	Middle	1.01	**16.19**	0.98	**−51.19**	1.11	**243.32**
	Far	1.43	**900.52**	1.34	**743.45**	1.15	**338.68**
Land use diversity	Low (single)	0.49	**−1303.88**	0.61	**−1005.50**	0.43	**−1524.74**
	Middle	1.10	**235.10**	1.07	**184.98**	1.14	**362.64**
	High (mixed)	1.73	**1123.48**	1.53	**855.28**	1.75	**1157.54**

(continued)

Table 10.9 (continued)

Residential land use growth		2004–2008		2008–2012		2012–2015	
Factor	Class	FR	*C/S* (*C*) (WoE)	FR	*C/S* (*C*) (WoE)	FR	*C/S* (*C*) (WoE)
Urban density	Low (density)	0.28	**−1481.52**	0.51	**−1087.37**	0.55	**−1005.10**
	Middle	1.35	**793.96**	1.24	**575.07**	1.21	**503.06**
	High (density)	1.26	**564.85**	1.17	**379.78**	1.17	**377.22**
Proximity to public transportation	Near	1.20	**437.10**	1.15	**339.15**	1.13	**289.05**
	Middle	1.07	**155.29**	1.09	**207.42**	1.10	**228.93**
	Far	0.73	**−613.55**	0.76	**−567.92**	0.78	**−538.17**
Proximity to water body	Near	1.20	**429.00**	1.13	**282.81**	1.11	**239.16**
	Middle	1.15	**318.69**	1.16	**367.72**	1.16	**360.93**
	Far	0.66	**−778.17**	0.72	**−681.05**	0.74	**−628.45**
Proximity to restricted area	Near	1.10	**215.03**	1.20	**440.34**	1.19	**425.68**
	Middle	1.14	**301.41**	1.05	**107.21**	1.05	**105.24**
	Far	0.77	**−533.78**	0.76	**−573.27**	0.77	**−556.78**
Proximity to flood zone	Near	1.19	**404.19**	1.08	**188.50**	1.06	**146.68**
	Middle	1.06	**134.72**	1.14	**318.09**	1.14	**311.17**
	Far	0.76	**−556.99**	0.78	**−526.34**	0.80	**−475.70**
Geological types	Acid	0.87	**−372.22**	0.91	**−270.96**	0.91	**−243.57**
	Quartz	0.87	**−38.18**	1.10	**29.08**	1.06	**17.59**
	Schist	0.72	**−61.72**	0.70	**−70.82**	0.67	**−78.08**
	Filit	1.13	**474.23**	1.09	**318.49**	1.09	**280.51**
Soil types	T-A-LA	1.30	**260.93**	1.45	**391.71**	1.43	**372.57**
	M-S	0.99	**−23.19**	0.97	**−52.34**	0.99	**−17.55**
	U-L	1.29	**743.07**	1.18	**467.27**	1.14	**365.07**
	M-L	1.22	**139.46**	0.94	**−39.58**	1.06	**42.05**
	R-J	0.40	**−1046.75**	0.63	**−677.06**	0.68	**−600.57**
	S-L	0.00	**−29.52**	0.00	**#NUM!**	0.00	**#NUM!**

Bold letters indicate the important factors

factors. The proximity to existing land-use types is one of the effective factors for growth of each land-use type. The maximum probability value of WoE for residential land-use is in the near class of proximity to residential area. The same conditions are valid for commercial and industrial growths, which mainly depend on the proximity to commercial and industrial areas respectively. In contrast, as distance increases from existing land-use type (e.g., third class of residential land-use), the probability of growth for the same land-use type (residential use) substantially decreases. Different land-use types affect the growth of other land uses as well. For example, residential and industrial areas tend to be far from each other as commonly expected. The near classes of proximity to industrial areas have negative probability value for residential growth. Similarly, as the distance from industrial areas to residential areas decreases, the probability of growth for industrial use significantly decreases. These effects are constant in all time periods. However, commercial land use has no direct influence on other categories. To illustrate, proximity to commercial area increases the probability of growth for residential land use. However, the middle class of residential proximity has the highest value of probability of growth for commercial land use. Similarly, the middle class of industrial proximity possesses higher probability value for commercial growth. However, the farthest distance from the commercial area has the highest value or probability for industrial growth. Therefore, commercial land-use type is mainly affected by other parameters, such as road and public transportation proximity.

Few important factors directly affect the growth of residential, commercial, and industrial land uses. The middle class of road proximity has a high probability of residential

Table 10.10 Frequency ratio and weights-of-evidence calculation results for commercial growth

Commercial land use growth		2004–2008		2008–2012		2012–2015	
Factor	Class	FR	*C*/*S* (*C*) (WoE)	FR	*C*/*S* (*C*) (WoE)	FR	*C*/*S* (*C*) (WoE)
Proximity to housing	Near	0.28	**–479.83**	0.58	**–362.83**	0.00	**#NUM!**
	Middle	2.25	**840.01**	1.56	**541.58**	1.56	**578.14**
	Far	0.36	**–442.58**	0.77	**–207.93**	1.15	**143.82**
Proximity to commercial	Near	3.03	**761.83**	2.05	**885.46**	3.06	**669.36**
	Middle	0.08	**–510.70**	0.22	**–648.64**	0.02	**–471.47**
	Far	0.00	**#NUM!**	0.77	**–217.00**	0.00	**–313.43**
Proximity to industrial	Near	0.84	**–118.86**	0.80	**–180.42**	0.59	**–369.49**
	Middle	1.34	**253.96**	1.44	**408.66**	1.32	**296.09**
	Far	0.82	**–139.06**	0.75	**–237.41**	1.08	**73.70**
Proximity to road network	Near	2.04	**709.66**	1.49	**440.88**	1.40	**366.40**
	Middle	0.81	**–145.82**	1.13	**125.84**	1.14	**134.85**
	Far	0.17	**–523.48**	0.39	**–544.39**	0.46	**–489.05**
Proximity to facility	Near	1.76	**530.18**	1.36	**327.13**	1.29	**264.09**
	Middle	1.00	**1.23**	0.84	**–149.48**	0.96	**–33.78**
	Far	0.27	**–497.25**	0.81	**–177.52**	0.76	**–229.12**
Proximity to recreation	Near	0.97	**–20.66**	0.82	**–166.38**	1.15	**137.84**
	Middle	1.35	**260.66**	1.26	**238.67**	1.17	**156.50**
	Far	0.68	**−240.90**	0.92	**−74.14**	0.68	**−293.08**
Proximity to infrastructure	Near	1.25	**182.27**	1.05	**48.63**	1.00	**3.84**
	Middle	1.10	**73.70**	1.03	**24.17**	1.12	**111.06**
	Far	0.66	**–253.04**	0.92	**–72.13**	0.88	**–115.11**
Proximity to agriculture	Near	0.38	**–435.41**	0.78	**–199.68**	0.75	**–232.67**
	Middle	0.69	**–234.96**	0.74	**–249.07**	1.25	**230.66**
	Far	1.93	**651.84**	1.48	**441.08**	1.00	**–0.12**
Land use diversity	Low (single)	0.10	**–575.85**	0.77	**–239.43**	0.51	**–504.89**
	Middle	1.15	**126.72**	0.87	**–143.13**	1.52	**537.28**
	High (mixed)	2.33	**699.04**	1.64	**437.70**	0.94	**–39.86**
Urban density	Low (density)	0.11	**–479.60**	0.41	**–486.00**	0.40	**–490.27**
	Middle	1.08	**63.41**	1.26	**255.37**	1.25	**246.30**
	High (density)	1.70	**505.29**	1.25	**228.45**	1.26	**242.31**
Proximity to public transportation	Near	1.61	**439.26**	1.38	**349.98**	1.37	**341.44**
	Middle	1.07	**55.47**	0.85	**–141.00**	0.87	**–125.28**
	Far	0.33	**–469.36**	0.77	**–211.60**	0.77	**–218.41**
Proximity to water body	Near	1.63	**452.26**	1.30	**279.07**	1.29	**270.71**
	Middle	0.96	**–27.24**	1.11	**99.78**	1.12	**114.47**
	Far	0.42	**–416.67**	0.60	**–373.66**	0.59	**–379.58**
Proximity to restricted area	Near	1.14	**103.24**	0.99	**–8.00**	1.01	**7.54**
	Middle	1.42	**311.55**	1.09	**82.48**	1.08	**74.25**
	Far	0.44	**–402.86**	0.92	**–74.89**	0.91	**–82.06**

(continued)

Table 10.10 (continued)

Commercial land use growth		2004–2008		2008–2012		2012–2015	
Factor	Class	FR	*C*/*S* (*C*) (WoE)	FR	*C*/*S* (*C*) (WoE)	FR	*C*/*S* (*C*) (WoE)
Proximity to flood zone	Near	1.49	**354.29**	1.22	**204.96**	1.21	**196.34**
	Middle	0.76	**–177.73**	0.83	**–154.57**	0.83	**–160.98**
	Far	0.76	**–180.25**	0.95	**–50.42**	0.96	**–35.28**
Geological types	Acid	1.04	**42.40**	0.63	**–470.23**	0.64	**–458.30**
	Quartz	2.00	**92.46**	1.60	**69.70**	1.59	**68.50**
	Schist	0.00	**#NUM!**	0.22	**–62.88**	0.22	**–63.22**
	Filit	0.95	**–50.07**	1.37	**466.75**	1.35	**455.05**
Soil types	T-A-LA	1.20	**59.85**	0.77	**–86.62**	0.76	**–89.29**
	M-S	0.39	**–327.86**	1.14	**98.69**	1.13	**92.60**
	U-L	1.59	**514.23**	1.37	**410.15**	1.36	**399.70**
	M-L	0.72	**–59.31**	0.29	**–179.42**	0.29	**–180.61**
	R-J	0.54	**–252.96**	0.44	**–379.02**	0.47	**–359.78**
	S-L	0.00	**#NUM!**	0.00	**#NUM!**	0.00	**#NUM!**

Bold letters indicate the important factors

Table 10.11 Frequency ratio and weights-of-evidence calculation results for industrial growth

Industrial land use growth		2004–2008		2008–2012		2012–2015	
Factor	Class	FR	*C*/*S* (*C*) (WoE)	FR	*C*/*S* (*C*) (WoE)	FR	*C*/*S* (*C*) (WoE)
Proximity to Housing	Near	0.04	**–881.21**	0.10	**–1085.65**	0.00	**#NUM!**
	Middle	1.07	**106.23**	0.79	**–359.45**	0.41	**–1068.34**
	Far	1.89	**1215.14**	2.09	**1552.90**	2.52	**2035.41**
Proximity to commercial	Near	0.54	**–655.84**	0.53	**–726.81**	0.50	**–802.85**
	Middle	1.10	**151.63**	1.07	**112.92**	1.13	**212.97**
	Far	1.33	**493.13**	1.38	**602.47**	1.36	**578.11**
Proximity to industrial	Near	3.07	**819.86**	2.60	**2012.03**	3.04	**832.73**
	Middle	0.01	**–642.12**	0.07	**–1097.10**	0.01	**–658.09**
	Far	0.00	**#NUM!**	0.39	**–941.69**	0.00	**#NUM!**
Proximity to road network	Near	0.85	**–218.87**	0.79	**–331.38**	0.87	**–219.94**
	Middle	1.17	**256.81**	1.03	**56.63**	1.17	**279.54**
	Far	0.97	**–42.42**	1.17	**269.17**	0.96	**–64.55**
Proximity to facility	Near	0.25	**–1004.87**	0.31	**–1022.94**	0.20	**–1154.55**
	Middle	0.98	**–35.52**	0.83	**–272.17**	0.78	**–367.16**
	Far	1.74	**1061.69**	1.83	**1258.34**	1.99	**1507.02**
Proximity to recreation	Near	0.74	**–386.34**	0.62	**–599.04**	0.21	**–1166.70**
	Middle	1.31	**455.74**	0.91	**–139.90**	0.68	**–537.18**
	Far	0.95	**–78.32**	1.46	**715.94**	2.11	**1649.83**
Proximity to infrastructure	Near	1.07	**101.10**	0.92	**–124.59**	0.79	**–341.68**
	Middle	1.17	**247.49**	1.04	**70.14**	1.02	**28.94**
	Far	0.77	**–349.23**	1.03	**52.55**	1.19	**306.33**
Proximity to agriculture	Near	0.78	**–333.11**	1.38	**587.29**	1.06	**102.37**
	Middle	1.70	**1016.00**	1.28	**444.17**	1.03	**46.49**
	Far	0.51	**–726.52**	0.34	**–1002.09**	0.91	**–149.24**

(continued)

Table 10.11 (continued)

Industrial land use growth		2004–2008		2008–2012		2012–2015	
Factor	Class	FR	*C/S* (*C*) (WoE)	FR	*C/S* (*C*) (WoE)	FR	*C/S* (*C*) (WoE)
Land use diversity	Low (single)	1.40	**672.66**	1.53	**900.49**	1.74	**1286.90**
	Middle	0.79	**–352.60**	0.76	**–448.23**	0.66	**–637.74**
	High (mixed)	0.65	**–396.33**	0.54	**–555.88**	0.30	**–822.20**
Urban density	Low (density)	1.70	**912.49**	1.79	**1096.34**	1.91	**1282.99**
	Middle	0.75	**–403.05**	0.72	**–483.26**	0.66	**–602.52**
	High (density)	0.67	**–503.79**	0.61	**–619.26**	0.57	**–703.70**
Proximity to public transportation	Near	0.81	**–279.09**	0.66	**–535.63**	0.62	**–622.42**
	Middle	1.00	**7.09**	0.74	**–411.55**	0.72	**–459.43**
	Far	1.18	**268.06**	1.59	**907.80**	1.65	**1029.98**
Proximity to water body	Near	0.98	**–34.91**	0.84	**–247.89**	0.79	**–346.58**
	Middle	1.08	**116.08**	0.81	**–309.39**	0.80	**–326.81**
	Far	0.95	**–81.50**	1.34	**543.59**	1.40	**652.96**
Proximity to restricted area	Near	0.18	**–1056.40**	0.28	**–1057.79**	0.26	**–1112.10**
	Middle	0.60	**–602.45**	0.80	**–325.15**	0.73	**–444.14**
	Far	2.20	**1602.78**	1.91	**1348.24**	1.99	**1500.02**
Proximity to flood zone	Near	1.30	**439.12**	1.02	**33.91**	0.95	**–74.91**
	Middle	0.62	**–565.93**	0.49	**–804.89**	0.59	**–671.33**
	Far	1.08	**125.09**	1.49	**760.50**	1.45	**727.70**
Geological types	Acid	0.90	**–210.51**	1.14	**304.75**	1.21	**466.01**
	Quartz	0.00	**#NUM!**	0.00	**#NUM!**	0.00	**#NUM!**
	Schist	2.09	**142.10**	1.77	**108.95**	1.65	**95.95**
	Filit	1.11	**223.59**	0.87	**–284.90**	0.80	**–444.46**
Soil types	T-A-LA	0.09	**–446.40**	0.10	**–482.50**	0.10	**–496.71**
	M-S	1.93	**990.78**	1.52	**605.19**	1.41	**492.59**
	U-L	0.93	**–132.13**	0.82	**–354.26**	0.76	**–472.27**
	M-L	1.73	**297.09**	1.57	**252.50**	1.47	**214.50**
	R-J	0.46	**–593.01**	0.74	**–309.28**	1.01	**8.53**
	S-L	0.00	**#NUM!**	4.18	**760.86**	3.90	**724.40**

Bold letters indicate the important factors

and industrial growths. However, in the case of industrial growth from 2008 to 2012, the farthest class was the more suitable area for industrial growth. Hence, the areas nearest to the main road networks appear to be the best location for commercial use.

In the case of community facilities, proximity to these locations is clearly expected to have a higher probability of growth for residential use and negative probability of growth for industrial use. In this case, proximity to facilities has the same effects for residential and commercial land uses. Nearly the same effect can be seen for the proximity to recreational facilities with respect to residential and industrial growths. However, the middle class of recreational proximity has a higher probability of commercial growth. Considering that infrastructures are well distributed in Kajang City and are extremely important for all types of developments, proximity to these utilities increases the probability of growth for all three land-use types. However, from 2008 to 2015, areas far from these utilities had higher probability of growth for industrial land use. This finding is due to the growth of industrial areas through agricultural fields, where only few utilities are generally provided. This growth can also be observed in 2008–2012 and in 2012–2015 in the near class of agricultural proximity factor, which has a higher value of probability of industrial growth. In contrast, as distance increases from agricultural fields, the probability of residential and commercial growths significantly increases. Interestingly, city compactness indicators have straightforward effects on the growth of these three land-use categories. Higher urban density and land-use diversity increase the growth of residential and commercial land use. In contrast, given that industrial areas are mainly located in single land-use and lower urban density environments, they have inverse results compared with the other two land-use types.

Similar to community facilities, public transportation also has the same effect on the growth of residential, commercial, and industrial land uses; proximity to these facilities increases the probability of growth for residential and commercial land uses and reduces the probability of growth for industrial land use.

Unlike other factors, water bodies, restricted areas (e.g., prisons, cemeteries, etc.), flood zones, soil, and geological types failed to provide understandable and acceptable effects with respect to the growth of selected land-use types. Owing to the improper amount and distribution of these factors and their classes, the results seem to be random and stochastic. However, implementing FR and WoE analyses on all factors provide a clearer view regarding the spatial influence of these factors with respect to the growth and changes of various land-use types.

All the calculated WoE values were standardized in the range of 1 as the least probability to 5 as the highest probability of growth. Next, these standardized values were applied to each class of factors. Furthermore, by aggregating all factors with respect to each land-use type, a projected probability of growth map for that corresponding land-use type was created. The FR and, consequently, the WoE model were applied for all the three time steps, namely, 2004–2008 to create the probability of growth map for 2012, 2008–2012 for 2016, and 2012–2015 for 2018, based on all derived factors for residential, commercial, and industrial land-use types. However, only the probability of growth for years 2012 and 2016 could be validated using the actual land-use map of 2012 and 2015. Hence, after validating each land-use growth probability map with the actual future land-use map and by considering all factors, less accuracy and similarity were achieved among actual and modeled land-use maps. After evaluating probability maps with the actual land-use maps, an optimized list of factors was extracted from all available derived factors. Among all the factors, only eight were considered the most effective factors for land-use change, specifically for this study area. Other factors were ignored from further processing because of data redundancy and/or random effects. These eight factors are as follows: proximity to residential, commercial, industrial, agricultural areas, including proximity to road network, public transportation, community facilities, and infrastructures.

Figures 10.8, 10.9 and 10.10 show the probability maps of residential, commercial, and industrial growth for all three time steps: 2004–2008 to create the probability map for 2012, 2008–2012 for 2016, and 2012–2015 for 2018. In general, central parts of Kajang City are found to have a higher probability of growth for residential and commercial uses, while the eastern and western sides have a higher probability of growth for industrial land use. However, residential areas have broader extensions compared with the commercial areas. The areas with higher probability of growth for commercial land use are mainly located along the main roads in the central parts and passing the main public transportation of Kajang City (southern parts). In contrast, residential growth has a higher probability in wider extensions mainly located in the central parts. Industrial land use has a higher probability of growth in the eastern regions, which are mainly covered by agricultural fields, and western parts near existing industrial buildings and open spaces. Note that unlike residential and commercial areas with nearly the same patterns for all the projected maps (2012, 2016, and 2018), industrial land use has a different growth pattern in 2016. The effects of road network mainly cause this different pattern. This effect can be noticed from Table 10.11 in factor no. 4 proximity to road network. For this factor, in the 2004–2008 and 2012–2015 time periods, only the middle class has a positive probability value. This exclusive limitation to the middle class consequently highlighted the roads in different colors in both time periods. In contrast, in the 2008–2012 time period, only the near class has a negative value. Hence, the effects of road network are not significantly visible.

The legends of these maps (Figs. 10.8, 10.9 and 10.10) are in the range of high, with a value of 40, and low, with a value of 8. As mentioned before, only eight factors were selected out of the 16 factors as the most important ones. All the WoE values were standardized into five classes as well. Hence, after aggregating all reclassified WoE maps, the areas with minimum probability of growth were assigned a value of 8; in contrast, the areas with the maximum probability of growth were assigned a value of 40. In general, the main differences among the probability maps (for each separate land-use type) for different time steps (2004–2008, 2008–2012, and 2012–2015) are the change in probability value, which can be observed from the intensity of colors in each of the three maps. For example, the increase in probability of growth for residential land use can be observed in the eastern parts during 2012–2018. The yellow color of this area in the first map gradually changes to green in the 2018 map. The same condition in the same area is happening for commercial land-use growth. However, in the case of

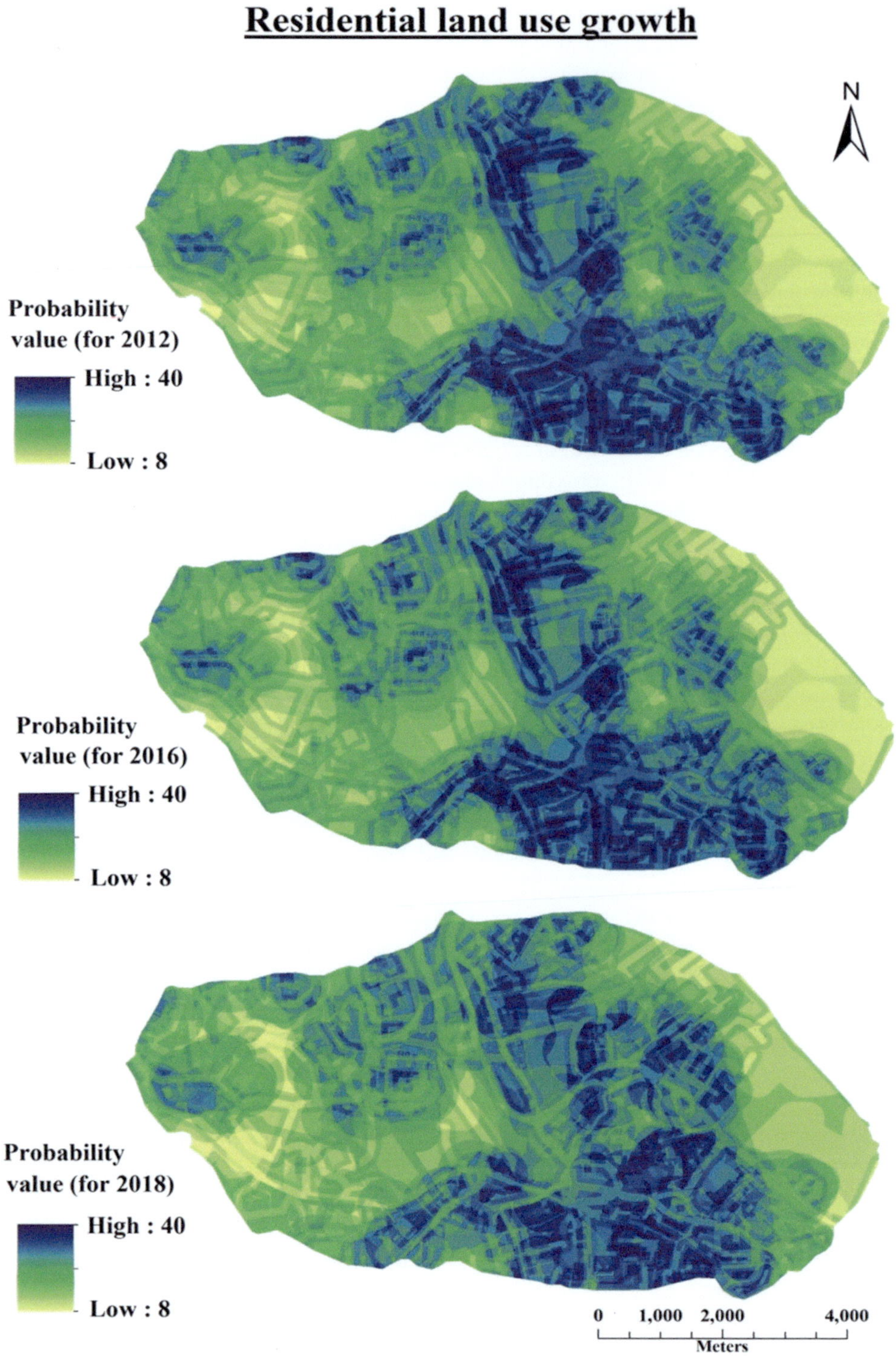

Fig. 10.8 Residential land use growth probability maps for years 2012, 2016 and 2018

industrial growth, 2016 has the highest intensity of growth (blue color) with respect to other land probability maps.

The predicted probability of growth maps of years 2012 and 2016 was validated by comparing these with actual land-use maps of 2012 and 2015 respectively using the area under the ROC curve. This validation technique measures the relationship and fitness among the real and projected maps. Figures 10.11 and 10.12 show the AUC graphs for both probability maps. In both graphs, industrial and commercial land uses have the highest and lowest accuracies

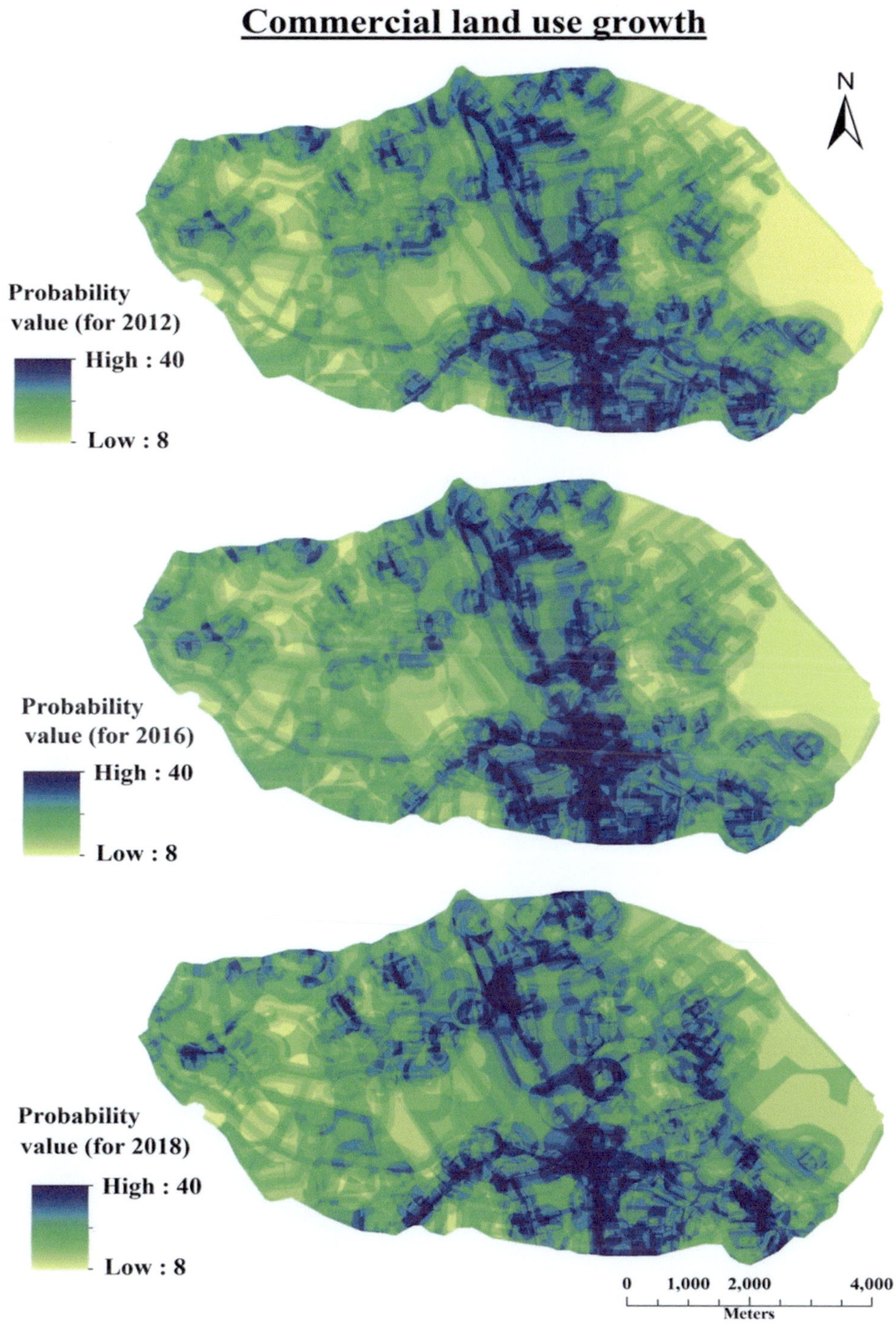

Fig. 10.9 Commercial land use growth probability maps for years 2012, 2016 and 2018

respectively. This finding is attributed to the small proportion of commercial areas rather than other factors, as well as the dependency on the additional number of variables and parameters. In contrast, industrial land use has a larger proportion than commercial land use, but mainly depends on the proximity to existing industrial areas. In both graphs, all land-use categories have high and acceptable similarities with actual land-use maps, indicating the reliability of the WoE land-use change modeling process.

After confirming the WoE performance accuracy to create the future probability of growth maps, this process was conducted to create the probability of growth maps for the next 11 years around 2026 based on the 2004–2015 time period. Table 10.12 presents the FR and WoE calculation

Fig. 10.10 Industrial land use growth probability maps for years 2012, 2016 and 2018

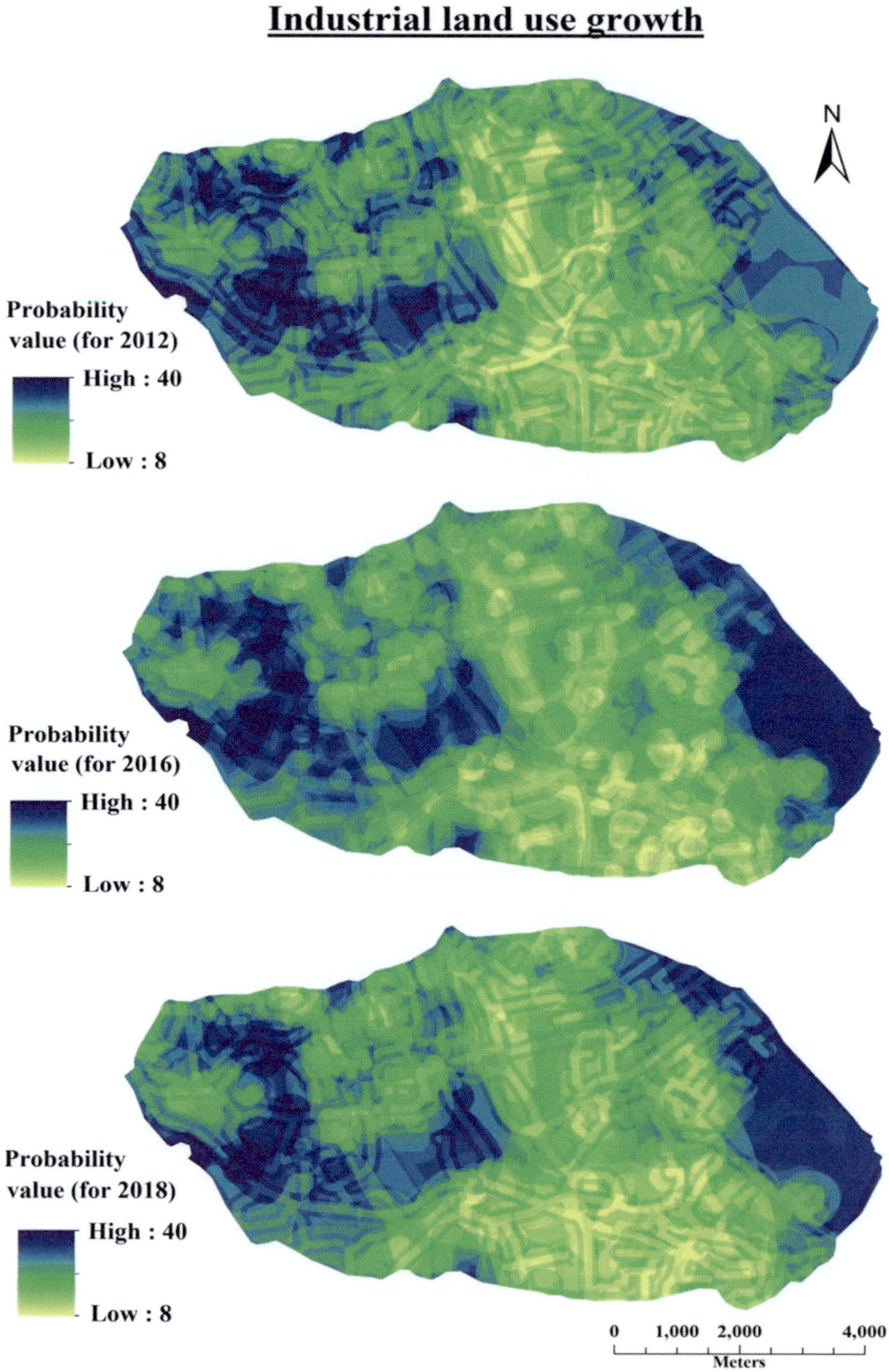

results for this time period on the basis of the most effective selected factors. Similar to previous time steps, all the WoE values were standardized in the range of 1 as minimum and 5 as maximum probability of growth values.

Figure 10.13 depicts the probability maps for residential, commercial, and industrial land-use growth for 2026. Similar to previous probability maps, residential growth has a wider extension in central parts, and commercial growth is

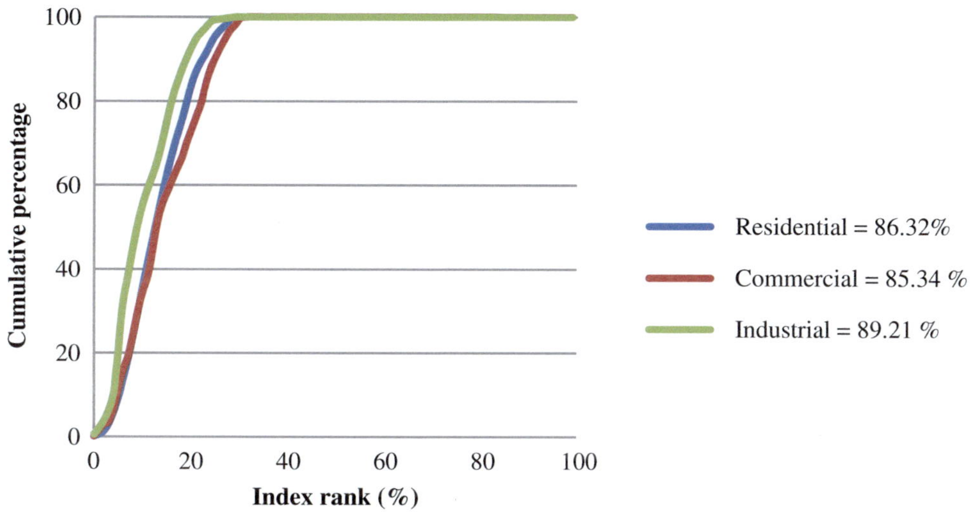

Fig. 10.11 AUC validation for growth probability map of 2012

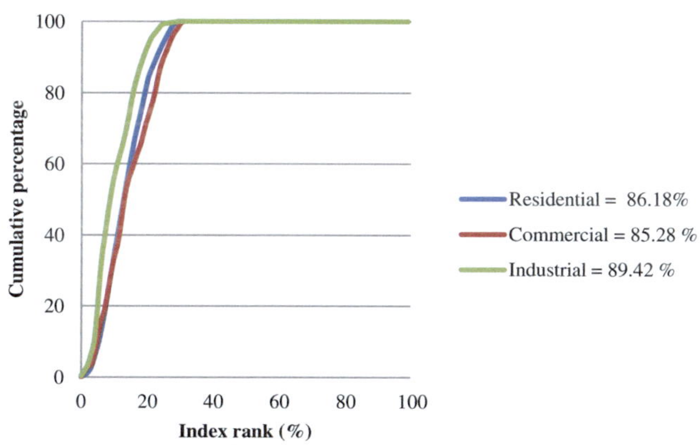

Fig. 10.12 AUC validation for growth probability map of 2016

mainly along the main roads. Furthermore, industrial land use has higher probability of growth in the eastern and western parts far from the central business districts of Kajang City.

10.6.2 Projected Land-Use Maps Using CA_WoE (Business-as-Usual Scenario)

Similar to WoE, the MC model evaluated the probability of changes among various land-use types, but without considering any driving forces. MC predicted land-use changes based on historical information and stochastic concept. This model produced two matrices for each time period: transitional area matrix, which shows the number of pixels expected to change to other land-use category; and transitional probability matrix, which shows the probability of changes to other land-use category. Transitional probability matrices were calculated using cross-tabulation of two land-use maps adjusted by proportional error. Transitional area matrices were calculated by multiplying each column in transitional probability matrices by the number of cells of the corresponding land-use in later land-use maps. Tables 10.13, 10.14, 10.15, 10.16, 10.17 and 10.18 present these matrices for the time periods 2004–2008, 2008–2012 and 2012–2015, respectively. Similar to cross-tabulation matrices, these matrices have rows representing the earlier land-use maps and columns representing later land-use maps. Next, this process was conducted to produce the transitional probability and area matrices for 2026 using the time period of 2004–2015 (Tables 10.19 and 10.20).

The MC model produced valuable quantitative information regarding the future changes of the study area; however, in the case of graphical illustration, the output maps failed to present a proper spatial location of land-use changes because of the lack of spatial dependency of this model. Hence, the produced matrices were used for further modeling process by integrating CA and WoE. Accordingly, land-use change occurrence was based on related evidences (the effects of selected factors and neighborhood conditions) and not entirely random. The spatial dependency and cellular basis of CA modeling and factor analysis of WoE provided a strong and reliable methodological approach to project future land-use growth and changes.

To run the CA_WoE integration approach, all the land-use growth probability maps produced from WoE modeling and other land-use type maps were merged and inserted into the CA modeling. This process calibrated the CA by integrating the influences of the driving forces to land-use change processing. CA_WoE was run three times: (1) for the 2004–2008 time step to project the land-use map for 2012 and validate the actual land-use map for 2012; (2) for the 2008–2012 time step to project the land-use map for 2016 and validate the actual land-use map for 2015; and finally (3) for the 2004–2015 time step to project the

Table 10.12 Frequency ratio and weights-of-evidence calculation results for 2004–2015 time periods

No.	Factor	Class	Residential		Commercial		Industrial	
			FR	*C/S* (*C*) (WoE)	FR	*C/S* (*C*) (WoE)	FR	*C/S* (*C*) (WoE)
1	Proximity to Housing	Near	**2.05**	**2009.06**	**0.61**	**–354.62**	**0.06**	**–1086.60**
		Middle	**0.59**	**–1036.66**	**1.57**	**542.88**	**0.97**	**–52.28**
		Far	**0.40**	**–1429.68**	**0.77**	**–211.37**	**1.97**	**1436.76**
2	Proximity to Commercial	Near	**1.17**	**372.52**	**2.04**	**872.15**	**0.53**	**–740.06**
		Middle	**1.11**	**260.46**	**0.28**	**–627.72**	**1.05**	**88.99**
		Far	**0.73**	**–661.25**	**0.75**	**–238.93**	**1.39**	**631.39**
3	Proximity to industrial	Near	**0.67**	**–790.82**	**0.81**	**–177.26**	**2.63**	**2088.13**
		Middle	**1.26**	**595.05**	**1.49**	**457.79**	**0.08**	**–1150.42**
		Far	**1.06**	**129.53**	**0.69**	**–290.96**	**0.36**	**–1021.50**
4	Proximity to road network	Near	**1.01**	**14.11**	**1.55**	**493.91**	**0.86**	**–233.38**
		Middle	**1.19**	**440.13**	**0.99**	**–9.15**	**1.06**	**92.00**
		Far	**0.80**	**–482.08**	**0.47**	**–478.79**	**1.08**	**138.32**
5	Proximity to facility	Near	**1.24**	**517.97**	**1.35**	**317.80**	**0.27**	**–1099.45**
		Middle	**1.11**	**256.62**	**0.81**	**–181.89**	**0.89**	**–179.13**
		Far	**0.66**	**–821.74**	**0.86**	**–135.44**	**1.81**	**1264.27**
6	Proximity to public transportation	Near	**1.13**	**289.05**	**1.38**	**343.99**	**0.62**	**–618.68**
		Middle	**1.10**	**228.93**	**0.86**	**–134.61**	**0.71**	**–475.07**
		Far	**0.78**	**–538.17**	**0.77**	**–211.63**	**1.66**	**1042.95**
7	Proximity to infrastructure	Near	**1.06**	**134.46**	**0.96**	**–39.45**	**0.97**	**–48.00**
		Middle	**1.11**	**241.13**	**1.06**	**54.02**	**0.97**	**–53.83**
		Far	**0.84**	**–386.07**	**0.98**	**–14.63**	**1.06**	**101.01**
8	Proximity to agriculture	Near	**0.78**	**–522.47**	**0.60**	**–363.93**	**1.32**	**515.83**
		Middle	**0.96**	**–104.69**	**0.93**	**–63.28**	**1.28**	**452.98**
		Far	**1.26**	**586.75**	**1.46**	**423.28**	**0.40**	**–957.03**

Bold letters indicate the important factors

land-use map for 2026. For each time step, the matrices produced from the MC model were also integrated to control the amount of changes and use the probability of changes from one land-use category to the other. Accordingly, CA reweighted each land-use growth map in each pass as a result of the contiguity filter on each current land-use type. Once reweighted, the revised suitability map was then run through the model to allocate 1/11 (1/11 for 2004–2015; 1/4 for 2004–2008, and 1/4 for 2008–2012) of the required land in the first run, 2/11 in the second run, and so on, until the full allocation of land for each land-use class is achieved. At the end of each run, all land-use types are masked, and the contiguity filter runs for them. Then, this result was multiplied to each land-use growth map to create input for the new run. Notably, the transitional area matrix created from the MC model aims to control how much land can be allocated to each land-use type over the future land-use maps. Figures 10.14 and 10.15 respectively illustrate the projected maps of years 2012 and 2016 using the CA_WoE integration approach.

The probability maps for various land-use types helped CA to define the transitional rules for land-use changes. Clearly, proximity to the same land-use types controlled the model more significantly than the other factors. This strong effect was expected because of the neighborhood effect of the CA model and the proximity factors from the growth probability maps. The growth of industrial buildings through agricultural fields and open spaces existing in their neighborhood can be observed in the central west areas from 2004 onwards. Residential land use also faced the same conditions. All the new residential areas are around or are linked to previous residential land uses. A considerable conversion of agricultural and open spaces to residential use can be observed in the entire study area. However, in the case of commercial land use, proximity to road has a higher influence than the proximity to existing commercial areas. Hence, in the case of land-use categories with small proportion, other parameters seem to play more effectively than the proximity to the same land-use type. In fact, open spaces with potential to change and located near the main roads and

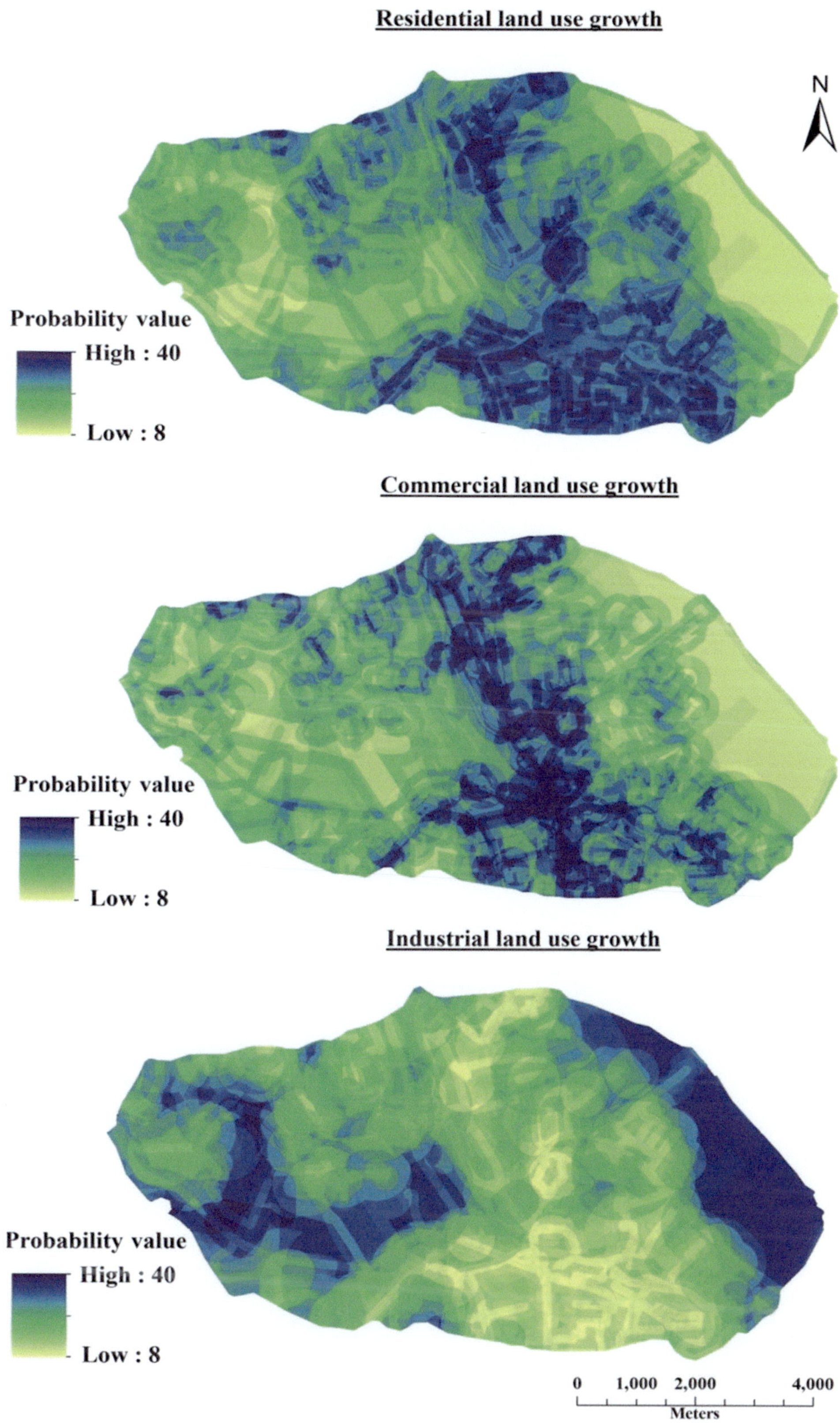

Fig. 10.13 Projected probability of growth maps for year 2026

Table 10.13 Transitional area matrix for 2004–2008 (m^2)

Class	1	2	3	4	5	6	7	8	9
1	**88,093**	355	90,381	8593	436	750	223	15,137	0
2	0	**37,285**	1147	149	0	0	0	5651	0
3	31,132	6196	**346,764**	108,075	13,100	8365	5016	52,875	5
4	43	4	85,561	**416,768**	0	0	4	14,771	0
5	21,447	13	3101	102	**162,669**	64	0	6521	0
6	12	0	3592	1509	4	**33,604**	0	3178	0
7	170	0	19,578	1683	0	11	**132,819**	4452	0
8	7096	4408	31947	33,041	1000	2001	563	**437,241**	31
9	0	0	0	0	0	0	0	1998	**11,268**

Note 1 = Agriculture, 2 = Commercial, 3 = Open spaces, 4 = Housing, 5 = Industry, 6 = Infrastructure, 7 = Community facility, 8 = Road network and 9 = Water body
Bold letters indicate the important factors

Table 10.14 Transitional probability matrix for 2004–2008

Class	1	2	3	4	5	6	7	8	9
1	**0.4319**	0.0017	0.4431	0.0421	0.0021	0.0037	0.0011	0.0742	0
2	0	**0.8429**	0.0259	0.0034	0	0	0	0.1278	0
3	0.0545	0.0108	**0.6067**	0.1891	0.0229	0.0146	0.0088	0.0925	0
4	0.0001	0	0.1654	**0.8059**	0	0	0	0.0286	0
5	0.1106	0.0001	0.016	0.0005	**0.8389**	0.0003	0	0.0336	0
6	0.0003	0	0.0857	0.036	0.0001	**0.802**	0	0.0759	0
7	0.0011	0	0.1234	0.0106	0	0.0001	**0.8368**	0.0281	0
8	0.0137	0.0085	0.0618	0.0639	0.0019	0.0039	0.0011	**0.8452**	0.0001
9	0	0	0	0	0	0	0	0.1506	**0.8494**

Bold letters indicate the important factors

Table 10.15 Transitional area matrix for 2008–2012 (m^2)

Class	1	2	3	4	5	6	7	8	9
1	**118,478**	2335	24,369	33,687	18,507	3716	370	23,089	218
2	264	**35,453**	3994	9260	4422	759	1573	13,948	0
3	49,929	14,310	**105,641**	66,146	28,543	7646	13,363	31,989	526
4	16727	21,524	58,263	**413,686**	11,645	10,725	9867	47,102	914
5	6855	2475	41,115	7317	**145,648**	2251	584	17,961	0
6	256	251	11,815	1098	869	**42,838**	361	3590	0
7	1376	10,491	10,544	7382	0	344	**146,349**	4154	556
8	8071	7623	34,693	50,042	17,723	9922	4591	**436,751**	6109
9	0	0	1284	89	0	0	0	1284	**14,609**

Bold letters indicate the important factors

Table 10.16 Transitional probability matrix for 2008–2012

Class	1	2	3	4	5	6	7	8	9
1	**0.5271**	0.0104	0.1084	0.1499	0.0823	0.0165	0.0016	0.1027	0.001
2	0.0038	**0.5088**	0.0573	0.1329	0.0635	0.0109	0.0226	0.2002	0
3	0.157	0.045	**0.3321**	0.2079	0.0897	0.024	0.042	0.1006	0.0017
4	0.0283	0.0365	0.0987	**0.7006**	0.0197	0.0182	0.0167	0.0798	0.0015
5	0.0306	0.011	0.1834	0.0326	**0.6496**	0.01	0.0026	0.0801	0
6	0.0042	0.0041	0.1934	0.018	0.0142	**0.7014**	0.0059	0.0588	0
7	0.0076	0.0579	0.0582	0.0407	0	0.0019	**0.8077**	0.0229	0.0031
8	0.014	0.0132	0.0603	0.0869	0.0308	0.0172	0.008	**0.7589**	0.0106
9	0	0	0.0744	0.0051	0	0	0	0.0744	**0.8462**

Bold letters indicate the important factors

Table 10.17 Transitional area matrix for 2012–2015 (m^2)

Class	1	2	3	4	5	6	7	8	9
1	**107,189**	0	38,012	4767	18,393	0	0	0	0
2	1317	**59,685**	1317	1317	1317	1317	1317	1317	1317
3	0	1934	**268,086**	65,117	0	167	0	36	0
4	0	0	91,979	**521,183**	0	0	0	0	0
5	0	0	37,012	0	**203,056**	0	0	0	0
6	1146	**1146**	1146	1146	1146	**51,956**	1146	1146	1146
7	3397	3397	3397	3397	3397	3397	**154,017**	3397	3397
8	0	0	69,067	17,267	0	0	0	**489,199**	0
9	324	324	324	324	324	324	324	324	**14,675**

Bold letters indicate the important factors

Table 10.18 Transitional probability matrix for 2012–2015

Class	1	2	3	4	5	6	7	8	9
1	**0.6367**	0	0.2258	0.0283	0.1092	0	0	0	0
2	0.0187	**0.85**	0.0187	0.0187	0.0187	0.0187	0.0187	0.0187	0.0187
3	0	0.0058	**0.7994**	0.1942	0	0.0005	0	0.0001	0
4	0	0	0.15	**0.85**	0	0	0	0	0
5	0	0	0.1542	0	**0.8458**	0	0	0	0
6	0.0187	0.0187	0.0187	0.0187	0.0187	**0.85**	0.0187	0.0187	0.0187
7	0.0187	0.0187	0.0187	0.0187	0.0187	0.0187	**0.85**	0.0187	0.0187
8	0	0	0.12	0.03	0	0	0	**0.85**	0
9	0.0187	0.0187	0.0187	0.0187	0.0187	0.0187	0.0187	0.0187	**0.85**

Bold letters indicate the important factors

Table 10.19 Transitional area matrix for 2004–2015 (m^2)

Class	1	2	3	4	5	6	7	8	9
1	**46,182**	945	31,840	28,944	28,141	3659	1360	27,089	81
2	298	**35,322**	3143	10,337	4993	313	1250	14,561	0
3	28,159	19,125	**110,979**	91,275	18,247	7666	14,261	45,110	496
4	10,546	26,689	42,217	**448,168**	6099	12,668	12,509	52,760	1459
5	7190	2004	42,018	8137	**159,345**	2119	673	18,581	0
6	361	164	4370	1211	1391	**46,932**	585	6109	0
7	1312	9602	9726	8323	0	341	**145,647**	5721	525
8	6220	4966	19,088	57,760	19,664	8800	5599	**445,501**	7857
9	0	0	1073	80	0	0	0	1511	**14,601**

Bold letters indicate the important factors

Table 10.20 Transitional probability matrix for 2004–2015

Class	1	2	3	4	5	6	7	8	9
1	**0.2745**	0.0056	0.1893	0.172	0.1673	0.0218	0.0081	0.161	0.0005
2	0.0042	**0.503**	0.0448	0.1472	0.0711	0.0045	0.0178	0.2074	0
3	0.084	0.057	**0.331**	0.2722	0.0544	0.0229	0.0425	0.1345	0.0015
4	0.0172	0.0435	0.0689	**0.731**	0.0099	0.0207	0.0204	0.0861	0.0024
5	0.03	0.0083	0.175	0.0339	**0.6637**	0.0088	0.0028	0.0774	0
6	0.0059	0.0027	0.0715	0.0198	0.0228	**0.7678**	0.0096	0.1	0
7	0.0072	0.053	0.0537	0.0459	0	0.0019	**0.8038**	0.0316	0.0029
8	0.0108	0.0086	0.0332	0.1004	0.0342	0.0153	0.0097	**0.7742**	0.0137
9	0	0	0.0622	0.0046	0	0	0	0.0875	**0.8457**

Bold letters indicate the important factors

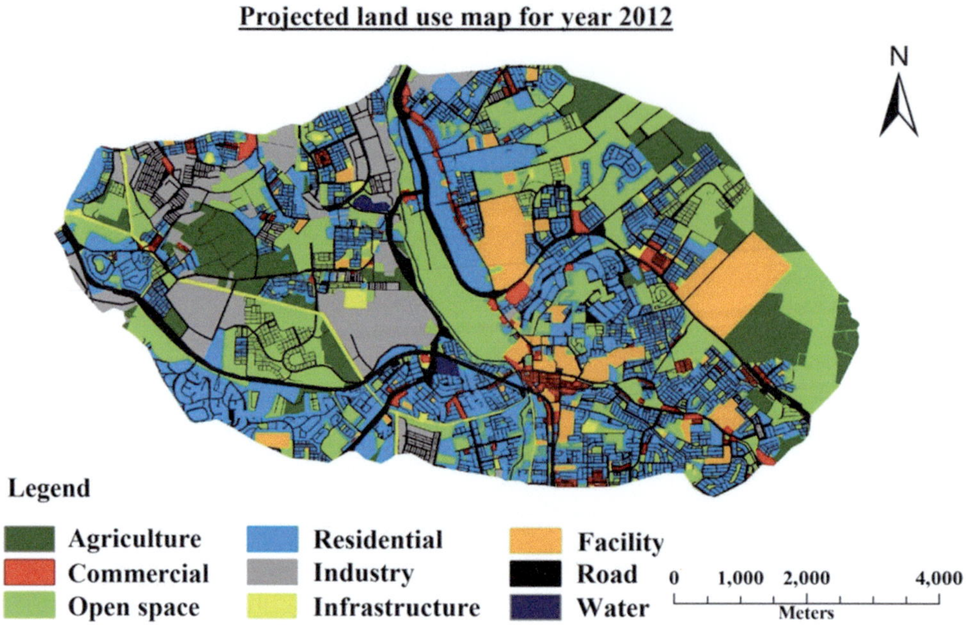

Fig. 10.14 Projected land use map for year 2012 using CA_WoE aggregation approach

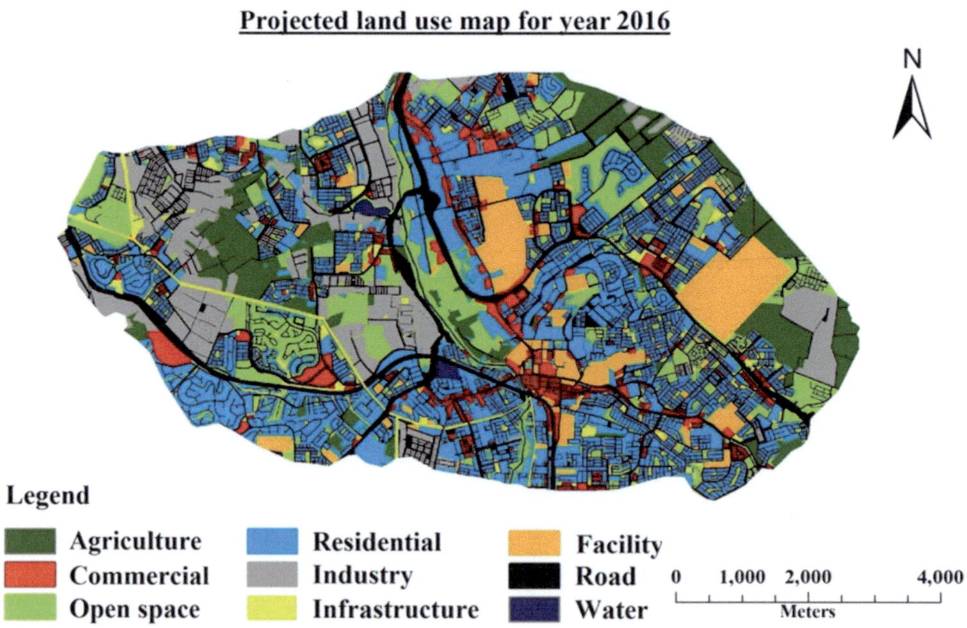

Fig. 10.15 Projected land use map for year 2016 using CA_WoE aggregation approach

public transportation stations are the main targets for growth of residential and commercial land-use types. Hence, apart from the proximity to the same land-use types, accessibility either by road network or public transportation is the controlling factor in land-use growth and changes. Another possible significant change is the conversion of agricultural land use into open spaces in the entire city, especially in agricultural environments located in the eastern regions. This conversion is mainly a land clearance process required for any type of new developments and/or different types of phonological and maintenance processes of agricultural fields. From the output maps, the main flow of changes in this study area can be observed as consisting of the conversion of agricultural fields into open spaces, and from open spaces into other land-use categories, such as residential, commercial, industrial, and transportation.

In the projected map of 2016 (Fig. 10.15), the growth of commercial area along the road network is more visible than that of the previous year.

Additionally, the growth of industrial areas through agriculture fields in central west can be clearly observed. Both projected maps were validated with the actual land-use maps of 2012 and 2015. This process was conducted using a contingency table based on two-by-two comparisons of projected and actual land-use maps (Tables 10.21 and 10.22). The diagonal values of these two tables are the areas modeled to change to one specific land-use type and are actually changed to that land-use type. Hence, the accuracies of both maps were calculated by the sum of diagonal values divided by the total area of Kajang City.

$$\text{Correct proportion for 2012 modelled map} = \frac{\mathbf{3907}}{\mathbf{5654}} = 70\%$$

$$\text{Correct proportion for 2016 modelled map} = \frac{\mathbf{5165}}{\mathbf{5654}} = 91\%$$

The projected map of the second time period (2008–2012) produced better results than that of the first time period (2004–2008). As shown in Tables 10.21 and 10.22, the last columns show the total areas of each land-use type in projected maps. The last rows of tables also present the total areas of each land-use type in actual land-use maps. These columns and rows are extracted from each map and presented in a separate table for a better view and judgment regarding these projected maps (Table 10.23).

The following differences were extracted from Table 10.23 between projected and actual land-use maps of year 2012:

- Open spaces changed significantly more into other land-use types than predicted;
- Agricultural loss was less than predicted;
- Industrial and commercial land-use growths were more than predicted; and

Table 10.21 Contingency table between projected land use map and actual land use map of 2012 (Hectare)

Land use		Actual (2012)									Total
		1	2	3	4	5	6	7	8	9	
		Agric.	Comm.	Open	House	Industry	Infra	Facility	Road	Water	
Model (2012)	1	**302.795**	4.2675	43.315	57.41	31.17	6.625	0.7	40.85	0.1325	487.265
	2	1.9475	**68.3175**	9.9275	13.9075	5.9775	1.0475	3.4625	22.675	0.09	127.3525
	3	184.0075	51.335	**499.6375**	236.7775	105.555	29.5125	46.1325	123.1125	1.8575	1277.928
	4	46.1925	32.215	119.8025	**1089.958**	19.73	18.5725	23.16	136.4875	1.8825	1488
	5	16.9825	3.9875	67.045	12.3825	**378.01**	3.4975	0.8575	29.0425	0.0775	511.8825
	6	0.4	0.2375	11.7125	1.06	0.9925	**82.445**	0.29	3.73	0	100.8675
	7	0.7725	5.79	5.915	3.825	0	0.28	**373.1475**	2.605	0.32	392.655
	8	8.68	7.9975	37.4225	60.465	18.9725	10.64	5.2225	**1079.855**	6.13	1235.385
	9	0	0	0.3825	0.005	0	0	0	0.075	**32.6725**	33.135
Total		561.7775	174.1475	795.16	1475.79	560.4075	152.62	452.9725	1438.433	43.1625	5654.47

Note 1 = Agriculture, 2 = Commercial, 3 = Open spaces, 4 = Housing, 5 = Industry, 6 = Infrastructure, 7 = Community facility, 8 = Road network and 9 = Water body
Bold letters indicate the important factors

Table 10.22 Contingency table between projected land use map of 2016 and actual land use map of 2015 (Hectare)

Land use		Actual (2015)									Total
		1	2	3	4	5	6	7	8	9	
		Agric.	Comm.	Open	House	Industry	Infra	Facility	Trans	Water	
Model (2016)	1	**347.02**	0.015	79.8975	9.6225	34.295	0	0	0.1525	0.0375	471.04
	2	4.9825	**167.9725**	28.26	30.265	1.6075	0.1575	23.3125	4.845	0	261.4025
	3	0.105	1.36	**648.135**	43.455	0	0.1175	0	0.0975	0.04	693.31
	4	45.8925	0.045	58.8425	**1434.345**	2.995	0.685	4.625	36.4625	0.055	1583.948
	5	22.8875	6.0875	23.1825	14.1075	**561.1875**	0.5425	0	11.1	0	639.095
	6	0.0025	0	0	0	0	**151.305**	0	0	0	151.3075
	7	0.0125	0	0	0	0	0	**425.055**	0	0.0025	425.07
	8	0	0.055	0.03	0.9625	0.085	0	0	**1386.108**	0	1387.24
	9	0	0.01	0	0.1475	0	0	0	0.0675	**43.0275**	43.2525
Total		420.9025	175.545	838.3475	1532.905	600.17	152.8075	452.9925	1438.833	43.1625	5655.665

Bold letters indicate the important factors

Table 10.23 The comparison of land use areas for actual and projected land use maps (Hectare)

Land use type	Projected map of 2012	Actual map 2012	Percentage of Error	Projected map of 2016	Actual map of 2015	Percentage of error
Agriculture	486.49938	562.1719	13.46	480.83	421.0981	14.18
Comm.	125.98022	174.41	27.77	257.5394	175.7851	46.51
Open	1283.1347	795.1358	61.37	704.7783	838.2406	15.92
Residential	1471.5896	1475.721	0.28	1560.539	1532.784	1.81
Industry	506.39061	560.5319	9.66	629.6502	600.2177	4.90
Infra	99.760198	152.8637	34.74	153.5961	152.981	0.40
Facility	390.98393	453.0455	13.70	421.601	453.0455	6.94
Transport	1259.2806	1438.603	12.47	1400.83	1438.612	2.63
Water	32.766378	43.2858	24.30	46.95813	43.2858	8.48
Sum	5656.39	5655.77		5656.32	5656.05	

Table 10.24 Kappa statistic index of agreement for validation of probability maps of 2012 and 2016

Kappa measurements		2012	2016
Kappa for no information	K_{no}	0.7496	0.9203
Kappa for location	$K_{location}$	0.7875	0.9487
Kappa standard	$K_{standard}$	0.7264	0.9129

- Infrastructures, facilities, and transportation had grown more than predicted.

These results mainly explain the lower accuracy for 2012 land-use prediction. Despite the one-year difference between the projected map 2016 and the actual land use 2015, this time period contrastingly has a higher accuracy for the projected map.

Apart from the contingency tables, projected maps were validated using the Kappa statistic index of agreement validation (Table 10.24). Similarly, from this assessment, the fitness of the projected probability map of 2016 has a higher accuracy with the actual land-use map 2015 than the projected probability map of 2012 with the actual land-use map 2012.

The growths and changes in other land-use types (other than residential, commercial, and industrial) seemed to be higher during 2004–2008 than in 2008–2012 and onwards. These growths were likewise more complex and dependent on an additional number of factors. However, after the optimization process in factor analysis, a higher accuracy could be achieved for the next time period (2008–2012 for the 2016 projection).

After confirmation regarding the model performance accuracy, the process was conducted again for the longer time period, that is, 2004–2015, to project the land-use map for 2026. Figure 10.16 depicts the projected map for 2026 based on the business-as-usual scenario. A significant growth of commercial land use along the main roads, especially near the train station (central areas of Kajang City) can be clearly seen. Industrial land use has grown in the vicinity of previous industrial buildings. The loss of agricultural fields in central west areas caused by the growth of industrial land use can be observed. In the north east of the study area, a gradual growth of industrial buildings is also noted. The same condition occurred for residential land use, which has growth near the existing residential area in the entire city.

The density and land-use diversity of the entire study area are increased, especially in the central regions. However, several open spaces still exist within the city. Confirming that these areas are abandoned land can enable a proper BR process to increase the compactness of the projected map.

The compactness of the projected map was evaluated using a similar DoC assessment based on urban density, intensity, and land-use diversity. Figures 10.17 and 10.18 depict the three compactness indicators and overall compactness maps of the projected land-use map of year 2026 respectively. The growth in urban density in the east and central west can be observed with respect to previous density maps. The intensity of the city center near the central business district has higher growth than the other regions; however, the growth through the eastern and western parts can also be noticed. Finally, land-use diversity of the study area also faced marginal growth during the 11 years' time period.

Visually, the overall compactness of the projected map is highly similar to the compactness map of 2012, with a fine

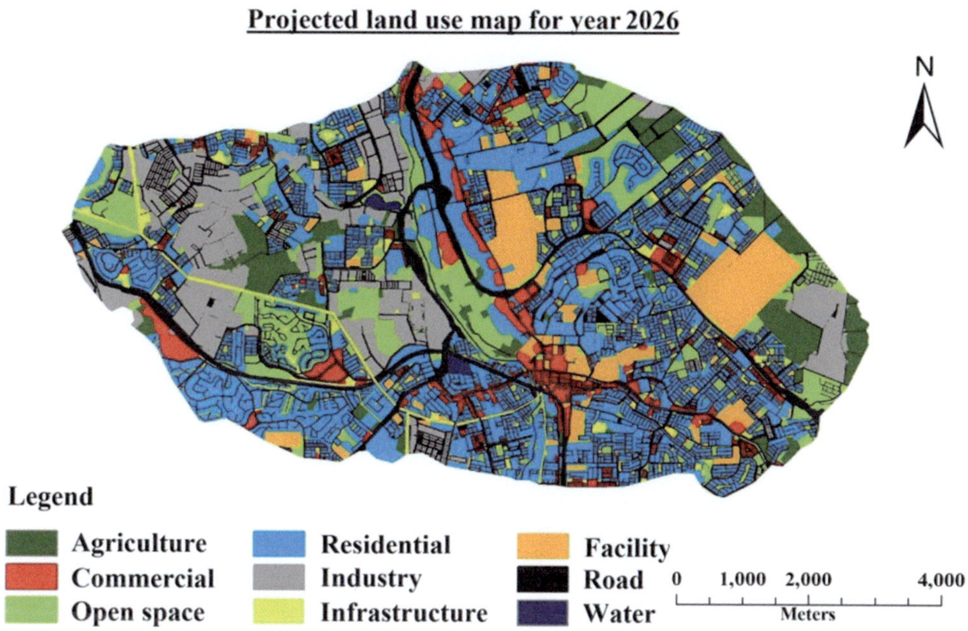

Fig. 10.16 Projected land use map for year 2026 using CA_WoE aggregation approach

distribution of compact centers in the entire city. However, a significant growth of compactness can be observed in the northern and southern regions, given that even small development changes (a single land-use change) can affect the compactness pattern of the neighborhood. For example, the conversion of a recreational field into residential land use in the residential neighborhood significantly reduces the DoC. Thus, expecting the growth of compactness by development growth during a specific time period is irrational. This finding can possibly explain the lower DoC in the northern east region of the study area in 2026 with respect to previous compactness maps. Figure 10.19 and Table 10.25 present the overall compactness of the projected map in quantitative perspectives. From Fig. 10.19, the amount of pixels with a high DoC value (near 50) can be seen to have increased to over 5% in 2026. Thus, a higher distribution of centers with high DoC value can be seen in this map. As shown in Table 10.25, the overall DoC value of the entire study area has a small reduction, which proved that the land-use growth of Kajang City requires an alternative development pattern.

10.6.3 Compact City Land Use Modeling Approach

In this study, the BR process was chosen as the most feasible and cost-effective approach to increase the DoC or compact pattern of Kajang City. First, existing BF sites were extracted from all open spaces to exclude the buffer zones of recreational and useful natural spaces. Figure 10.20 illustrates the spatial location of the existing BF sites and all other open spaces.

In general, small BF sites and open spaces are located in the south and south eastern parts near the dense residential and commercial areas, where the compactness pattern is higher than in the other regions. In contrast, BF sites with larger areas are located in the east, west, and central west near the industrial and agricultural fields or less compact regions.

Overlaying the compactness map of year 2026 with the BF sites of year 2015 reveals that, most of the BF sites are located in areas with lower DoC value (Fig. 10.21). Hence, a proper land-use type for these BF sites based on local neighborhood demands was proposed to increase the compactness pattern of the study area. Proper land-use categories were proposed for each BF site for land-use map 2015 through the analysis and evaluation of the proximity of existing various land-use types, requirements of different community facilities for each neighborhood, master plan, and probability of growth maps (produced from the WoE modeling). Next, using the new land-use map of 2015, the CA_WoE integration land-use modeling approach was run one more time to project a more compact land-use map. This study mainly aimed to evaluate the BR process as a city intensification approach, whether this process will increase the city compactness or not. Figure 10.22 illustrates the projected land-use map according to the proposed compact land-use modeling approach.

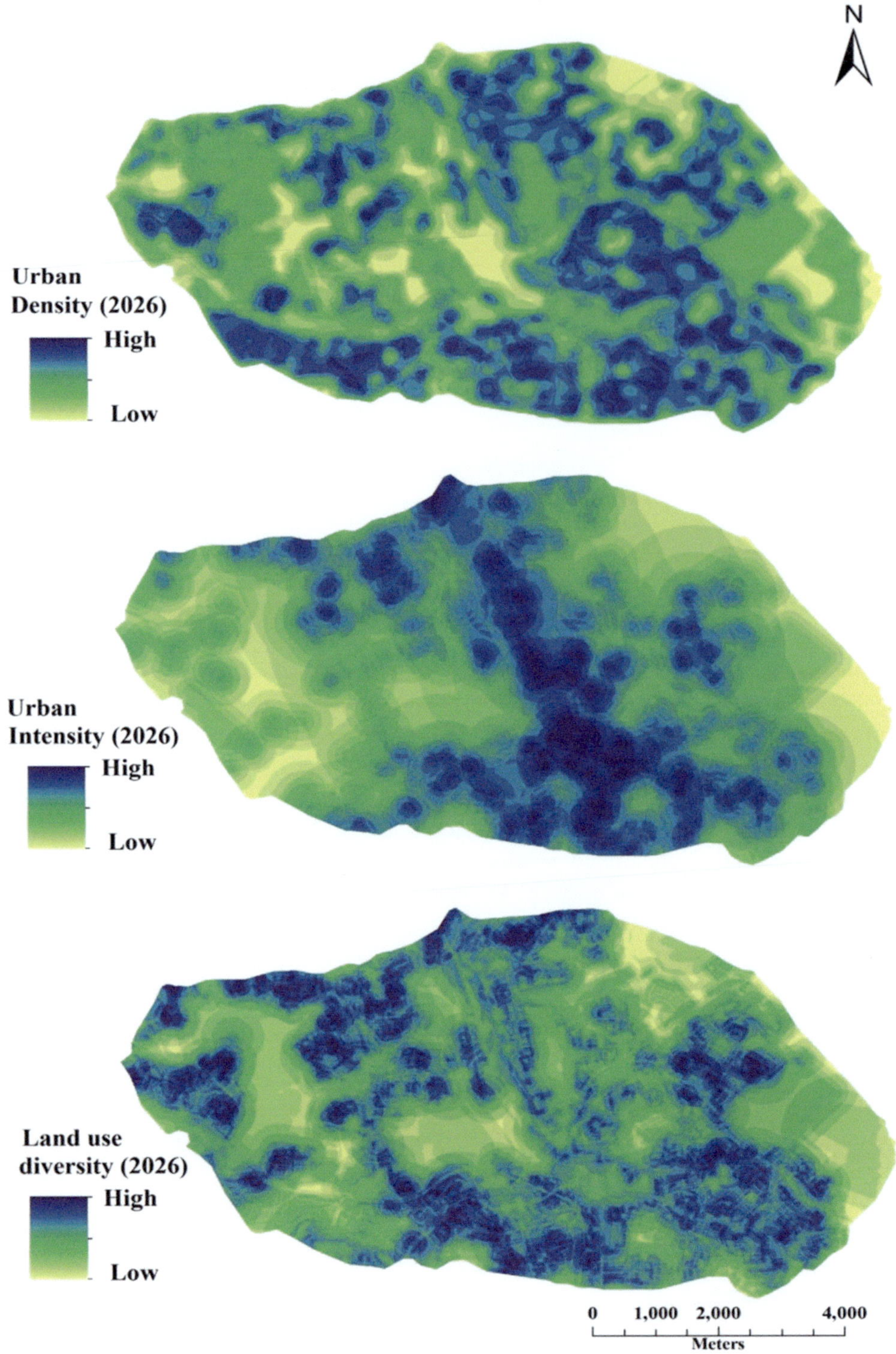

Fig. 10.17 City compactness indicators for projected map of year 2026 (business-as-usual scenario)

Hence, after running the land-use modeling process one more time, the projected map was assessed with respect to city compactness. Figure 10.23 depicts the overall compactness of the projected land-use map based on the compact land-use form scenario. At first glance, a slight difference can be noticed between this map and the projected map based on the business-as-usual scenario. However, through subtle investigation (especially quantitative assessment), the second scenario can be observed to have a higher degree of compactness. The degree of compactness of BF sites shown

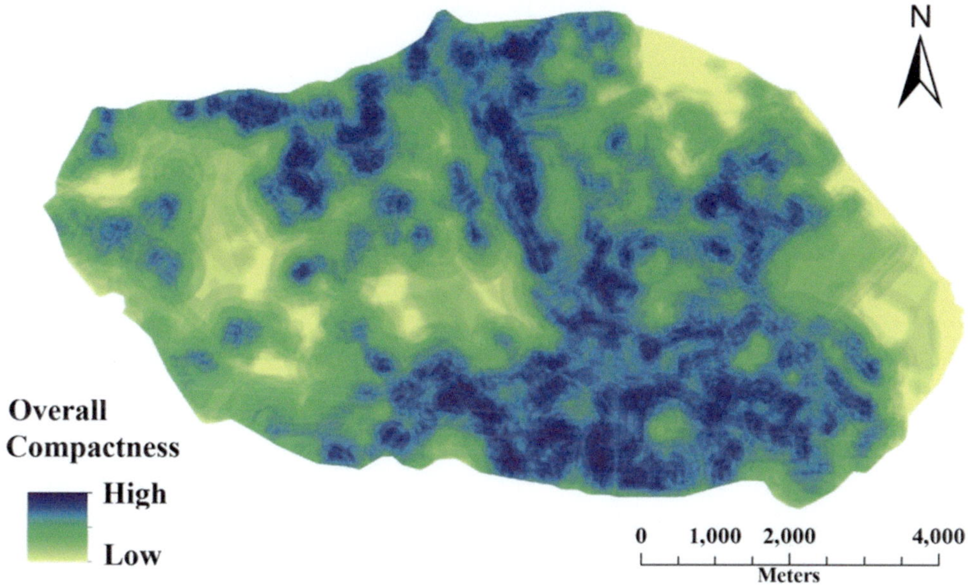

Fig. 10.18 Overall city compactness for projected map of year 2026 (business-as-usual scenario)

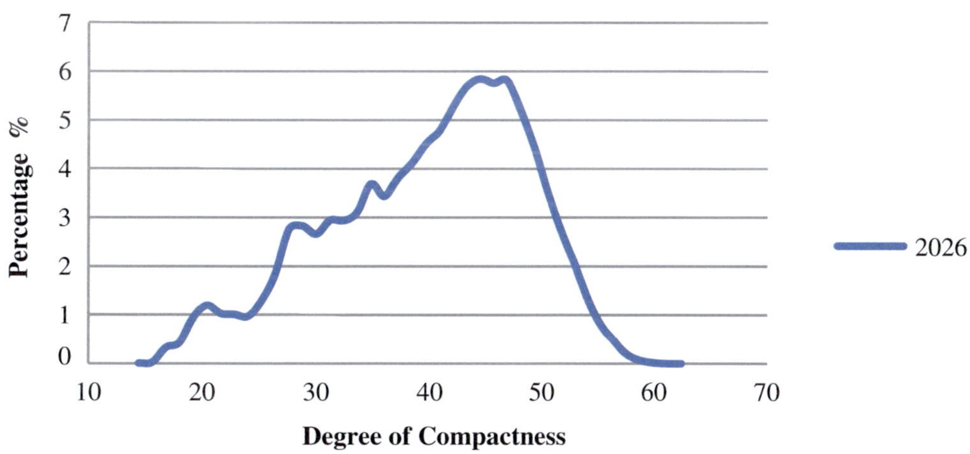

Fig. 10.19 Graphical presentation of compactness value of year 2026 (business as usual scenario) with respect to area percentage

Table 10.25 Quantitative assessment of overall compactness for year 2026 land use map

DoC	No. of cells	Percentage	Perc. × DoC
14	161	0.01	0.00
16	626	0.03	0.00
17	7493	0.33	0.06
18	9813	0.43	0.08
19	21,322	0.94	0.18
20	27,039	1.20	0.24
22	23,187	1.03	0.22
23	22,741	1.01	0.23
24	21,937	0.97	0.23
25	29,283	1.30	0.33
26	41,762	1.85	0.49
28	62,237	2.75	0.76
29	63,821	2.82	0.81

(continued)

Table 10.25 (continued)

DoC	No. of cells	Percentage	Perc. × DoC
30	60,079	2.66	0.80
31	66,485	2.94	0.92
32	66,317	2.94	0.95
34	70,258	3.11	1.04
35	83,315	3.69	1.28
36	77,510	3.43	1.23
37	86,281	3.82	1.42
38	93,074	4.12	1.58
40	102,028	4.52	1.79
41	108,145	4.79	1.95
42	119,198	5.28	2.22
43	128,531	5.69	2.46
44	132,013	5.84	2.59
46	130,036	5.76	2.62
47	131,198	5.81	2.72
48	117,306	5.19	2.49
49	100,731	4.46	2.19
50	80,516	3.56	1.80
52	62,429	2.76	1.43
53	47,582	2.11	1.11
54	30,701	1.36	0.73
55	18,029	0.80	0.44
56	10,536	0.47	0.26
58	3991	0.18	0.10
59	1485	0.07	0.04
60	256	0.01	0.01
61	6	0.00	0.00
62	10	0.00	0.00
Sum	2,259,468	100	39.82

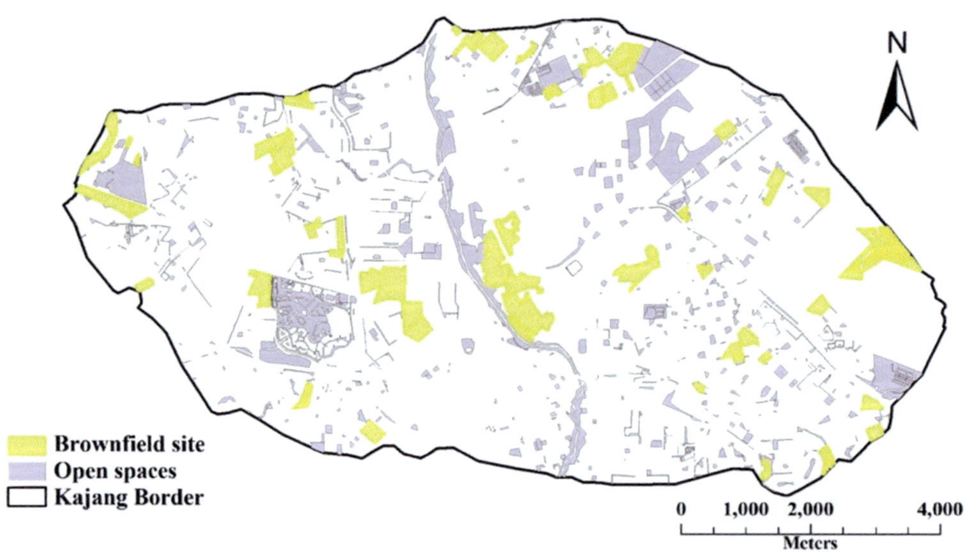

Fig. 10.20 Existing BF sites of Kajang City (2015)

Fig. 10.21 The spatial relation of BF sites with compactness map (projected map of year 2026)

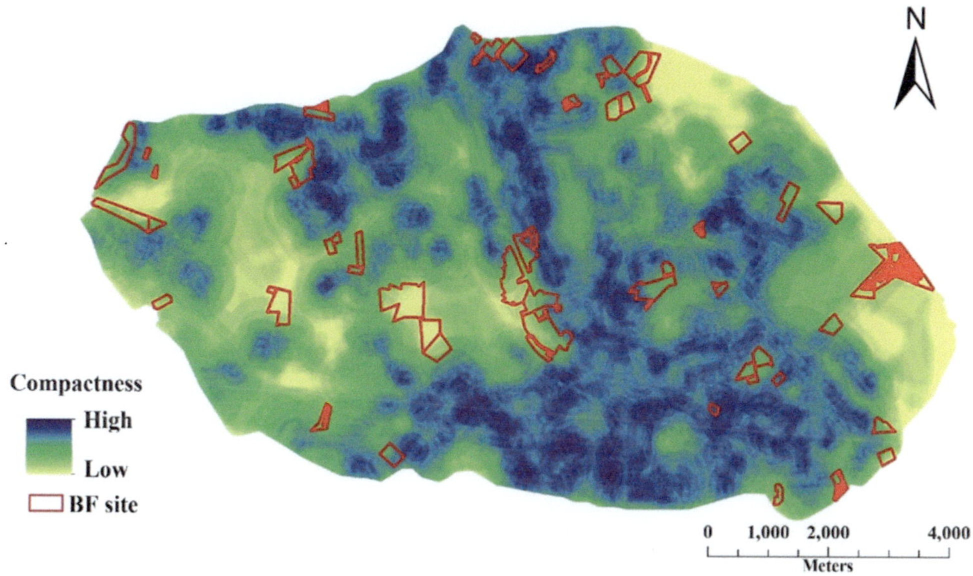

Fig. 10.22 Projected compact land use map for year 2026 using CA_WoE aggregation approach

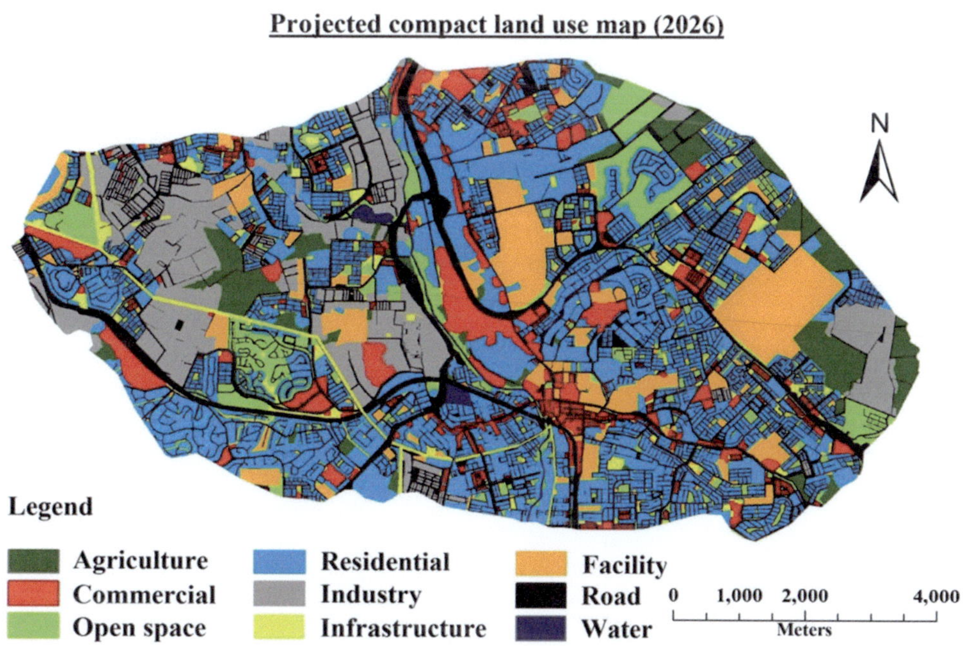

in Fig. 10.21 is also higher in this scenario, given that these areas now have higher building density (in some cases, residential density as well), higher urban intensity, and land-use diversity rather than the previous time. Figure 10.24 illustrates the percentage of area bearing various degrees of compactness value for both scenarios. Obviously, the compact land-use form scenario has a higher proportion of area, as it has a higher DoC value with respect to the business-as-usual scenario. Finally, Table 10.26 shows the quantitative assessments of city compactness for the second scenario. This scenario has a higher overall DoC value with respect to the business-as-usual scenario, and even higher than the other historical land-use maps (2004, 2008, 2012, and 2015). Hence, through a proper BR as a city intensification process, the compactness and, eventually, the urban sustainability of a region can be improved.

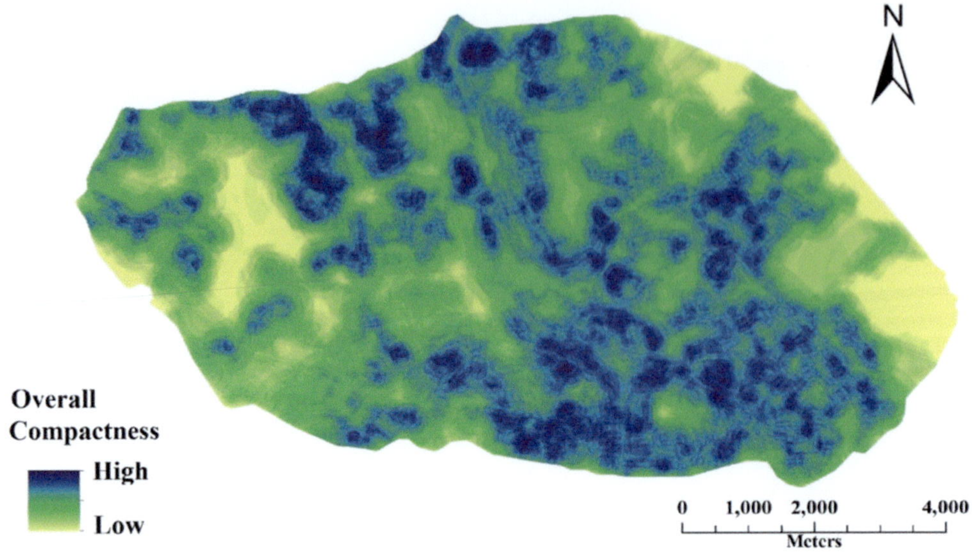

Fig. 10.23 Overall city compactness for projected map of year 2026 (compact land use form scenario)

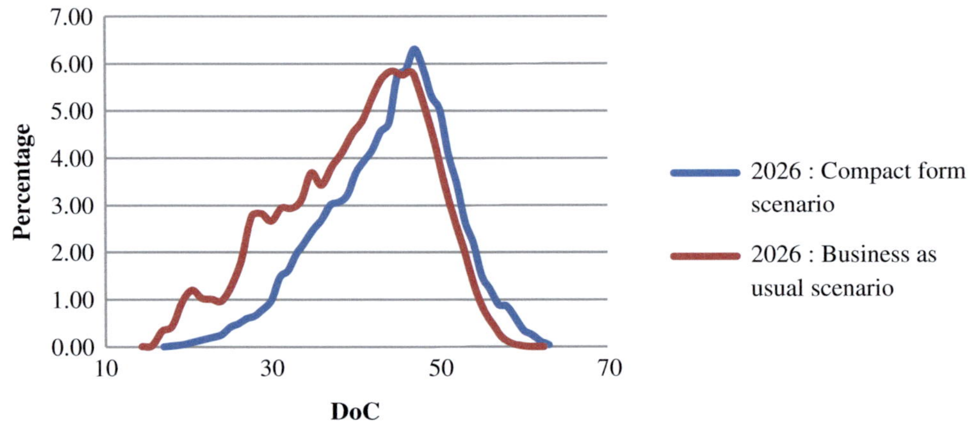

Fig. 10.24 Graphical presentation of compactness value of year 2026 (business-as-usual vs. compact land use map scenarios) with respect to area percentage

Table 10.26 Quantitative assessment of overall compactness for year 2026 land use map for compact land use map

DoC	No. of cells	Percentage	Perc. × DoC
17	43	0.00	0.00
18	494	0.02	0.00
19	905	0.04	0.01
20	1566	0.07	0.01
21	2879	0.13	0.03
22	3567	0.16	0.03
23	4607	0.20	0.05
24	5835	0.26	0.06
25	9358	0.41	0.10
26	11,060	0.49	0.13
27	13,521	0.60	0.16
28	14,825	0.66	0.18
29	18,121	0.80	0.23
30	22,352	0.99	0.30
31	33,243	1.47	0.46

(continued)

Table 10.26 (continued)

DoC	No. of cells	Percentage	Perc. × DoC
32	36,430	1.61	0.52
33	44,442	1.97	0.65
34	49,810	2.20	0.75
35	55,909	2.47	0.87
36	60,598	2.68	0.97
37	67,912	3.00	1.11
38	69,368	3.07	1.17
39	72,776	3.22	1.26
40	83,355	3.69	1.48
41	89,115	3.94	1.62
42	94,373	4.18	1.75
43	103,065	4.56	1.96
44	107,236	4.74	2.09
45	129,868	5.75	2.59
46	132,582	5.87	2.70
47	142,542	6.31	2.96
48	133,727	5.92	2.84
49	119,940	5.31	2.60
50	113,235	5.01	2.50
51	92,171	4.08	2.08
52	78,245	3.46	1.80
53	59,707	2.64	1.40
54	49,484	2.19	1.18
55	33,839	1.50	0.82
56	27,494	1.22	0.68
57	20,264	0.90	0.51
58	19,406	0.86	0.50
59	13,914	0.62	0.36
60	7917	0.35	0.21
61	5826	0.26	0.16
62	2647	0.12	0.07
63	838	0.04	0.02
Sum	2,260,411	100	43.92

10.7 Conclusion and Future Recommendations

In recent decades, one of the most important issues for urban planners and scientists is achieving the objectives of sustainable urban development. Various aspects of sustainable urban development environmental protection (especially agricultural and forest conservations) are dominant in tropical countries, such as Malaysia. Developing cities and towns (comprising huge agricultural and natural spaces with a high potential for growth due to their proximity to large metropolitan cities) require regulation to avoid large horizontal urban expansion and destruction of valuable natural and agricultural fields. Hence, proposing various development scenarios based on objectives of urban sustainability is highly important to avoid negative consequences of sprawl urban development. In this regard, one of the most promising and widely accepted solutions for this purpose is developing urban areas in a more compact form, with high density built-up, mixed land-use development, and intensified neighborhoods. Apart from the environmental perspectives, a compact city increases the quality of life of the local residents, as in the case of social perspectives.

This study utilized geospatial data within the GIS environment to analyze the urban growth process and its pattern with respect to the compact city paradigm. This study focused on the spatial pattern, growth, and changes of urban land-use types of Kajang City from 2004 to 2015, which can be useful in evaluating the city compactness of urban plans and urbanization policies for Kajang City. The analyses and modeling approaches used in this study can be employed to guide the identification and measurements of the changes and growths likely to happen in urban areas. The analysis produced several figures and tables to understand and assess urban land-use growths and changes in Kajang City. Results also confirmed that the proposed modeling approaches, geospatial data, statistical techniques, and GIS are highly practical in identifying future urban growth and land-use change patterns and their general trends.

This study specifically proposed a land-use change modeling approach by combining the WoE model with the CA framework to predict future spatial patterns. Using this innovative integration modeling approach enabled the simultaneous application of the following two aspects of urban land-use changes: self-organization as a consequence of neighborhood effects, and external driving forces affecting land-use changes. Integrating the surrounding environment effects using the CA model and external driving forces was considered by using the WoE model. This integration model provided detailed information on land-use change patterns, such as the effectiveness of various related factors or influence of various land-use categories.

Through validation techniques, the resultant maps from the model were verified to fall into reasonable accuracies. The output probability maps for the main land-use types (residential, commercial, and industrial) revealed that some land-use category growths are mainly affected by the proximity to the same land-use or the distance from other land-use categories. New residential buildings tend to be developed near the existing residential area and be far from existing industrial buildings. Similar conditions are generally happening for industrial land-use types. In contrast, the growth of other categories, such as commercial land use, is mainly affected by the accessibility of the sites. Hence, proximity to main roads and public transportation control the growth of such land-use types. Along with these neighborhood land-use effects, the linkages among different factors and each land-use types were also precisely investigated.

The WoE allowed us to identify the influence of spatial determinants on the analyzed transitions. As a regression analysis, WoE provided better explanatory power and outperformed some methods, such as ANNs. Furthermore, in the case of functional relationships among the dependent and independents variables, WoE provided clearer and more informative results. The direct or indirect relationships of land-use types were revealed with respect to selected factors, especially proximity- and ordinal-based factors. For example, residential growth is directly linked to the proximity to community facilities. Higher proximity to facilities increases the growth of residential land use. In contrast, residential growth has an inverse relation to agricultural fields. Higher proximity to agricultural environment reduces the probability of residential growth. Hence, implementing the WoE analysis on all factors provides a clearer view regarding the spatial influence of these factors with respect to growth and changes of the different land-use types.

Apart from WoE, utilizing the MC model controlled how much land-use types can be allocated to other types over future land-use maps based on a historical trend. The growth of other land-use types (with their probability of growth not analyzed using WoE, such as infrastructure, facilities, etc.) was high during the initial years (2004–2008). When urban developments of the study area were saturated, these land-use growths were also significantly reduced after 2008. Thus, an insignificant growth of these land-use categories was observed in the final projected map.

As a second scenario, a compact land-use pattern modeling approach was proposed by combining the land-use modeling process used in the previous objective and the concept of urban intensification. This task was done to increase the city compactness level of the study area by redeveloping the existing BF sites. The DoC and ToC evaluations indicated the compactness characteristics of the various parts of Kajang City. The local planning authority can evaluate the study area using these compactness maps and propose different scenarios and solutions to increase the compactness pattern of the various parts of the city. Redeveloping and revitalizing the existing BF sites are considered the most time- and cost-effective approach to deal with this issue. In fact, developing an integrated modeling approach for compact land-use modeling aimed to evaluate the BR process, whether or not it increases the city compactness of local neighborhoods, is an effective task. Regardless of the results, considering that even a negligible development change (single parcel land-use change) can affect the compactness pattern of the neighborhood is extremely important. For example, although converting open spaces into residential land use within a residential neighborhood increases the urban density, it generally reduces the degree of compactness significantly. This finding is attributed to the decrease in land-use diversity and urban intensity. Hence, expecting an increase in compactness level by any type of development implementation is illogical. Furthermore, an extremely high level of compactness pattern means a region without any agricultural and forest areas with several negative impacts of extremely high urban density. Hence, a DoC value of over 50% of the total range is an acceptable pattern of urban development, which can be gradually improved by various proper development scenarios based on urban sustainability perspectives.

Some limitations remain within the analyzed and proposed approaches, mainly relying on the physical aspect of land-use categories and their interaction among one another and the external driving forces. Although the master plan of the study area was utilized in the prediction modeling stages, political issues are important aspects of urban growth and the changes mainly proposed and implemented by the government. Hence, refining these modeling approaches and implementing the assessment on the basis of the updated strategies and policies could extend the applicability and reliability of these approaches to achieve sustainable urban development. This study analyzed the probability of growth of the three main land-use types (residential, commercial, and industrial), which are commonly developed by private agencies. However, considering the growth of all land-use types (road network, infrastructures, and community facilities) improves the reliability of the projected maps and increases the accuracy of the modeling approaches.

Although these challenges exist, this study has shown promising approaches for city compactness assessment, namely, business-as-usual and compact land-use pattern modeling for researchers aiming to study compact urban development as an effective task to achieve sustainable urban development.

References

Abdullahi, S., & Pradhan, B., (2015). Sustainable brownfields land use change modeling using GIS-based weights-of-evidence approach. *Applied Spatial Analysis and Policy*, 1–18.

Abdullahi, S., Pradhan, B., & Jebur, M. N. (2015a). GIS-based sustainable city compactness assessment using integration of MCDM, Bayes theorem and RADAR technology. *Geocarto International, 30*(4), 365–387.

Abdullahi, S., Pradhan, B., Mansor, S., & Shariff, A. R. M. (2015b). GIS-based modeling for the spatial measurement and evaluation of mixed land use development for a compact city. *GIScience & Remote Sensing, 52*(1), 18–39.

Arsanjani, J. J., Kainz, W., & Mousivand, A. J. (2011). Tracking dynamic land-use change using spatially explicit Markov Chain based on cellular automata: the case of Tehran. *International Journal of Image and Data Fusion, 2*(4), 329–345.

Bohman-Carter, G. (1994). Geographic information systems for geoscientists: Pergamon. Oxford.

Chen, J., Chang, K.-T., Karacsonyi, D., & Zhang, X. (2014). Comparing urban land expansion and its driving factors in Shenzhen and Dongguan, China. *Habitat International, 43,* 61–71.

Collins, F. P. (2002). Small business liability relief and Brownfields Revitalization Act: A critique. *Duke Envtl. L. & Pol'y F., 13,* 303.

Domencich, T. A., & McFadden, D. (1975). Urban travel demand-a behavioral analysis.

Gross, M. (2010). Ignorance, research and decisions about abandoned opencast coal mines. *Science and Public Policy, 37*(2), 125–134.

Hassan, A., Sarah, H., & Algadi, M. (2013). *Kajang sustainable urban design*. Malaysia: University Kebangsaan Malaysia.

Kang, J. (2012). *A study on the future sustainability of Sejong, South Korea's multifunctional administrative city, focusing on implementation of transit oriented development.* Master, Uppsala University, Sejong city, South Korea. (93).

Li, X., & Yeh, A. G.-O. (2000). Modelling sustainable urban development by the integration of constrained cellular automata and GIS. *International Journal of Geographical Information Science, 14*(2), 131–152.

Oliver, L., Ferber, U., Grimski, D., Millar, K., & Nathanail, P. (2005). *The scale and nature of European brownfields.* Paper presented at the CABERNET 2005-International Conference on Managing Urban Land LQM Ltd., Nottingham, UK, Belfast, Northern Ireland.

Pontius, R. G., & Schneider, L. C. (2001). Land-cover change model validation by an ROC method for the Ipswich watershed, Massachusetts, USA. *Agriculture, Ecosystems & Environment, 85* (1), 239–248.

Pradhan, B., & Lee, S. (2010). Delineation of landslide hazard areas on Penang Island, Malaysia, by using frequency ratio, logistic regression, and artificial neural network models. *Environmental Earth Sciences, 60*(5), 1037–1054.

Pradhan, B., Sezer, E. A., Gokceoglu, C., & Buchroithner, M. F. (2010). Landslide susceptibility mapping by neuro-fuzzy approach in a landslide-prone area (Cameron Highlands, Malaysia). *IEEE Transactions on Geoscience and Remote Sensing, 48*(12), 4164–4177.

Schädler, S., Morio, M., Bartke, S., Rohr-Zaenker, R., & Finkel, M. (2011). Designing sustainable and economically attractive brownfield revitalization options using an integrated assessment model. *Journal of Environmental Management, 92*(3), 827–837.

Thomas, M. R. (2002). A GIS-based decision support system for brownfield redevelopment. *Landscape and Urban Planning, 58*(1), 7–23.

Van Eck, J. R., & Koomen, E. (2008). Characterising urban concentration and land-use diversity in simulations of future land use. *The Annals of Regional Science, 42*(1), 123–140.

City Intensification Process Using Brownfields Land Use Change Modeling

11

Saleh Abdullahi and Biswajeet Pradhan

11.1 Background

Brownfields are abandoned or underused properties that should be redeveloped or reused because of the real or suspected presence of hazardous substances, pollutants, or containment (Collins 2002; Oliver et al. 2005). Rapid urban growth and economic restructuring has led to increase in number of brownfields in urban region. Redevelopment of existing brownfields is one of the important objectives to enhance the sustainable urban development theory (Nijkamp et al. 2002; De Sousa 2008) and reduce urban sprawl (Nuissl and Schroeter-Schlaack 2009). However, a brownfield redevelopment needs a comprehensive effort to resolve and negotiate among several stakeholders with different interests (Bardos 2003; Gross 2010). In addition, this task is complicated by environmental contamination such as heavy metal, mineral oil, and hydrocarbon (Bleicher and Gross 2010). These complexities have caused many brownfields remain undeveloped, especially in developing countries. De Sousa (2008) had estimated that total number of contaminated sites in the USA, Canada, and European Union is about 450,000; 30,000, and 250,000, respectively. That is why most of researches regarding brownfield redevelopment in 1970s and 1980s were related to contamination and pollution caused by these sites. However, in the 1990s after publication of Brundtland Report in 1987, other aspects such as economic, environment, and social which were the main principles of sustainable development were also involved in brownfield redevelopment research objectives. The redevelopment of these abandoned land according sustainable development principles become the main action of the governments in order to reduce land consumption and rural and natural destruction. Thus, since 1990s, several research projects such as CABERNET, RESCUE, and SMART-e have been proposed and implemented for this objective. The main goal of these projects was to develop policies for sustainable brownfield development and remediation. However, scientists and urban planners realized that it is insufficient to only consider policies and technologies, and more complex and decision-making process is required in this field. For instance, Bleicher and Gross (2010) presented and discussed about the megasite management system based on SAFIRA II Program as shown in Fig. 11.1. This support system guides the decision makers in the development of efficient remediation strategies for the future use to contaminated site.

On the other hand, according to the literature, there is a variety of approaches for various aspects of brownfield redevelopment such as risk assessment (Carlon et al. 2008), policy analysis (Linkov et al. 2006), optimization of remediation (Bürger et al. 2007), remediation cost assessment (Kaufman et al. 2005), general success factors for brownfield redevelopment (Lange and McNeil 2004), infrastructure redevelopment (Attoh-Okine and Gibbons 2001), urban planning, and site prioritization under budget constraints (Alvarez-Guerra et al. 2009), etc. (Schädler et al. 2011). Hence, strong approaches are required to integrate these aspects and manage complexities of information and results (Bardos et al. 2000; Agostini et al. 2007).

Land use change modeling is important for various urban planning and management issues. It is a suitable approach to deal with redevelopment and revitalization of existing brownfields. Modeling of land use changes not only improve and select various land development scenarios, but also to evaluate the impact of development alternatives. For example, proper analysis and prediction of urban growth may prevent many social and environmental problems caused by the urban sprawl (Hayek et al. 2011), suburbanization process (Helbich and Leitner 2009) and unorganized land developments. The main environmental problems that can be prevented are encroachment on valuable agricultural, forest, and natural areas. In addition, land use change modeling can help local planning authorities to provide better community facilities and services to sustain developments (Hathout 2002). In fact, most of the urban development scenarios are an act to achieve urban sustainability. Compact development, transit-oriented development, and smart city are good examples of development scenarios that are based on

B. Pradhan (ed.), *Spatial Modeling and Assessment of Urban Form*,
DOI 10.1007/978-3-319-54217-1_11

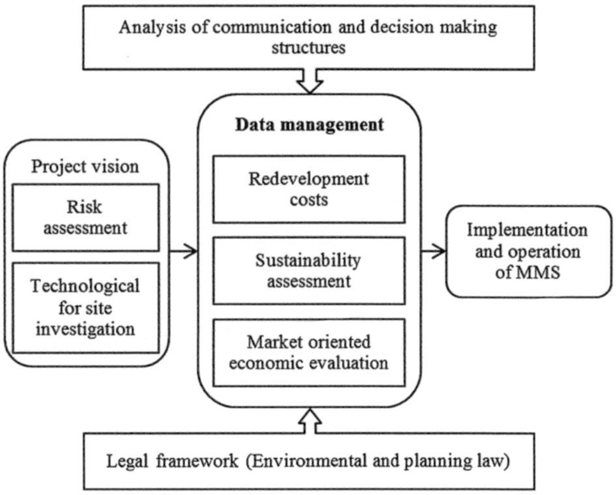

Fig. 11.1 The megasite management system of the SAFIRA II Program (modified after Bleicher and Gross 2010)

sustainable development principles (Livingstone and Rogers 2003). One of the environmental perspectives of compact urban development is to emphasize on containment of rural developments and revitalization of central areas (Jenks et al. 1996; Lin and Yang 2006). The rationale behind this concept is to try redeveloping existing brownfields inside the cities instead of growing built up areas through rural environments. Cho et al. (2011) attempted to solve this problem by evaluating the hypothesis that land value tax contained rural area development and encouraged compact and development closer to and within built up areas. Rall and Haase (2011) assessed the brownfield revitalization program of the City of Leipzig in the context of urban sustainability. The assessment was performed through a triangular integrated evaluation method combined with site surveys and interviews, as well as expert knowledge. However, these assessments and analyses can be improved significantly by modeling the land use changes in order to predict and propose a proper land use types for each brownfield site. Schädler et al. (2012) described and proposed a framework which integrates a GIS-based identification of areas to be remediated, an estimation of associated clean-up costs, and an assessment of the planned future land use's contribution to sustainable urban development. Furthermore, Schädler et al. (2013) proposed a scheme to transfer the evaluation of site-specific sets of sustainability indicators into automated quantitative and spatially explicit assessments, which can be integrated into multidisciplinary spatial optimization algorithms. On the other hand, the main aim of their study was to gain a site-specific understanding of sustainable land use planning, and of the potential advantages that mixed land use development may have over uniform use with only one single land use type for a brownfield site. However, the current chapter attempts to deal with brownfield redevelopment based on compact development paradigms through land use change modeling to achieve more sustainability environment.

11.2 Land Use Change Modeling

According to the literature, following four core principles are the bases of all land use change simulation models; historical evidence based, suitability bases, neighborhood bases, and actor interaction bases (Verburg et al. 2004a). The logic behind historical evidence based is that, "past is the key for future." Therefore, background information can be helpful in predicting future land use change as demonstrated by Kuijpers-Linde et al. (2007). Suitability bases may consist of several factors in a land parcel in order to evaluate for an allocation of specific purpose (Abdullahi et al. 2014a). Therefore, the underlying premise is to achieve maximum profit and minimize liability. Neighborhood bases deal with neighborhood interaction cells that affect the transition of one land use to another (Li et al. 2008). Actor interaction bases assume that land use change is the result of an interaction of several actors or agents. The agents can be one or a group of factors. This core principle is a promising research tool for land use change modeling (Matthews et al. 2007; Jjumba and Dragićević 2012).

There are a few main concepts of land use changes such as Markov chains, economic-based concept, agent-based systems, statistical analysis, cellular automata, and artificial neural networks. The Markov chain concept is based on a continuation of historical trend of development. This concept calculates a probability matrix of changes of one land use type to another. The main disadvantage of this model is the lack of spatial bases of results therefore additional assumptions are required for allocation (Verburg et al. 2004a; Al-sharif and Pradhan 2014). The economic-based concept is also an important reason for land use changes and is mainly based on the suitability of a land, although the core principle of continuation of historical development can also be included. In general, economic-based is not exactly a concept, however, cannot be left out of the list of concepts of land use change (van Schrojenstein Lantman et al. 2011). More recent applications of economic-based have been reported by Nelson and Hellerstein (1997) and Walker (2004) which all use as the based theory to explain tropical deforestation. An agent-based systems of land use change modeling which is based on the core principle of actor interaction, consists of two main components: a map of a study area and a model with agents that represent human decision-making (Parker et al. 2003). An agent is a representation of actors important in the process with their own preferences (Grimm et al. 2006). These preferences can be defined by expert knowledge, using questionnaires, or using artificial neural networks technique (van Schrojenstein Lantman et al. 2011).

Various kinds of statistical computation can be derived from land use maps. For example, logistic regression, frequency ratio, and weights-of-evidence techniques can be used to analyze the probability of occurrence of a dependent variable on each class of independent variables (Verburg et al. 2004a). The coefficients of each variable can be calculated from historical land use changes. Furthermore, they can be projected for future land use changes. Other statistical modeling approaches such as logit modeling in the planning and policy environment are also so common like land use scanner (Hilferink and Rietveld 1999), which has been used in producing sustainability outlooks for the Netherlands (Kuijpers et al. 2007). The Cellular Automata (CA) is the most well-known techniques in modeling of land use changes (White and Engelen 1993). The main logic behind CA modeling for land use changes is the current state of each cell and its interaction with neighborhood cells. Therefore, this model is based on core principles of historical trend and neighborhood interaction. However, CA does not necessarily consider the relationship and interaction among the related parameters. Thus, usually, the CA is integrated with other techniques such as Markov Chain (Al-sharif and Pradhan 2014), Fuzzy Theory (Al-Ahmadi et al. 2009), etc., to increase the strength of the modeling. The use of artificial neural networks (ANN) has increased significantly due to advances in computing performance and flexibility of software (Skapura 1996). The pattern recognition capability (Pijanowski et al. 2002), that makes a relationship between past and future land use and suitability maps (Verburg et al. 2004b), are important parameters that can emphasize the strength of ANN models in land use change modeling. The first to apply ANNs to a computer simulation model was Pijanowski et al. (2002). The model trains itself on a dataset and the corresponding land use maps of different years enabling it to recognize and reproduce the pattern of land use categories (Mas et al. 2004; Pijanowski et al. 2005). Appropriate knowledge about the land use change modeling based on their concepts allows modelers to select the most appropriate model for area of investigation.

Land use change modeling requires availability of rich spatial data, spatial analysis tools, and displaying capability to illustrate the output maps. Geographic information system (GIS) and remote sensing are the most useful tools to support modeling. GIS can provide a proper environment to store, manage, analyze, manipulate, and display spatial data associated with the models. In addition, GIS can aid modelers to define and create spatial variables for the models (Openshaw and Clarke 1996), predict land use changes based on several independent spatial variables (Mertens and Lambin 2000), and evaluate predicted changes in a spatial pattern. However, some capability of GIS is also questioned regarding the extent and type of GIS used in the planning practice (Olafsson and Skov-Petersen 2014). Hence, updated knowledge regarding the performance and limitation of GIS, as well as integration with other models, will improve the strength of the analysis. There are numerous studies on land use change modeling using integration of GIS technology (Li and Yeh 2002; Pijanowski et al. 2002; Verburg et al. 2004a). For instance, Thomas (2002) stated that to assess land use modeling performance with respect to the redevelopment of brownfields, accessibility of information such as land capability, environmental concerns, public preferences, etc., for both governmental agencies and decision makers are required. He discussed a GIS-based decision support system to provide access to geospatial data in various scales for better understanding of the brownfields redevelopment issue.

Although several applications of global parametric models have been used in land use change modeling (Tayyebi et al. 2014), very few urban related studies have considered city compactness as an objective of the land use change modeling process to achieve urban sustainability. The lack of this application specifically for brownfields redevelopment planning is the motivation to investigate the potential and capability of this integration modeling for existing brownfields of Kajang City, Malaysia. In addition, the current study is an attempt to analyze the urban land use changes and spatial patterns of the study area in a quantitative manner. This will benefit Malaysian case studies as these effects are usually explained without quantitative perspectives in the country (Nourqolipour et al. 2014). Therefore, the main objective is to integrate the land use change modeling concept with the brownfield redevelopment plan on the bases of city compactness paradigms. Specifically, this chapter illustrates how statistical-based weights-of-evidence (WoE) approach within GIS aids in the understanding of the process of land use changes. WoE was used to measure the extent and direction of various land use growth based on temporal datasets for the year 2008 and 2012. In addition, the model was utilized to apply and evaluate the driving forces responsible for the change of land use types in a compact pattern. One benefit of using the model is the ability to extract and utilize the most effective factors from all the selected factors before evaluating the probability of land use growth.

By integrating WoE process with brownfield redevelopment strategy, one of the environmental objectives of compact development can be fulfilled. Hence, this study initially predicts land use changes of Kajang City using the probability of growth of each land use according to compact development evidences. Consequently, the extent and direction of each land use types were projected. After validating the process, the created probability maps and the master plan of the study region was used to assess the existing brownfields land use types. It should be mentioned that due to utilization of the standard and common urban related parameters as well as statistical-based methodology, this process can be easily replicated in other international cities for implementing brownfields redevelopment strategies.

11.3 Study Area

The proposed modeling approach was used to predicting the multiple land use changes for Kajang City, Malaysia (3° 00′ 19″N, 101° 46′ 42″E). The study area is located 21 km from Kuala Lumpur, the capital city of Malaysia (Fig. 11.2). According to the 2010 census data, the city has population of 246,618, with an estimated population growth of 9% per annum. The study area covers approximately 60 km^2. The west parts of the city are mainly covered by agriculture and forest lands. Recent urban sprawl developments have mushroomed Kajang City because of its proximity to three main cities of Malaysia. Although there are many abandoned and brownfields in the city (Fig. 11.3), most of these new developments have been constructed at the outskirts of the agricultural and forest lands. For this reason, the present research attempts to assess the brownfields land use changes according to the city compactness paradigm to make Kajang City more environmentally sustainable.

11.4 Data and Methodology

In the first stage, a contextualized definition of the compact urban development and its indicators was needed. Generally, in a large scale study area, urban compactness is measured based on the cellular bases and the concentration of the built up cells in a specific area as in the study conducted by Li and Yeh (2004) which assesses the urban compactness using entropy and compactness index methods. Due to availability of several urban detail layers, the current study performed a compactness assessment in more accurate and detail bases. There are three main city compactness indicators: urban density, land use diversity, and urban intensity (Burton 2002; Abdullahi et al. 2014b). Each of these indicators is divided into several parameters according to the availability of data and the objective of the research. In addition to city compactness indicators, other related urban parameters that are important for land use changes and some physical properties of the sites were included in the analysis, as shown in Table 11.1. The overall methodology flowchart of the process is shown in Fig. 11.4.

Most of the data, such as the land use map of year 2008 and 2012, the road network, the soil map, etc., were collected from the local planning authority of Kajang City. Other layers were also extracted or created from existing layers. It was essential to select the most important parameters among others, which have a significant effect on the land use conversion for the specific study area. Therefore, an optimization process was applied to select the most effective parameters. This process was performed by the frequency ratio (FR) approach, which is the initial step of running the weights-of-evidence technique (Pradhan et al. 2010; Pourghasemi et al. 2013).

The FR model has the ability to examine the existence and changes (the increase or decrease) of land use types with respect to each class of all parameters. In this manner, the effectiveness of each parameter could be assessed by investigating the trend of land use changes based on their classes. This process also assessed the spatial dependency of the factors. The classification of the parameters was defined according to their types. For instance, a proximity analysis was performed for the distance-based parameters. Then these distances were divided into classes, which include their spatial extent. Every cell are in a distance class: “near” to, “middle”, and “far” from land uses or points of interest. For ordinal

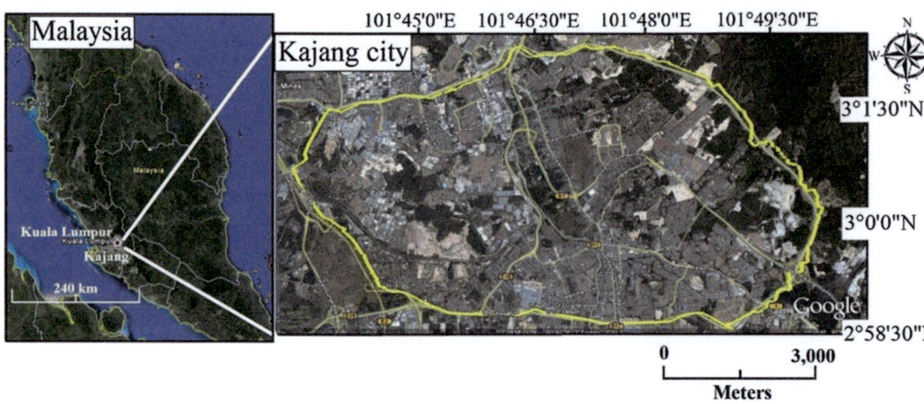

Fig. 11.2 The map of Malaysia, Kajang city

Fig. 11.3 An abandoned land in Kajang; RGB photo (Taken by Author) and satellite image (Google Earth)

Table 11.1 Land use change modeling parameters

No	Parameters	Categories
1	Population density	City compactness
2	Built up density	City compactness
3	Residential density	City compactness
4	Land use diversity	City compactness
5	Proximity to public transportation facilities	Site-specific/City compactness
6	Proximity to recreation facilities	Site-specific/City compactness
7	Proximity to community facilities	Site-specific/City compactness
8	Proximity to infrastructure	Site-specific/City compactness
9	Proximity to road networks	Site-specific/City compactness
10	Proximity to same land use types	Site-specific
11	Distance from agricultural fields	Site-specific
12	Soil and geology properties	Physical properties
13	Distance from flood zones	Physical properties

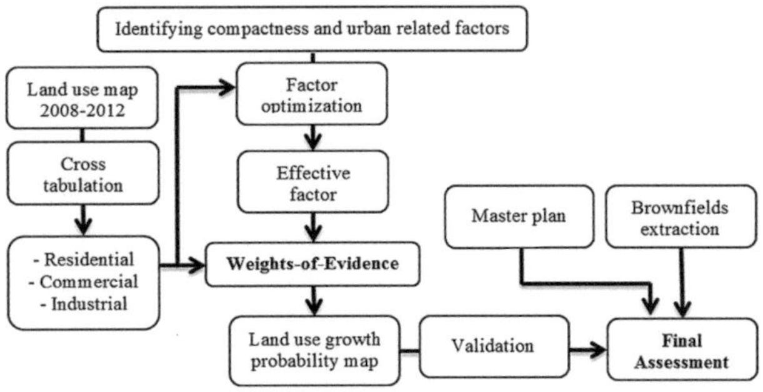

Fig. 11.4 Proposed brownfield redevelopment methodological flowchart

parameters, such as the land use diversity and urban population, "high", "moderate", and "low" mixed or density were applied respectively. In the case of nominal parameters such as soil types or geology types, each type was used as one class.

The entire base layer of all factors was converted into a grid cell to assess the growth of each land use type in their classes. For instance, the proximity to industrial land use causes a reduction in existence of residential land use cells. In contrast, in areas near to recreational facilities, more number of residential cells can be observed. However, various geological types do not have any significant effect on existence or absence of residential cells. Hence, the proximity to industrial and recreational facilities selected as important parameters. Moreover, geological characteristic assumed as not an effective factor, hence were removed from the process.

As previously mentioned, historical evidence bases are one of the core principles of land use change modeling. In order to investigate the trend of the land use changes during a four year period, a cross-tabulation analysis of land use maps was performed between the years 2008 and 2012. A cross-tabulation enabled the observation of the significant growth of the main land use types. This process is a mathematical matrix, which gives unbiased information concerning the entire area of interest, to derive unbiased summary statistics (Pontius and Millones 2011). For this case study, the matrix gave unbiased information concerning the relationship between the land use maps of 2008 and 2012. It showed that only three main land use types (residential, commercial and industrial) were growing and changing significantly than others. New-build gentrification literature also proved that the land use change process is mainly from pre-industrial or brownfields to residential, commercial, or institutional uses (Davidson and Lees 2005; Sabri et al. 2012). Furthermore, the growth of these three land use types resulted in the reduction of agricultural fields. Hence, it is decided to focus on residential, commercial, and industrial land use types, to evaluate and project their growth through other land use types.

For the proposed land use change modeling, Bayesian theorem was applied, with an update of prior probabilities through the weights-of-evidence (WoE) approach (Bonham-Carter 1994; Pradhan et al. 2010). Dempster–Shafer theory of evidence developed by Dempster (1967) and then by Shafer (1976) is a spatial integration model with mathematical representations (Carranza 2009; Althuwaynee et al. 2012).

The selected parameters from the optimization process were utilized as evidence in order to evaluate the probability of growth for each main land use type. The WoE allowed the ability to assess and combine evidences according to variation of the land use changes. The advantage of this theory is its flexibility to compute uncertainty and to combine evidences from different sources of data (Thiam 2005; Bui et al. 2012). The model created an opportunity to analyze land use changes according to the city compactness paradigm. In general, WoE evaluates the degree to which evidences support the hypothesis, in this case the land use change occurrence, and the degree to which those evidences do not refute the hypothesis (Dempster 1967; Shafer 1976). The WoE has been widely applied in the literature in a variety of applications such as geological mapping (Chen et al. 2013), and natural disaster management (Althuwaynee et al. 2012; Bui et al. 2012; Pourghasemi et al. 2013). However, a few studies have utilized this approach in urban applications such as land use dynamic modeling by Maria de Almeida et al. (2003) and mixed land use development probability mapping by Abdullahi et al. (2015). As an example, the WoE of residential land use growth with respect to the proximity to the road is shown in Table 11.2. The value of C was calculated by subtracting W+ (natural logarithm of occurrence) and W− (natural logarithm of non-occurrence). This value represents the spatial association of each land use pixel and each class of evidence. A positive value represents a higher number of specific land use pixels occurring in this class. In contrast, a negative value represents a lesser number of land use pixels occurring in this class. S2(W+) and S2(W−) are variances of W+ and W−, respectively, and S(C) is the standard deviation of the contrast. Finally, C/S(C) is the standardized value of C which represents the significance of the spatial association and measures the relative certainty of the posterior probability (Bonham-Carter 1994).

Further detailed description of the mathematical formulation is available in Maria de Almeida et al. (2003), Pradhan et al. (2010) and Regmi et al. (2010). The probability value of the land use growth for every cell of the study area is calculated by considering the prior probability of occurrence

Table 11.2 Weights-of-evidence for residential land use growth with respect to proximity to road, proximity to public transportation and facilities and population density evidences

Factor	Class	No. cells	No. deposit	FR	W+	W-	C	S2(W+)	S2(W−)	S(C)	C/S(C)
Proximity to road	Near	745,948	215,086	**1.08**	0.08	−0.04	0.12	0.000007	0.000004	0.0032	**38.23**
	Middle	752,993	209,126	**1.04**	0.04	−0.02	0.07	0.000007	0.000004	0.0032	**20.49**
	Far	762,663	176,854	**0.87**	−0.14	0.06	−0.20	0.000007	0.000004	0.0033	**−59.78**
	Total	2,261,604	601,066								
Proximity to public transportation	Near	736,018	215,086	**1.10**	0.09	−0.05	0.14	0.000007	0.000004	0.0032	**44.33**
	Middle	769,543	209,126	**1.02**	0.02	−0.01	0.03	0.000007	0.000004	0.0032	**10.43**
	Far	756,043	176,854	**0.88**	−0.13	0.06	−0.19	0.000007	0.000004	0.0033	**−55.82**
	Total	2,261,604	601,066								
Proximity to public facility	Near	723,298	243,621	**1.27**	0.24	−0.13	0.37	0.000006	0.000004	0.0032	**114.96**
	Middle	769,661	240,939	**1.18**	0.16	−0.10	0.26	0.000006	0.000004	0.0032	**80.78**
	Far	768,645	116,506	**0.57**	−0.56	0.20	−0.76	0.000010	0.000003	0.0036	**−207.46**
	Total	2,261,604	601,066								
Population density	Low	668,052	111,643	**0.63**	−0.46	0.14	−0.61	0.000011	0.000003	0.0037	**−162.34**
	Middle	832,707	265,225	**1.20**	0.18	−0.12	0.30	0.000006	0.000005	0.0032	**93.32**
	High	760,845	224,198	**1.11**	0.11	−0.06	0.16	0.000006	0.000004	0.0032	**50.51**
	Total	2,261,604	601,066								

Bold letters indicate the significant/important values

and non-occurrence of land use types in each class of evidence. The majority of the evidences are distance-based or accessibility. Hence, assessing the weights across the different distance ranges is possible. The other two evidences were the main city compactness characteristics and characterized in ordinal bases. The transitional probability was computed according to the proportion of the observed transition in each predefined class of evidences.

The output of this process is the three land use growth probability maps which show the probability of each land use growth according to the selected evidences separately. Each map was later classified into three classes: areas with very high, moderate, and low probability of growth of a specified land use type. Finally, all three maps as first scenario which was equal priority scenario were aggregated with the same weights. The first scenario was to illustrate the overall view of the study area regarding the growth of each single land use types, as well as mixture of them. The other scenarios were defined according to the master plan of the study area, where each land use growth map had different priority values. Hence, the site potential and suitability, the local demands of the neighborhood, and the local planning and development policy can assist to assign a proper priority to each land use type.

The next step was to extract the existing brownfields of Kajang City. For this process, the site indicators and criteria as listed in study conducted by Thomas (2002) were tested. All open spaces such as the buffer zone around rivers and highways, recreational play grounds, and natural landscapes were excluded from the analysis. As shown in Fig. 11.5, most of the small brownfield sites are located in the center, south, and

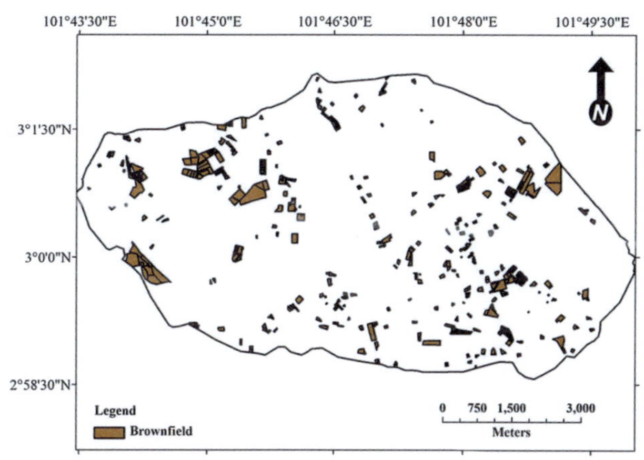

Fig. 11.5 Existing brownfields of Kajang city

Table 11.3 Most effective parameters for land change modeling

Important factors	Land use type
Proximity to residential	Residential, commercial, industrial
Proximity to commercial	Residential, commercial, industrial
Proximity to industrial	Residential, commercial, industrial
Proximity to roads	Residential, commercial, industrial
Proximity to public transportation	Residential, commercial, industrial
Proximity to community facilities	Residential, industrial
Proximity to recreational facilities	Residential, industrial
Proximity to infrastructure	Residential, industrial
Proximity to agricultural fields	Residential, commercial, industrial
Population density	Residential, commercial, industrial
Land use diversity	Residential, commercial, industrial

southeast. They are near dense residential and commercial areas or compact regions. In contrast, brownfields with larger area are located in the east, west, and central west near to industrial and agricultural fields or less compact regions.

11.5 Results and Discussion

In the first step to understand the current trend of land use change of the study area, a cross-tabulation process was run between the land use map of year 2008 and 2012. This process revealed that residential land use attempts to capture almost all types of activities. However, this growth is more noticeable through open spaces and agricultural fields. In fact, the growth in main land use types through brownfields and abandoned land is desirable. However, loss of 345 ha of agricultural fields in 4 years only from residential land use development is an unsustainable problem that should be avoided. Unfortunately, loss of agricultural fields can be seen from the growth of commercial and industrial land use as well. By running a cross-tabulation process, the total growth and total loss of each land use types were computed. Residential, commercial, and industrial land use had growth values of 367, 72, and 75 ha, respectively, during four years period. Moreover, in overall Kajang City has lost more than 348 ha area of its agricultural fields in same period.

After evaluation of the land use growth with respect to all selected parameters, it was noticed that some of them do not have influence on these growths. Therefore, by running the optimization process, the most effective factors as shown in Table 11.3 were extracted. Residential and industrial land use types apparently are more similar in factor effectiveness rather than commercial land use. However, it should be noted that most the factors have an inverse relationship with these two land use types: the proximity to community and recreational facilities, the population density, and the land use diversity. In contrast, commercial and residential land uses have a direct relationship in case of the most factors.

The probability value of land use growth (C/S(C) value) for every cell of study area was calculated considering the prior probability of occurrence and non-occurrence of land use types in each class of evidences. A summarized weights-of-evidence calculation for each land use growth is given in Table 11.4.

A majority of evidences were based on distance or accessibility and it was possible to examine the probability of growth of land use types across the different distance classes. On Table 11.4 and as shown in Fig. 11.6, the probability of growth for each land use types is higher in close proximity to the same land use types. In contrast, the residential and industrial land uses tend to keep distances from each other. In general, proximity to recreational and community facilities offer advantages for the housing environment. In this specific case study, the proximity caused positive probability values for residential and negative values for industrial land use growth. This confirms the inverse relationship of the residential and industrial land uses theory.

Accessibility to main roads and public transportation facilities is another important characteristic of site suitability. Having proper accessibility is a positive factor for most of new developments. However in Kajang City, most of the industrial land uses are located near agricultural fields or rural areas, which not much urban development is observed.

Table 11.4 Summarized weights-of-evidence for main three land use growths.

Factors	Class	C/S(C)	C/S(C)	C/S(C)
		Residential	Commercial	Industrial
Proximity to residential	Near	364	−14	−365
	Middle	−264	41	330
	Far	−156	−45	126
Proximity to roads	Near	53	167	−137
	Middle	29	−88	−40
	Far	−86	−89	172
Proximity to recreational facilities	Near	129	–	−225
	Middle	42	–	−85
	Far	−188	–	300
Population density	Low	−158	−99	251
	Moderate	112	40	−114
	High	18	60	−142
Proximity to commercial	Near	14	198	−163
	Middle	60	−131	−6
	Far	−77	−70	159
Proximity to public transportation	Near	63	117	−172
	Middle	−3	−99	−6
	Far	−62	−17	174
Proximity to infrastructure	Near	40	–	−70
	Middle	60	–	30
	Far	−105	–	37
Land use diversity	Low	−79	−120	216
	Moderate	74	73	−76
	High	2	56	−156
Proximity to industrial	Near	−206	−60	380
	Middle	108	65	−241
	Far	76	−6	−166
Proximity to community facilities	Near	72	–	−222
	Middle	88	–	−75
	Far	−174	–	287
Proximity to agricultural fields	Near	−132	−42	2
	Middle	21	6	75
	Far	100	36	−78

This issue can be seen from the proximity to agricultural fields as well. Consequently, in the case of industrial land use, negative values are for classes near to main roads and public transportation facilities.

Population and land use diversity evaluation were in the range of high and low population and single to mixed land use, respectively. These two compactness-based evidences revealed straightforward effects on the land use growth. Higher population and higher land use diversity resulted positive values of C/S(C) for residential and commercial land uses. In contrast, the area in single land use and lower population density has a higher probability for industrial land use growth.

Land use growth map for residential, commercial, and industrial are given in Fig. 11.6, respectively. The white areas are constraint areas such as transportation, flood zone, and water bodies, which have been removed from the analysis.

For the validation of land use change simulation, it is desirable to quantitatively evaluate the degree of fitness or

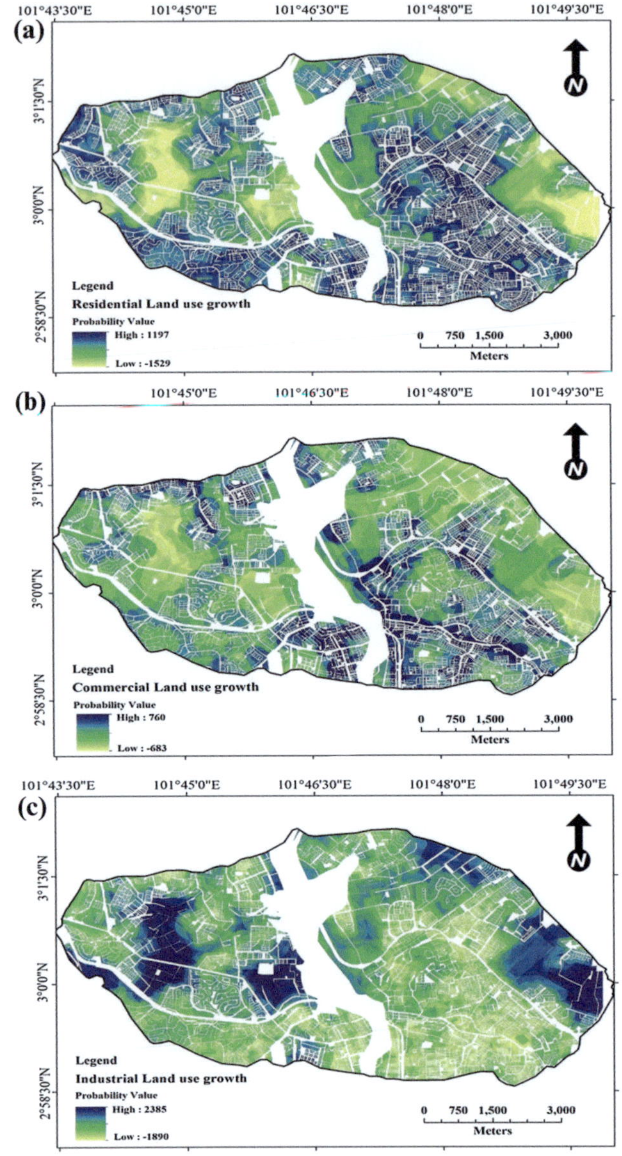

Fig. 11.6 **a** Residential, **b** commercial and **c** industrial land use growth map from WoE model

similarity between the projected land use and the actual land use map. This similarity assessment was performed using the relative operating characteristic (ROC) based area under curve (AUC) to evaluate the probability of growth maps created by WoE approach (Pontius and Schneider 2001; Van Eck and Koomen 2008; Chen et al. 2013). To run AUC, WoE was applied on the land use map of 2008 in order to create the main land use (residential, commercial and industrial) probability of growth for the future. Subsequently, the similarity of these three maps was assessed by the actual residential, commercial and industrial land use of year 2012. In this manner, the process determined how well the method and parameters produced the land use growth map. The AUC of 50% indicates random results. The AUC of residential, commercial and industrial land use were gained as 77.4, 78, and 67%, respectively. Lower AUC value of industrial land use indicated that industrial land use growth is more depend on economical perspectives rather than physical or proximity to same land use type parameters.

After the aggregation of land use growth maps as the first scenario (equal priority), with the brownfields map, from a total area of brownfields (2,908,550 m^2), 400,000 m^2 was assigned as the single land use growth (Fig. 11.7 and Table 11.5). The rest of them were projected for mixture of two or three land use types.

Single land use developments are assigned to brownfields located in an area with high probability of growth for only one land use type. For instance, as shown in Fig. 11.6, the central west of Kajang City is only suitable for the industrial land use type. Therefore, most of the brownfields located in these areas were automatically assigned as a single land use development for industrial purposes. However, according to Fig. 11.6a, b, most of the commercial land use growth are suitable for residential as well. Residential–commercial mixed land use area covered more than one third of whole city (1,107,877). This large area could be expected due to the high similarity of the C/S(C) value and the direct relationship of residential and commercial land use with respect to all evidences. These areas can be developed in vertical mixed use, which means the basement floors are for commercial use and upper floors for housing purposes. In addition, significant differences of the C/S(C) values of commercial and industrial caused only 136,000 m^2 area to become a mixture of these two land use types. Areas with mixture of residential–industrial are mainly located in the borders of these land use types. It should be mentioned that the industrial use which assigned for these areas are mainly light industry. The rest of the area was assigned as a mixture of all three land use types. Considering the size, potential, and suitability of these locations, these brownfields can serve the neighborhood as mixed land use development. Further scenarios can be defined by consideration of local expert knowledge in order to give different priority values for each land use type.

11.6 Conclusion

Rapid urban growth has resulted in intensive loss of natural and valuable agricultural lands especially in fast growing regions. Therefore, the simulation or projection of the future urban growth and land use changes provide very beneficial information for local planners and decision makers. However, like other urban related issues, land use change modeling is often difficult

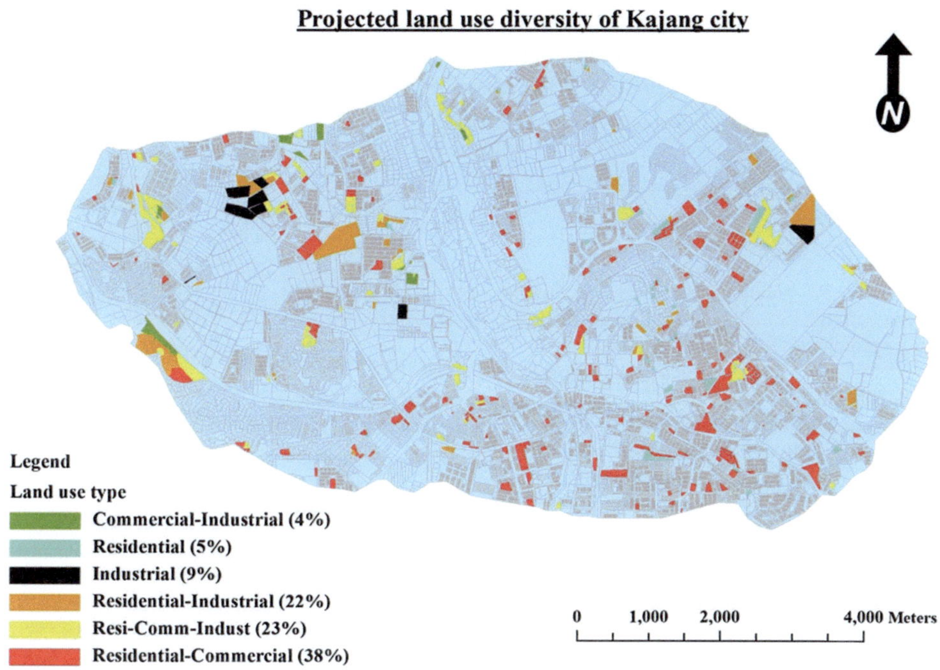

Fig. 11.7 Projected land use types for existing brownfields

Table 11.5 Projected land use types for existing brownfields

Land use diversity	Land use type	Area (m^2)
Single land use development	Residential	140,222.3
	Commercial	0
	Industrial	258,892.5
	Total	399,114.8
Mixed land use development	Residential–commercial	1,107,877.2
	Residential–industrial	625,962.9
	Commercial–industrial	136,064.8
	Residential, commercial, industrial	639,531.7
	Total	2,509,436.6
Total		2,908,551.4

to simulate due to its complexities, uncertainties, and several numbers of involved parameters and/or stakeholders. These difficulties need to be dealt with multidisciplinary geospatial techniques and other systematic procedures.

This chapter illustrates the application of GIS-based WoE for modeling the brownfields land use changes. The future land use type of each existing brownfield could be identified. This projection can be in single land use or a mixture of two or more land use types depending on brownfield properties, potential, and surrounding environment conditions. The model process was based on the trend and historical land use changes of the study area. Furthermore, the projected land use changes were based on city compactness paradigms such as urban density, urban intensity, and land use diversity in order to develop the city according to sustainable development theory. Several other urban related factors were involved in the analysis as well. However, the results show that one of the main controlling factors for these changes was based on spatial autocorrelation of land use types.

The WoE model is a statistical-based model. Hence, the parameters were evaluated statistically instead of subjective choice of weighing technique by expert knowledge, which is the main source of uncertainty. For this reason, it can be concluded that the model revealed reliable and promising results for brownfields land use change modeling. The final outputs provide valuable land use growth maps and information about the future of existing brownfields of the city. Redeveloping and revitalizing these areas according to compact development concept will make Kajang City environmentally more sustainable.

References

Abdullahi, S., Mahmud, A. R. B., & Pradhan, B. (2014a). Spatial modelling of site suitability assessment for hospitals using geographical information system-based multicriteria approach at Qazvin city, Iran. *Geocarto International, 29*(2), 164–184.

Abdullahi, S., Pradhan, B., & Jebur, M. N. (2014b). GIS-based sustainable city compactness assessment using integration of MCDM, Bayes theorem and RADAR technology. *Geocarto International* (just-accepted), 1–39.

Abdullahi, S., Pradhan, B., Mansor, S., & Shariff, A. R. M. (2015). GIS-based modeling for the spatial measurement and evaluation of mixed land use development for a compact city. *GIScience & Remote Sensing, 52*(1), 18–39.

Agostini, P., Carlon, C., Critto, A., & Marcomini, A. (2007). A step toward contaminated megasites management: Six European experiences at comparison. In *Landfill research trends* (pp. 47–75). New York: Nova Science Publishers Inc.

Al-Ahmadi, K., Heppenstall, A., Hogg, J., & See, L. (2009). A fuzzy cellular automata urban growth model (FCAUGM) for the city of Riyadh, Saudi Arabia. Part 1: Model structure and validation. *Applied Spatial Analysis and Policy, 2*(1), 65–83.

Al-sharif, A. A., & Pradhan, B. (2014). Monitoring and predicting land use change in Tripoli Metropolitan City using an integrated Markov chain and cellular automata models in GIS. *Arabian Journal of Geosciences, 7*(10), 4291–4301.

Althuwaynee, O. F., Pradhan, B., & Lee, S. (2012). Application of an evidential belief function model in landslide susceptibility mapping. *Computers & Geosciences, 44,* 120–135.

Alvarez-Guerra, M., Viguri, J. R., & Voulvoulis, N. (2009). A multicriteria-based methodology for site prioritisation in sediment management. *Environment International, 35*(6), 920–930.

Attoh-Okine, N. O., & Gibbons, J. (2001). Use of belief function in brownfield infrastructure redevelopment decision making. *Journal of Urban Planning and Development, 127*(3), 126–143.

Bardos, P. (2003). *Sharing experiences in the management of megasites: Towards a sustainable approach in land management of industrially contaminated areas.* Paper presented at the Final Report of the NICOLE Workshop.

Bardos, R., Nathanail, C., & Weenk, A. (2000). *Assessing the wider environmental value of remediating land contamination.* Paper presented at the Proceedings 7th International FZK/TNO Conference on Contaminated Soil.

Bleicher, A., & Gross, M. (2010). Sustainability assessment and the revitalization of contaminated sites: Operationalizing sustainable development for local problems. *International Journal of Sustainable Development and World Ecology, 17*(1), 57–66.

Bonham-Carter, G. (1994). *Geographic information systems for geoscientists: Modelling with GIS.* Amsterdam: Elsevier.

Bui, D. T., Pradhan, B., Lofman, O., Revhaug, I., & Dick, O. B. (2012). Spatial prediction of landslide hazards in Hoa Binh province (Vietnam): A comparative assessment of the efficacy of evidential belief functions and fuzzy logic models. *Catena, 96,* 28–40.

Bürger, C. M., Bayer, P., & Finkel, M. (2007). Algorithmic funnel-and-gate system design optimization. *Water Resources Research, 43*(8).

Burton, E. (2002). Measuring urban compactness in UK towns and cities. *Environment and Planning B, 29*(2), 219–250.

Carlon, C., Pizzol, L., Critto, A., & Marcomini, A. (2008). A spatial risk assessment methodology to support the remediation of contaminated land. *Environment International, 34*(3), 397–411.

Carranza, E. J. M. (2009). Controls on mineral deposit occurrence inferred from analysis of their spatial pattern and spatial association with geological features. *Ore Geology Reviews, 35*(3), 383–400.

Chen, C., He, B., & Zeng, Z. (2013). A method for mineral prospectivity mapping integrating C4. 5 decision tree, weights-of-evidence and m-branch smoothing techniques: A case study in the eastern Kunlun Mountains, China. *Earth Science Informatics*, 1–12.

Cho, S.-H., Kim, S. G., & Roberts, R. K. (2011). Measuring the effects of a land value tax on land development. *Applied Spatial Analysis and Policy, 4*(1), 45–64.

Collins, F. P. (2002). Small business liability relief and brownfields revitalization act: A critique. *The Duke Envtl. L. & Pol'y F., 13,* 303.

Davidson, M., & Lees, L. (2005). New-build 'gentrification' and London's riverside renaissance. *Environment and Planning A, 37* (7), 1165–1190.

De Sousa, C. A. (2008). *Brownfields redevelopment and the quest for sustainability* (Vol. 3). Emerald Group Publishing.

Dempster, A. P. (1967). Upper and lower probabilities induced by a multivalued mapping. *The Annals of Mathematical Statistics*, 325–339.

Grimm, V., Berger, U., Bastiansen, F., Eliassen, S., Ginot, V., Giske, J., et al. (2006). A standard protocol for describing individual-based and agent-based models. *Ecological Modelling, 198*(1), 115–126.

Gross, M. (2010). Ignorance, research and decisions about abandoned opencast coal mines. *Science and Public Policy, 37*(2), 125–134.

Hathout, S. (2002). The use of GIS for monitoring and predicting urban growth in East and West St Paul, Winnipeg, Manitoba, Canada. *Journal of Environmental Management, 66*(3), 229–238.

Hayek, U. W., Jaeger, J. A., Schwick, C., Jarne, A., & Schuler, M. (2011). Measuring and assessing urban sprawl: What are the remaining options for future settlement development in Switzerland for 2030? *Applied Spatial Analysis and Policy, 4*(4), 249–279.

Helbich, M., & Leitner, M. (2009). Spatial analysis of the urban-to-rural migration determinants in the Viennese metropolitan area. A transition from suburbia to postsuburbia? *Applied Spatial Analysis and Policy, 2*(3), 237–260.

Hilferink, M., & Rietveld, P. (1999). Land use scanner: An integrated GIS based model for long term projections of land use in urban and rural areas. *Journal of Geographical Systems, 1*(2), 155–177.

Jenks, M., Burton, E., & Williams, K. (1996). *Compact cities and sustainability: An introduction*. London: E & FN Spon.

Jjumba, A., & Dragićević, S. (2012). High resolution urban land-use change modeling: Agent iCity approach. *Applied Spatial Analysis and Policy, 5*(4), 291–315.

Kaufman, M. M., Rogers, D. T., & Murray, K. S. (2005). An empirical model for estimating remediation costs at contaminated sites. *Water, Air, and Soil pollution, 167*(1–4), 365–386.

Kuijpers, M. A. J., Bouwman, A., & Natuurplanbureau, M.-E. (2007). *Nederland later: Tweede duurzaamheidsverkenning, deel: Fysieke leefomgeving Nederland.* Bilthoven: Milieu en Natuur Planbureau.

Lange, D. A., & McNeil, S. (2004). Brownfield development: Tools for stewardship. *Journal of Urban Planning and Development, 130*(2), 109–116.

Li, X., Yang, Q., & Liu, X. (2008). Discovering and evaluating urban signatures for simulating compact development using cellular automata. *Landscape and Urban Planning, 86*(2), 177–186.

Li, X., & Yeh, A. G.-O. (2002). Neural-network-based cellular automata for simulating multiple land use changes using GIS. *International Journal of Geographical Information Science, 16*(4), 323–343.

Li, X., & Yeh, A. G.-O. (2004). Analyzing spatial restructuring of land use patterns in a fast growing region using remote sensing and GIS. *Landscape and Urban Planning, 69*(4), 335–354.

Lin, J., & Yang, A. (2006). Does the compact-city paradigm foster sustainability? An empirical study in Taiwan. *Environment and Planning B: Planning and Design, 33*(3), 365.

Linkov, I., Satterstrom, F., Kiker, G., Batchelor, C., Bridges, T., & Ferguson, E. (2006). From comparative risk assessment to multi-criteria decision analysis and adaptive management: Recent developments and applications. *Environment International, 32*(8), 1072–1093.

Livingstone, K., & Rogers, R. (2003). *Housing for a compact city*. Greater London authority.

Maria de Almeida, C., Batty, M., Vieira Monteiro, A. M., Câmara, G., Soares-Filho, B. S., Cerqueira, G. C., et al. (2003). Stochastic cellular automata modeling of urban land use dynamics: Empirical development and estimation. *Computers, Environment and Urban Systems, 27*(5), 481–509.

Mas, J.-F., Puig, H., Palacio, J. L., & Sosa-López, A. (2004). Modelling deforestation using GIS and artificial neural networks. *Environmental Modelling and Software, 19*(5), 461–471.

Matthews, R. B., Gilbert, N. G., Roach, A., Polhill, J. G., & Gotts, N. M. (2007). Agent-based land-use models: A review of applications. *Landscape Ecology, 22*(10), 1447–1459.

Mertens, B., & Lambin, E. F. (2000). Land-cover-change trajectories in southern Cameroon. *Annals of the Association of American Geographers, 90*(3), 467–494.

Nelson, G. C., & Hellerstein, D. (1997). Do roads cause deforestation? Using satellite images in econometric analysis of land use. *American Journal of Agricultural Economics, 79*(1), 80–88.

Nijkamp, P., Rodenburg, C. A., & Wagtendonk, A. J. (2002). Success factors for sustainable urban brownfield development: A comparative case study approach to polluted sites. *Ecological Economics, 40*(2), 235–252.

Nourqolipour, R., Shariff, A. R. B. M., Balasundram, S. K., Ahmad, N. B., Sood, A. M., & Buyong, T. (2014). Predicting the effects of urban development on land transition and spatial patterns of land use in Western Peninsular Malaysia. *Applied Spatial Analysis and Policy*, 1–19.

Nuissl, H., & Schroeter-Schlaack, C. (2009). On the economic approach to the containment of land consumption. *Environmental Science & Policy, 12*(3), 270–280.

Olafsson, A. S., & Skov-Petersen, H. (2014). The use of GIS-based support of recreational trail planning by local governments. *Applied Spatial Analysis and Policy, 7*(2), 149–168.

Oliver, L., Ferber, U., Grimski, D., Millar, K., & Nathanail, P. (2005). *The scale and nature of European brownfields*. Paper presented at the CABERNET 2005-International Conference on Managing Urban Land LQM Ltd, Nottingham, UK, Belfast, Northern Ireland, UK.

Openshaw, S., & Clarke, G. (1996). Developing spatial analysis functions relevant to GIS environments. In *Spatial analytical perspectives on GIS* (pp. 21–37).

Parker, D. C., Manson, S. M., Janssen, M. A., Hoffmann, M. J., & Deadman, P. (2003). Multi-agent systems for the simulation of land-use and land-cover change: A review. *Annals of the Association of American Geographers, 93*(2), 314–337.

Pijanowski, B. C., Brown, D. G., Shellito, B. A., & Manik, G. A. (2002). Using neural networks and GIS to forecast land use changes: A land transformation model. *Computers, Environment and Urban Systems, 26*(6), 553–575.

Pijanowski, B. C., Pithadia, S., Shellito, B. A., & Alexandridis, K. (2005). Calibrating a neural network-based urban change model for two metropolitan areas of the Upper Midwest of the United States. *International Journal of Geographical Information Science, 19*(2), 197–215.

Pontius, R. G., Jr., & Millones, M. (2011). Death to Kappa: Birth of quantity disagreement and allocation disagreement for accuracy assessment. *International Journal of Remote Sensing, 32*(15), 4407–4429.

Pontius, R. G., Jr., & Schneider, L. C. (2001). Land-cover change model validation by an ROC method for the Ipswich watershed, Massachusetts, USA. *Agriculture, Ecosystems & Environment, 85*(1), 239–248.

Pourghasemi, H. R., Pradhan, B., Gokceoglu, C., Mohammadi, M., & Moradi, H. R. (2013). Application of weights-of-evidence and certainty factor models and their comparison in landslide susceptibility mapping at Haraz watershed, Iran. *Arabian Journal of Geosciences, 6*(7), 2351–2365.

Pradhan, B., Oh, H.-J., & Buchroithner, M. (2010). Weights-of-evidence model applied to landslide susceptibility mapping in a tropical hilly area. *Geomatics, Natural Hazards and Risk, 1*(3), 199–223.

Rall, E. L., & Haase, D. (2011). Creative intervention in a dynamic city: A sustainability assessment of an interim use strategy for brownfields in Leipzig, Germany. *Landscape and Urban Planning, 100*(3), 189–201.

Regmi, N. R., Giardino, J. R., & Vitek, J. D. (2010). Modeling susceptibility to landslides using the weight of evidence approach: Western Colorado, USA. *Geomorphology, 115*(1), 172–187.

Sabri, S., Ludin, A. N. M. M., & Ho, C. S. (2012). Conceptual design for an integrated geosimulation and analytic network process (ANP) in gentrification appraisal. *Applied Spatial Analysis and Policy, 5*(3), 253–271.

Schädler, S., Finkel, M., Bleicher, A., Morio, M., & Gross, M. (2013). Spatially explicit computation of sustainability indicator values for the automated assessment of land-use options. *Landscape and Urban Planning, 111*, 34–45.

Schädler, S., Morio, M., Bartke, S., & Finkel, M. (2012). Integrated planning and spatial evaluation of megasite remediation and reuse options. *Journal of Contaminant Hydrology, 127*(1), 88–100.

Schädler, S., Morio, M., Bartke, S., Rohr-Zaenker, R., & Finkel, M. (2011). Designing sustainable and economically attractive brownfield revitalization options using an integrated assessment model. *Journal of Environmental Management, 92*(3), 827–837.

Shafer, G. (1976). *A mathematical theory of evidence* (Vol. 1). Princeton: Princeton University Press.

Skapura, D. M. (1996). *Building neural networks*. Addison-Wesley Professional.

Tayyebi, A., Pijanowski, B. C., Linderman, M., & Gratton, C. (2014). Comparing three global parametric and local non-parametric models to simulate land use change in diverse areas of the world. *Environmental Modelling and Software, 59*, 202–221.

Thiam, A. K. (2005). An evidential reasoning approach to land degradation evaluation: Dempster-Shafer theory of evidence. *Transactions in GIS, 9*(4), 507–520.

Thomas, M. R. (2002). A GIS-based decision support system for brownfield redevelopment. *Landscape and Urban Planning, 58*(1), 7–23.

Van Eck, J. R., & Koomen, E. (2008). Characterising urban concentration and land-use diversity in simulations of future land use. *The Annals of Regional Science, 42*(1), 123–140.

van Schrojenstein Lantman, J., Verburg, P. H., Bregt, A., & Geertman, S. (2011). Core principles and concepts in land-use modelling: A

literature review. In *Land-use modelling in planning practice* (pp. 35–57). Berlin: Springer.
Verburg, P. H., Schot, P. P., Dijst, M. J., & Veldkamp, A. (2004a). Land use change modelling: Current practice and research priorities. *GeoJournal, 61*(4), 309–324.
Verburg, P. H., Van Eck, J. R., de Nijs, T. C., Dijst, M. J., & Schot, P. (2004b). Determinants of land-use change patterns in the Netherlands. *Environment and Planning B, 31*(1), 125–150.
Walker, R. (2004). Theorizing land-cover and land-use change: The case of tropical deforestation. *International Regional Science Review, 27*(3), 247–270.
White, R., & Engelen, G. (1993). Cellular automata and fractal urban form: A cellular modelling approach to the evolution of urban land-use patterns. *Environment and Planning A, 25*(8), 1175–1199.

12 Extraction and Modeling of Urban Sprawl Development in Karbala City Using VHR Satellite Imagery

Amer D. Salman Aal-shamkhi, Hossein Mojaddadi, Biswajeet Pradhan and Saleh Abdullahi

12.1 Background

Currently, half of the world's population lives in urban areas. The rate is expected to significantly increase by the middle of the present century. Unfortunately, this rapid urban expansion has been strongly associated with poverty and slum growth. The increasing concentration of urban population in slum areas generally indicates increasing urban poverty. This process is recognized as the urbanization of poverty. According to new estimates reported by UN-HABITAT, over 200 million people in the developing countries are expected to be lifted out of slum conditions between the year 2000 and 2010. However, in the course of the same years, the number of slum residents is estimated to increase by six million every year. Based on these trends, the world's slum population will continue to grow if no proper action is taken in the coming years (UN 2009). Proper urban planning and development activities for improving the living conditions worldwide depend on the use of strong and comprehensive spatial data (Mason and Fraser 1998); such data can be obtained by remote sensing (RS) technology (Hofmann 2001; Bhatta 2010; Bhatta et al. 2010). Given that traditional methods are costly and time consuming, alternative approaches, such as sophisticated techniques, must be used to extract information from remotely sensed data.

Mapping and monitoring of unorganized urban expansion are important concerns for many national and international initiatives. Monitoring settlements is beneficial in acquiring information on various phenomena, such as illegal immigration, and is important in the current political agenda. Mapping and monitoring techniques are widely applied and particularly relevant for developing countries. Usually, no real subdivision of the land is performed, and sprawl developments are characterized by rapid, unstructured, and unplanned developments. Spatial technology may assist in analyzing the patterns of these settlements by forecasting their possible changes and providing information on the way to improve the living conditions in these areas from their present status. However, for its effective application in unorganized urban settlement, spatial technology must provide low cost data acquisition and processing. The technology must also be as automated as possible to achieve fast and reliable results, simple to use, and largely based on tested routines and algorithms. Field survey and visual interpretation of satellite data are traditionally used to produce reliable information. Nevertheless, these methods are manually operated and require considerable expertise. Apart from the operator skill bias, these methods are time consuming and cannot cope with the rapid development. In general, remote sensing and geographic information system are useful tools for providing input data and proper environment for analysis and visualization of various urban applications (Bhatta 2009, 2010; Al-sharif et al. 2013; Abdullahi et al. 2014; Abdullahi et al. 2015), especially in monitoring of urban sprawl development (Shekhar 2012; Hegazy and Kaloop 2015). In recent decades, several high spatial resolution satellite data, such as IKONOS (1999), EROS (2000), QuickBird (2001), SPOT-5 (2002), ALOS (2006), and GeoEye-1 (2008), have been applied in various aspects. These satellite data are recommended for various mapping applications because they can clearly identify many details and other elements of phenomena. Urban geographers have recognized the potential of satellite data for urban applications, including updating of maps, extraction of urban features (e.g., road networks and other engineering and social infrastructure), generation of urban models, and land use mapping. For the purpose of the current study, the challenge lies in obtaining appropriate methods to reliably detect and monitor the spatial behavior of unorganized settlements (Lemma et al. 2006). In an ideal case, these methods can be applied without expert knowledge and human interaction. In practice, ease of use and the degree of automation for information extraction from RS imagery depend on the data and phenomena to be extracted from the

B. Pradhan (ed.), *Spatial Modeling and Assessment of Urban Form*,
DOI 10.1007/978-3-319-54217-1_12

image. In this context, unorganized settlements show a relatively high inner structural heterogeneity. Consequently, describing patterns in image becomes difficult, thereby hampering the generation of an automated and easy-to-use detection process.

Considerable research on the effects of urban sprawl development and expansion on environmental, social, and economic aspects (Squires 2002; Grant 2006; Abdullahi et al. 2015; Dadi et al. 2016) clearly indicates the necessity for serious monitoring and controlling of urban growth through proper policies and strategies. Therefore, analysis, modeling, and prediction of land use change and growth are essential tasks and provide the necessary baseline information and environment for dealing with the resulting complex problems (Li and Yeh 2002). Land use changes can be modeled and predicted using several approaches, such as Markov chain (Koomen and Borsboom-van Beurden 2011), agent-based modeling (Parker et al. 2003; Grimm et al. 2006), statistical approaches (Verburg et al. 2004; Abdullahi and Pradhan 2015), artificial neural network (Pijanowski et al. 2002, 2014) and cellular automata (Li and Yeh 2000; Li et al. 2008). Among these approaches, cellular automata (CA) is the most widely applied in this field because of basing on the neighborhood interaction of surrounding cells that affect the transition and conversion of land use types. Integrating this model with other approaches, such as ANN, multi-criteria decision analysis, Bayes rule, and Markov chain, can provide a strong and comprehensive methodological approach for modeling and predicting future land use pattern.

This study dealt with the common problems of urban sprawl development and the expansion of Karbala City (Iraq) through four main objectives: (1) to develop rule sets in object-based classification for extracting land covers of the study area, including unorganized built-up area; (2) to detect and analyze land use changes in 2002, 2007, and 2013, particularly on urban growth based on the structural plan of Karbala; (3) to predict the future pattern of land use growth and changes using the CA method for the year 2024; and (4) to extract and count the houses in a part of the study area using eCognition segmentation-based methods and produce a map of unorganized settlements. This study can assist in comprehensively understanding urban pattern and behavior for effective decision making and urban planning and overcoming the present problems and limitations on sprawl development.

12.2 Study Area

The surrounding regions (northern and eastern parts) of Karbala City in Iraq are selected as the study area. The study area is an important part of Karbala City for future development. This area covers approximately 62 km^2 and is located between 32.37–32.35 N and 44.2–44.4 E (Fig. 12.1). This area is fertile and includes agricultural land and orchards rich with palms, citrus, and various fruit trees. Karbala is an important province in Iraq and is 100 km away from the capital city (Baghdad). The entire area of Karbala is 5228 km^2. This province is one of the holy cities in Iraq and has an estimated population of 1,013,500 in 2008.

Karbala has experienced a steady growth over the last two decades. Although urban growth is perceived as necessary for a sustainable economy, uncontrolled or sprawling urban growth results in various problems. Urban sprawl not only rapidly consumes the precious rural land resources at the outskirt of the city, but also results in landscape alteration, environmental pollution, traffic congestion, infrastructure pressure, rising taxes, and neighborhood conflicts. Unfortunately, urban growth prediction at the regional level in the entire Karbala City has not been conducted. Without information generated from reliable predictions, discussions or debates on these issues will remain at a superficial level.

Iraq has been subjected to harsh conditions of international and internal wars. These conditions have significantly affected the economic situation of the country. Consequently, the quality of life has been seriously affected, resulting in deterioration of agriculture and industrial sectors and difficulty in residential development for poor citizens. These issues have increased rural–urban migration for job opportunities to fulfil people's basic needs. However, the high cost of buying or renting existing houses has stimulated the growth and expansion of improper and unorganized residential environments around the city centers. After the US-led war against Iraq and the fall of the previous government, the local owners of orchard fields near the city centers began to divide their vast lands into small residential parcels. They then sold these portions to poor families without obtaining proper approval from the local planning authorities. The most recent war in Iraq has resulted in a substantial wave of internal and external displacement along with increased sectarian violence and ethnic tension. The subsequent conflict has exacerbated conditions within the nation and further increased displacement. Karbala Province received a large number of these displaced families after the deterioration of the security situation in Baghdad. Therefore, proposing a robust technique for modeling land use changes and extracting urban sprawl development is highly required for Karbala City and other cities in Iraq.

12.3 Data and Methodology

The methodological processes of this study (Fig. 12.2) are listed below.

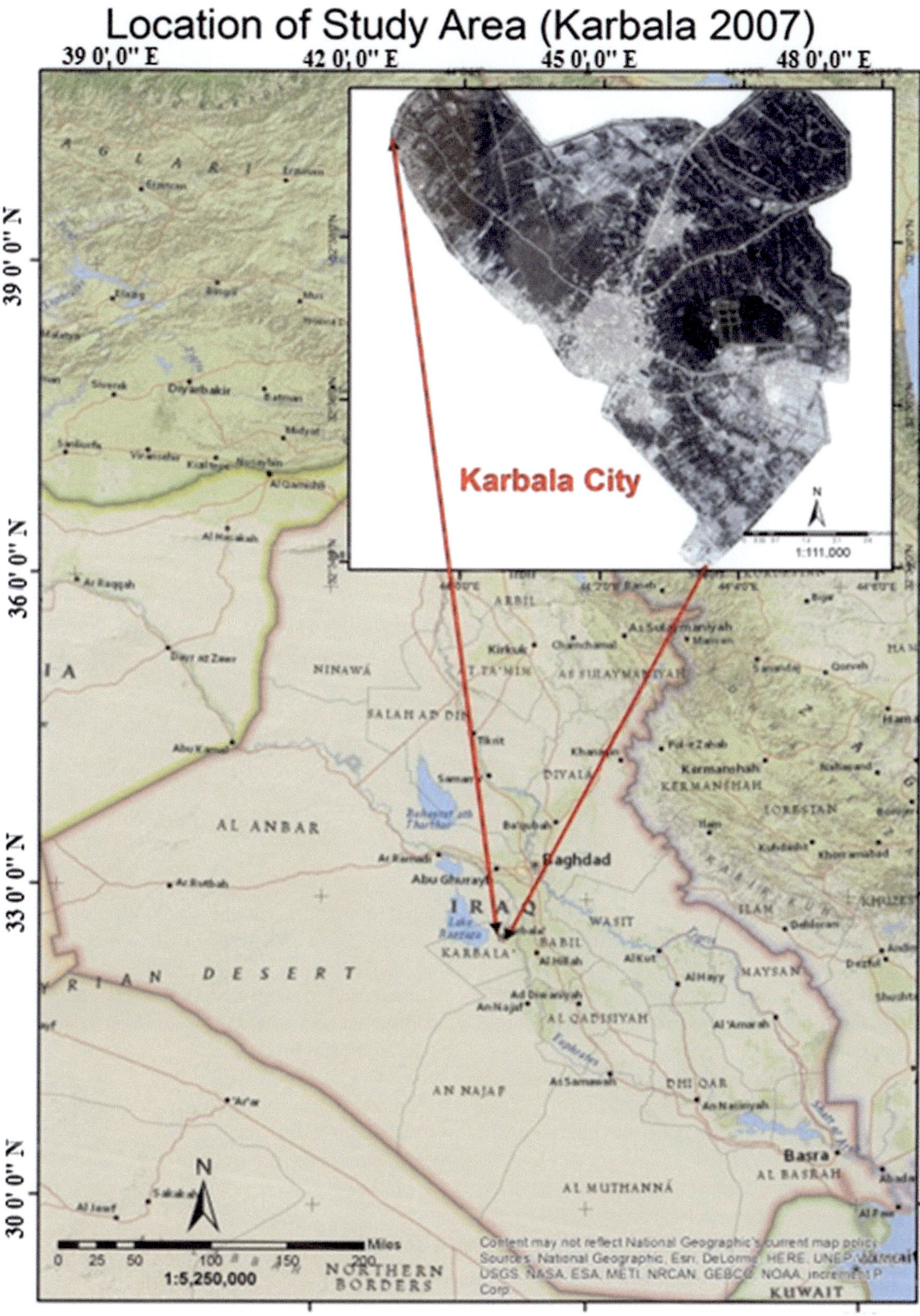

Fig. 12.1 Location of study area (Karbala city)

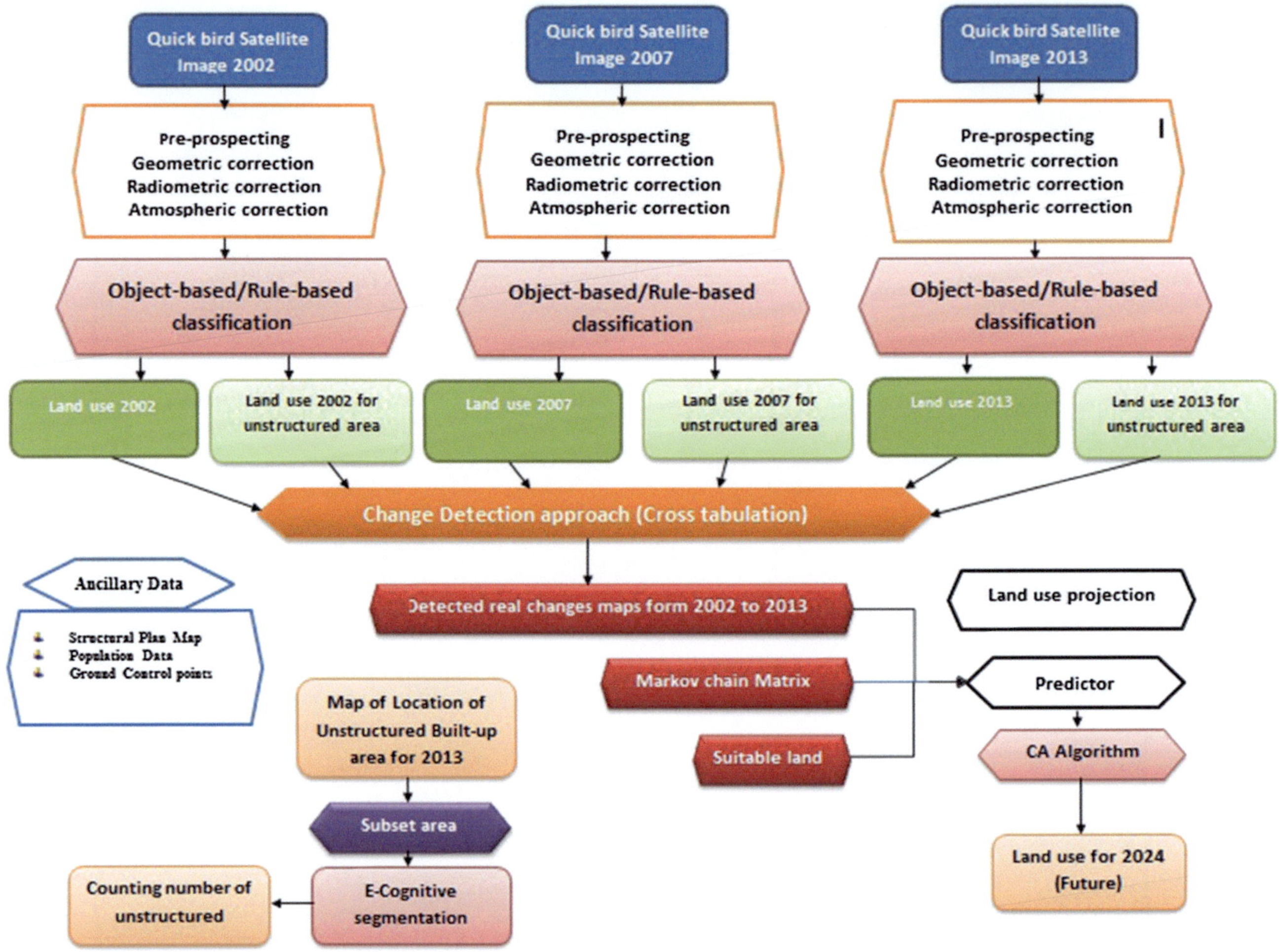

Fig. 12.2 Overall methodological process

- Image processing and analysis
    i. Pre-processing: to correct the geometric, radiometric, and atmospheric errors.
    ii. Processing: to subset the images and apply the image classification method and change detection in three scenarios.
    iii. Post-processing: to produce a map, layout the results, and check accuracy assessment.
- Land use change modeling: to predict the future growth and change pattern.
- Counting the unorganized houses inside a subset of the study area using object-based image analysis (OBIA) and eCognition segmentation method.

Table 12.1 and Fig. 12.3 show the image processing steps (QuickBird image). Three very high spatial resolution images (0.6 m) were utilized. These images provided detailed information of the location and distribution of roads, buildings, orchards, and other structures. They contained the ground information and the spatial location of developments in 2002, 2007, and 2013 of all centers and outskirts of Karbala City in Iraq.

In addition to these satellite images, the following set of ancillary datasets was utilized to improve the accuracy and reliability of the process:

- Structural plan of the city,
- Population Data, and
- GCPs Ground Control point Points 106 points during two weeks.

The structural plan of Karbala City was collected from the ministry of municipalities. This plan was approved in 2006 by the prime minister of Iraq. Population data were used to evaluate the urban growth with population and socioeconomic factors. Table 12.2 presents the total population in Iraq and Karbala City in 2002, 2007, and 2013.

Table 12.1 Characteristics of Quickbird images (2002, 2007, and 2013)

Spatial resolution	Panchromatic	Multispectral
	0.6 m GSD	2.4 m GSD
Spectral range	445–900 nm	(blue) 450–520 nm (green) 520–600 nm (red) 630–690 nm (near IR) 760–900 nm

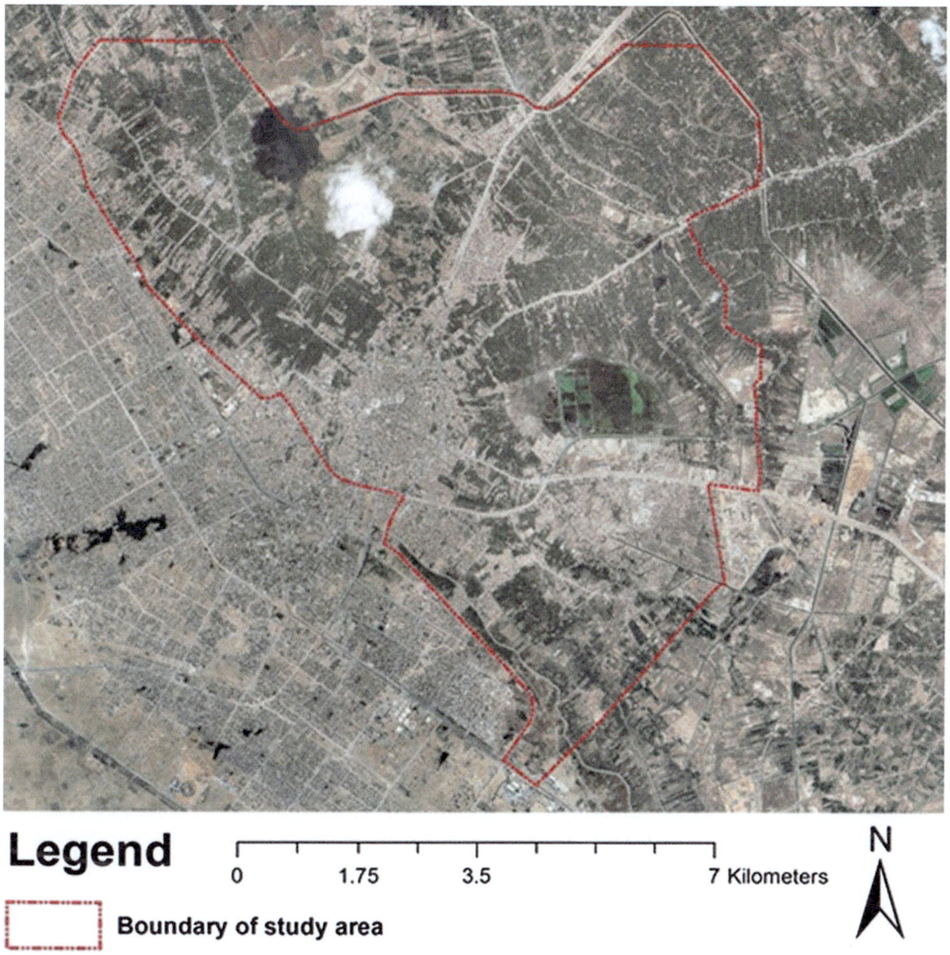

Fig. 12.3 Study area overlapped on Quickbird image (2013)

Table 12.2 Population data of Karbala

No.	Year	Population (Iraq)	Population (Karbala)
1	2002	34,208,000	755,995
2	2007	25,565,000	1,013,254
3	2013	29,682,000	1,663,500 (1,113,500 original + 550,000 displaced)

Source the statistical center of Karbala

12.3.1 Image Preprocessing and Classification

Prior to data analysis, initial processing on the raw data, such as radiometric, atmospheric, and geometric correction, was performed to correct all the distortion and errors attributed to the characteristics of the imaging system and conditions. In addition, the cloud-covered area was corrected through two steps: the spectral range of the cloud was analyzed using the image histogram first, and then a threshold was applied to extract the cloud-covered area and results were further refined by manual editing. The resulting image used for the processing was cloud free. Moreover, the image was compared with the closest image available (2009) to extract the cloud-covered area and replace it with the included features from the 2009 image.

Numerous image processing and analysis techniques have been developed to aid the interpretation of RS images and extract as much information as possible from the images. The choice of specific techniques or algorithms depends on the goals of each individual project. For the current study, several procedures commonly used in analyzing and interpreting Quickbird images were examined for extracting and monitoring much information on the sprawl development in Karbala City. The rule-based algorithm of the object-oriented classification method was applied on three available images (2002, 2007, and 2013). This algorithm is reported as the best method for this purpose because its spectral and spatial properties are more accurately than those of the pixel-based algorithm. Regarding the occurrences of settlement areas in RS data, pixel-based approach on a high-resolution image cannot represent the heterogeneity of complex urban environments. Hence, sophisticated method and data for slum analysis must be used. Using multi-resolution segmentation, initial objects were created according to texture, geometry, and contextual characteristics of the image objects and were classified into slum and non-slum areas.

OBIA is being developed recently contrary to traditional pixel-based image analysis. Pixel-based image analysis is based on the information in each pixel, whereas OBIA is based on information from a set of similar pixels called objects (Hamedianfar and Shafri 2015). More specifically, image objects are groups of pixels that are similar to one another based on a measure of spectral properties (i.e., color, size, shape, and texture) and a context from a neighborhood surrounding the pixels. In the present study, the image pixels were grouped on the basis of the spectral and spatial characteristics. This process requires a set of parameters, such as scale, shape, and compactness, which need to be specified for the process. Using this classification process, seven land covers (road, slough, water body, agriculture, built-up, orchard, and wasteland) were extracted from the utilized images.

Common accuracy measurements (overall accuracy and kappa coefficient) were implemented to evaluate the classification process. These measurements require training data from the field or other reliable sources, such as reference maps from agencies. The training data used in the current study were from the field and were collected by GCPs in 106 samples. These data were equally distributed to all classes together with other important locations from Google Earth.

12.3.2 Land Use Change Analysis and Modeling

Change detection analysis was implemented on classified images to describe and quantify differences between images of the same scene at different times. For this purpose, cross tabulation process was applied to individually extract conversion among various land use categories in the period from 2002 to 2013. The classified images of the three dates were used to calculate the area of different land covers and observe the changes that occurred in the span of the data. By comparing the three classified images, the unorganized settlements and number of houses in the study area were extracted and calculated.

Markov chain module was used to analyze the pair of land cover images. As a result, transition probability, transition area matrices, and a set of conditional probability images were produced. The transition probability matrix records the probability that each land cover category will change to every other category. The transition area matrix records the number of pixels that are expected to change from each land cover type to another over the specified number of time units. In these matrices, the rows represent the older land cover categories and the columns represent the newer categories. The conditional probability images report the probability that each land cover type will be found at each pixel after the specified number of time units. In the present study, these images were calculated as later projections of the two input land cover images. The output conditional probability images can be used as direct input for specification of the prior probabilities in maximum likelihood classification of remotely sensed imagery. On the basis of these outputs, a CA model was designed in this study for urban growth modeling to simulate the process of urbanization in a hypothetical region. This model comprises a set of rules that describe the spatial interaction of cells and a set of parameters that indicate different urban forms. Furthermore, different results of the model can be evaluated throughout by fractal analysis and the estimation of the fractal dimension. A significant connection was observed between the parameters of the model and the value of the fractal dimension. Several important factors were applied into the model to aid its feasibility in examining real urban patterns, such as landscape constraints, transportation networks, protected areas, and physical geographies (Apostolos 2010).

The design of the CA algorithm comprises defining the transition rules that control the urban growth, calibrating these rules, and evaluating the results for prediction purpose. Defining the transition rules is the most important phase in CA model design because these rules translate the effect of input data on the urban process simulation. Therefore, an accurate and realistic definition of the rules must be realized.

In this study, transitional rules were designed as a function of land use effect on the urban process, growth constraints, and population density. The transition rules were defined over the 3 × 3 neighborhood of a pixel to minimize the number of input variables to the model. The rules identify the necessary neighborhood for the tested cell to remain the same or change to another land use category. The growth constraints must reflect the conservation strategy adopted in the study area for certain land uses. For example, conservation of certain species of natural resources can be considered through these rules (Alkheder et al. 2006). Thus, the process was applied for land use maps of years 2002, 2007, and 2013 to project the future land use map for the year 2024.

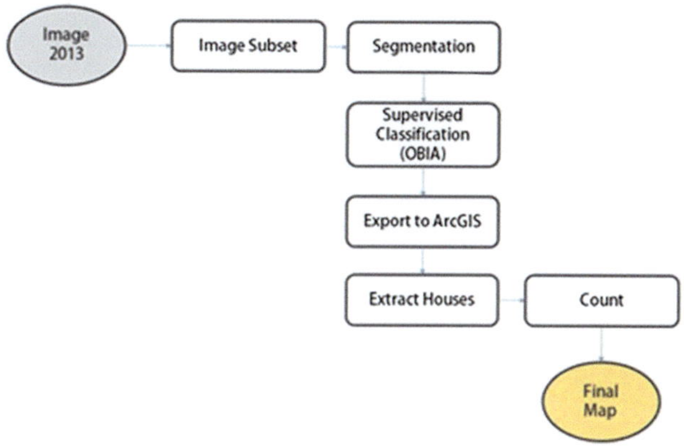

Fig. 12.4 Methodological flowchart for extraction of unorganized houses

12.3.3 Counting of Unorganized Houses Using OBIA

The process of counting unorganized houses starts from selecting a subset of the original image to show the efficiency of the developed method. First, the image was segmented using the trial and error approach and with the following parameters: scale = 10, shape = 0.9, and compactness = 0.5. The segmentation was classified using the supervised OBIA method with the spectral and spatial features. The result of the classification was then converted to ArcGIS data for further analysis. In ArcGIS, the houses were first extracted using the selection by attribute and SQL structure. The filtering was then applied to the area of the segments for counting, thereby generating several logical segments representing houses. In this step, the rule "Area >40 AND Area <500" was used. Using this rule, the original houses were filtered and the final segments were exported into a new dataset. Using the new dataset, the houses were counted on the basis of the number of segments left. The final map was produced by selection and preparation of the final houses as segments (Fig. 12.4).

12.4 Results and Discussion

12.4.1 Rule Sets Development for Land Use Classification

A set of rule sets were developed in ENVI software using the trial and error approach combined with the knowledge of the authors regarding the study area. The rules were developed to classify the satellite images into seven land covers identified using the Anderson scheme (Table 12.3).

Using the rule sets developed in this study, the satellite images were classified into seven classes (Fig. 12.5). The built-up areas were mostly clustered in the city center and the surrounding areas were mostly covered by agriculture and orchards. The sprawl developments were clearly observed within orchard areas. This finding is attributed to that the farmers started dividing their farms into small patches from 40 to 500 m^2. In the 2007 image and classification map, the built-up areas were expanded toward the south of the study area of approximately 2.24 km^2. In 2013, the built-up areas were distributed mainly in the center, north, and southeast of the city. In addition, the built-up areas were expanded toward the southeast and north of the study area of approximately 6.5 km^2. In 2013, the percentage of the agriculture and orchard classes was reduced because of the expansion of built-up areas. In other words, the government also expanded the development from the center toward all other directions. In addition, owners of orchards kept on dividing their lands into unorganized houses and selling these portions to poor citizens and displaced people.

In 2002, the study area was mostly covered by orchards and agriculture areas containing palm dates and other seasonal crops (covering more than 80%). The other components of built-up areas and other classes are shown in Table 12.4. This table shows the areas in square meters as determined from the land classification map of 2002, 2007, and 2013. The observed areas of built-up land cover class were 6,794,506.8, 9,038,500.2, and 15,521,743 m^2 in 2002, 2007, and 2013, respectively. The built-up areas included the proper development by the government and unorganized houses developed by local citizens. In addition, the spatial distribution of the organized development was found clustered in the city center. However, the unorganized houses were distributed in the green lands around the city center.

One of the important topics in RS and image classification is the accuracy assessment. In this study, the accuracy of the land cover classification was determined using four common accuracy measures (overall, kappa, and user and producer accuracy). The results of the accuracy assessment are shown in Table 12.5. The table shows that the overall

Table 12.3 Rules of each class in rule-based classification approach

Class		Rules	Description
1	Road	If avgband_1 <330 AND avgband_4 >374 AND Length ≥ 33	Roads have low values of NDVI, low reflectance in the blue band but a high reflectance in the NIR band
2	Orchard	If avgband_3 <360.4 AND tx_range >28.2 AND avgband_4 >2.4	Orchards have high values of NDVI and highly textured
3	Built-up	If avgband_1 <355.1 AND tx_range >15.5 AND tx_variance <167.9 AND avgband_4 >242.8 AND Area >50	Built-up areas have low values of NDVI and low reflectance in the blue band. The texture variance is small, but the range is big. Most buildings are higher than 2.8 m
4	Water bodies	If tx_range <12.4 AND avgband_4 < 460.3	Water has low values of NDVI, low reflectance in the NIR band and a small texture range
5	Agriculture	If avgband_3 >420 AND avgband_1 >310.6 AND avgband_2 <480	Land has low values of NDVI and high reflectance in blue and almost NIR bands
6	Wasteland	If tx_range <13.5 AND avgband_4 <460	Wasteland has high values of NDVI, but not highly textured
7	Slough	If avgband_4 >420 AND avgband_1 <360 tx_mean >245.6	Slough and wetland have high reflectance in the NIR band, but low reflectance in the other bands. They are highly textured

Table 12.4 Area of each class of land cover

ID	Class	2002	2007	2013
1	Road	2,322,735.48	2,481,872.04	3,218,462
2	Slough	1,374,043.68	2,090,996.64	1,406,288
3	Water Body	1,200,949.56	1,206,203.04	988,364.2
4	Agriculture	16,695,924.2	10,956,547.8	13,213,710
5	Built-up Area	6,794,506.8	9,038,500.2	15,521,743
6	Orchard	33,696,323.3	36,308,532.24	27,735,324
7	Wasteland	390,343.32	390,346.92	394,275.2
Total		62,474,826.4	62,472,998.88	62,478,166

Table 12.5 Accuracy assessment for classification of all three land covers

Land use map	Overall accuracy (%)	Kappa coefficient
2002	83.77	0.79
2007	84.62	0.80
2013	80.51	0.75

accuracies of the land classification map for the year 2002, 2007, and 2013 were 83.77, 84.62, and 80.51%, respectively. Meanwhile, the user accuracies of the built-up areas were 79.39, 83.56, and 80.79% for the three land use maps.

The area of organized housing development was excluded from the original land cover map to check the distribution and area of the classes without the organized houses, especially the sprawl development. Most of the houses observed in built-up areas were significantly distributed in the green land compared with the organized houses. Therefore, the expansion of the houses in Karbala City is insufficient and strategic planning is necessary for further development to aid in better planning and managing of the city in the future. Figure 12.5b, d, and f show the land cover classification map excluding the proper housing development in 2002, 2007, and 2013.

The surface areas of land covers without city illustrations are shown in Table 12.6. This table shows the areas in square meters as determined from the land classification map for the three land use maps. The areas of built-up land cover class were 4,806,455, 6,324,836, and 11,556,451 m^2 for 2002, 2007, and 2013, respectively. The built-up areas only

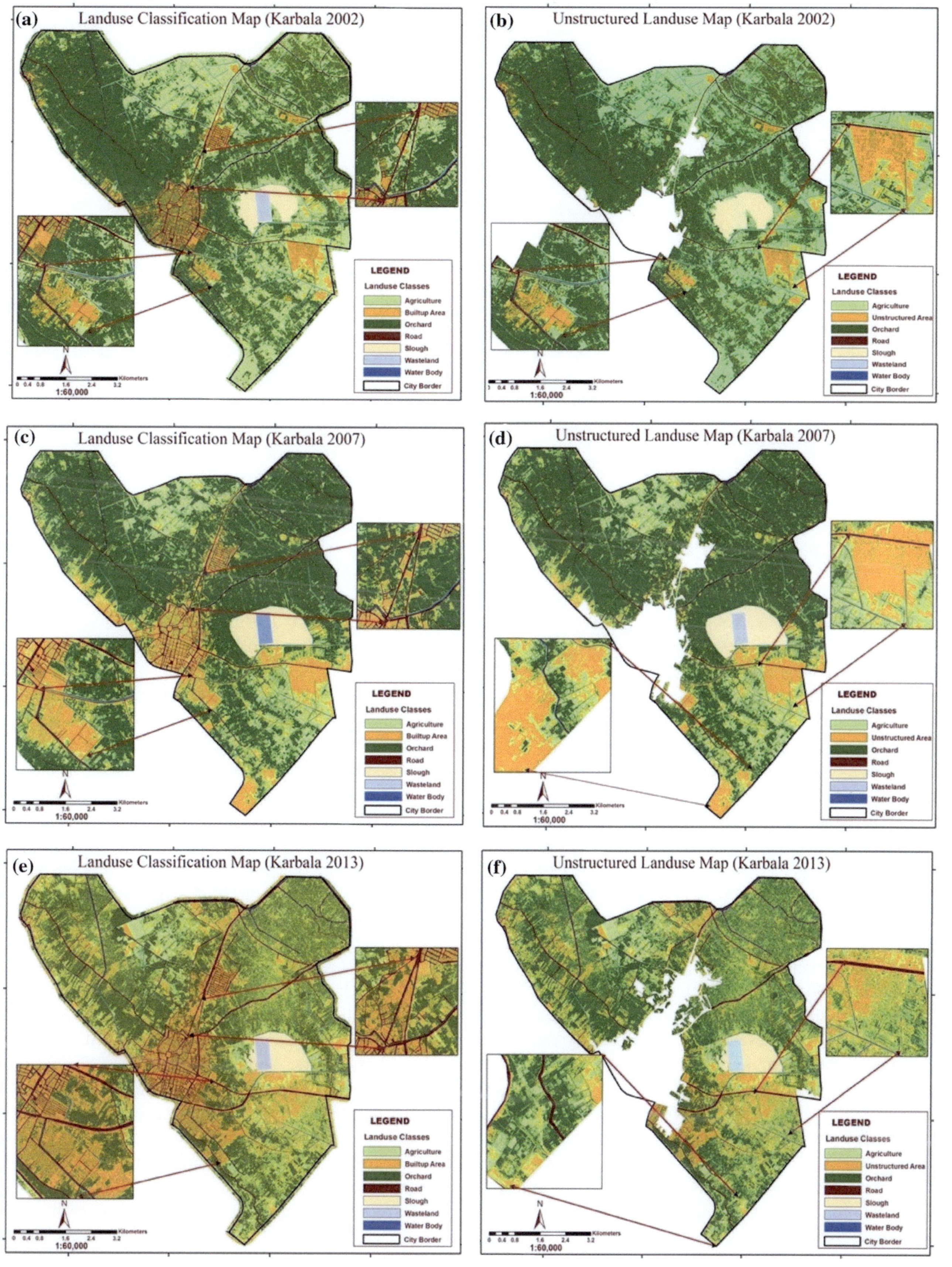

Fig. 12.5 Land use classification map for years 2002, 2007 and 2013; and unstructured land use maps (unorganized urban development) for corresponding classified images

Table 12.6 Area of each class of land covers 2002 in m^2

ID	Class	2002	2007	2013
1	Road	1,495,122	1,502,163	1,881,840
2	Slough	1,374,044	2,090,997	1,406,288
3	Water body	1,200,040	1,187,180	943,501
4	Agriculture	16,646,498	10,109,991	12,949,531
5	Built-up	4,806,455	6,324,836	11,556,451
6	Orchard	32,291,047	35,605,549	25,844,092
7	Wasteland	390,343.3	390,346.9	394,275.2
Total		58,203,549	57,211,062	54,975,979

included sprawl housing development. In addition, the spatial distribution of the unorganized houses was distributed in the green lands around the city center. The growth of these unplanned built-up areas in the farmlands and the reduced orchards was roughly the same as that of the organized built-up areas. This phenomenon is attributed to the approximately the same percentage of agriculture and orchards in the study area in both years.

12.4.2 Change Detection and Analysis

The trend of land use change and urban growth of the study area from 2002 to 2013 were analyzed to achieve the second objective. The analysis was based on thematic change maps, which show the conversion of various land cover classes into other classes. In the 2002–2007 analysis, 1,383,356.52 m^2 from the agricultural field were converted into built-up areas. This conversion indicates that orchards were converted into built-up areas (as shown before in the area of the classes in three years). The agriculture areas were also converted, which were most likely transformed into sprawl development. Most of these transformations occurred in the southeast and south parts of the study area. The other transformations among the classes are shown in Table 12.7.

The 2007–2013 change map shows that the expansion of built-up areas increased from 2007 to 2013 by exactly 4,904,657.28 m^2 (from the agriculture class). The conversion of 2,115,230.04 m^2 of orchards into built-up areas shows a substantial expansion of built-up areas, indicating the growth of sprawl development in the study area. The other transformations among other classes are presented in Table 12.8

The spatial distribution of land cover change shows that the expansion of built-up areas was generally toward the southeast and northwest of the study area. Most of the sprawl urban growth and expansion occurred toward and close to the central parts. Significant numbers of orchard areas were converted to agriculture and built-up areas from 2007 to 2013. The analysis of long-term change detection from 2002 to 2013 (Table 12.9) shows that 2,289,665 m^2 of orchard and 1,245,898 m^2 of agriculture were converted into built-up areas, and most of the built-up areas were unorganized development. This expansion mostly occurred toward the center, northwest, and southeast of the city.

By specifically performing change detection analysis on sprawl development, nearly 531 ha of observed orchards were replaced by built-up areas, followed by 443 ha of agricultural lands converted to housing areas. Moreover, 653 ha of orchard areas were replaced by agricultural lands. However, wasteland slough and water bodies did not show any substantial changes during these years.

12.4.3 Land Use Prediction Analysis

CA_MARKOV is usually used in modeling and predicting future land use change and growth. This method is a combination of CA, Markov chain, multi-criteria, and multi-objective land allocation. The land cover of the study area of the year 2024 was predicted using this integrated approach. This prediction was conducted to analyze the future trend of built-up area expansion in the study area, especially the unorganized and sprawl development. The land cover map of the year 2024 is presented in Fig. 12.6. This map shows the spatial distribution of the seven land classes for 11 years, from 2013 to 2024. The prediction was generated on the basis of the analysis of the land covers of the years 2002 and 2013.

The prediction of land cover of the year 2024 shows that agriculture and orchard classes will be reduced by 22.8 and 3.3 km^2. Meanwhile, the built-up area will be significantly increased by 5.31 km^2 from 2013 to 2024. This analysis indicates that the city will lose considerable orchard lands and most of the transformations will occur to the built-up class. Therefore, strategic planning and comprehensive management are required to avoid these conversions.

With regard to area comparison of land cover classes for the selected period of time, Table 12.10 and Fig. 12.7 illustrate the growth and loss in the areas of all the classes.

Table 12.7 Change detection analysis between 2002 and 2007 (m^2)

2002									
		Road	Slough	Water body	Agriculture	Built-up	Orchard	Wasteland	Total
2007	Road	2255947	0	15700.32	36622.8	37977.84	111874.32	0	2481801.5
	Slough	0	1373310	442.08	445054.68	36023.04	234574.56	346.32	2090996.6
	Water body	14962.68	262.44	1161933.12	3307.68	1618.2	12405.6	0	1205973
	Agriculture	10781.28	15.48	5662.44	5247970.2	983356.52	4301472.96	0	10956296
	Built-up	21741.84	19.8	3599.28	3259065.96	4663033.64	2780222.04	0	9038161.4
	Orchard	12695.76	45	10152.36	7665167.52	1065648.92	26241102	0	36304292
	Wasteland	0	349.92	0	0	0	0	389997	390346.92
	Total	2322735	1374044	1200949.56	16665923.52	6794506.8	33696322.56	390343.32	

Table 12.8 Change detection analysis between 2007 and 2013 (m^2)

2007									
		Road	Slough	Water	Agric.	Built-up	Orchard	Wasteland	Total
2013	Road	2170825.2	3.6	249324.5	177068.5	295888.32	318048.12	0	3218462
	Slough	0	1401570.36	552.6	5.04	3.24	12.96	4121.28	1406288
	Water	58757.76	1986.12	856323	19741.68	24913.08	24213.6	0	988364.2
	Agriculture	34494.12	293962.68	23296.32	3319155	1394705.04	7842802.68	0	13213709
	Built-up	126042.84	224265.6	38919.6	3811339	6104657.28	6397741.8	0	15521743
	Orchard	72924.48	160915.32	27263.88	3626787	1215230.04	21717690.12	0	27735204
	Wasteland	0.36	8049.6	0	25.56	60.84	33.48	386225.64	394395.5
	Total	2481801.48	2090996.64	1205973	10956296	9038161.44	36304295.04	390346.92	

Table 12.9 Change detection analysis between 2002 and 2013 (m^2)

2002									
		Road	Slough	Water body	Agriculture	Built-up	Orchard	Wasteland	Total
2013	Road	2051690	0	243383.8	358689.2	180643.3	357718.3	0	3218462
	Slough	0	951319.4	600.84	335339.3	35510.4	78885.72	4048.2	1406288
	Water body	52091.28	1682.28	869104.1	16444.08	4646.16	36606.96	0	988364.2
	Agriculture	30286.8	169414.2	20536.2	5027251	945898	6715347	0	13213709
	Built-up	109921	130855	35922.24	4901886	4236170	7291858	0	15521743
	Orchard	66079.44	112389.1	23833.8	6022176	1389665	19209284	0	27735204
	Wasteland	0	7980.12	0	9.36	41.76	69.12	386295.1	394395.5
	Total	2322735	1374044	1200950	16665924	6794507	33696323	390343.3	

Figure 12.7 clearly shows that most transformations and changes occurred in agriculture, orchard, and built-up classes. As discussed earlier, the main conversion is the growth of built-up areas from orchard fields. This analysis also confirms that the expansion of the sprawl development is annually increasing in the city.

12.4.4 Counting the Unorganized Residential Buildings

On the basis of the image segmentation analysis, a step was further applied to count the number of unorganized residential buildings in a subset selected from the study area.

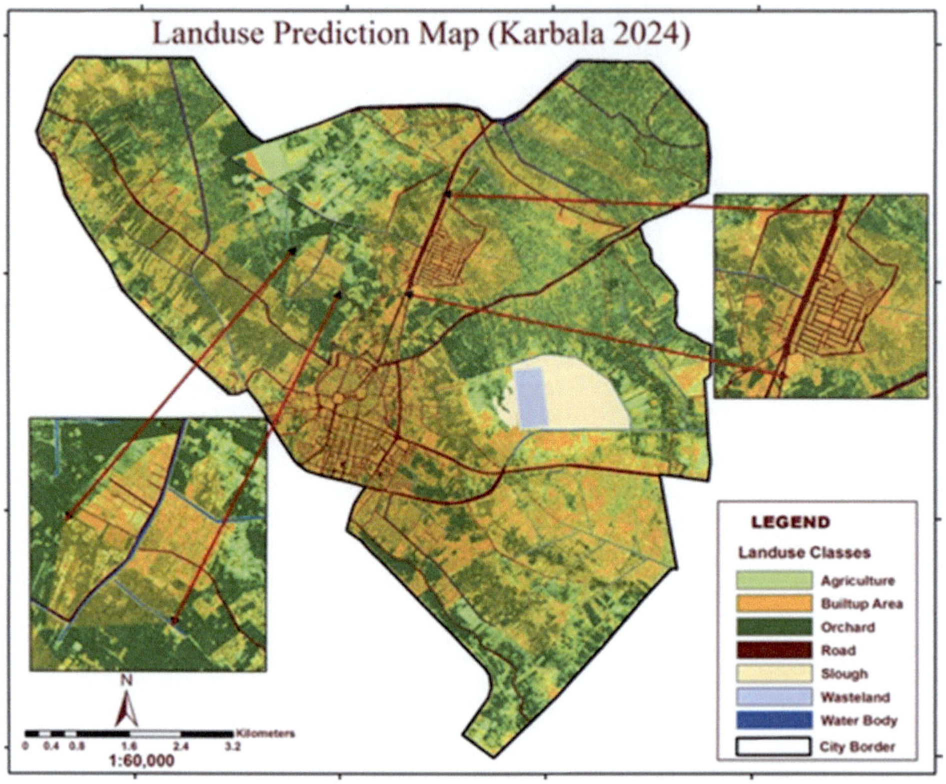

Fig. 12.6 Projected land use map for year 2024 using CA_Markov

Table 12.10 Comparison between land covers during the 22 years (2002–2024)

Class	2002	2007	2013	2024
Road	2,322,735.48	2,481,872	3,218,462	3,459,505.3
Slough	1,374,043.68	2,090,997	1,406,288	1,410,465.8
Water Body	1,200,949.56	1,206,203	988,364.2	999,835
Agriculture	16,695,924.24	10,956,548	13,213,710	10,933,905.8
Built-up Area	6,794,506.8	9,038,500	15,521,743	20,840,306.8
Orchard	33,696,323.28	36,308,532	27,735,324	24,437,469.3
Wasteland	390,343.32	390,346.9	394,275.2	390,345.8
Total	62,474,826.36	62,472,999	62,478,166	62,471,833.5

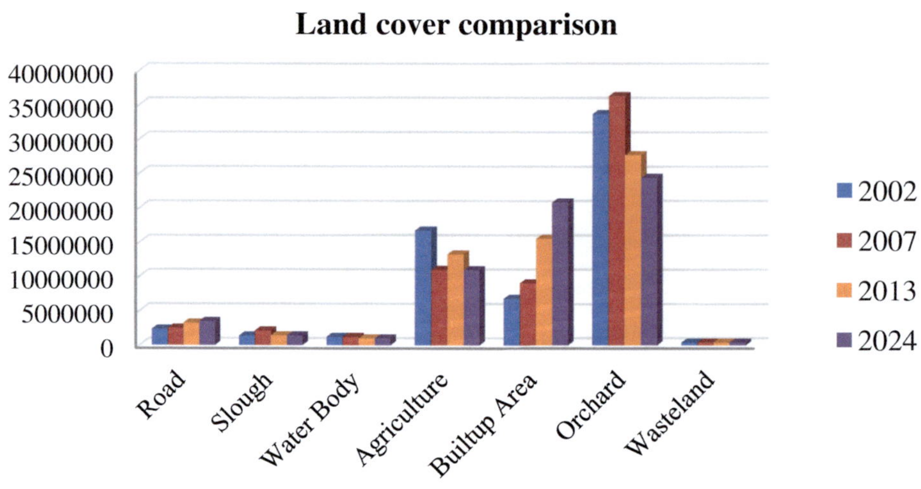

Fig. 12.7 Surface areas of land cover classes in different times (2002, 2007, 2013, and 2024)

The step of segmentation grouped the pixels of the image and created segments from these groups. The segments were then converted to vectors in ArcGIS software (Fig. 12.8). The conditional rule of "40< AREA AND AREA <500" was applied to extract unorganized houses. The lower limit of 40 m^2 was selected because this area is the minimum possible area that can be sold by the owners. The upper limit of 500 m^2 was selected because this area is the largest area used to build a house. Given that the remainder of this area may be used for agricultural purposes, the owners normally divide their land parcels to less than 500 m^2. In addition, selling these lands for the built-up is much costly than selling them as agriculture fields.

According to the rules and conditions of selling agricultural lands for housing, a filter was designed to improve the result of house counting from image segmentation. The filter is a structure of SQL language that can be used in ArcGIS software by applying the selection by attribute tool.

The image segmentation and designed filter were applied on the 2013 satellite image to count the number of houses in a subsetted image. The results of this application are presented in Fig. 12.9. This figure shows the houses in red color generated by the image segmentation process in the eCognition software. After the application of the filter, the houses were extracted individually (right-hand side of the figure).

Figure 12.9 also shows the image segments before and after the filtering process. The map in Fig. 12.10 shows the difference between image segments before and after the filtering process. The importance of the designed filter for house counting is also reflected in the map.

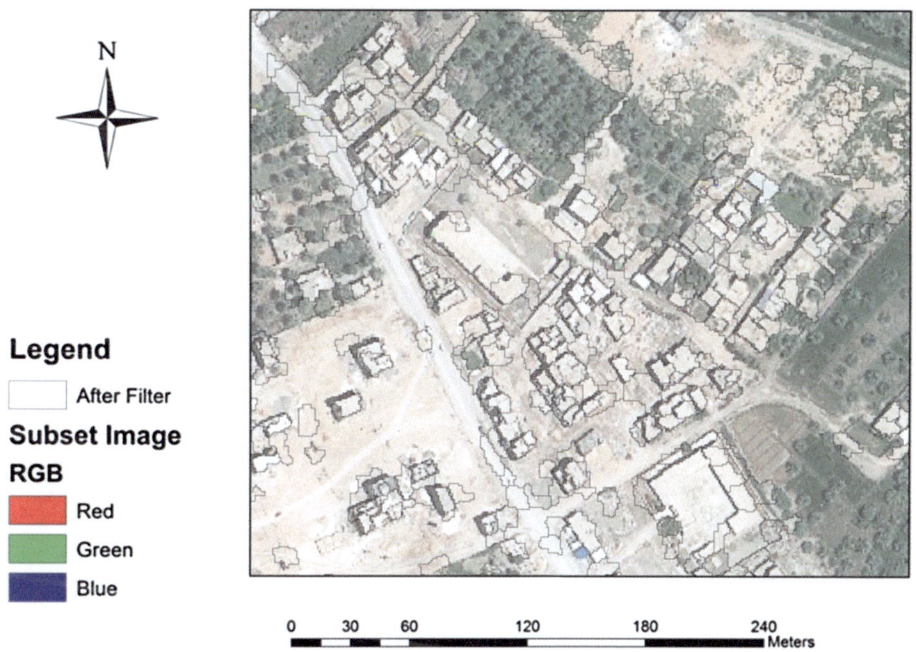

Fig. 12.8 Subset from the study area shows the segmentation of unorganized housing development

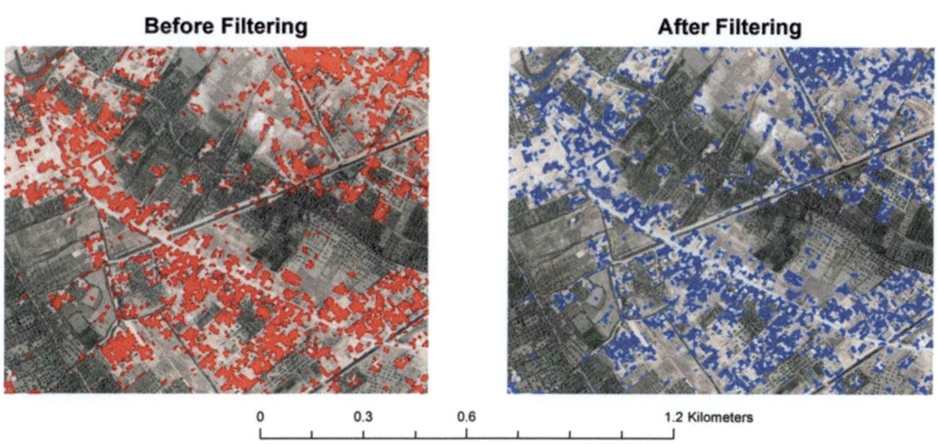

Fig. 12.9 Results of detected unorganized houses before and after filtering

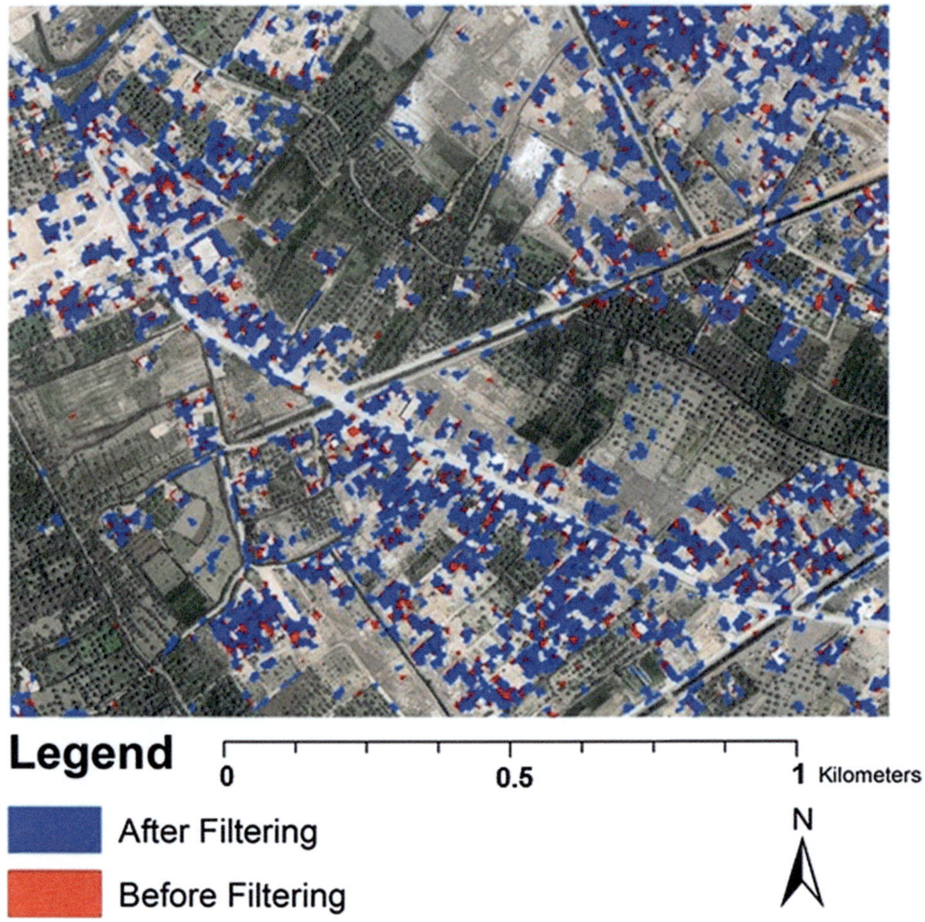

Fig. 12.10 Result of detected unorganized houses of both before and after filtering combined together

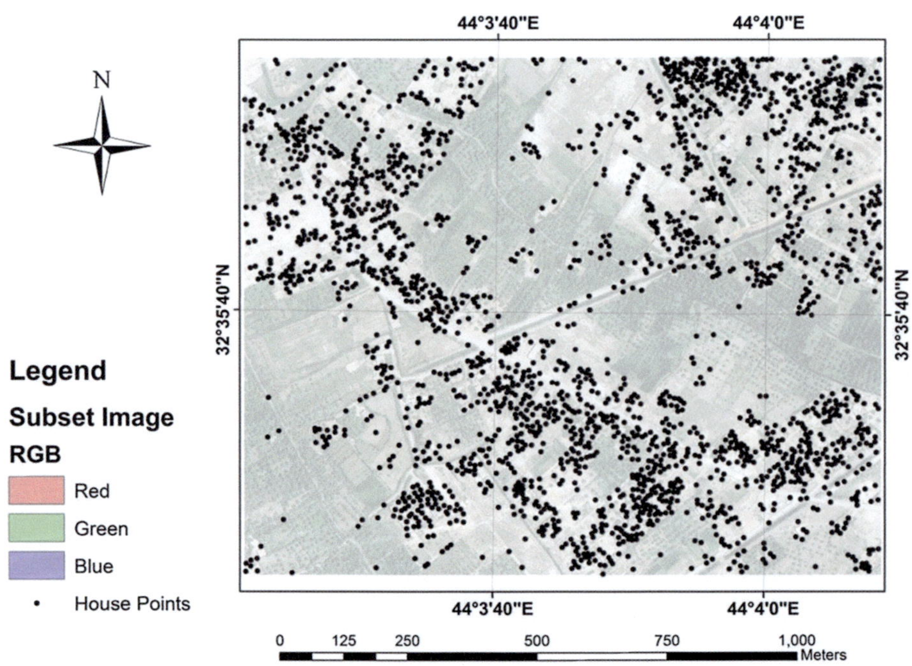

Fig. 12.11 Detected houses as points created from the polygons

The vector format of the housing layer was converted to a point format to count the number of points that are similar to the number of unorganized houses in the subset of the study area (Fig. 12.11). The result of the house extraction process shows that the number of houses in the subset selected from the study area was 2,217 in the year 2013.

12.5 Conclusion

The growth of unorganized housing development commonly known as urban sprawl development creates several problems in various social, environmental, and economic aspects. Particularly, unorganized housing expansion has inappropriate infrastructure and utilities, thereby destroying valuable green and natural environments and creating several environmental issues attributed to lack of proper sewerage and waste management system. Thus, monitoring and controlling unorganized expansions is essentially important for governments and local authorities to avoid these issues and problems. Using RS technology and its related methodologies and approaches in extracting information from remotely sensed data provides strong assistance to the authorities in controlling and managing urban growth and expansion.

The environmental degradation in several cities of Iraq was a result of internal and external wars. Particularly, the war in 2003 has forced people to migrate to safe cities, such as Karbala, to live and protect themselves from the terrorism threats. These situations have increased the conversion of agricultural fields to informal residential buildings in Karbala City. In addition, the local government could not control the sprawl development and expansion because of internal and external wars; therefore, this phenomenon has significantly increased in recent years.

This study aimed to extract, analyze the changes, and predict the urban expansion in Karbala City, Iraq. The most effective classification approach (rule-based object-oriented technique) for land cover classification was used to extract various land cover classes from high spatial resolution imagery. In general, the accuracy assessment process indicates that uniform land covers, such as water bodies and wastelands, were extracted with high accuracy. Meanwhile, built-up areas attributed to several distinction properties, such as geometrical shape and format, could be extracted by defining proper rules. In terms of the second objective, this study successfully highlighted and discovered the spatiotemporal change pattern of urban land cover in Karbala City. The process shows a significant growth of built-up areas in the orchard and agricultural fields. In addition, the projected map for the year 2024 provides valuable information on the future land cover growth and changes, especially the expansion rate and direction of sprawl development. This information can aid the local government in controlling and managing these growths and additional provisions of proper infrastructures and facilities. Thus, this process provides a future vision of urban pattern for the government. The information provided by this study also provides beneficial ideas to the decision makers and warns them regarding the consequences of sprawl development in Karbala City. The results indicate that, if the current rate of the trend of land cover change continues, then Karbala City will soon suffer from lack of green environments and orchard fields.

Acknowledgements The authors would like to thank the Information Center of Karbala, the Thu Shaan Office in Karbala, and the Joint Coordination Center of Police Directorate of Karbala for providing parts of data required to complete this work.

References

Abdullahi, S., Mahmud, A. R. B., & Pradhan, B. (2014). Spatial modelling of site suitability assessment for hospitals using geographical information system-based multicriteria approach at Qazvin city Iran. *Geocarto International, 29*(2), 164–184.

Abdullahi, S., & Pradhan, B. (2016). Sustainable brownfields land use change modeling using GIS-Based weights-of-evidence approach. *Applied Spatial Analysis and Policy, 9*(1), 21–38.

Abdullahi, S., Pradhan, B., & Jebur, M. N. (2015a). GIS-based sustainable city compactness assessment using integration of MCDM, Bayes theorem and RADAR technology. *Geocarto International, 30*(4), 365–387.

Abdullahi, S., Pradhan, B., Mansor, S., & Shariff, A. R. M. (2015b). GIS-based modeling for the spatial measurement and evaluation of mixed land use development for a compact city. *GIScience & Remote Sensing, 52*(1), 18–39.

Al-sharif, A. A., Pradhan, B., Shafri, H. Z. M., & Mansor, S. (2013). Spatio-temporal analysis of urban and population growths in Tripoli using remotely sensed data and GIS. *Indian Journal of Science and Technology, 6*(8), 5134–5142.

Alkheder, S., Wang, J., & Shan, J. (2006). Change detection—cellular automata method for urban growth modeling. In *Proceedings of International Society of Photogrammetry and Remote Sensing Mid-term Symposium*.

Apostolos, L. (2010). Urban growth simulation through cellular automata: A way to explore the fractal nature of cities. *Scienze regionali*.

Bhatta, B. (2009). Analysis of urban growth pattern using remote sensing and GIS: A case study of Kolkata India. *International Journal of Remote Sensing, 30*(18), 4733–4746.

Bhatta, B. (2010). *Analysis of urban growth and sprawl from remote sensing data*. Springer Science & Business Media.

Bhatta, B., Saraswati, S., & Bandyopadhyay, D. (2010). Urban sprawl measurement from remote sensing data. *Applied Geography, 30*(4), 731–740.

Dadi, D., Azadi, H., Senbeta, F., Abebe, K., Taheri, F., & Stellmacher, T. (2016). Urban sprawl and its impacts on land use change in Central Ethiopia. *Urban Forestry & Urban Greening, 16*, 132–141.

Grant, J. (2006). *Planning the good community: New urbanism in theory and practice* (Vol. 9). Taylor & Francis.

Grimm, V., Berger, U., Bastiansen, F., Eliassen, S., Ginot, V., Giske, J., et al. (2006). A standard protocol for describing individual-based and agent-based models. *Ecological Modelling, 198*(1), 115–126.

Hamedianfar, A., & Shafri, H. Z. M. (2015). Detailed intra-urban mapping through transferable OBIA rule sets using WorldView-2 very-high-resolution satellite images. *International Journal of Remote Sensing, 36*(13), 3380–3396.

Hegazy, I. R., & Kaloop, M. R. (2015). Monitoring urban growth and land use change detection with GIS and remote sensing techniques in Daqahlia governorate Egypt. *International Journal of Sustainable Built Environment, 4*(1), 117–124.

Hofmann, P. (2001). Detecting informal settlements from IKONOS image data using methods of object oriented image analysis-an example from Cape Town (South Africa). In C. Jürgens (Ed.), *Remote Sensing of Urban Areas/Fernerkundung in urbanen Räumen*, 41–42.

Koomen, E., & Borsboom-van Beurden, J. 2011. *Land-use modelling in planning practice* (Vol. 101). Springer Science & Business Media.

Lemma, T., Sliuzas, R., & Kuffer, M. (2006). A participatory approach to monitoring slum conditions: an example from Ethiopia. *Participatory Learning and Action, 54*(1), 58–66.

Li, X., Yang, Q., & Liu, X. (2008). Discovering and evaluating urban signatures for simulating compact development using cellular automata. *Landscape and Urban Planning, 86*(2), 177–186.

Li, X., & Yeh, A. G.-O. (2000). Modelling sustainable urban development by the integration of constrained cellular automata and GIS. *International Journal of Geographical Information Science, 14*(2), 131–152.

Li, X., & Yeh, A. G.-O. (2002). Neural-network-based cellular automata for simulating multiple land use changes using GIS. *International Journal of Geographical Information Science, 16*(4), 323–343.

Mason, S., & Fraser, C. (1998). Image sources for informal settlement management. *Photogrammetric Record, 16*(92), 313–330.

Parker, D. C., Manson, S. M., Janssen, M. A., Hoffmann, M. J., & Deadman, P. (2003). Multi-agent systems for the simulation of land-use and land-cover change: a review. *Annals of the Association of American Geographers, 93*(2), 314–337.

Pijanowski, B. C., Brown, D. G., Shellito, B. A., & Manik, G. A. (2002). Using neural networks and GIS to forecast land use changes: a land transformation model. *Computers, Environment and Urban Systems, 26*(6), 553–575.

Pijanowski, B. C., Tayyebi, A., Doucette, J., Pekin, B. K., Braun, D., & Plourde, J. (2014). A big data urban growth simulation at a national scale: Configuring the GIS and neural network based land transformation model to run in a high performance computing (HPC) environment. *Environmental Modelling and Software, 51*, 250–268.

Shekhar, S. (2012). Detecting slums from QuickBird data in Pune using an object-oriented approach. *International Archives of the Photogrammetry, Remote Sensing and Spatial Information Sciences, 39* (8), 519–524.

Squires, G. D. (2002). *Urban sprawl and the uneven development of metropolitan America* (pp. 1–22). Urban sprawl: Causes, consequences, and policy responses.

UN. (2009). *The millennium development goals report 2009* United Nations Publications; Department of Economic and Department of Public Information.

Verburg, P. H., Schot, P. P., Dijst, M. J., & Veldkamp, A. (2004). Land use change modelling: current practice and research priorities. *Geo Journal, 61*(4), 309–324.

Application of GIS and RS in Urban Growth Analysis and Modeling

13

Saleh Abdullahi and Biswajeet Pradhan

13.1 Introduction

Modeling of complex systems involves spatial and geographic aspects, such as urban areas. Such modeling requires approaches with capabilities of spatial and geographical analyses. Understanding urban growth patterns, urban dynamic aspects, and their relationships is the most important objective in this field. In addition, comprehensive data and information regarding historical and current urban patterns and processes are necessary for predicting future growth and changes. Critical issues regarding the spatial organization of urban areas, such as location, reason, time of developments, and positive and negative consequences must be addressed properly (Koomen and Borsboom-van Beurden 2011). Answering all these issues can help in preparing an appropriate urban development, thereby achieving sustainable development. Hence, professional data collection, data management, and data processing tools are necessary to deal with these complexities (Koomen and Borsboom-van Beurden 2011). Specifically, modeling and simulation of land use change require the availability of rich spatial data, spatial analysis tools, and displaying capability to illustrate output maps (Bhatta 2010; Bhatta et al. 2010b). Geographic information system (GIS) and remote sensing (RS) are the most beneficial tools to support these models; a general flowchart of their application in modeling process is shown in Fig. 13.1 (Basse et al. 2016; Karteris et al. 2016). Forster (1984) reported that the recent improvements in RS and GIS technologies provide a unique perspective on the processes of urban expansion and urban land use change.

Complex urban information systems incorporate traditional data (e.g., reports and analog maps), digital data (e.g., attribute databases, digital maps, and RS imageries), and ground measurements. Therefore, the integration of these sources of data in a proper environment (such as GIS) provides improved analysis methods, thereby resulting in enhanced urban area management and good urban planning. However, integrated approaches are inefficiently used in urban management and planning in developing countries because of deficient data, communication, and interaction among citizens, politicians, and planners (Bhatta et al. 2010a).

In the past decades, GIS techniques and remotely sensed data have been extensively employed for urban monitoring (to understand the urban development processes), urban modeling (to simulate the urban land use/land cover changes and urban sprawl), measuring (to analyze and assess the urban systems), and mapping (to highlight and understand the spatial urban patterns). The spatial patterns and physical expression of urban development and sprawl on the landscape can be investigated, detected, analyzed, and mapped by adopting GIS technology and RS data (Kumar et al. 2007; Bhatta et al. 2010b). Using multiagent evaluations and decision support systems available within the GIS enables scholars to assess geospatial and RS datasets (Parker et al. 2003; Bhatta 2010) and predict the prospects using the historical and current datasets. These systems are effectively applied to analyze, detect, and model the dynamics of urban expansions (Dadhich and Hanaoka 2011; Al-shalabi et al. 2013; Arsanjani et al. 2013).

13.2 Remote Sensing in Urban Application

RS technologies can reasonably and cost-effectively provides spatial data with the desired coverage. RS is a suitable source and data collection method for supporting and improving urban-related studies (Donnay et al. 2003; Dadhich and Hanaoka 2011; Wakode et al. 2014). This technology can be used to collect various essential data, such as detailed temporal and spatial information of land use patterns, infrastructures, and urban morphologies, and effective factors of the changes and population growth (Bhatta 2010). The information extracted from remotely sensed data is extremely beneficial for modeling, managing, and describing the urban environment (Longley et al. 2003; Al-shalabi et al. 2013). Remotely sensed data are powerful tools for measuring and detecting several elements related to the urban morphology of metropolises, such as density, amount,

B. Pradhan (ed.), *Spatial Modeling and Assessment of Urban Form*,
DOI 10.1007/978-3-319-54217-1_13

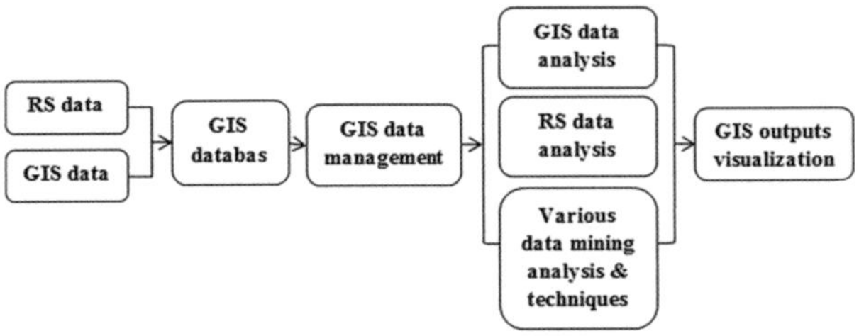

Fig. 13.1 Conceptual application of GIS and RS in modeling process

textural form, shape, and diffusion of built-up areas (Bhatta 2010). Moreover, RS data are particularly important in urban areas that face rapid land use changes and when updating information using traditional mapping and surveying approaches is time consuming and tedious. Monitoring of urban expansion can identify and determine the amount, location, and type of land conversion (Arsanjani et al. 2011; Guan et al. 2011). RS techniques are also effective for extracting and investing the interaction between urban environments and people (Hu and Lo 2007).

Advances in land surface mapping using RS techniques have contributed to the generation of significantly detailed urban maps, thereby providing detailed understanding on urban change dynamics (Wilson et al. 2003; Chen et al. 2014). In summary, applying RS in urban analysis is advantageous because of its capability to obtain temporal datasets in a large coverage and short time, and conduct digital analysis and processing within GIS environment (Bhatta 2009; Bhatta et al. 2010b). Compared with other techniques, RS of urban environments, particularly with space-borne platforms, is a fairly modern topic for geographers and the RS community. Space-borne satellite data are primarily valuable for developing countries because traditional urban survey methods are time consuming and costly. The reliance on using remotely sensed data in urban studies is gradually increasing. In addition, continuous advancements in software, hardware, and RS technologies have increased the demand and application of remotely sensed data in urban analysis.

Many sources of remotely sensed data are available in various scales, capabilities, efficiencies, and resolutions. Remotely sensed data are mainly selected and characterized on the basis of four resolutions, namely, spectral, temporal, radiometric, and spatial resolutions (Jensen 2009). Spectral resolution defines the intensity of interaction between electromagnetic radiation and surface features; temporal resolution refers to the length of time a satellite takes to complete one entire orbit cycle; radiometric resolution refers to the capability to discriminate slight differences in energy; spatial resolution of the sensor refers to the size of the smallest possible feature that can be detected (Jensen 2009). In the context of urban analysis, spatial and spectral resolutions are important and effective because the RS technology in these areas is challenged by spatial and spectral heterogeneities (Herold et al. 2004; Jensen and Im 2007; Eyoh et al. 2012). Hence, sensors with high spatial resolution, such as Quick Bird, IKONOS, and World View 2 and 3, and sensors with high spectral resolution (hyperspectral sensors), such as Hyperion, AVIRIS, and MERIS, have attracted considerable attention among urban scientists. However, urban growth and development perspectives in spatial resolution play a more important role than those in spectral resolution (Jensen 2009; Bhatta 2010).

Apart from RS technology resolutions, the type of sensors in data capturing (active or passive) also affects the capability and applicability of remotely sensed data. Passive sensors use the sun as the source of energy and radiation. The energy of the sun is either reflected as for visible wavelength or absorbed and then reemitted as for thermal infrared wavelengths. Notably, passive sensors can only be used to detect energy when the naturally occurring energy is available (Jensen 2009). Meanwhile, active sensors provide their own source for illumination. These sensors emit radiation, which is directed toward the target (surface features) to be investigated. They also detect and record the reflected radiation from the target. Contrary to optical RS, radar imagery can capture images and obtain measurements regardless of the time of day or season (Jensen 2009). Specifically, synthetic aperture radar (SAR) data are particularly suited for updating urban structure extraction because these data can pass through cloud cover and acquire data on every satellite pass (Abdullahi et al. 2015a). Di- and trihedral corner reflectors in an urban environment are other important features of radar data collection, thereby creating more bright areas than non-urban areas (Esch and Roth 2004). In addition, the built-up area can be accurately highlighted in SAR imagery because of the backscatter of radar wave, which is largely determined by the physical properties of surface objects, such as shape and surface roughness (Henderson and Xia 1997). Considerable research has been conducted on the application of radar imagery in urban studies, such as detection and extraction of urban settlement and growth detection, by various classification analyses (Fatone et al. 2001; Corr et al. 2003; Stasolla and Gamba 2008; Esch et al. 2010).

Many techniques have been developed and adopted to preprocess, interpret, and extract information from remotely sensed data for urban analysis (Jensen 2009; Bhatta 2010). However, the selection of available techniques is based on the application and objective of the research. For example, multitemporal RS data are particularly essential for measuring the structural differences of urban land use/land cover pattern (Liu and Phinn 2003). The information extracted from multitemporal data is beneficial to avoid the cumulative and irreversible effects of urban development (Yuan 2008) and support the optimization of urban service allocation (Barnsley and Barr 1996). Land use/land cover data generated from RS approaches are also important for devising sustainable urban development and environmental planning strategies (Alphan 2003; Jensen and Im 2007).

RS techniques have also been applied in a wide range of urban applications and decision making. Particularly, several urban planning studies have been conducted using RS images, especially in urban change modeling, urban expansion analysis, and urban land use/cover assessment.

13.2.1 Application of RADAR Imagery in Building Extraction

The total amount of urban areas covers an insignificant percentage of the Earth's land surface; however, these man-made regions represent the most complex and multidimensional-based environments for various evaluation, analysis, and modeling objectives (Henderson and Xia 1997). The continuous growth of urban areas has strengthened these complexities and enhanced the necessity of powerful tools and approaches for regular updating of information in urban-related fields. This information is necessary for scientists, planners, resource managers, and conservationists to plan for sustainable environments (Dewan and Yamaguchi 2009; Abdullahi et al. 2015b). The availability of accurate and up-to-date information regarding urban pattern is necessary to achieve this ultimate goal. Specifically, understanding the growth and changes in different land use and land covers are the main interests of urban planners and scientists (Rimal 2011; Abdullahi and Pradhan 2015). In addition, identifying the existence, size, density, pattern, and distribution of various urban features, such as buildings and roads, provides valuable information for analysis and modeling in urban-related applications. On this basis, remote sensed data have been extensively utilized for extracting and monitoring urban land use and land cover changes. Among all RS technologies, optical sensors have received most of the attention in this field and thus offered the most widely used approaches (Goodman and Ustin 2007; Roberts et al. 2008; Tehrany et al. 2014). Unlike optical sensors, SAR is well suited for urban detection and analysis especially in tropical regions because of its several advantages, such as cloud cover penetration (Grey et al. 2003; Jebur et al. 2014; Abdullahi et al. 2015a). Optical sensors measure the reflectance based on molecular resonances of surface materials, whereas radar sensors evaluate the physical characteristics (such as surface roughness and dielectric constant) of surface objects from backscattering (Henderson and Xia 1997). SAR technology has several other applications in the studies of agriculture, forestry, soil, oceans, geomorphology, and particularly in distinguishing among different land cover categories (Dell'Acqua and Gamba 2003; Tison et al. 2004; Bonci et al. 2006; Gamba et al. 2007).

SAR imagery has also been applied in urban studies, especially in settlement detection (Henderson and Xia 1997; Stasolla and Gamba 2008; Esch et al. 2010), urban changes mapping (Grey et al. 2003), and urban classification (Lombardo et al. 2001; Corr et al. 2003; Esch and Roth 2004). SAR technology is applied in this field to mainly detect urban structures and features and investigate their interaction with radar signals. The intensity and quality of SAR backscattering from urban areas generally depends on radar system properties, feature characteristics, and other environmental factors. For example, building size and orientation properties strongly affect the backscatter of radar wave because of its relationship with the azimuth angle of the sensors (Grey et al. 2003).

In recent years, several models have been developed to describe the interaction between radar signals and urban objects for improving the knowledge and interpretation of SAR images (Delliere et al. 2007; Franceschetti et al. 2007). Although the proposed models have provided certain benefits, most practical knowledge regarding SAR imagery properties is based on empirical observation. Speckle is a challenging issue in SAR imagery because it reduces the capability of SAR data in various applications, such as image classification, change detection, biomass estimation, and interpretation by degradation in appearance, quality, and recorded power of backscattering (Ali et al. 2008; Lee and Pottier 2009). Hence, speckle effects must be considered prior to processing and analyzing SAR images for precise applications, especially pixel-based analyses. The effects of this noise can be reduced mainly by the use of image filtering coupled with edge and texture preservation (Domg and Milne 2001; Xiao et al. 2003). However, selecting the ideal filter to reduce speckles for all SAR image data is a challenging process; hence, the selection is usually an application-dependent task.

Speckle is a grainy "salt and pepper" appearance in radar images and is attributed to random constructive and destructive interference from the multiple scattering returns. Given its effect on visual interpretation, speckle must be reduced prior to any analysis and interpretation. Adaptive filters such as Lee, enhanced Lee, Frost, enhanced Frost, Kuan, and Gamma filters are the most common approaches for reducing speckle while preserving the high-frequency features (Esch and Roth 2004).

Texture refers to the spatial variation of image tone as a function of scale. To be considered as a distinct textural area, the gray levels within the area must be more homogeneous as a unit than the areas with different textures. ENVI software provides various textural filters based on occurrence and cooccurrence measures, such as data range, mean, variance, and entropy homogeneity. Cooccurrence is a matrix of relative frequencies in which pixel values occur in two neighboring processing windows separated by a specified distance and direction. Cooccurrence shows the number of occurrences of the relationship between a pixel and its specified neighbor.

Texture property is one of the image interpretation elements and is an important factor recognized by the human visual system for recognizing features or area of interests in an image. Texture analysis is important for pixel and segment-based classification schemes, especially in SAR images with noise, such as speckle (Ulaby et al. 1986). In addition to texture properties, some of the other important elements of image interpretation in this field are context, edges, and tonal variation. This process can be compared to computer processing, in which only tonal information is often used. Texture filters are often included in the process of image classification to create computer interpretation of images similar to man-made interpretations, thereby improving the classification accuracy (Jensen 2009). Among the various approaches applied for image texture analysis, GLCM-based methods are the most common approaches for remotely sensed images (Kandaswamy et al. 2005).

Radar technologies, such as ENVISAT, ALOS PALSAR, RADARSAT, TerraSAR-X, and TanDEM-X, provide effective and useful images for land cover classification analysis. Classification of these images into a thematic map is a complex and challenging process and depends on several parameters, such as landscape heterogeneity, sensor selection, and adopted analysis and classification techniques (Mishra et al. 2011). Image classification of SAR images is traditionally based on a pixel-based concept. The classification accuracy of SAR images has been enhanced by integrating texture information (Dekker 2003; Dell'Acqua et al. 2006). These classifications are mainly in binary format between the two classes of the urban and non-urban environment. In addition, supervised and unsupervised classification techniques have been widely examined (Gomez-Chova et al. 2006; He et al. 2006; Chamundeeswari et al. 2007). Nevertheless, recent studies have concentrated on object-based techniques because of their utilization of geometrical, textural, and contextual properties of the objects in the classification procedure (Abdullahi et al. 2015a). This process segments the images into various objects containing similar pixels. Object-based classifiers typically incorporate spatial and spectral information. Contrary to pixel-based approaches that consider spectral and textural information only, the object-based approaches incorporate shape characteristics and neighborhood relations to the classification in addition to the abovementioned information (Shackelford and Davis 2003; Ban and Hu 2007). These tasks assist in extracting urban settlements by applying spectral, geometrical, and textural characteristics along with other information of the surrounding area (Esch et al. 2010). Object-based classifications also produce homogenous products with high detail and accurate mapping (Ban and Hu 2007).

Pixel-based and object-based classification techniques differ mainly in image segmentation process, which refers to assigning image pixels (homogeneous pixels) into different classes. Image segmentation groups the identical pixels of remotely sensed images into classes by matching the informational properties of user interest through comparing pixels to one another and those of known identity (Perumal and Bhaskaran 2010). In this process, the image is divided into unclassified "object primitives." Accordingly, various image objects are created for further analysis. This process is based on various properties of surface features, such as shape, size, color, and pixel topology, which are controlled by the parameters defined by the analysts. The selected and defined parameters indicate the effects of spectral and spatial properties of the image layers on the shape and size of the image objects. The analysts edit the parameters on the basis of the objective of the research and data quality and resolution. The multiscale image segmentation process is usually known as region-based segmentation because it is a bottom-up region merging approach. This process assumes each pixel as an object first and then starts to merge the initial objects for the purpose of creating large objects according to their similarity and homogeneity in terms of color and shape properties. In addition, a merging cost is assigned to each merging stage. Pixel-based and object-based classification methods have been compared in several studies (Wei et al. 2010; Duro et al. 2012; Tehrany et al. 2014). For instance, Abdullahi et al. (2015a, b) compared pixel-based and object-based classification methods for building extraction to assess city compactness (Fig. 13.2). Notably, object-based classification approaches perform better than object-based ones because of their inherent properties.

In summary, radar images are particularly difficult to interpret because of various properties and complexities of the active sensor performance, data collection, and interaction with the Earth's surface. However, useful results and information can be extracted from these images especially from urban environments by applying various techniques and models. Notably, the implementation of fully polarimetric, interferometric, and ancillary data can increase the accuracy and improve the information extraction performance. These ancillary data can help resolve confusion and inaccuracies by defining additional features of the study area, such as shape and size. Contrary to extracting other classes, such as vegetation and water, building extraction is more difficult because of its several categories, including single or multi-story buildings, various rooftop materials, interaction with surrounding vegetation covers, and effects of look direction during data collection. Hence, the integration of other

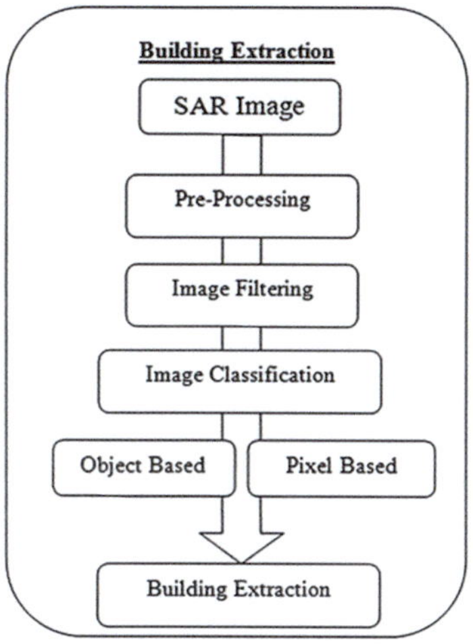

Fig. 13.2 RADAR image processing flowchart for building extraction (Abdullahi et al. 2015a, b)

information, especially from more than one look direction, can significantly increase the performance accuracy.

13.3 GIS in Urban Planning and Development Application

GIS can provide a proper environment for storing, managing, analyzing, manipulating, and displaying spatial data corresponding to the applied models. GIS has developed parallel to the advancement of other data capturing and analysis technologies for spatial data processing. GIS can provide a consistent visualization environment for displaying the input data and outputs of the models; this feature is beneficial in various applications (Weng 2001). Apart from its data processing capability, GIS also provides functional tools for problem solving and decision making. GIS can assist modelers in defining and creating spatial variables for the models (Openshaw and Clarke 1996), predicting land use changes based on several independent spatial variables (Mertens and Lambin 2000), and evaluate predicted changes in spatial pattern (de Koning et al. 1999). Simulation of growth patterns can also be conducted in a GIS environment using several kinds of constraints (Yeh and Li 1998). The evolution of RS and GIS and digital computing technologies has provided powerful tools to deal with very complex mathematical urban models. GIS can be employed to model urban expansion with a high level of spatial accuracy (Xie et al. 2005).

GIS is a digital map-based technology that provides a database management system to store, analyze, and display geographical-based information. GIS can manipulate spatial data from digitizing to editing and processing raw data. GIS comprises five main components, which are as shown in Fig. 13.3.

- Hardware: The hardware includes computers, such as PCs and workstation, operating systems, and additional equipment, including monitors, digitizers, and scanners.
- Software: The software includes the source code and user interface. The code may be written in C++, Visual Basic, or Python.
- People: GIS professionals define the purpose and objectives and provide the reason and justification for using GIS.
- Data: Data and information are the initial input for processing and analyzing the objectives, and are mainly obtained using RS and traditional in situ data collection.
- Methods/Infrastructure: The infrastructure refers to the necessary physical, organizational, administrative, and cultural environments that support GIS operations. The infrastructure includes requisite skills, suitable methods, data standards, data clearinghouses, and general organizational patterns.

The critical function of GIS is, by design, the analysis of spatial data. GIS is not a new invention because geographic information processing has already been applied in various disciplines. Particularly, natural resource specialists and environmental scientists have been actively processing graphical data and promoting their techniques since the 1960s. At present, generic GIS is distinguished from the previous geoprocessing because of its use of computer automation to integrate geographic data processing tools in a friendly and comprehensive environment. The advent of sophisticated computer techniques has proliferated along with the multidisciplinary application of geoprocessing methodologies, thereby providing data integration capabilities that are logistically impossible.

Fig. 13.3 Five main components of GIS

The basic data type in a GIS reflects traditional data found on a map. GIS technology uses the following two types of data:

- Spatial data, which describe the absolute and relative location of geographic features; and
- Attribute data, which describe the characteristics of the spatial features that can be quantitative or qualitative in nature. Attribute data are often referred to as tabular data.

GIS is an ideal tool for analyzing and solving multiple criteria problems because of its characteristics as follows:

- GIS database combines spatial and nonspatial information;
- GIS has ideal data viewing capabilities, thereby allowing efficient and effective visual examinations of solutions;
- GIS allows users to interactively modify solutions to perform sensitivity analysis; and
- GIS, by definition, must contain spatial query and analytical capabilities, such as area measurement, distance measurement, overlay capabilities, and corridor analysis.

Using GIS in urban studies has numerous advantages. GIS is a database management system that provides data mapping operations for viewing geographical information and data retrieval operations for map inquiries (Almeida et al. 2005). These functions allow planners and urban analysts to communicate, display, and manage information effectively (Weng 2001). These GIS functions are augmented by methods of data modeling (e.g., data conversion routines and cartographic analysis), which can improve land use and transportation analysis (Stanilov 2003). However, regardless of its significant benefits, the urban analyst cannot avoid the complexities, nuances, and subtle interactions intrinsic in the spatial data usually used in urban studies with GIS alone (Paez et al. 2001). Although GIS is a beneficial tool in decision support technologies, it is not a decision support system by itself and cannot handle multiple decision factors (Turskis et al. 2006). Predictive models have been implemented within GIS environment to analyze and project various future development scenarios and land use conversion (Wu 1998; Li and Yeh 2002; Pijanowski et al. 2002; Verburg et al. 2004). However, simulating these changes is a challenging process. GIS also cannot fulfill all the requirements of these simulations because of the localized, dynamic, and complex nature of the modeling. Therefore, GIS is integrated with other techniques to provide a strong simulation approach. Cellular automata (CA), multicriteria decision making (MCDM), and statistical-based approaches are examples of models integrated to GIS.

The rapid advancement of GIS improves the application of CA in urban modeling. Cell-based GIS provides a beneficial tool, including analysis and assessments in urban-related projects. Integrating CA and GIS can overcome several drawbacks and requirements in urban systems because of the combined capabilities, such as fast iterative computation of CA and the beneficial information provided by GIS in defining transition rules (Li and Yeh 2000). In addition, CA can serve as an analytical engine to provide a flexible framework for the programming and running of dynamic spatial models. With this integration, various development scenarios can be proposed and evaluated to achieve sustainable urban development. These GIS scenarios can go beyond theoretic constructs and become realistic by considering real-world data, factors, and constraints in their modeling. The heterogeneity of the geographical spaces can be easily investigated by developing spatial difference equations in the context of GIS (Batty and Longley 1994). CA models can be developed and processed in a GIS environment using stored data and information in GIS database and available analytical tools in GIS for geospatial data. In addition to GIS, RS data can benefit CA modeling in defining constraints of sustainable urban development modeling (Li and Yeh 2000). Moreover, temporal data captured and prepared by RS technology can be used to calibrate the neighborhood functions of CA modeling. RS provides beneficial data regarding land use and land covers as input to CA modeling. Therefore, the integration of RS data for CA modeling within GIS environments can significantly enhance the capability of dynamic spatial modeling of urban growth applications.

13.3.1 GIS in Urban Land Use Site Selection Process

Site suitability analysis is the process of determining the fitness of a provided tract of land for a defined use and is a fundamental processing in urban planning and design (Steiner et al. 2000). In other words, site suitability analysis is the process of determining whether the available lands are suitable for a few specific requirements and determining the suitability level. Establishing and integrating appropriate suitability factors related to location, development activities, socioeconomic, and environmental aspects are the essential requirements of suitability analysis. These techniques assist planners, landscape architects, and local decision makers in analyzing the factor interactions in various ways. In processing a certain development, integrating these factors can assist in possibly identifying the most and least suitable sites for desired proposals. If the location and proposed development are known, then the potential conflicts can be qualitatively determined.

In case of site suitability for a specific purpose, such as site selection for a hospital in an urban environment, the selection of an optimum site can be influenced by the uncertainty inherent in describing and ranking available alternatives based on effective criteria. These criteria include physical, socioeconomic, and environmental quality and amenities. In the site suitability example, the following criteria must be defined: positive and negative effects of the new hospital on its surrounding area and positive and negative effects of the environment on the new hospital during either the construction phase or the operation phase. The necessary in situ or remotely sensed data for meeting all criteria must be carefully collected and managed and organized in a GIS environment. Criteria are selected to evaluate potentials of the existing hospitals and support decisions regarding the location of the additional hospitals in various zones. The criteria must be selected according to literature, planning guidelines, and government regulations of the study area. For example, the effect of a new hospital as a new resource on the local community can be divided into long-term and short-term effects. Short-term effect includes the construction phase, while long-term effect includes the operation phase. All developments have an environmental effect during the construction phase, such as increased heavy traffic (noise and pollution), building noise, antisocial aspects of building (dust), and potential disruption to local services (short-term cutting off of utilities, including water, gas, and electricity). For the most part, these effects are considered in conditions attached to planning consent for the eventual option. All developments during the operation phase can also result in positive or negative effects. For example, the positive economic effects a new hospital as a new resource in a community include providing local jobs and staff spending money in local shops. Meanwhile, the negative environmental effects include overspill parking on residential roads and the effect of people, staff, patient, ambulances, and visitors traveling to the site.

Site selection process in the urban area considers the following objectives:

- Technical aspects,
- Environmental issues,
- Social aspects,
- Economic issues,
- Biological aspects,
- Physical aspects.

Under these objectives, several criteria can be identified:

- Population density,
- Land use/land cover,
- Distance from existing resources,
- Distance from highways or intersections,
- Land cost,
- Land area,
- Travel time,
- Distance from polluted areas,
- Distance from rivers and canals,
- Distant to main roads,
- Safety aspect.

These criteria have different effects on the suitability of the location for a new resource or development. Some of these criteria are effects of the new resource on the environment, while others are the effect of the environment on the new resource. For example, the development of a hospital can negatively affect its surrounding area, and a polluted environment can negatively affect a hospital as a new resource. Therefore, a city planner must consider the mutual relationship between the effects of the environment and new resource on each other.

Several main points are qualitatively and quantitatively assessed in site selection, and the degree of fitness is identified to propose the following requirements:

- The degree to which the site supports the function requirement and capabilities of human beings;
- The degree to which the site is suitable for construction of the proposed project, in terms of materials, workers, time, and money;
- The degree to which the site fulfills the requirements with regard to form and capacity of space; and
- The degree to which the site has access to main resources, such as human and natural resources.

Given that most of the related issues and data in site selection and suitability process are based on geographical information, GIS allows for the consideration and combination of these data and variables to deal with the modeling process. GIS can apply various data regarding geology, topography, land use/land cover, demography, natural resources, and transportation networks.

Using GIS in site suitability analysis and urban planning process is advantageous because GIS can develop alternative scenarios of urban development. GIS techniques and methods can be used to assist in developing and designing a growth management plan for urban areas. GIS can also analyze existing zoning and land use plan and discuss future projection build-out scenario in accordance with existing plans and regulations. Alternative development scenarios and 3D visualizations can be developed and comprehensively compared. Such comparison can be conducted by collecting, compiling, and editing database developments and maps for all the base layers and identified constraint layers for any future development of townships, such as unbuildable parcels, agricultural district, airport zones, and floodplains. The areas suitable for development are assigned a value of one, while the areas unsuitable for development are assigned a value of zero. Each of these layers is assigned a specific weight first and then combined using the raster calculator. The weighting stage is a complex mathematical calculation based on the decision-making process.

Decision making is the study of identifying and choosing alternatives based on the value and preferences of the decision makers. Making a decision implies that alternative choices are available for consideration. Decision making requires identifying as many of these alternatives as possible and choosing the one that best fits with the goals, desires, aims, and values. Most high equity decision problems have strong spatial connotations, and geography is an inherent component of these problems because of the spatial arrangement of infrastructure, population density, settlements, and patterns of the physical world. Thus, geography cannot be ignored in finding solutions to such problems.

Spatial decision making faces the following decision complexities:

- Spatial nature and temporal development of phenomena and process;
- Complex multidimension and heterogeneous data describing decision situations;
- Large or extremely large datasets that include data in numerical, map, image, text, and other forms;
- Substantial number of available alternatives or a need to generate decision alternatives "on the fly" according to the changing situation;
- Multiple participants with different and often conflicting interests; and
- Multiple categories of knowledge involved, including expert and layman knowledge.

Spatial decision support is the computational or information assistance for making well-informed decisions regarding problems with a geographic or spatial component. This support assists with the development, evaluation, and selection of proper policies, plans, projects, scenarios, interventions, or solution strategies.

GIS can provide the following tools for assisting in the decision-making process:

- Maps/display as means of visualizing the problem,
- Overlay as means of defining relationships, and
- Modeling as means of predicting outcomes.

Although GIS is a beneficial tool for spatial analysis issues, whether the GIS decision support capabilities are sufficient is unclear (Jankowski 2006). In fact, GIS is a decision support system and not a decision-making system. GIS can provide certain tools for assisting in the decision-making process. GIS approaches are incapable of processing multiple criteria and conflicting objectives (Carver 1991). GIS approaches are also limited in integrating geographical information with subjective values/priorities imposed by the decision maker (Malczewski 1999).

In making a good decision, the following requirements must be provided:

- Knowledge and foresight,
- Insight and intelligence, and
- Expertise and others.

Although GIS does not provide the preceding requirements, this system can fulfill the important role in decision making by providing decision support. Therefore, integrating different techniques with GIS is required to increase the analysis capability in site suitability process. One of the main techniques integrated to GIS is MCDM. For the last three decades, the integrated GIS and MCDM technique has been used in solving site selection problems.

Decision analysis is a set of systematic procedures for analyzing complex decision problems. These procedures include dividing the decision problems into small and highly understandable components, analyzing each component, and logically integrating the components to produce a meaningful solution. MCDM incorporates an explicit statement of preferences of decision makers. Such preferences are presented by various quantities, weighting schemes, constraints, goals, utilities, and other parameters. These preferences analyze and support decision through formal analysis of alternative options, attributes, evaluation criteria, goals or objectives, and constraints. However, they assume homogeneity within the study area. Such assumption is unrealistic for site selection problems.

In the MCDM approach, the decision maker must define the criterion preferences. MCDM problems frequently involve criteria of varying importance to decision makers, thereby resulting in policies, established hierarchies, cause–effect relationships, and subjective preferences. Preferences are expressions of the values of the decision maker. In the MCDM context, preference represents the varying degrees of importance assigned to criteria.

A weight is a numeric value assigned to an evaluation criterion that indicates its importance relative to other criteria in the decision situation. The criterion with a large the weight is considered the most important. The weights are usually normalized such that their sum for all n-criteria considered in a given decision situation equals 1. Different weighting techniques are available, such as ranking procedures, rating, and pair-wise comparison.

Any spatial decision problem can be structured into the following three major phases (Fig. 13.4):

- Intelligence, which examines the existence of a problem or opportunity for change;
- Design, which determines the alternatives; and
- Choice, which decides the best alternative.

The framework for spatial multicriteria decision analysis is provided in Fig. 13.5.

Problem Definition: The decision problem is the difference between the desired and existing state of the real world. This stage involves searching the decision environment for conditions, processing, and examining the raw data to identify the problems. The GIS capability for storage, management, manipulation, and analysis are used in this stage, thereby providing a major support.

Evaluation Criteria: This stage involves specifying a comprehensive set of objectives that reflect all concerns relevant to the decision problem and determines the achievement of those objectives defined as attributes. GIS data handling and analyzing capabilities are used to generate inputs for spatial decision-making analysis.

Criterion Weights: Weight is the value assigned to an evaluation criterion that indicates its importance relative to other criteria under consideration.

Decision Rules: The criterion map layers and weightings must be integrated to provide an overall assessment. This task is accomplished by an appropriate decision rule or aggregation function.

Sensitivity Analysis: In real-world situation, analysis must be made to investigate whether the preliminary conclusion is robust or not. Sensitivity analysis aims to identify the effect of the changes in the inputs, such as geographical data and the preferences of the decision maker on the outputs. If the changes insignificantly affect the outputs, then the ranking is assumed as robust and satisfactory.

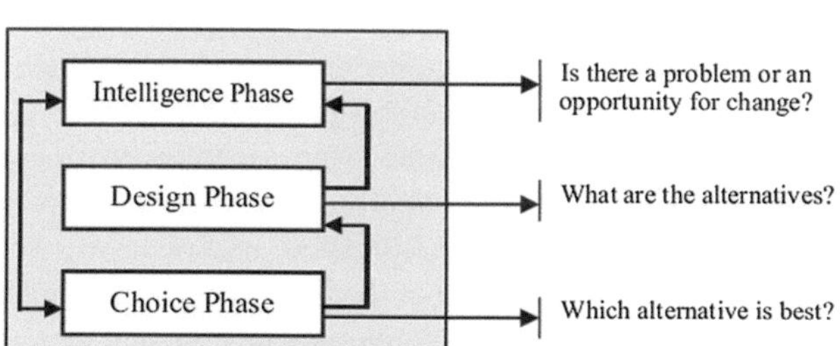

Fig. 13.4 Three phase of decision-making process (Malczewski 1999)

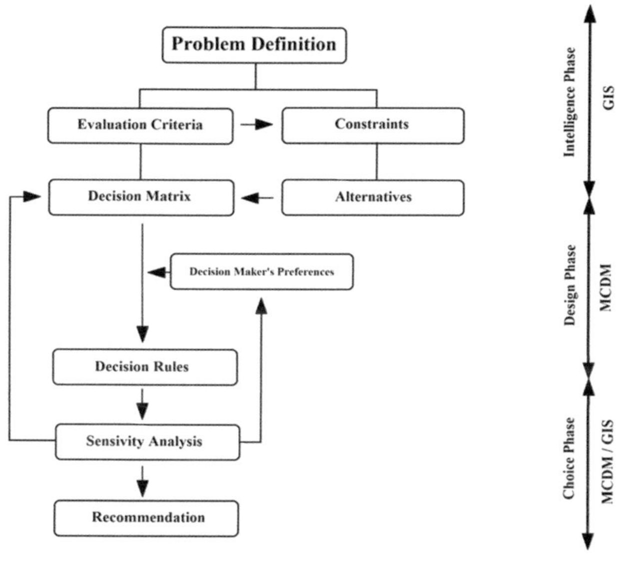

Fig. 13.5 Framework for spatial multicriteria decision analysis (Malczewski 1999)

The choice of the MCDM method is important because it has a significant effect on the outcome. The characteristics and properties of the MCDM method must be compatible with the specific nature of the decision problem (Laaribi et al. 1996).

Among various MCDM methods, analytical hierarchy process (AHP) is common used in site selection process. The pair-wise comparison technique is developed by Thomas Saaty in the 1970s and 1980s in the context of the MCDM method called AHP (Saaty 1980), This technique represents a theoretically based approach to computing weights representing the relative importance of criteria. Weights are not directly assigned; however, they represent a "best fit" set of weights derived from the eigenvector of the square reciprocal matrix used to compare all possible pairs of criteria.

The technique comprises using pairs of criteria C_i and C_j and asking the two following questions:

(1) Which criterion is more important, C_i or C_j?
(2) How much/How many more times is said criterion more important relative to the lesser important criterion? Typically answered as "about the same" or "strongly more important" and subsequently scored on a one to nine scale (Table 13.1).

Answers to these two questions are used to generate the cell values in a square matrix A, where i is a row and j is a column. Since each factor is of equal importance to itself, the diagonal in A matrix is filled with 1s. If C_i (row element) and C_j (column element) are if equal importance, then a_{ij} (the value in the matrix A at the intersection of row i and column j)

Table 13.1 Pair-wise comparison matrix (Saaty 1980)

Intensity of importance	Definition
1	Equal importance
2	Equal to moderate importance
3	Moderate importance
4	Moderate to strong importance
5	Strong importance
6	Strong to very strong importance
7	Very strong importance
8	Very to extremely strong importance
9	Extremely importance

equals 1; if C_j is more important than C_i, then a_{ij} is set equal to the importance score and will be >1; finally, if C_i is more important than C_j, then a_{ij} is set equal to the reciprocal of the importance score (i.e., 1/score) and will be <1. The structure of the matrix A can be presented as follows: C_1, C_2, C_3, C_n;

$$A = \left[a_{ij}\right] = \begin{matrix} C_1 \\ C_2 \\ C_3 \\ \vdots \\ C_n \end{matrix} \begin{bmatrix} 1 & a_{12} & a_{13} & \cdots & a_{1n} \\ 1/a_{12} & 1 & \cdots & \cdots & a_{2n} \\ 1/a_{13} & \cdots & 1 & \cdots & \cdots \\ 1/a_{14} & \cdots & \cdots & 1 & \cdots \\ 1/a_{1n} & 1/a_{2n} & \cdots & \cdots & 1 \end{bmatrix}$$

where A is the reciprocal and square pair-wise comparison matrix. $a_{ij} = 1$, and $a_{ji} = 1/a_{ij}$, $i, j = 1, 2, 3, \ldots, n$. In matrix A, the problem becomes one of assigning to the n elements, $C_1, C_2, C_3, \ldots C_n$, a set of numerical weights $W_1, W_2, \ldots, W_n$ that reflect the recorded judgments. If A is a consistency matrix, the relation between weights W_i and judgments a_{ij} are simply given by $W_i/W_j = a_{ij}$ (for $i, j = 1, 2, \ldots, n$) and $C_1, C_2, C_3, \ldots C_n$

$$A = \begin{matrix} C_1 \\ C_2 \\ C_3 \\ \vdots \\ C_n \end{matrix} \begin{bmatrix} W_1/W_1 & W_1/W_2 & \vdots & \vdots & W_1/W_n \\ W_2/W_1 & W_2/W_2 & \vdots & \vdots & W_2/W_n \\ \vdots & \vdots & 1 & \cdots & \vdots \\ \vdots & \vdots & \cdots & 1 & \vdots \\ W_n/W_1 & W_n/W_2 & \cdots & \cdots & W_n/W_n \end{bmatrix}$$

The consistency test is one of the essential features of the AHP method and aims to eliminate the possible inconsistency presented in the weights by computing the consistency level of each matrix (Saaty 2000). The degree of consistency achieved in the ratings is determined by a CR, which induces the probability of the matrix ratings randomly generated because individual judgment can never be perfectly agreed. The rule of thumb is that a CR less than or equal to 0.10 indicates an acceptable reciprocal matrix A, while ratios over 0.10 indicate that the matrix must be revised.

Contrary to ranking and rating, pair-wise comparison has a solid theoretical foundation based on ratio-scale judgments regarding pairs of criteria and the properties of the theoretical matrix of pair-wise comparisons. This technique has the disadvantage of the need to conduct numerous judgments when the numbers of criteria are large.

Using these techniques in site suitability analysis can result in a model for evaluating the suitable location for a specific purpose, such as building sites to support the decision making in locating additional housing areas. Owing to the complexity of the site selection process, integrating several decision support tools, such as high spatial resolution remotely sensed data and GIS and multicriteria analysis using AHP, is essential in this process. This integration can benefit urban planners and decision makers. GIS is used based on a set of criteria derived from the spatial aspects, environment, policies, and national and local physical plan.

The normal AHP technique of Saaty can be used to form main criterion factors and subcriterion factors. The basic assumption is that the weightings derived from the hierarchical comparison in normal AHP can be influenced by the preferences provided to a particular criterion factor. Therefore, a sensitivity test is performed on the criterion preferences and is evaluated based on various preference factors thought to influence weightings. Separate hierarchical pair-wise comparisons of main criterion factors are performed for each preference to analyze the sensitivity of the weights obtained. The pair-wise comparisons of criteria and subcriterion factors are independently performed and similar judgments are provided for all the preferences. Definite to very strong preferences are provided to the factor in their pair-wise comparison to reflect the preferences toward a certain factor. Then, the consistency ratio must be analyzed to verify the reliability of the judgment of the decision maker.

Site selection analysis can be performed through GIS spatial analysis and 3-D analysis using ArcView or ArcMap Model Builder. Models are represented as sets of spatial processes, such as buffer, classification, and reclassification and overlay techniques. Each of the input themes is assigned a weight influence based on its importance first, and then the result is successively multiplied by each of the constraints. Next, the GIS overlay process is used to combine the factors and constraints in the form of a weighting overlay process. Finally, the result is summed up to produce a suitability map.

Performing site selection process using site screening method and AHP demonstrates how evaluation criteria, such as physical, socioeconomical, technical, environmental, and their regulatory subcriteria can be introduced into an over layer technique to screen limited appropriate zones in the area. Utilizing an MCDM method is recommended for hierarchy computations of the process to find the optimal site among the primarily screened site. Using the introduced method, an accurate sitting procedure for urban and environmental planning in an area can be enabled. As a result, several suitable zones can be screened in the area and the most suitable site can be chosen as the optimal site for locating the proposed facility or development using AHP.

AHP is an efficient method for solving multiobjective decision-making problems and can be used to locate the optimal site among primarily selected zones. As the alternatives in hierarchy process, the sites are screened in the first phase to satisfy all involved considerations and the standards are utilized. Integrating GIS and AHP for site suitability process can provide the correct solution to assist the decision maker in determining appropriate values for physical suitability criteria.

In conclusion, the integration and improvements in GIS and RS technology provide a unique perspective on urban expansion and land use change process (Forster 1984). The information collected from remotely sensed data within a GIS environment are extremely beneficial to modeling, managing, and describing urban growth behavior (Longley et al. 2003; Al-shalabi et al. 2013). Moreover, the capability of GIS to be integrated with other evaluation and decision support systems enables urban planners to assess geospatial and remotely sensed datasets (Parker et al. 2003; Bhatta 2010; Chen et al. 2014) and predict and model future patterns by considering historical and current datasets.

13.4 GIS in Urban Planning from the Malaysian Perspective

In recent decades, GIS has been used in many countries for several applications, such as disaster monitoring and urban growth management in China, business application through the Integrated Land Use System in Singapore, preparing development plan including the framework of future land use in Calcutta, tourism planning in Canada, and studying the environmental effect of tourism on fragile reefs in the Cayman Islands (Selamat et al. 2012).

Historically, Malaysian governments have used old and traditional approaches (such as blueprint) of preparation and monitoring for urban growth and development (Samat 2006). However, given their several drawbacks, such as difficulties in monitoring uncontrolled urban growth, complexity, and time-consuming process (Yusoff et al. 2010), these approaches are replaced by GIS, RS, and other advanced techniques to deal with geographical-based data and issues in urban applications. The advance growth in information technology development has improved and increased the application of GIS in urban planning in this country. GIS was first used in Malaysia in the 1980s using the digital cadastral database and the National Topographic Database developed by the Department of Survey and Mapping (Selamat et al. 2012).

GIS has also been applied in the following applications:

- Education: GIS provides an attractive environment and further develops the creative, critical, and innovative thinking of student (Lateh and Muniandy 2011);
- Medicine: GIS has been used in mapping the distribution of health facilities and dengue disease in affected areas. Moreover, GIS has been used in the research on dengue fever in Bandar Baru Bangi and Kajang and in health care database (Shaharudin et al. 2002);
- Geology: GIS has been used to generate digital elevation models stored in the geological map to produce 3D-shaped display in the Klang Valley (Abd Manap et al. 2009);
- Crime analysis: GIS has emerged as decision support system in crime prevention projects. GIS allows the integration of crime information systems with spatial data and assists in the production of accurate and high-quality maps that clearly show the locations of different kinds of crimes as crime spots (Suryavanshi 2001);
- Health studies: GIS has been used to determine geographic accessibility for various hospitals for patients, determine geographic access to pharmacies for the community, and describe the geographic pattern of specific illness or accidents based on emergency department reports. GIS can also be used as a decision support tool to allocate health services such that they are geographically accessible to the population they intend to serve. Determining the distance is typically performed with a "buffering" operation in the GIS, wherein buffers at various straight-line distances from a particular point are generated. However, the method is not the most effective in measuring actual travel distance or travel time. Therefore, network analysis is usually better because of its sophisticated basis related to urban road network behavior.

GIS is used by the local planning authorities of Malaysia to improve the planning system, particularly in controlling the development in the area of jurisdiction, creating and developing geospatial and urban-related database, running and analyzing the proposed planning scenarios, and evaluating the consequences (Yaakup 2004). Samat (2006) believed that Malaysia is a rapidly developing country, and this rapid growth results in high land consumption rate. Accordingly, strong and powerful land use management systems, such as GIS, are needed to control and monitor the utilization of land resources and prevent conversion of valuable green and natural environments to urban land uses. GIS provides a proper data management and retrieval system related to land resources and a mechanism that can be used to implement the planning function involving the daily administrative management operations. Zaini (2007) used GIS in preserving historical heritage buildings in Taiping, Perak because of its capability to manipulate data and produce various development scenarios required by the local planning agencies in monitoring the development of Taiping. Abdullahi and Pradhan (2015) used a GIS environment to propose a brownfield land use change model based on the compact city paradigm. Abdullahi et al. (2015a, b) analyzed and evaluated mixed land use development of Kajang City as an objective to achieve sustainable urban development using GIS database and analytical tools.

One of the important roles of GIS at the national level is site selection projects for various purposes, such as community facilities (medical, educational, and transportation), government buildings, and developing entirely new cities (Putrajaya and Cyberjaya). Given that various parameters, including social, environmental, and physical parameters, must be considered in site selection analysis, the capability of GIS in integrating various data sources with different formats and scales is beneficial in this analytical process (Abdullahi et al. 2014). Thus, GIS is applicable in a wide range of land management, land use planning, and suitability, including the interpretation and formulation of land use policy (Selamat et al. 2012). Several areas and divisions need GIS for management and performing processes for urban applications, such as Corporate Plan, National Physical Development Plan, Regional Planning, Legal and Regulatory Planning, Legal Unit, National Land Use

Table 13.2 Development of geographical information system from 1980 to present (Selamat et al. 2012)

Period	Application
1980s	• Digital Cadastral Database (DCDB) • National Topographic Database • Geographic Distribution Information System (GDIS) for gas Malaysia
1990s	• Penang Geographic Information System (PEGIS) • Department of Agriculture • Sabah Agriculture Department • Computerized Planning Information System (Melaka City Council) • Forest Department of Sarawak • National Infrastructure for Land Information System (NaLIS) • Department of Survey and Mapping (JUPEM) • Malaysian Center for Remote Sensing (MACRES) • Geological Survey Department • Valuation and Property Services Department • Public Work Department • Economic Planning Unit
2000s and onward	• MACRES Ground Remote Sensing System (MGRS) • Posse 2, Vehicle Satellite-Tracking System • MiniLBS Application Suite in B200 Handheld Communicator • Satellite Image Map (SIM) • National Geospatial Data Network (myGDI) • Kajang GIS • Safe City Monitoring System (SCMS) • Commercial Vehicle Licensing Board (LPKP) • Melaka Structure Plan • Planning Information System for Johor Bharu Tengah Municipal Council (SIMAP-MPJBT) • GIS Database Development for Nalis System of Alor Gajah Municipal Council • Planning Information Application Process for Batu Pahat Municipal Council • Development Control System for Planning of DBKL • GIS System for Taiping Landscape Master Plan • GIS System for Cameron Highland Landscape Master Plan • Impervious Surface Estimation Model for Housing Estate/Subdivisions Using GIS and Remote Sensing • Land Use Monitoring System, Selangor • Integrated Land Use, Landscape and Building Design Guidelines as Flood Mitigation Tools • Integrated GIS Database Development for Ministry of Federal Territories and Urban Well-Being and etc.

Information, Internal Audit Unit, and the Division of Research and Development. Table 13.2 lists the various applications and stages of GIS utilization in the town planning field in Malaysia.

The utilization of GIS in these governmental and urban-related tasks has obtained good level of improvements and achievements, such as (1) AGISwlk of GIS for Klang Valley Region; (2) GIS9, which is the Negeri Sembilan planning system and acts as a manual system to monitor the structure plan document; and (3) PEGIS, which is an application of GIS to Penang which plays a key role in providing information to the Economic Planning Unit of Penang (Selamat et al. 2012). The integration of GIS has provided a tool that can contribute to a clear understanding of real planning problems and prescriptive planning scenarios for enhancing the quality of urban planning and management.

References

Abd Manap, M., Firuz Ramli, M., & Redzwan, G. (2009). The application of digital elevation model for the interpretation of Klang Valley geological structure. *Disaster Prevention and Management: An International Journal, 18*(5), 504–510.

Abdullahi, S., Mahmud, A Rb, & Pradhan, B. (2014). Spatial modelling of site suitability assessment for hospitals using geographical information system-based multicriteria approach at Qazvin city, Iran. *Geocarto International, 29*(2), 164–184.

Abdullahi, S., & Pradhan, B. (2015). Sustainable brownfields land use change modeling using GIS-based weights-of-evidence approach. *Applied Spatial Analysis and Policy*, 1–18.

Abdullahi, S., Pradhan, B., & Jebur, M. N. (2015a). GIS-based sustainable city compactness assessment using integration of MCDM, Bayes theorem and RADAR technology. *Geocarto International, 30*(4), 365–387.

Abdullahi, S., Pradhan, B., Mansor, S., & Shariff, A. R. M. (2015b). GIS-based modeling for the spatial measurement and evaluation of mixed land use development for a compact city. *GIScience & Remote Sensing, 52*(1), 18–39.

Al-shalabi, M., Billa, L., Pradhan, B., Mansor, S., & Al-Sharif, A. A. (2013). Modelling urban growth evolution and land-use changes using GIS based cellular automata and SLEUTH models: The case of Sana'a metropolitan city, Yemen. *Environmental Earth Sciences, 70*(1), 425–437.

Ali, S. M., Javed, M. Y., Khattak, N. S., & Mohsin, A. (2008). Despeckling of synthetic aperture radar images using inner product spaces in undecimated wavelet domain.

Almeida, C., Monteiro, A. M. V., Câmara, G., Soares-Filho, B. S., Cerqueira, G. C., Pennachin, C. L., et al. (2005). GIS and remote sensing as tools for the simulation of urban land-use change. *International Journal of Remote Sensing, 26*(4), 759–774.

Alphan, H. (2003). Land-use change and urbanization of Adana, Turkey. *Land Degradation & Development, 14*(6), 575–586.

Arsanjani, J. J., Helbich, M., & de Noronha Vaz, E. (2013). Spatiotemporal simulation of urban growth patterns using agent-based modeling: The case of Tehran. *Cities, 32,* 33–42.

Arsanjani, J. J., Kainz, W., & Mousivand, A. J. (2011). Tracking dynamic land-use change using spatially explicit Markov Chain based on cellular automata: The case of Tehran. *International Journal of Image and Data Fusion, 2*(4), 329–345.

Ban, Y., & Hu, H. (2007). *RADARSAT fine-beam SAR data for land-cover mapping and change detection in the rural-urban fringe of the Greater Toronto Area.* Paper presented at the Urban Remote Sensing Joint Event, 2007.

Barnsley, M. J., & Barr, S. L. (1996). Inferring urban land use from satellite sensor images using kernel-based spatial reclassification. *Photogrammetric Engineering and Remote Sensing, 62*(8), 949–958.

Basse, R. M., Charif, O., & Bódis, K. (2016). Spatial and temporal dimensions of land use change in cross border region of Luxembourg. Development of a hybrid approach integrating GIS, cellular automata and decision learning tree models. *Applied Geography, 67,* 94–108.

Batty, M., & Longley, P. A. (1994). *Fractal cities: A geometry of form and function*. Academic Press.

Bhatta, B. (2009). Modelling of urban growth boundary using geoinformatics. *International Journal of Digital Earth, 2*(4), 359–381.

Bhatta, B. (2010). *Analysis of urban growth and sprawl from remote sensing data.* Berlin: Springer Science & Business Media.

Bhatta, B., Saraswati, S., & Bandyopadhyay, D. (2010a). Quantifying the degree-of-freedom, degree-of-sprawl, and degree-of-goodness of urban growth from remote sensing data. *Applied Geography, 30*(1), 96–111.

Bhatta, B., Saraswati, S., & Bandyopadhyay, D. (2010b). Urban sprawl measurement from remote sensing data. *Applied Geography, 30*(4), 731–740.

Bonci, E., Frate, F., & Solimini, D. (2006). *Monitoring Urban changes in Rome, Italy by multi-temporal ERS-SAR images.* Paper presented at the 2006 IEEE International Symposium on Geoscience and Remote Sensing.

Carver, S. J. (1991). Integrating multi-criteria evaluation with geographical information systems. *International Journal of Geographical Information System, 5*(3), 321–339.

Chamundeeswari, V. V., Singh, D., & Singh, K. (2007). *Unsupervised land cover classification of SAR images by contour tracing.* Paper Presented at the Geoscience and Remote Sensing Symposium, 2007. IGARSS 2007. IEEE International.

Chen, J., Chang, K.-T., Karacsonyi, D., & Zhang, X. (2014). Comparing urban land expansion and its driving factors in Shenzhen and Dongguan, China. *Habitat International, 43,* 61–71.

Corr, G., Walker, A., Benz, U., Lingenfelder, I., & Rodrigues, A. (2003). *Classification of urban SAR imagery using object oriented techniques.* Paper Presented at the Geoscience and Remote Sensing Symposium, 2003. IGARSS'03. Proceedings. 2003 IEEE International.

Dadhich, P. N., & Hanaoka, S. (2011). Spatio-temporal urban growth modeling of Jaipur, India. *Journal of Urban Technology, 18*(3), 45–65.

de Koning, G. H., Verburg, P. H., Veldkamp, A., & Fresco, L. (1999). Multi-scale modelling of land use change dynamics in Ecuador. *Agricultural Systems, 61*(2), 77–93.

Dekker, R. (2003). *Texture analysis and classification of SAR images of urban areas.* Paper Presented at the Remote Sensing and Data Fusion over Urban Areas, 2003. 2nd GRSS/ISPRS Joint Workshop on.

Dell'Acqua, F., & Gamba, P. (2003). Texture-based characterization of urban environments on satellite SAR images. *IEEE Transactions on Geoscience and Remote Sensing, 41*(1), 153–159.

Dell'Acqua, F., Gamba, P., & Trianni, G. (2006). Semi-automatic choice of scale-dependent features for satellite SAR image classification. *Pattern Recognition Letters, 27*(4), 244–251.

Delliere, J., Maitre, H., & Maruani, A. (2007). *SAR measurement simulation on urban structures using a FDTD technique.* Paper Presented at the Urban Remote Sensing Joint Event, 2007.

Dewan, A. M., & Yamaguchi, Y. (2009). Land use and land cover change in Greater Dhaka, Bangladesh: Using remote sensing to promote sustainable urbanization. *Applied Geography, 29*(3), 390–401.

Domg, Y., & Milne, A. K. (2001). Toward edge sharpening: A SAR speckle filtering algorithm. *IEEE Transactions on Geoscience and Remote Sensing, 39*(4), 851–863.

Donnay, J.-P., Barnsley, M. J., & Longley, P. A. (2003). *Remote sensing and urban analysis: GISDATA 9*. CRC Press.

Duro, D. C., Franklin, S. E., & Dubé, M. G. (2012). A comparison of pixel-based and object-based image analysis with selected machine learning algorithms for the classification of agricultural landscapes using SPOT-5 HRG imagery. *Remote Sensing of Environment, 118*, 259–272.

Esch, T., & Roth, A. (2004). *Semi-automated classification of urban areas by means of high resolution radar data*. Paper Presented at the ISPRS 2004 Congress.

Esch, T., Thiel, M., Schenk, A., Roth, A., Muller, A., & Dech, S. (2010). Delineation of urban footprints from TerraSAR-X data by analyzing speckle characteristics and intensity information. *IEEE Transactions on Geoscience and Remote Sensing, 48*(2), 905–916.

Eyoh, A., Olayinka, D. N., Nwilo, P., Okwuashi, O., Isong, M., & Udoudo, D. (2012). Modelling and predicting future urban expansion of Lagos, Nigeria from remote sensing data using logistic regression and GIS. *International Journal of Applied Science and Technology, 2*(5), 116–124.

Fatone, L., Maponi, P., & Zirilli, F. (2001). *Fusion of SAR/optical images to detect urban areas*. Paper Presented at the Remote Sensing and Data Fusion over Urban Areas, IEEE/ISPRS Joint Workshop 2001.

Forster, B. (1984). Remote sensing and its application to urban studies. *Urban Policy and Research, 2*(4), 25–32.

Franceschetti, G., Guida, R., Iodice, A., Riccio, D., Ruello, G., & Stilla, U. (2007). *Simulation tools for interpretation of high resolution SAR images of urban areas*. Paper Presented at the Urban Remote Sensing Joint Event, 2007.

Gamba, P., Dell'Acqua, F., & Trianni, G. (2007). *Satellite SAR and human settlement detection*. Paper Presented at the Urban Remote Sensing Joint Event, 2007.

Gomez-Chova, L., Fernández-Prieto, D., Calpe, J., Soria, E., Vila, J., & Camps-Valls, G. (2006). Urban monitoring using multi-temporal SAR and multi-spectral data. *Pattern Recognition Letters, 27*(4), 234–243.

Goodman, J., & Ustin, S. L. (2007). Classification of benthic composition in a coral reef environment using spectral unmixing. *Journal of Applied Remote Sensing, 1*(1), 011501–011517.

Grey, W., Luckman, A., & Holland, D. (2003). Mapping urban change in the UK using satellite radar interferometry. *Remote Sensing of Environment, 87*(1), 16–22.

Guan, D., Li, H., Inohae, T., Su, W., Nagaie, T., & Hokao, K. (2011). Modeling urban land use change by the integration of cellular automaton and Markov model. *Ecological Modelling, 222*(20), 3761–3772.

He, C., Xia, G.-S., & Sun, H. (2006). An adaptive and iterative method of urban area extraction from SAR images. *IEEE Geoscience and Remote Sensing Letters, 3*(4), 504–507.

Henderson, F. M., & Xia, Z.-G. (1997). SAR applications in human settlement detection, population estimation and urban land use pattern analysis: A status report. *IEEE Transactions on Geoscience and Remote Sensing, 35*(1), 79–85.

Herold, M., Roberts, D. A., Gardner, M. E., & Dennison, P. E. (2004). Spectrometry for urban area remote sensing—Development and analysis of a spectral library from 350 to 2400 nm. *Remote Sensing of Environment, 91*(3), 304–319.

Hu, Z., & Lo, C. (2007). Modeling urban growth in Atlanta using logistic regression. *Computers, Environment and Urban Systems, 31*(6), 667–688.

Jankowski, P. (2006). 12 Integrating geographical information systems and multiple criteria decision-making methods. *Classics from IJGIS: Twenty years of the International Journal of Geographical Information Science and Systems, 9*(3), 265.

Jebur, M. N., Pradhan, B., & Tehrany, M. S. (2014). Detection of vertical slope movement in highly vegetated tropical area of Gunung pass landslide, Malaysia, using L-band InSAR technique. *Geosciences Journal, 18*(1), 61–68.

Jensen, J. R. (2009). *Remote sensing of the environment: An earth resource perspective* (2nd ed.). Pearson Education India.

Jensen, J. R., & Im, J. (2007). Remote sensing change detection in urban environments. In *Geo-spatial technologies in urban environments* (pp. 7–31). Berlin: Springer.

Kandaswamy, U., Adjeroh, D., & Lee, M.-C. (2005). Efficient texture analysis of SAR imagery. *IEEE Transactions on Geoscience and Remote Sensing, 43*(9), 2075–2083.

Karteris, M., Theodoridou, I., Mallinis, G., Tsiros, E., & Karteris, A. (2016). Towards a green sustainable strategy for Mediterranean cities: Assessing the benefits of large-scale green roofs implementation in Thessaloniki, Northern Greece, using environmental modelling, GIS and very high spatial resolution remote sensing data. *Renewable and Sustainable Energy Reviews, 58*, 510–525.

Koomen, E., & Borsboom-van Beurden, J. (2011). *Land-use modelling in planning practice* (Vol. 101). Berlin: Springer Science & Business Media.

Kumar, J. A. V., Pathan, S., & Bhanderi, R. (2007). Spatio-temporal analysis for monitoring urban growth—A case study of Indore city. *Journal of the Indian Society of Remote Sensing, 35*(1), 11–20.

Laaribi, A., Chevallier, J., & Martel, J.-M. (1996). A spatial decision aid: A multicriterion evaluation approach. *Computers, Environment and Urban Systems, 20*(6), 351–366.

Lateh, H., & Muniandy, V. (2011). GIS dalam pendidikan geografi di Malaysia: Cabaran dan potensi (GIS in the Malaysian geography education: Challenges and potentials). *Geografia: Malaysian Journal of Society and Space, 7*(1), 42–52.

Lee, J.-S., & Pottier, E. (2009). *Polarimetric radar imaging: From basics to applications*. CRC press.

Li, X., & Yeh, A. G.-O. (2000). Modelling sustainable urban development by the integration of constrained cellular automata and GIS. *International Journal of Geographical Information Science, 14*(2), 131–152.

Li, X., & Yeh, A. G.-O. (2002). Neural-network-based cellular automata for simulating multiple land use changes using GIS. *International Journal of Geographical Information Science, 16*(4), 323–343.

Liu, Y., & Phinn, S. R. (2003). Modelling urban development with cellular automata incorporating fuzzy-set approaches. *Computers, Environment and Urban Systems, 27*(6), 637–658.

Lombardo, P., Pellizzeri, T. M., & Tomasuolo, A. (2001). *Classification of polarimetric SAR images of suburban areas using joint annealed segmentation and "H/A/α" decomposition*. Paper Presented at the Remote Sensing and Data Fusion over Urban Areas, IEEE/ISPRS Joint Workshop 2001.

Longley, P. A., Barnsley, M. J., & Donnay, J.-P. (2003). Remote sensing and urban analysis: A research agenda. In M. J. B. J. P. Donnay & P. A. Longley (Eds.), *Remote sensing and urban analysis: GISDATA 9* (pp. 245–258). London: Taylor and Francis.

Malczewski, J. (1999). *GIS and multicriteria decision analysis*. Wiley.

Mertens, B., & Lambin, E. F. (2000). Land-cover-change trajectories in southern Cameroon. *Annals of the Association of American Geographers, 90*(3), 467–494.

Mishra, P., Singh, D., & Yamaguchi, Y. (2011). Land cover classification of PALSAR images by knowledge based decision tree classifier and supervised classifiers based on SAR observables. *Progress in Electromagnetics Research B, 30,* 47–70.

Openshaw, S., & Clarke, G. (1996). Developing spatial analysis functions relevant to GIS environments. In *Spatial analytical perspectives on GIS* (pp. 21–37).

Paez, A., Uchida, T., & Miyamoto, K. (2001). Spatial association and heterogeneity issues in land price models. *Urban Studies, 38*(9), 1493–1508.

Parker, D. C., Manson, S. M., Janssen, M. A., Hoffmann, M. J., & Deadman, P. (2003). Multi-agent systems for the simulation of land-use and land-cover change: A review. *Annals of the Association of American Geographers, 93*(2), 314–337.

Perumal, K., & Bhaskaran, R. (2010). Supervised classification performance of multispectral images. *arXiv preprint* arXiv:1002.4046

Pijanowski, B. C., Brown, D. G., Shellito, B. A., & Manik, G. A. (2002). Using neural networks and GIS to forecast land use changes: A land transformation model. *Computers, Environment and Urban Systems, 26*(6), 553–575.

Rimal, B. (2011). Urban growth and land use/land cover change of Biratnagar sub-metropolitan city, Nepal. *Applied Remote Sensing Journal, 2*(1), 6–15.

Roberts, J. W., Van Aardt, J., & Ahmed, F. (2008). Assessment of image fusion procedures using entropy, image quality, and multispectral classification. *Journal of Applied Remote Sensing, 2*(1), 023522–023528.

Saaty, T. L. (1980). *The analytic hierarchy process: Planning, priority setting, resources allocation.*

Saaty, T. L. (2000). *Fundamentals of decision making and priority theory with the analytic hierarchy process* (Vol. 6). Rws Publications.

Samat, N. (2006). *Applications of geographic information systems in urban land use planning in Malaysia.* Paper Presented at the 4th Taipei International Conference on Digital Earth, Taiwan.

Selamat, M. H., Selamat, A., Othman, M. S., Shamsuddin, N., & Zukepli, N. (2012). A review on geographical information system (GIS) in town planning: Malaysia experience. *Geoinformatica: An International Journal (GIIJ), 2*(2), 27–38.

Shackelford, A. K., & Davis, C. H. (2003). A combined fuzzy pixel-based and object-based approach for classification of high-resolution multispectral data over urban areas. *IEEE Transactions on Geoscience and Remote Sensing, 41*(10), 2354–2363.

Shaharudin, I., Shamsul, A., Tahir, A., Mariam, M., Azah, D., & NI, N. S. (2002). Sistem maklumat geografi (GIS) dan sektor kesihatan awam: kajian demam denggi di Bandar Baru Bangi dan Kajang. *Jurnal Kesihatan Masyarakat, 8*(S), 34–42.

Stanilov, K. (2003). Accessibility and land use: The case of suburban Seattle, 1960–1990. *Regional Studies, 37*(8), 783–794.

Stasolla, M., & Gamba, P. (2008). Spatial indexes for the extraction of formal and informal human settlements from high-resolution SAR images. *IEEE Journal of Selected Topics in Applied Earth Observations and Remote Sensing, 1*(2), 98–106.

Steiner, F., McSherry, L., & Cohen, J. (2000). Land suitability analysis for the upper Gila River watershed. *Landscape and Urban Planning, 50*(4), 199–214.

Suryavanshi, V. M. (2001). *Land use and opportunities for crime: Using GIS as an analysis tool.* Virginia Polytechnic Institute and State University.

Tehrany, M. S., Pradhan, B., & Jebuv, M. N. (2014). A comparative assessment between object and pixel-based classification approaches for land use/land cover mapping using SPOT 5 imagery. *Geocarto International, 29*(4), 351–369.

Tison, C., Nicolas, J.-M., Tupin, F., & Maître, H. (2004). A new statistical model for Markovian classification of urban areas in high-resolution SAR images. *IEEE Transactions on Geoscience and Remote Sensing, 42*(10), 2046–2057.

Turskis, Z., Zavadskas, E. K., & Zagorskas, J. (2006). Sustainable city compactness evaluation on the basis of GIS and Bayes rule. *International Journal of Strategic Property Management, 10*(3), 185–207.

Ulaby, F. T., Kouyate, F., Brisco, B., & Williams, T. L. (1986). Textural infornation in SAR images. *IEEE Transactions on Geoscience and Remote Sensing, 2,* 235–245.

Verburg, P. H., Van Eck, J. R., de Nijs, T. C., Dijst, M. J., & Schot, P. (2004). Determinants of land-use change patterns in The Netherlands. *Environment and Planning B, 31*(1), 125–150.

Wakode, H. B., Baier, K., Jha, R., & Azzam, R. (2014). Analysis of urban growth using Landsat TM/ETM data and GIS—A case study of Hyderabad, India. *Arabian Journal of Geosciences, 7*(1), 109–121.

Wei, S., Chao, Z., Jianyu, Y., Honggan, W., Minjie, C., Anzhi, Y., et al. (2010). Knowledge-based object oriented land cover classification using SPOT5 imagery in forest-agriculture ecotones. *Sensor Letters, 8*(1), 22–31.

Weng, Q. (2001). A remote sensing? GIS evaluation of urban expansion and its impact on surface temperature in the Zhujiang Delta, China. *International Journal of Remote Sensing, 22*(10), 1999–2014.

Wilson, E. H., Hurd, J. D., Civco, D. L., Prisloe, M. P., & Arnold, C. (2003). Development of a geospatial model to quantify, describe and map urban growth. *Remote Sensing of Environment, 86*(3), 275–285.

Wu, F. (1998). SimLand: A prototype to simulate land conversion through the integrated GIS and CA with AHP-derived transition rules. *International Journal of Geographical Information Science, 12*(1), 63–82.

Xiao, J., Li, J., & Moody, A. (2003). A detail-preserving and flexible adaptive filter for speckle suppression in SAR imagery. *International Journal of Remote Sensing, 24*(12), 2451–2465.

Xie, C., Huang, B., Claramunt, C., & Chandramouli, C. (2005). *Spatial logistic regression and GIS to model rural-urban land conversion.* Paper Presented at the Proceedings of PROCESSUS Second International Colloquium on the Behavioural Foundations of Integrated Land-use and Transportation Models: Frameworks, Models and Applications.

Yaakup, A. (2004). Sistem maklumat perancangan bandar dan wilayah: kronologi pembangunan dan penggunaannya. *Jurnal Alam Bina, Jilid, 6,* 1–26.

Yeh, A., & Li, X. (1998). Sustainable land development model for rapid growth areas using GIS. *International Journal of Geographical Information Science, 12*(2), 169–189.

Yuan, F. (2008). Land-cover change and environmental impact analysis in the Greater Mankato area of Minnesota using remote sensing and GIS modelling. *International Journal of Remote Sensing, 29*(4), 1169–1184.

Yusoff, Z. M., Maassoumeh, B., Omar, D., Latif, Z. A., & Samad, A. M. (2010). *Applicability of geographical information system in assessing the accessibility and mobility of urban lower-income family living.* Paper Presented at the Signal Processing and Its Applications (CSPA), 2010 6th International Colloquium on.

Zaini, Z. (2007). *Sistem maklumat Geografi dalam Penentuan Kawasan Pemeliharaan Warisan. Kajian Kes: Bandar Taiping, Perak.* Thesis Ijazah Sarjana Sains Perancangan (Teknologi Maklumat), Fakulti Alam Bina, Universiti Teknologi Malaysia.

14 Quantifying Spatiotemporal Urban Sprawl Patterns in the City of Tripoli Metropolis (Libya) Over the Past Four Decades Using Satellite Data Sets

Abubakr A.A. Al-sharif and Biswajeet Pradhan

14.1 Introduction

Worldwide urban population was approximately 3% of the total population in the 1800s, reached nearly 30% in the 1950s, and became 50% by the end of the last century. The United Nations (UN) estimated that 60% of global inhabitants would be living in urbanized areas by 2025 (Al-sharif and Pradhan 2013b). In addition, urbanization is accelerating worldwide. Uncontrolled urbanization induces significant effects on ecosystems and landscapes in metropolis cities and neighboring areas (Xian et al. 2007; Li et al. 2010). Moreover, rapid population increase has profoundly affected socioeconomics in urban growth centers in developing countries (Kong et al. 2012). In such countries, where the most rapid urban expansions are occurring, urban policies and plans are frequently weaker than those in developed countries. The reasons behind this situation include lack of expertise, the absence of holistic environmental considerations, and the need for the integration of urban scientific manners in the decision-making process (Li et al. 2010). However, urban expansion is a permanent phenomenon that includes transformations in large areas of land cover that are associated with progressive population density. Spontaneously urbanizing landscapes lead to unbalanced urban population density, unplanned infrastructure, and a significant lack of basic necessity facilities (Ramachandra et al. 2012b; Sun et al. 2013).

The phenomenon of urban sprawl occurs in a variety of forms that are strongly connected with geographical, economic, and institutional contexts, and can be defined in many different ways. In recent years, urban sprawl detection has been conducted using maps and satellite images from different periods (Bhatti and Tripathi 2014). The main spatial indicator in sprawl investigation is the overall total built-up area and time series analysis of such indicator (Verbeek et al. 2014). In several developing countries, urban expansion process has different drivers and appears in different guises. Hence, the characteristics of considered regions must be accounted for when studying and investigating urban sprawl (Epsteln et al. 2002; Jaeger et al. 2010). Landscape changes associated with urbanization, mainly urban sprawl, continued to remain significant over the past five decades and its significance is expected to continue over the next decades. The spatial landscape indices are valid in assessing and measuring different landscape changes and patterns with various ecological meanings (Tang et al. 2006). Interest in using landscape metric techniques to analyze urban environments has grown in recent years. Remotely sensed data, such as satellite systems data, is a valuable resource for mapping urban areas. It provides a comprehensive and synoptic view, which, in the case of large study area, is not possible through field survey. Another important benefit of using remote sensing data is the availability of temporal records that help in understanding and mapping urban sprawl over time periods (Bhatti and Tripathi 2014; Abdullahi et al. 2015a). As reported, combining remote sensing and spatial metrics is a good step for improving the analysis and modeling of urban growth and land-use change, leading to an improved understanding and representation of urban dynamics and contributing to the development of alternative concepts of urban spatial structure and change (Su et al. 2010). However, rapid urbanization leads to haphazardly dispersed urban development in city fringes that can be considered as sprawl. Hence, urban sprawl phenomena are scattered and uncontrolled suburban expansions that deplete resources as a consequence of considerable land-use change (conversion of green lands, water bodies, parks, etc.) (Zhang et al. 2015). Nevertheless, the term "urban sprawl" has no clear consistent definition yet among urban researchers, that is, this term has different definitions (Siedentop 2005; Zhang et al. 2014). According to Pumain (2004), urban sprawl is the spatial expansion of urbanized settlements into countryside areas along with the deconcentration of urban functions. Other researchers consider the

B. Pradhan (ed.), *Spatial Modeling and Assessment of Urban Form*,
DOI 10.1007/978-3-319-54217-1_14

dominance of low-density urban settlements as urban sprawl (Taubenböck et al. 2009a). Meanwhile, a number of urban activists defined urban sprawl as the transformation of previously monocentric and compacted cities into dispersed, discontinuous, and fragmented polycentric urban patterns (Bhatta 2010; Abdullahi et al. 2015b). Another definition of urban sprawl is urban fragmented developments that contradict the objectives and concepts of spatial urban plans (Al-sharif and Pradhan 2013a). Urban sprawl can also mean large imponderables between population increase and spatial urban expansion (Brueckner 2001). Urban sprawl typically occurs in suburban areas, but not all suburban growths can be regarded as urban sprawl (Yue et al. 2012). Moreover, urban sprawl processes entail the expansion of urbanized areas from a city center toward the fringes of the city. These small, scattered urban pockets in the outskirts of a city typically need basic amenities such as electricity, treated water, and sanitation facilities (Ramachandra et al. 2012b). From another perspective, urban sprawl occurs if land-use rate is faster than population increase (Fulton et al. 2001). Correspondingly, some researchers reported that unlimited "leapfrog" urban expansions with low density can also be considered as urban sprawl (Burchell et al. 2005; Yue et al. 2012). In general, the common pattern of the urban sprawl process is that progressively urban landscapes become geometrically fragmented and complex (Ramachandra et al. 2012b). However, without a universal description of urban sprawl, modeling, and quantifying it is difficult. Hence, the urban sprawl phenomenon should be analyzed from multidimensional perspectives (Taubenböck et al. 2009a).

Monitoring urban changes by mapping landscapes at a temporal scale enhances planning sustainability (Al-sharif et al. 2013). However, the dynamicity of landscape diversity resulting from the urban expansion process has not been satisfactorily discussed within the context of urban areas (Yeh and Huang 2009; Abdullahi et al. 2015a; Abdullahi and Pradhan 2016). Moreover, the urban landscape is dynamic and constantly changing as a result of urban development. Consequently, studying and analyzing spatial change in urban landscape patterns at a single temporal point cannot identify the real dynamics behind urban landscape transformation (Xian et al. 2007; Bagan and Yamagata 2015). Thus, using historical spatiotemporal data to analyze and visualize urban expansions is helpful in identifying suitable and probable areas of intense urban growth and sprawl (Alsharif and Pradhan 2013a). One important means to address urbanization problems and their side effects is by analyzing the urban sprawl process and by applying appropriate urban strategies that facilitate sustainable urban development (Jaeger et al. 2010; Kong et al. 2012). In this context, socioeconomic measures, such as the number of commercial establishments, employment opportunity, population growth, and so on, are used to detect and identify urban sprawl indirectly (Brueckner 2000; Ramachandra et al. 2012a). Nevertheless, these measurement techniques cannot efficiently identify the spatial impacts of urban sprawl (Ramachandra et al. 2012a). Some researchers have reported that urban sprawl can be characterized by considering various indicators, such as density, accessibility, growth, decentralization, socioeconomic, open space, cost of open space, aesthetic, etc. (Bhatta 2010). Developing quantitative techniques to identify urban growth/sprawl patterns is crucial to assist regional and local urban planners in addressing and understanding issues attributed to urban sprawl (Sun et al. 2013). From this perspective, accessibility of spatial remotely sensed data at numerous time intervals supports identifying and monitoring rapid land-use changes (Chen et al. 2000; Epsteln et al. 2002; Dietzel et al. 2005). Furthermore, remote sensing images can easily explain the quantitative physical spatial formation of urban environments. By combining geographic information system (GIS) data, satellite imageries, digital datasets, and spatial metrics, the landscape patterns can be conveniently described, analyzed, and estimated (Herold et al. 2003; Yeh and Huang 2009; Sun et al. 2013). Recent advances in landscape ecology discipline provide helpful tools in monitoring, modeling, quantifying, and predicting urban growth (Sudhira et al. 2004; Kong et al. 2012). Landscape metrics are generally used in ecological studies (Li et al. 2005; Peng et al. 2010); however, landscape metrics have been recently extended and used in urban studies to improve understanding on different urban forms and the urbanization process at a landscape level (Peng et al. 2010; Kong et al. 2012).

Dietzel et al. (2005) and Herold et al. (2003) demonstrated that landscape metrics can be used to trace the spatiotemporal trends of landscape patterns. Landscape metrics can effectively categorize and quantify complex landscapes; thus, these metrics can be used to determine many spatial properties that are not directly observable (Ramachandra et al. 2012b). Several studies used spatial landscape metrics to study dynamic patterns, as well as socioeconomic and political factors that underlie urbanization (Schneider et al. 2005). Landscape metrics can be considered as reliable approaches to quantify urban spatial patterns, as well as to detect and provide in-depth information on urban sprawl.

The main objectives of this study are to assess, investigate, and measure quantitatively the spatiotemporal patterns of urban sprawl in the Tripoli metropolitan area using a series of landscape metrics. This work investigates the urban sprawl patterns of Tripoli metropolitan city and its districts in different temporal points, which gives good guidance to urban planner and decision maker. This study highlights different behaviors of urban expansions, which are helpful for further researches and deeper understanding of various urban sprawl patterns.

14.2 Study Area

Tripoli is the capital of Libya, as well as the political, financial, and commercial center of the country. Tripoli City is thousands of years old. It is located along the Mediterranean coast in the northwestern part of Libya between latitudes (32° 36′ 18″ N and 32° 54′ 17″ N) and longitudes (12° 54′ 04″ E and 13° 26′ 38″ E). Tripoli covers an area of approximately 1143.73 km^2, with a population of more than 1.3 million. This city is divided into nine districts, namely, Central Tripoli, Suq Aljumma, Hey Alandalus, Abuslim, Tajoura, Ainzara, Janzour, Kaser Ben Ghashir, and Alswany (Fig. 14.1).

The Tripoli metropolitan area has become economically active, particularly during the last decade after international sanctions were lifted. Despite the presence of government urban plans, urban growth, and development are spontaneous, uncontrolled, and haphazard. Implementing plans was affected by corruption, political situations, and economic conditions, thus resulting in urban sprawl becoming primarily dependent on citizen trend regardless of plans in most situations.

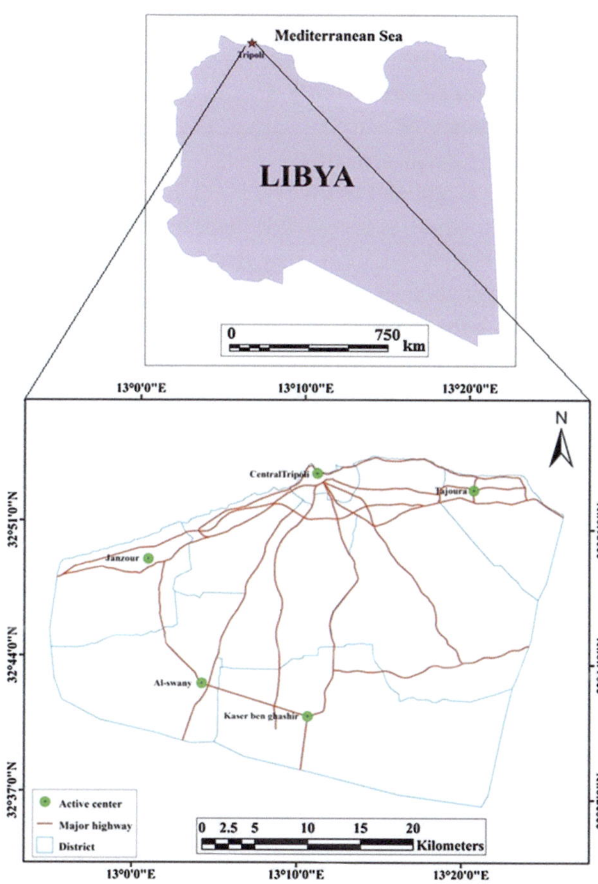

Fig. 14.1 Study area location map

The major results of rapid and uncontrolled urban expansions in the study area are the conversion of fertile and green lands, the destruction of environmental reserve areas, and the formation of illegal spontaneous settlements that generate socioeconomic and physical problems.

14.3 Data Description and Methods

Four satellite images were used in this study (Table 14.1). Landsat satellite images and SPOT 5 image from 2002 were obtained from the Biruni Remote Sensing Center, Libya. A SPOT 5 image from 2010 was also obtained from the Libyan Center for Remote Sensing and Space Science. Available data are limited, and there was no access to imageries for the same periods.

Numerous techniques have been developed to analyze, process, and extract information from remotely sensed data. Selecting specific algorithms or methods that will be employed depends on the objectives of the study (Zhang et al. 2015). ARC/INFO GIS software package was used for image processing, classified land cover map generation, spatial analysis, and map preparation.

The images used were collected as standard products, and were then radiometrically and geometrically corrected. The standards used by the agencies that provided the images are different, thus resulting in misregistration and low accuracy of images co-alignment. To resolve this problem, images were rectified and geo referenced during the preprocessing step, such that the high accuracy of the overlay was matched.

In this study, the images used had different spatial resolutions. The simple approach used to fix this problem was to resample high-resolution images to match them with low-resolution imageries.

Subsequently, a maximum likelihood supervised classification method was applied on the images during the classification process phase. All images were classified by selecting samples (polygons) as training areas to present different classes (2). This work focused on urban land cover and urban expansion; consequently, classification was performed by considering only two important classes: nonurban area and urban area; such classification is considered to be sufficient (Bhatta 2009).

Table 14.1 List of used satellite imageries

Sensor type	Acquisition year	Spatial resolution (m)
LANDSAT–TM	Sep. 1984	30
LANDSAT–TM	Aug. 1996	30
SPOT 5	Oct. 2002	2.5
SPOT 5	Aug. 2010	5

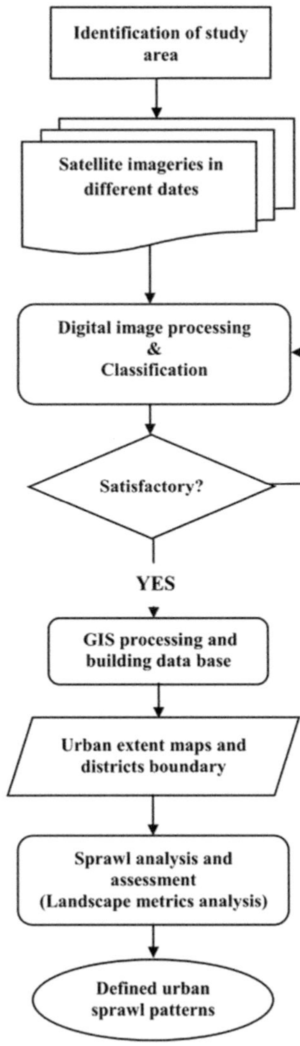

Fig. 14.2 Overall methodology flow chart

As a final step in data preparation, all maps were clipped with the study area boundary vector map and resampled to a grid size of 30 m × 30 m. Figure 14.2 demonstrates the overall steps flow chart. All the prepared data were converted to ASCII formats to be further used in analysis using the FRAGSTATS software.

14.4 Analyzing Urban Sprawl Using Landscape Metrics

During the last three decades, many landscape metrics have been developed, tested, and used to analyze landscape structure and composition (Peng et al. 2010). In general, the urban sprawl process changes and modifies landscape compositions over time by increasing landscape fragmentation and generating small urban patches. In this study, six landscape metrics were applied to investigate and analyze the spatiotemporal patterns of urban sprawl in the study area and each individual district, as well as to assess sprawl from different perspectives. The applied landscape metrics can measure clumpiness, aggregation, complexity, and level of dispersion of urban class in the landscape of the study area. The metrics used are: 1-edge density (ED) to report edge length per unit area basis that facilitates comparison among landscapes of varying size. ED = 0 when there is no class (urban) edge in the landscape; 2-largest patch index (LPI) to quantify the percentage of total landscape area comprised by the largest patch. As such, it is a simple measure of dominance (i.e., it equals the percentage of the landscape comprised by the largest patch), LPI approaches 0 when the largest patch of the corresponding patch type is increasingly small. LPI = 100 when the entire landscape consists of a single patch of the corresponding patch type; that is, when the largest patch comprises 100% of the landscape; 3-shape index (SHAPE) to measure the complexity of patch shape compared to a standard shape (square) of the same size; 4-landscape shape index (LSI) to provide a standardized measure of edge density that adjusts for the size of the landscape. Because it is standardized, it has a direct interpretation, in contrast to total edge, for example, that is only meaningfully relative to the size of the landscape; 5-patch density (PD) is a limited, but fundamental aspect of landscape pattern. PD has the same basic utility as number of patches as an index, except that it expresses number of patches on a per unit area basis that facilitates comparisons among landscapes of varying size. PD is ultimately constrained by the grain size of the raster image, because the maximum PD is attained when every cell is a separate patch. Therefore, ultimately cell size will determine the maximum number of patches per unit area. Large PD reflects high dispersed urban patterns; 6-Simpson's evenness index (SIEI) expresses such that an even distribution of area among patch types results in maximum evenness. As such, evenness is the complement of dominance, the higher SIEI means higher diverse distribution of considered urban landscape (Taubenböck et al. 2009). To perform urban landscape analysis, the statistical package FRAGSTATS version 4 (McGarigal et al. 2002) was employed to calculate all the aforementioned landscape quantitative measures.

14.5 Results

The classification of the multi-temporal satellites images into nonurban area and urban area has resulted in abstracted and simplified representation of study area as shown in Fig. 14.3. The four classified maps demonstrate the spatiotemporal patterns of urban expansions in the study area (La Rosa and Wiesmann 2013). The classified maps were assessed using the confusion matrix method. Real ground reference

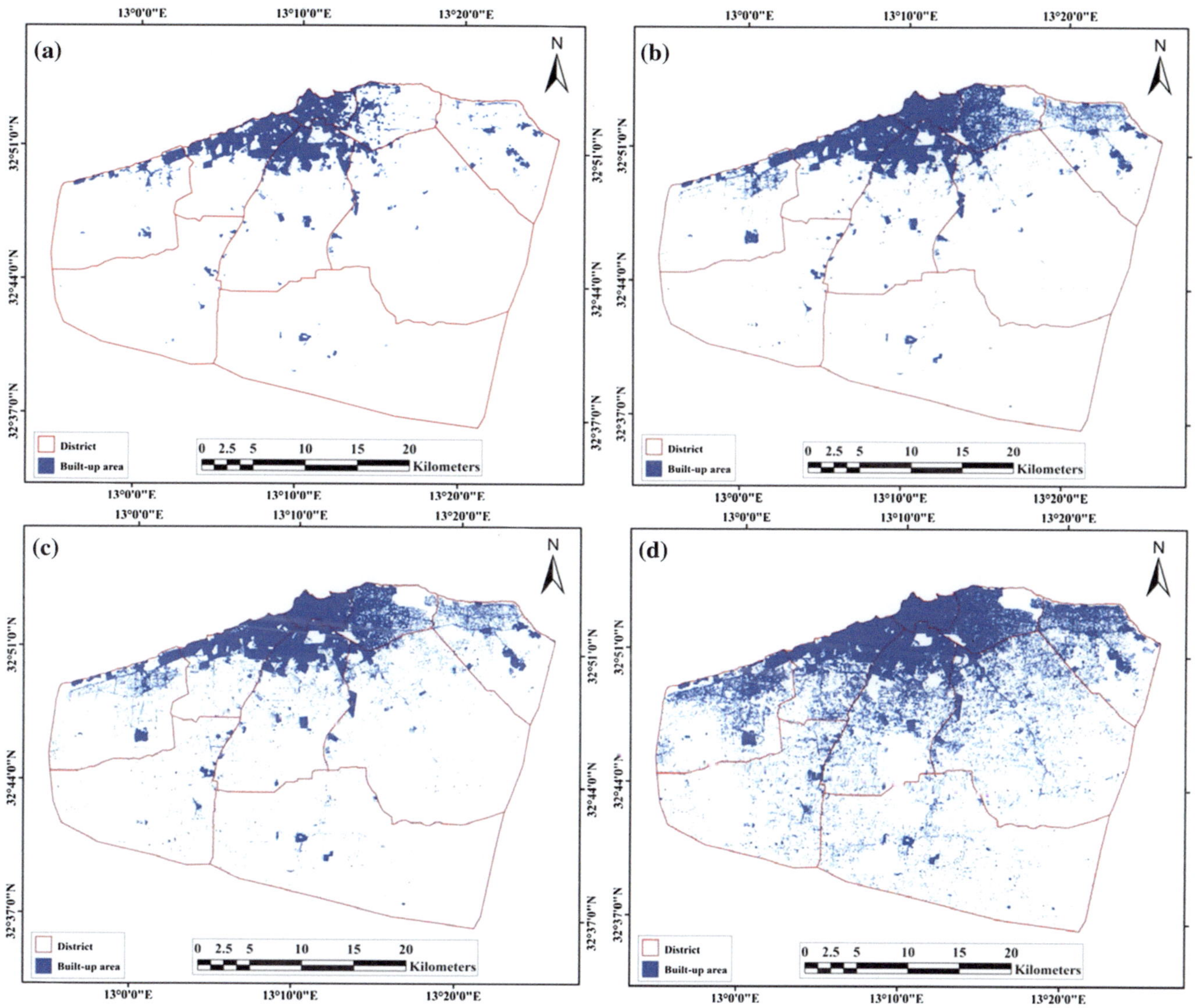

Fig. 14.3 Built-up area extent in different years; **a** 1984; **b** 1996; **c**; 2002; **d** 2010

polygons were compared with the classified output maps to assess the accuracy. Overall accuracy values of 91, 93.2, 95.7, and 94% and Kappa coefficients values of 0.89, 0.93, 0.94, and 0.93 were achieved for the classified maps for the years 1984, 1996, 2002, and 2010, respectively. In Fig. 14.3, dispersed rapid urban expansions in the Tripoli metropolis over the period of 1984 to 2010 are easily noted. However, it is significant to analyze and assess these urban growth maps with suitable statistical evidences to identify and describe the different urban development patterns that have happened in the study area. This will allow comparing and understanding the different urban patterns over time quantitatively. The aforementioned spatial landscape metrics were used to assess and compute only the spatiotemporal trends of urban areas. According to different definitions of urban sprawl and the analysis results collected from applied landscape metrics, the presence of sprawl was identified and assessed quantitatively.

The results of the synoptic analysis of the implemented spatial metrics provided an overall demonstration of urban sprawl spatiotemporal patterns (Figs. 14.4, 14.5, 14.6, 14.7, 14.8, and 14.9). The SIEI measure illustrated that the overall diversity level of the study area increased between 1984 and 2010. The increase rate of the SIEI measure was most obvious from 2002 to 2010. The SIEI index has decreased dramatically in central Tripoli region after 1984 and became very low in 2010. This finding reflect the low diversity of the landscape in this district, i.e., almost whole landscape was urbanized in such zone. This high urban compactness is because of that the Central Tripoli district includes the central business of district (CBD) of the study area.

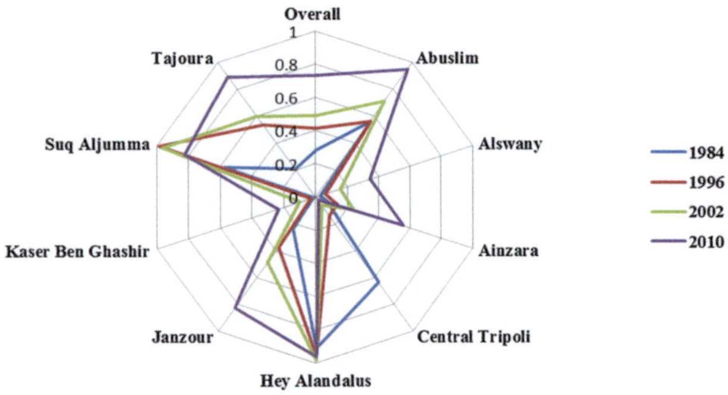

Fig. 14.4 Variation of SIEI measure in different time periods (without unit)

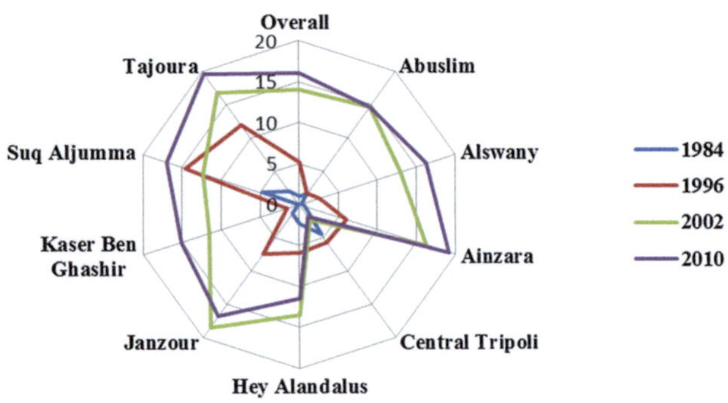

Fig. 14.5 Variation of PD measure in different time periods (Number per 100 ha)

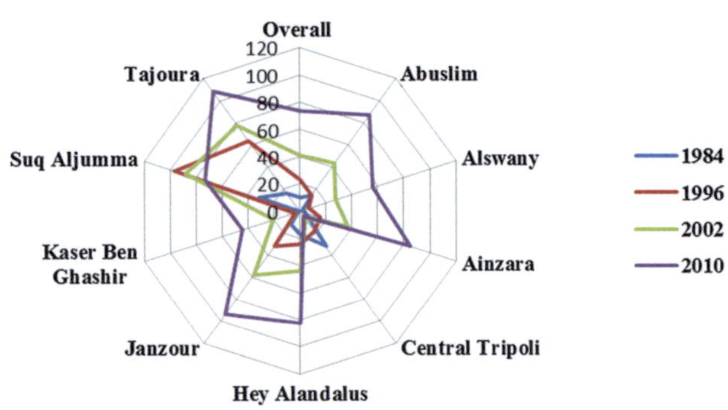

Fig. 14.6 Variation of ED measure in different time periods (m/ha)

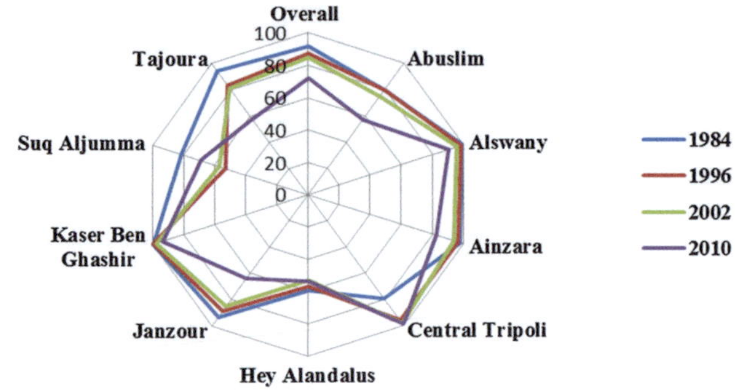

Fig. 14.7 Variation of LPI measure in different time periods (%)

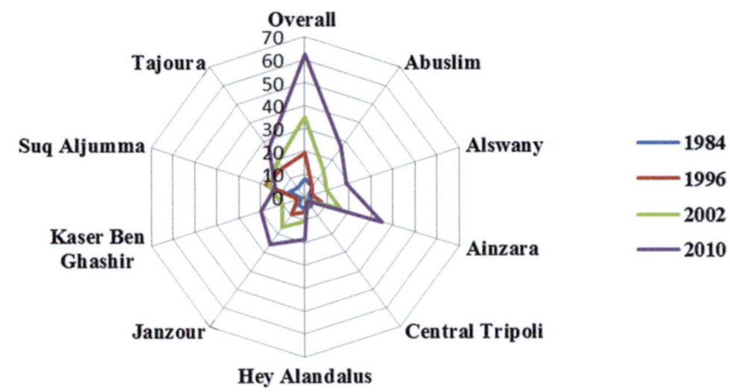

Fig. 14.8 Variation of LSI measure in different time periods (without unit)

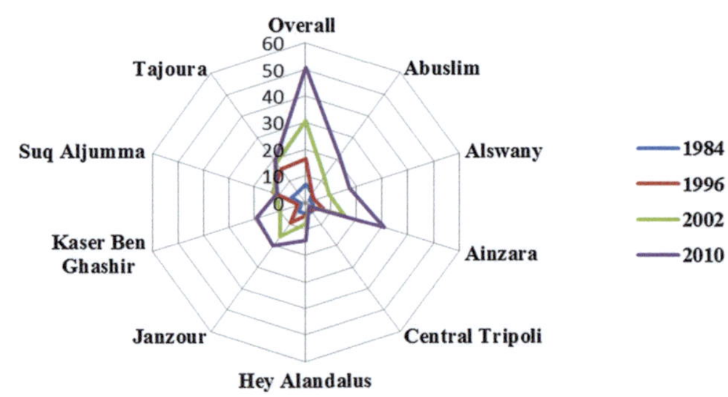

Fig. 14.9 Variation of SHAPE measure in different time periods (without unit)

The districts of Hey Alandalus and Suq Aljumma showed gradual decrease of SIEI in the last decade; that means the both regions are going to be condensed with built-up areas. In contrast, the other six districts showed different behavior and had high increased SIEI values; i.e. one can say these six areas have speedy urban expansion. The landscape metrics of PD and ED increased remarkably in the entire study landscape. Such large increases reflect the increase in isolated urban patches and the irregular formations of these patches. The highest increase of PD (i.e., increase of urban patches) in overall study area was from 1996 to 2002; the PD was increased further in 2010. However, districts of Central Tripoli, Hey Alandalus and Janzour showed decreased level of PD (i.e., decreasing in urban dispersion) in the last decade. On the other hand, the other districts encountered high PD levels after 1996 (i.e., the presence of urban leapfrog dispersed patterns). The spatial patterns of urban areas in overall study area in 1984 were more regular than other recent time periods based on ED measure. In 2010 the ED value was the greatest which reflects the very high irregular and complex urban expansions. By reading the analysis findings of ED measure for each individual zone, one can note that only two districts of Central Tripoli and Suq Aljumma showed decreased ED levels (i.e., more regular and dense urban patterns) with time progress. On the other hand, the all other districts had increased irregular rate of urban growth. These findings indicate high urban fragmentation as well as increasing overall urban sprawl process. However, the overall LPI metric has decreased since 1984, which is the largest decrease recorded in the last decade, that is, the largest urban patch in the study area has become increasingly small. This situation is another sign of uncontrolled and fragmented urban sprawl increase. However, after 1996 only the districts of Hey Alandalus, Central Tripoli, and Suq Aljumma had increased LPI values, this result confirms that coastal zones that near to CBD in study area have lower sprawl rates. The findings of the LPI investigation revealed that the Central Tripoli district presented an increase in LPI at all times. Meanwhile, the largest LPI in Suq Aljumma and Hey Alandalus occurred in 1984, and then, LPI declined from 1996 to 2002. After 2002, however, LPI increased to reflect the growing urban compactness of these districts. By contrast, the other six districts increasingly faced formations of smaller urban patches (i.e., dispersed expansions), and exhibited dramatic degradations in LPI values. Therefore, the occurrence of urban sprawl is obvious and easily detectable.

The analysis findings showed that LSI and SHAPE are increasing continuously based on the urban expansion history of the study area. Such increases indicate that the urban area aspect progressively has become irregular. Hence, we can consider that Tripoli is experiencing unplanned urban growth. The outcomes of ED, LSI, and SHAPE analyses confirmed the presence of highly fragmented complex urban patches in the districts of Hey Alandalus, Abuslim, Tajoura, Ainzara, Janzour, Kaser Ben Ghashir, and Alswany. The increase rate of these metrics was gradual from 1984 to 1996, but the trend of the three measures changed entirely after 1996. The remarkable increases in ED, LSI, and SHAPE values are noticeable; this indicating deteriorated urban expansion patterns and the presence of uncontrolled sprawl. Moreover, the SIEI measures exhibited highly diverse values in most districts, particularly in 2010, except in Central Tripoli, Suq Aljumma, and Hey Alandalus. This result confirmed the findings of the analyses of the other landscape metrics, and thus, can be considered as a strong evidence of urban sprawl.

14.6 Discussions

Since, understanding the socioeconomic and environmental consequences of urban expansion require quantifying the spatiotemporal patterns of urban landscapes, this study has illustrated that the spatiotemporal patterns of urban sprawl can be investigated and quantified using a combination of spatial landscape metrics. The study used a combined analysis of both the spatial and temporal changes of landscape pattern in response to urban expansion process. Thus, the results of this study can adequately address several important points such as overall urban sprawl presence, districts that contributed significantly to sprawl occurrence, and variation of sprawl over the time periods.

In the study area, an extensive urban sprawl occurred between 2002 and 2010. Another important trend of the recent urban growth process in Tripoli is the decline of urban densities, with most districts exhibiting a trend toward scattered low densities. The general pattern of urban expansions in Tripoli was that of an increasingly urbanized landscape becoming geometrically complex, spatially dispersed, and fragmented. Hence, when the degree of urban sprawl is high, both the density and shape complexity of urban patches will increase, whereas the size of the urban patches will decrease. Thus, further studies are necessary to confirm these findings.

The findings of the study demonstrated that urban expansions in the Tripoli metropolitan area have resulted in dramatic increases in PD, ED, LSI, SHAPE, and SIEI, and a sharp decrease in LPI. Consequently, the change rates of applied metrics indicate that the intensity of landscape changes increased because of rapid urban sprawl. In districts such as Central Tripoli and Suq Aljumma, the rapid urban expansion caused dramatic alterations in the landscape patterns; however, the rate of urbanization process gradually slowed down because of the lack of space for development. Hence, the degree of landscape diversity and fragmentation gradually decreased when urban land-use types became dominant. The infill of urban developments can help reduce PD and increase LPI in the city, enabling it to achieve the goal of a compact form of urban development and decrease the degree of fragmentation. The regions near CBD were assumed to already have been developed to almost its full capacity at the beginning of the time period investigated; hence, less urban expansions in landscape pattern occurred.

By contrast, the districts further away from CBD, including the urban fringes are places where the most rapid transformation of the landscape occurred because of the urban sprawl. Therefore, the old urban center gradually became stabilized whereas the urban fringes experienced the most rapid transformation of landscape patterns. In other words, a highly urbanized landscape is more homogenous and better connected than a rapidly urbanizing landscape.

However, based on results of the analysis and considering the continuous development policy from the local government and citizens, we predict that rapid urban sprawl will create significant environmental impacts in the coming years. This study showed that landscape metric analysis does not necessarily have to rely on complex mathematical computation; rather, easy to calculate and comprehensible metrics can provide important information on urban landscape patterns. The advantage of landscape metric analysis includes quantification of complex landscape into numerical indices, which can serve as criteria to evaluate different planning scenarios. Understanding how urban land use is expanded is more complex than simply mapping the urban cover. Measures such as PD, ED, LSI, SHAPE, SIEI, and LPI can quantitatively evaluate and compare urban sprawl to assist in the decision-making process. Moreover, as this study has demonstrated, using the landscape metrics at districts level can help in identifying areas with the most intensive urban development pressure.

14.7 Conclusion

This study investigated urban sprawl and its spatial patterns in the Tripoli metropolis area from 1984 to 2010. The spatial landscape metrics have assessed the dispersion, aggregation, diversity, complexity, and shape of urban areas to measure quantitatively urban sprawl in the study area from different perspectives. The analysis results of all applied metrics demonstrated a generally clear and uncontrolled urban dispersion pattern and urban sprawl in Tripoli City. Sprawl is increasing continuously as time progresses. Nevertheless, only two districts, namely, Central Tripoli and Suq Aljumma, exhibited non-sprawling urban growth during the last two decades. Furthermore, the outcomes illustrated that Hey Alandalus experienced a clear sprawl but would be more compact in the future based on landscape metrics results. All the metrics results indicated that all the other zones in Tripoli are experiencing extremely high sprawl levels, which reflect rapid urban expansions and the absence of clear and effective urban policies. The techniques used in this work can provide guidance in recognizing and assessing changes that are likely to occur if the trends exhibited in urban history persist. Finally, this research offers good insight into the urban expansion behaviors of the study area. The findings of the study can be useful in directing prospective urban plans and urbanization policies in Tripoli. Moreover monitoring and assessing effects of urban sprawl and its patterns provides important information and knowledge that support urban planning and establishing regional development policy. Wise urban land management legislation and clear strategy implementation should be adopted to protect fertile lands from being converted to unplanned built-up lands. However, this study has only examined the historical records of urban landscape changes. Therefore, analytical techniques need to be developed which can predict urban future trends and monitor sprawl changes.

References

Abdullahi, S., & Pradhan, B. (2016). Sustainable brownfields land use change modeling using GIS-Based weights-of-evidence approach. *Applied Spatial Analysis and Policy, 9*(1), 21–38.

Abdullahi, S., Pradhan, B., & Jebur, M. N. (2015a). GIS-based sustainable city compactness assessment using integration of MCDM, Bayes theorem and RADAR technology. *Geocarto International, 30*(4), 365–387.

Abdullahi, S., Pradhan, B., Mansor, S., & Shariff, A. R. M. (2015b). GIS-based modeling for the spatial measurement and evaluation of mixed land use development for a compact city. *GIScience & Remote Sensing, 52*(1), 18–39.

Al-Sharif, A. A., & Pradhan, B. (2013a). Monitoring and predicting land use change in Tripoli Metropolitan City using an integrated Markov chain and cellular automata models in GIS. *Arabian Journal of Geosciences, 7*(10), 4291–4301. doi:10.1007/s12517-013-1119-7

Alsharif, A. A., & Pradhan, B. (2013b). Urban sprawl analysis of Tripoli Metropolitan city (Libya) using remote sensing data and multivariate logistic regression model. *Journal of the Indian Society of Remote Sensing, 42*(1), 149–163. doi:10.1007/s12524-013-0299-7

Al-Sharif, A. A., Pradhan, B., Shafri, H. Z. M., & Mansor, S. (2013). Spatio-temporal analysis of urban and population growths in Tripoli using remotely sensed data and GIS. *Indian Journal of Science and Technology, 6*(8), 5134–5142.

Bagan, H., & Yamagata, Y. (2015). Analysis of urban growth and estimating population density using satellite images of nighttime lights and land-use and population data. *GIScience & Remote Sensing, 52*(6), 765–780.

Bhatta, B. (2009). Analysis of urban growth pattern using remote sensing and GIS: A case study of Kolkata India. *International Journal of Remote Sensing, 30*(18), 4733–4746.

Bhatta, B. (2010). *Analysis of urban growth and sprawl from remote sensing data*. Springer.

Bhatti, S. S., & Tripathi, N. K. (2014). Built-up area extraction using Landsat 8 OLI imagery. *GIScience & Remote Sensing, 51*(4), 445–467.

Brueckner, J. K. (2000). Urban sprawl: Diagnosis and remedies. *International Regional Science Review, 23*(2), 160–171. doi: 10.1177/016001700761012710

Brueckner, J. K. (2001). Urban sprawl: Lessons from urban economics. *Brookings-Wharton papers on urban affairs, 2001*(1), 65–97. doi:10.1353/urb.2001.0003

Burchell, R., Downs, A., McCann, B., & Mukherji, S. (2005). *Sprawl costs: Economic impacts of unchecked development*. Island Press.

Chen, S., Zeng, S., & Xie, C. (2000). Remote sensing and GIS for urban growth analysis in China. *Photogrammetric Engineering and Remote Sensing, 66*(5), 593–598.

Dietzel, C., Herold, M., Hemphill, J. J., & Clarke, K. C. (2005). Spatio-temporal dynamics in California's Central Valley: Empirical links to urban theory. *International Journal of Geographical Information Science, 19*(2), 175–195. doi:10.1080/13658810410001713407

Epsteln, J., Payne, K., & Kramer, E. (2002). Techniques for mapping suburban sprawl. *Photogrammetric Engineering & Remote Sensing, 63*(9), 913–918.

Fulton, W. B., Pendall, R., Nguyen, M., & Harrison, A. (2001). *Who sprawls most?: How growth patterns differ across the US*. Washington DC: Brookings Institution Center on Urban and Metropolitan Policy.

Herold, M., Goldstein, N. C., & Clarke, K. C. (2003). The spatiotemporal form of urban growth: Measurement, analysis and modeling. *Remote Sensing of Environment, 86*(3), 286–302.

Jaeger, J. A., Bertiller, R., Schwick, C., & Kienast, F. (2010). Suitability criteria for measures of urban sprawl. *Ecological Indicators, 10*(2), 397–406. doi:10.1016/j.ecolind.2009.07.007

Kong, F., Yin, H., Nakagoshi, N., & James, P. (2012). Simulating urban growth processes incorporating a potential model with spatial metrics. *Ecological Indicators, 20,* 82–91. doi:10.1016/j.ecolind.2012.02.003

La Rosa, D., & Wiesmann, D. (2013). Land cover and impervious surface extraction using parametric and non-parametric algorithms from the open-source software R: An application to sustainable urban planning in Sicily. *GIScience & Remote Sensing, 50*(2), 231–250.

Li, X., Jongman, R. H., Hu, Y., Bu, R., Harms, B., Bregt, A. K., et al. (2005). Relationship between landscape structure metrics and wetland nutrient retention function: A case study of Liaohe Delta. *China. Ecological Indicators, 5*(4), 339–349. doi:10.1016/j.ecolind.2005.03.007

Li, Y., Zhu, X., Sun, X., & Wang, F. (2010). Landscape effects of environmental impact on bay-area wetlands under rapid urban expansion and development policy: A case study of Lianyungang China. *Landscape and Urban Planning, 94*(3), 218–227. doi:10.1016/j.landurbplan.2009.10.006

McGarigal, K., Cushman, S. A., Neel, M. C., & Ene, E., (2002). *FRAGSTATS: Spatial pattern analysis program for categorical maps*. Computer software program produced by the authors at the University of Massachusetts, Amherst.

Peng, J., Wang, Y., Zhang, Y., Wu, J., Li, W., & Li, Y. (2010). Evaluating the effectiveness of landscape metrics in quantifying spatial patterns. *Ecological Indicators, 10*(2), 217–223. doi:10.1016/j.ecolind.2009.04.017

Pumain, D. (2004). *Urban sprawl: Is there a french case?* (pp. 137–157). Aldershot: Urban Sprawl.

Ramachandra, T. V., Aithal, B. H., & Sanna, D. D. (2012a). Insights to urban dynamics through landscape spatial pattern analysis. *International Journal of Applied Earth Observation and Geoinformation, 18,* 329–343. doi:10.1016/j.jag.2012.03.005

Ramachandra, T. V., Aithal, B. H., & Sreekantha, S. (2012b). Spatial metrics based landscape structure and dynamics assessment for an emerging Indian Megalopolis. *International Journal of Advanced Research in Artificial Intelligence, 1*(1), 48–57. doi:10.14569/IJARAI.2012.010109#sthash.IGWxFf8x.dpuf

Schneider, A., Seto, K. C., & Webster, D. R. (2005). Urban growth in Chengdu, Western China: Application of remote sensing to assess planning and policy outcomes. *Environment and Planning B: Planning and Design, 32*(3), 323–345. doi:10.1068/b31142

Siedentop, S. (2005). Urban Sprawl—verstehen, messen, steuern: Ansatzpunkte für ein empirisches Mess-und Evaluationskonzept der urbanen Siedlungsentwicklung. *disP-The Planning Review, 41*(160), 23–35. doi:10.1080/02513625.2005.10556903

Su, W., Gu, C., Yang, G., Chen, S., & Zhen, F. (2010). Measuring the impact of urban sprawl on natural landscape pattern of the Western Taihu Lake watershed. *China. Landscape and Urban Planning, 95* (1–2), 61–67. doi:10.1016/j.landurbplan.2009.12.003

Sudhira, H., Ramachandra, T. V., & Jagadish, K. (2004). Urban sprawl: Metrics, dynamics and modelling using GIS. *International Journal of Applied Earth Observation and Geoinformation, 5*(1), 29–39. doi:10.1016/j.jag.2003.08.002

Sun, C., Wu, Z.-F., Lv, Z.-Q., Yao, N., & Wei, J.-B. (2013). Quantifying different types of urban growth and the change dynamic in Guangzhou using multi-temporal remote sensing data. *International Journal of Applied Earth Observation and Geoinformation, 21,* 409–417. doi:10.1016/j.jag.2011.12.012

Tang, J., Wang, L., & Yao, Z. (2006). Analyzing urban sprawl spatial fragmentation using multi-temporal satellite images. *GIScience & Remote Sensing, 43*(3), 218–232.

Taubenböck, H., Wegmann, M., Roth, A., Mehl, H., & Dech, S. (2009). Urbanization in India-Spatiotemporal analysis using remote sensing data. *Computers, Environment and Urban Systems, 33*(3), 179–188.

Taubenböck, H., Wegmann, M., Roth, A., Mehl, H., & Dech, S. (2009a). Analysis of urban sprawl at mega city Cairo, Egypt using multisensoral remote sensing data, landscape metrics and gradient analysis. In *Proceedings of the ISRSE, Conference at Paper presented* Stresa, Italy.

Verbeek, T., Boussauw, K., & Pisman, A. (2014). Presence and trends of linear sprawl. Explaining ribbon development in the north of Belgium. *Landscape and Urban Planning, 128,* 48–59. doi:10.1016/j.landurbplan.2014.04.022

Xian, G., Crane, M., & Su, J. (2007). An analysis of urban development and its environmental impact on the Tampa Bay watershed. *Journal of Environmental Management, 85*(4), 965–976. doi:10.1016/j.jenvman.2006.11.012

Yeh, C.-T., & Huang, S.-L. (2009). Investigating spatiotemporal patterns of landscape diversity in response to urbanization. *Landscape and Urban Planning, 93*(3), 151–162.

Yue, W., Liu, Y., & Fan, P. (2012). Measuring urban sprawl and its drivers in large Chinese cities: The case of Hangzhou. *Land Use Policy*. doi:10.1016/j.landusepol.2012.07.018

Zhang, C., Chen, Y., & Lu, D. (2015). Detecting fractional land-cover change in arid and semiarid urban landscapes with multitemporal Landsat Thematic mapper imagery. *GIScience & Remote Sensing, 52*(6), 700–722.

Zhang, Q., He, C., & Liu, Z. (2014). Studying urban development and change in the contiguous United States using two scaled measures derived from nighttime lights data and population census. *GIScience & Remote Sensing, 51*(1), 63–82.

15 Effect of Urban Expansion on Land Surface Temperature in Putrajaya City, Malaysia

Marziyeh Zahabi and Biswajeet Pradhan

15.1 Introduction

In recent years, changes in land use (LU), climate, economic, and population have been observed owing to high urbanization range (IPCC 2007) With the rapid growth of urbanization and urban warming, climatologists have focused on the phenomenon called urban heat island (UHI) (Tran et al. 2006; Peng et al. 2011; Fujibe 2011). The UHI effect is the difference in solar radiation reflectivity (albedo), and thermal conductivity and thermal storage capacity between surfaces that are classified as rural and urban environments (Mitchell 2011).

Among different causative factors related to atmospheric UHI, the difference in land surface temperature (LST) depending on land cover (LC) type is the most fundamental (Landsberg 1981; Voogt and Oke 2003).

Urban climatology studies mostly require accurate spatial information to monitor the occurring changes (Voogt and Oke 2003; Weng 2009). Remote sensing technologies can obtain practical and modern data over large areas. Furthermore, such technologies can provide remarkable information when integrated with various methodologies, such as image classification, statistical analysis, and change detection. Applying remote sensing systems in urban climatology studies is advantageous because of their multispectral and multi-temporal capabilities and ease of integrating their data with geographic information systems (GISs) (Quattrochi and Luvall 1999; Weng et al. 2004; Weng 2009).

Concerning temporal and spatial resolution, a high temporal resolution is an advantage in the climatologic studies of UHIs (Tran et al. 2006). Climatologic studies investigate issues with a long data record; however, these studies have disadvantage of poor spatial resolution. Remote sensing views have been utilized for detecting the environmental change and analyzing UHI characteristics. UHI studies extensively use remotely sensed thermal infrared (TIR) images (Gallo et al. 1993; Roth et al. 1989; Streutker 2003) because these images have high spatial coverage and can provide information of the urban canopy layer.

LST is the critical component of the atmospheric UHI and can be observed using satellite or aerial remote sensing technology. Voogt and Oke (2003) proposed that the urban LST heating pattern observed by remote sensing is the "surface UHI" or SUHI. This phenomenon is an indicator of the energy stored by natural and built surfaces that are radiated to the surrounding air, thereby affecting the temperature of the lower atmosphere.

In this study, remote sensing techniques were utilized to retrieve LST using Landsat Thematic Mapper (TM) and Enhanced TM Plus (ETM+) satellite imagery. LST can be an appropriate indicator to demonstrate the thermal changes within LC changes. Zonal statistics techniques, as a method of GIS, were also utilized to achieve mean LST for each LC type in 2002 and 2009. Beneficial information for studying the urban LC change effects in Putrajaya City was obtained from the combined results of LC analysis and mean LST. The results show that the urban growth in Putrajaya City results in the increase in mean LST during the study period. This observation verifies the positive correlation between normalized difference built-up index (NDBI) and LST and the negative relationship between normalized difference vegetation index (NDVI) and LST.

15.2 Previous Studies on RS-Based UHI

Remote sensing data related to LST, vegetation indicator, and other surface factors have been used to characterize UHI phenomenon (Gallo and Owen 1999; Gallo et al. 1993; Weng et al. 2004). Hence, relations between LST and other factors and indices have been investigated in numerous studies. Weng (2001) investigated the direct effect of urban development on the increase in LST using GIS and remote sensing methods. Tehrany et al. (2013) explored LST

B. Pradhan (ed.), *Spatial Modeling and Assessment of Urban Form*,
DOI 10.1007/978-3-319-54217-1_15

variation and its relation with LC changes in Klang Valley in Malaysia. In addition, Zhou et al. (2014) evaluated the effects of LU/LC variables on LST across seasons using Landsat images acquired from various seasons.

NDVI is the main index of urban climate and has been investigated in the recent thermal studies. Xian and Crane (2006) and Huang et al. (2005) only performed correlation between changes in LST and LU/LC types caused by human activities. They also documented the effect of vegetation cover type and NDVI. Bokaie et al. (2016) estimated the mean temperature of LST over different LU/LC classes in Tehran City and used NDVI to study the distribution of vegetation spaces in the region. Gallo et al. (1993) analyzed the influence of NDVI on surface temperature over rural and urban areas, and their results show a direct relation between LST and NDVI.

Buyantuyev and Wu (2010) concluded that pavements and vegetation covers are the main factors for surface temperature variation. Thus, NDBI has been utilized to represent LC changes attributed to rapid urbanization (Xiong et al. 2012; Rinner and Hussain 2011). Weng et al. (2004) found a positive correlation between LST and impervious surface whereas a negative correlation between LST and vegetation LC class. Ma et al. (2016) demonstrated the influence of impervious surface on LST during various seasons in China to emphasize the importance of urban climatology. According to UHI research findings, LST is susceptible to other factors such as density and soil moisture (Weng 2009; Mallick et al. 2008). Subsequently, Sun et al. (2012) examined the influence of other indices, such as NDWI and normalized difference barren index, on LST. Ogashawara and Bastos (2012) observed a negative correlation between LST and NDWI. Literature review shows that UHI studies can be established by examining the relationship between different indices and temperature.

15.3 Study Area

The study area is situated in Putrajaya, Malaysia (2° 56′ 35.14″N and 101° 41′ 57.74″E); Putrajaya is a planned city and is the federal administrative center of Malaysia located at 25 km south of Kuala Lumpur (Fig. 15.1). The total area of this city is approximately 49 km^2 and its population is 67,964 in 2010. Malaysia has faced rapid urbanization growth during these periods. With this regard, Putrajaya is observed with 100% level in urbanization compared with other states for the period of 2000–2010.

Recently, the highest recorded temperature is 39 °C for this city, while the lowest recorded temperature is 21 °C.

15.4 Data Used

Two scenes of Landsat ETM+ and TM images acquired on February 11, 2002 and January 22, 2009 were effectively used to recognize the spatial distribution characteristics of surface temperature. NDVI was assigned to vegetation presence while NDBI was assigned to the built-up area in Putrajaya, Malaysia. Thermal bands of TM and ETM+ were analyzed in terms of surface temperatures. The relationship between LC types and temperature patterns in the city was

Fig. 15.1 Map of the study area

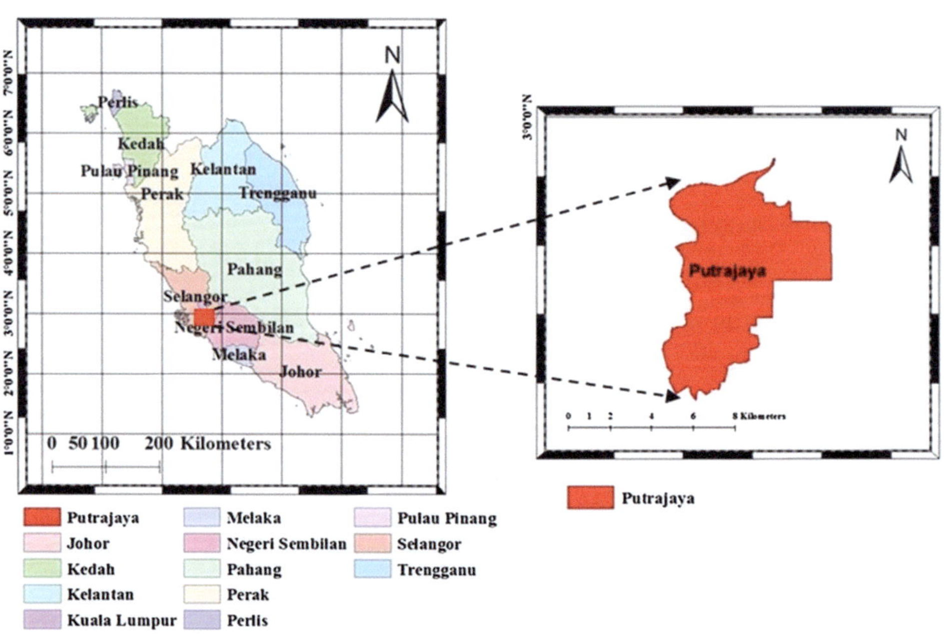

then determined by comparing the LC classification image and the surface temperature images.

15.5 Methodology

Various remote sensing data processing methods, such as image classification, LST retrieval, and NDVI and NDBI retrieval, were used to perform this research. The methodology comprised three main parts: data collection and preprocessing, data processing, and data analysis (Fig. 15.2).

15.5.1 Data Preprocessing

Data preprocessing is a notable step of satellite imagery processing and analysis and has an effect on all further steps and the final result quality (Bobrinskaya 2012). The purpose of image preprocessing is to restore proper image data from distorted raw data. The principal steps of this part include radiometric correction, geometric correction, and image registration. Usually, the responsible company for satellite data distribution applies several of these stages before handing the data to the user (Lillesand et al. 2007).

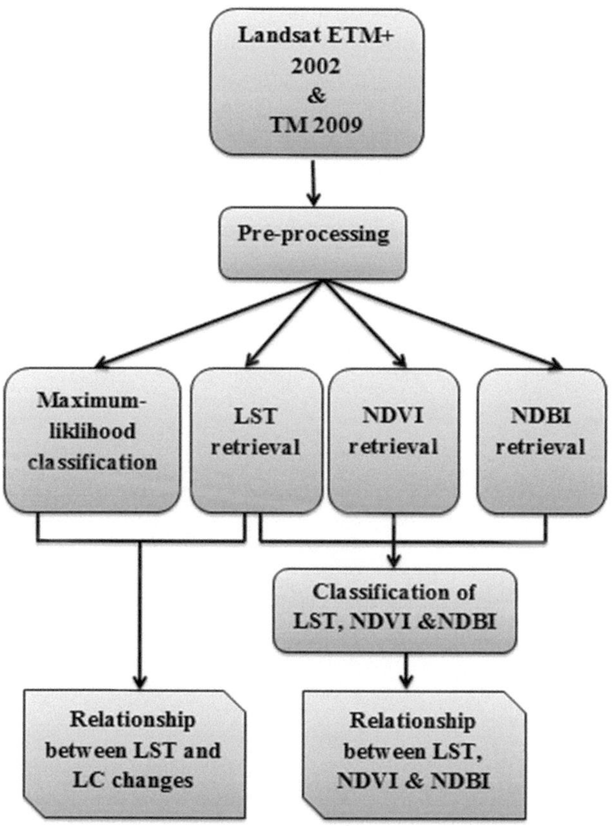

Fig. 15.2 Methodology used in the study

In this study, the Landsat ETM+ and TM images were collected under a clear atmospheric situation. The images were preprocessed by the USGS to rectify any geometric or radiometric distortions of the image to a level of 1G or 1T product. The USGS further rectified the images to the WGS1984 datum and Universal Transverse Mercator zone 29°N coordinate system (Landsat Project Science Office 2001; USGS 2010b). Original scenes of regions covering the case city and subsets from these original scenes were acquired.

15.5.2 Image Classification

The main purpose of image classification process is to automatically classify all pixels in an image into LU/LC classes or patterns. Digital classification is categorized into supervised and unsupervised classification. The main classifiers include minimum distance to mean classifier, parallelepiped classifier, and maximum likelihood classifier (MLC). Compared with other supervised methods, MLC is the most generally used supervised method and can provide better results (Foody et al. 1992).

In this study, images of two years (2002 and 2009) were classified using ENVI 4.8 program into five LC classes: (1) vegetation, (2) urban area, (3) water, (4) asphalt, and (5) bare area. These classes were chosen because the city has shown a dramatic increase in urban expansion in recent years.

15.5.3 List of Retrieval of Remote Sensing Data

LST is the principal factor that specifies surface radiation and energy exchange (Xiao et al. 2008). In this study, LST was derived from ETM+ band 61 (10.4–12.5 μm) and TM thermal band 6 with spatial resolutions of 60 and 120 m, respectively.

The DNs were converted to spectral radiance first and then the surface temperature was computed under the assumption of uniform emissivity (Ahmed et al. 2013). The calculation was implemented using the following equation:

$$T = \frac{k_2}{\ln\left(\frac{k_1 * \varepsilon}{\text{CVR}} + 1\right)}, \tag{15.1}$$

where T = the degree in Kelvin (−273.15 °C); CVR = the cell value as radiance; E = emissivity (typically 0.95); K_1 = 666.09 and 607.76 in mW cm^{-2} sr^{-1} μm^{-1} for Landsat ETM+ and TM, respectively; and K_2 = 1282.71 and 1260.56 in Kelvin for Landsat ETM+ and TM, respectively.

The resulting images were then classified into different classes of temperature using density slicing. The classes of temperature were 291–295, 296–300, 301–305, and >305 K.

15.5.4 Retrieval of NDVI

NDVI is one of the most widely used vegetation indices for demonstrating vegetation information (Lu et al. 2009). Vegetation information can be obtained by ratio calculation between the near-infrared (NIR) band and the red (R) band; this ratio is the main idea of NDVI (Lu et al. 2009). Equation 15.2 was used to calculate the NDVI.

$$\mathrm{NDVI} = \frac{\mathrm{NIR} - R}{\mathrm{NIR} + R} \tag{15.2}$$

Classification was performed on the NDVI images obtained in 2002 and 2009 using density slice on ENVI 4.8. Each NDVI image was divided into five classes by 0.15 intervals (>−0.3, −0.3 to −0.15, −0.15 to 0, 0 to 0.15, and >0.15).

15.5.5 Retrieval of NDBI

NDBI is one of the widely applied indices for reinforcing building information and extracting built-up land from urban areas (Lu et al. 2009). Moreover, NDBI is a reflectance of urban building and is higher in the fifth band than the fourth band; therefore, NDBI can be computed by the equation below

$$\mathrm{NDBI} = \frac{\mathrm{MIR} - \mathrm{NIR}}{\mathrm{MIR} + \mathrm{NIR}}. \tag{15.3}$$

The index was developed on the basis of the unique spectral response of built-up lands, that is, a higher reflectance in middle infrared wavelength range than in NIR wavelength range. However, this response is not always true (Xu 2007).

The same NDVI procedure was applied on extracted NDBI values. Six classes were assigned to each NDBI by 0.1 intervals using density slice on ENVI 4.8 (>−0.2, −0.2 to −0.1, −0.1 to 0, 0 to 0.1, 0.1 to 0.2, and >0.2).

15.6 Results and Discussion

15.6.1 Land Cover Classification Results

Five LC classification results were acquired from Landsat images (Fig. 15.3). Therefore, according to the classification of the 2002 image (with overall accuracy classification of 92.16% and Kappa value of 0.88) and the 2009 image (with overall accuracy classification of 90.41% and Kappa value of 0.87), urban area increases by 15.86 km^2 and shows a slight increase in water area and asphalt cover. Inversely, vegetation cover declines by 5 km^2 and bare area dramatically decreases by 12.64 km^2 (Fig. 15.4).

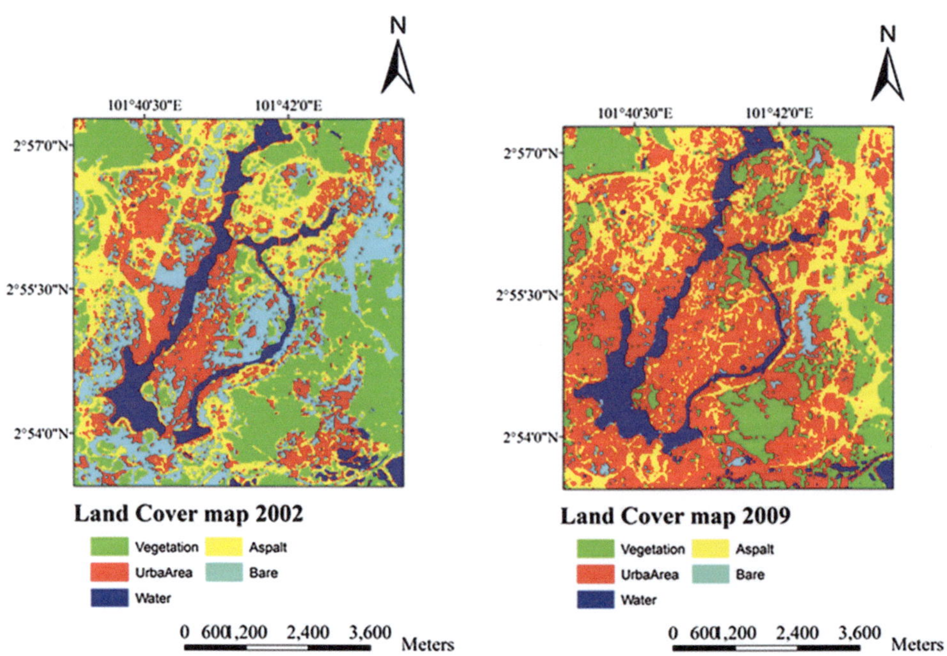

Fig. 15.3 Maximum likelihood classification results

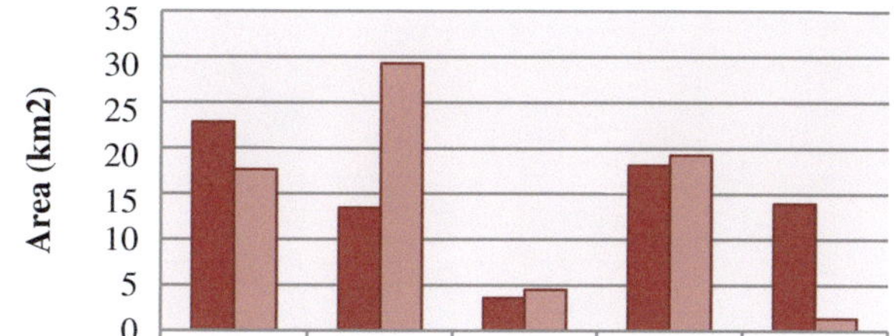

	Vegetati on	Urban Area	Water	Asphalt	Bare Area
2002/ km2	22.89	13.44	3.59	18.15	13.98
2009/ km2	17.64	29.3	4.5	19.27	1.34

Fig. 15.4 Changes of each land cover area from 2002 to 2009

The statistical results indicate that the urbanization trend in Putrajaya region has increased throughout the period of 9 years.

15.6.2 Relationship Between LST and Land Cover Types

The specific LST themes are related to the thermal characteristics of the LC types. To study the effect of urbanization on the local thermal environment, the changes in temperature over LU/LC must be investigated. As shown in Fig. 15.5, most areas of the 2002 image are observed in yellow and green colors, which indicate significantly low mean temperature. Inversely, most areas of the 2009 image are observed in red color, which indicates significantly high mean temperature.

The mean of temperature over each class was derived using zonal statistics from GIS operation to better analyze the LST changes in each LC type during the study period. The results of GIS zonal statistics are shown in Fig. 15.6.

Over the study period of 2002–2009, urban area shows the highest mean surface temperature, followed by asphalt which is still considered a part of the urban area. Bare area and vegetation exhibit the lowest degree of mean temperature in 2002 (295.17 and 295.63 °K); however, a significant increase in mean temperature is observed for both LCs in 2009 (300.66 °K over bare area and 299.15 °K over vegetation). The results show the occurrence of an ascending trend of mean temperature for all LC types, except water during the study period.

Urbanization growth in Putrajaya has engendered the enhancement of mean temperature in this city. The increased warming is attributed to the decrease in vegetation coverage

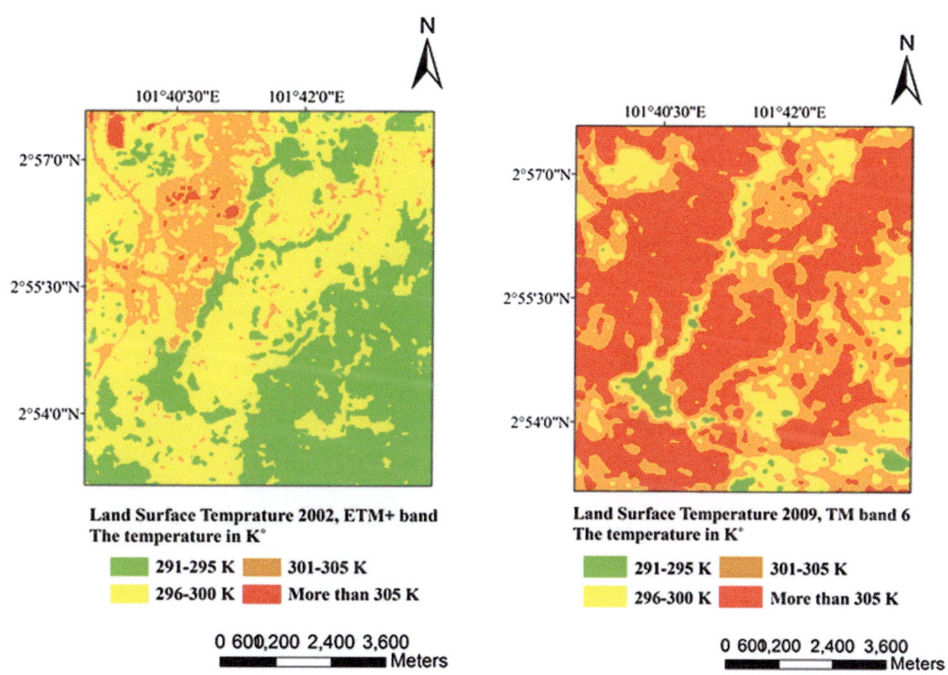

Fig. 15.5 Landsat thermal classification related to the period from 2002 to 2009

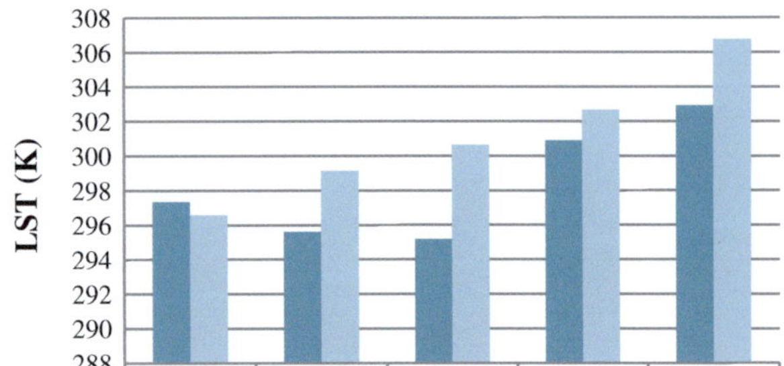

	Water	Vegetation	Bare Area	Asphalt	Urban Area
Mean/2002	297.35	295.63	295.17	300.9	302.93
Mean/2009	296.59	299.15	300.66	302.69	306.76

Fig. 15.6 Change of mean temperature over each LC type in the period of 2002–2009

and the replacement of bare area with urban and asphalt areas.

However, diversities in LST values show the effects of LC changes on the thermal environment. With regard to UHI expansion in the period from 2002 to 2009, substantial cover changes in urban area and major satellite towns intensify the regional UHI effect.

15.6.3 Relationship Between LST and NDVI

The urban thermal environment is closely associated with the decline in surface transpiration caused by a decrease in vegetation coverage. Based on previous studies, NDVI can be an indicator of the representative vegetation cover and land surface radiance temperature (Xiong et al. 2012)

Figure 15.7 provides the visual presentation of different NDVI ranges over different LC types in the period from 2002 to 2009. In this study, NDVI and LST show a close correlation in several LC categories, particularly in vegetated areas for both years. NDVI values were queried out by 0.15 intervals for each year, and then mean of temperature over each range was calculated by zonal statistics. Generally, urban areas show smaller NDVI values than those of nonurban areas. A consistent decrease in the NDVI value as the mean LST increases is also observed. Essentially, an increase in urban growth results in a decreased NDVI.

The negative NDVI ranges (<−0.3, −0.3 to −0.15, and −0.15 to 0) indicate a high mean temperature for the year 2002. These ranges correspond to the urban area, asphalt, and water. The highest values (0–0.15 and >0.15) are expressed by the vegetated area, and the mean temperature of this area is lower than that of the non-vegetated area (Fig. 15.8).

The same results are derived for mean temperature over each NDVI intervals in 2009. In particular, the mean temperature significantly declines over the vegetated area by positive and highest ranges compared with that of the non-vegetated area by the negative interval. The results in Fig. 15.9 represent the lowest mean temperature of 297.73 °K at >0.15, followed by 298.02 °K at 0–0.15.

According to the obtained results and illustrated charts of mean temperature changes over different NDVI ranges for each year (Figs. 15.8 and 15.9), the inverse relationship between LST and NDVI is obviously inferable in Putrajaya City in 2002 and 2009. Thus, the obtained results are in good agreement with those reported in the literature. Specifically, the presence of vegetation cover effects reduces the surface temperature.

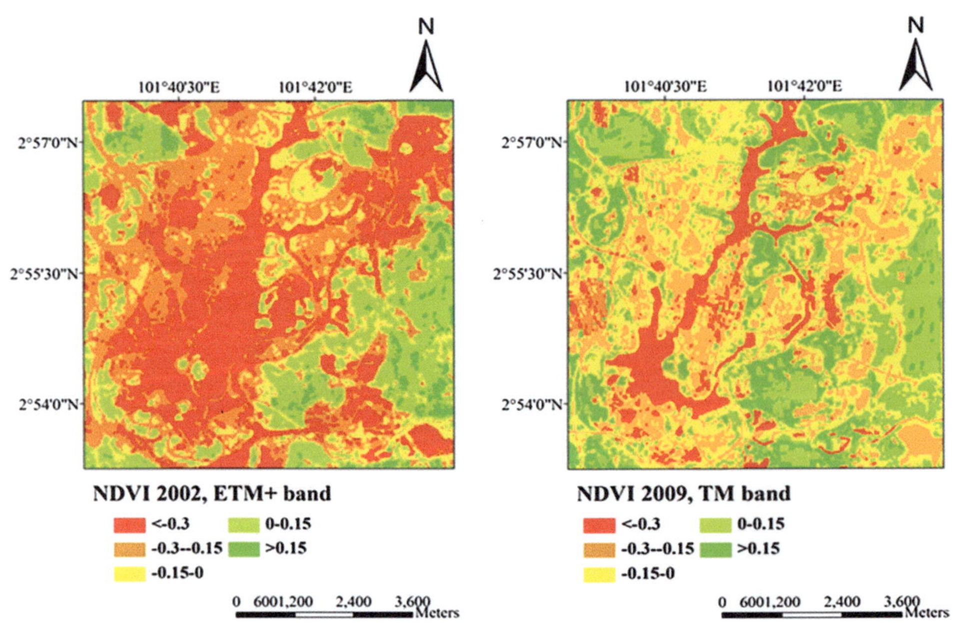

Fig. 15.7 Visual representation of classified NDVI in the period from 2002 to 2009

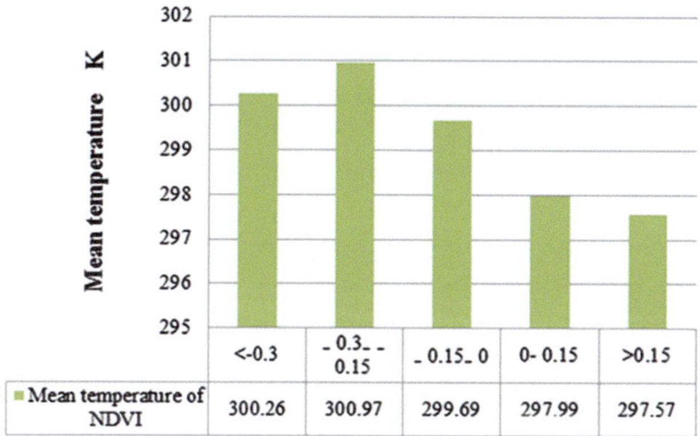

Fig. 15.8 Changes of mean temperature over NDVI intervals in 2002

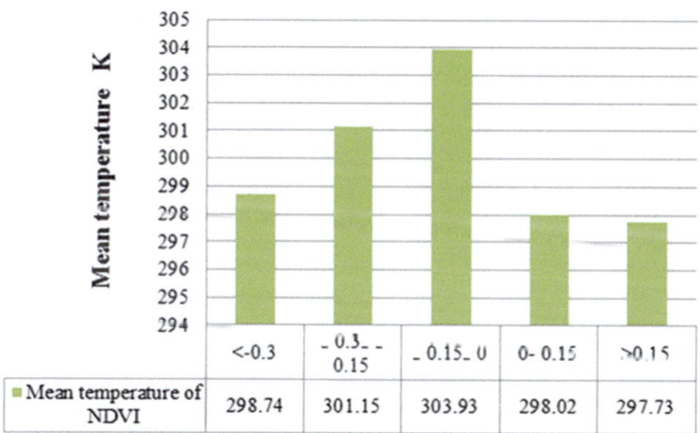

Fig. 15.9 Changes of mean temperature over NDVI intervals in 2009

15.6.4 Relationship Between LST and NDBI

NDBI is a good determinant of urban development for a specific region of interest, thereby providing new approaches for future research on UHI effects (Xiong et al. 2012; Chen et al. 2006; Liu and Zhang 2011). NDBI images of Putrajaya City related to 2002 and 2009 were analyzed in this study to achieve correlation between LST and NDBI during these years.

NDBI images were classified into six ranges using density slice method to obtain the relationship between LST and NDBI. The equal interval between ranges (divided by 0.1 intervals) was considered. Thereafter, mean temperature degrees over each interval were determined using zonal statistics in ArcGIS.

The analyses of NDBI in both years demonstrate that the negative ranges are assigned to the nonurban areas, such as vegetation and water, and the positive ranges display built-up areas, including asphalts and buildings.

As shown in Fig. 15.10, distinguishing between bare areas and built-up areas (in red and orange colors) in 2002 was difficult with NDBI because of their similar reflectance in the TM/Landsat five bands. Thus, the bare area was represented as built-up area. The negative ranges allocated to nonurban area present lower mean temperatures that those of the built-up area. The highest mean temperature degrees are dedicated to ranges of 0.1–0.2 by 300.71 °K and >0.2 by 301.09 °K.

Fig. 15.10 Visual representation of NDBI classified image in 2002

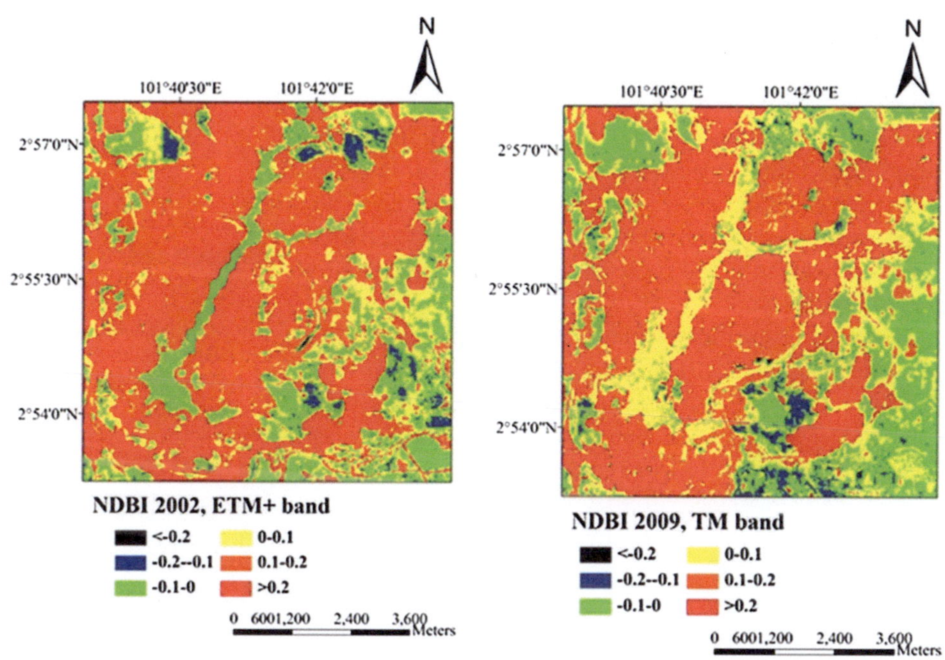

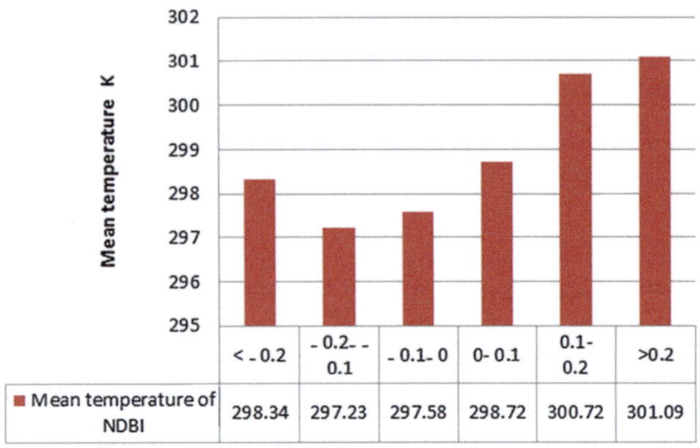

Fig. 15.11 Changes of mean temperature over NDBI intervals in 2002

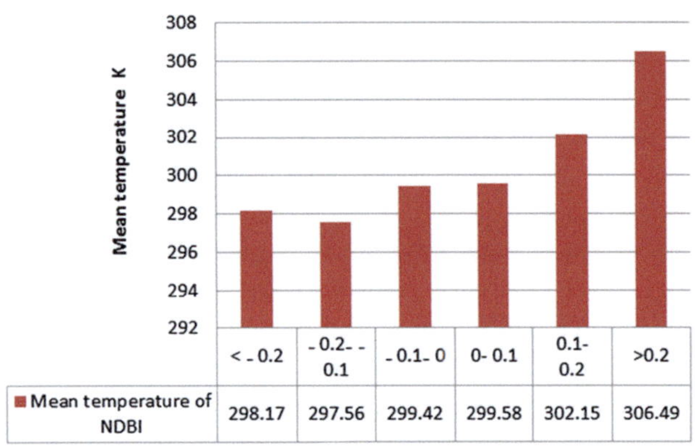

Fig. 15.12 Changes of mean temperature over NDBI intervals in 2009

These results demonstrate that the LST displays a positive relationship with NDBI during 2002 in Putrajaya. As shown in Fig. 15.11, the positive ranges exhibit higher temperature degrees than those of the negative intervals.

The same analysis in 2002 was performed for NDBI image from 2009. During this year, positive intervals demonstrate built-up areas similar to 2002. These areas are represented by red and orange colors. Most vegetation covers are located in the range of −0.1 to 0 represented by green color.

Figure 15.12 displays various mean temperatures over equal intervals of NDBI in 2009. The highest mean temperature degrees are allocated to the highest values of NDBI. This finding indicates an uptrend in NDBI ranges from negative to positive and high values. The highest mean temperature is assigned to >0.2 by 306.49 °K after the interval between 0.1 and 0.2 by 302.15 °K. The latter is considered as the highest temperature compared with other NDBI intervals during 2009.

Based on the analysis of NDBI results in 2009, an obvious direct relationship between LST and NDBI exists. Hence, the mean temperature enhances whereas the NDBI value increases because of the high values of NDBI associated to the built-up area with high degrees of temperature.

Although LST and NDBI show a direct relationship in Putrajaya during both years, the uptrend temperature in the built-up area in 2009 is contrary to that in 2002. This finding proves that urban expansion directly correlates with temperature. Thus, UHI has an ascending trend in urban growth in Putrajaya during these years.

15.7 Conclusion

This research investigated the relationship of LC classes and LSTs and the effects of urban expansion on LST changes. Putrajaya City was the chosen study area. According to the results, the temperature of the urban area approximately increases by 3.8 °K. Within the Putrajaya City, LST and NDVI share a noticeable inverse relationship. This result is attributed to that increasing vegetation abundance can generally reduce surface temperatures that cause severe UHI. In addition, NDBI reveals a direct relationship with LST, indicating that increasing urbanization (an increase of buildings and impervious surface, such as asphalt) in this city directly correlates with rising temperature. However, a high accuracy for NDBI must be considered in future studies because of similar reflectance; therefore, NDBI is mixed with the bare area. In recent years, increasing LST in Putrajaya has become a significant factor related to UHI, thereby accelerating the growth of the UHI phenomenon. This trend will rise along with urban expansion in this urbanized city.

The research findings can be beneficial for urban decision makers and urban ecological planning in Putrajaya with regard to clarifying which types of performance may be the most advantageous.

References

Ahmed, B., Kamruzzaman, M., Zhu, X., Rahman, M. S., & Choi, K. (2013). Simulating land cover changes and their impacts on land surface temperature in Dhaka Bangladesh. *Remote Sensing, 5*(11), 5969–5998.

Bobrinskaya, M. (2012). Remote sensing for analysis of relationships between land cover and land surface temperature in ten mega cities.

Bokaie, M., Zarkesh, M. K., Arasteh, P. D., & Hosseini, A. (2016). Assessment of Urban Heat Island based on the relationship between land surface temperature and land use/land cover in Tehran. *Sustainable Cities and Society, 23,* 94–104.

Buyantuyev, A., & Wu, J. (2010). Urban heat islands and landscape heterogeneity: Linking spatiotemporal variations in surface temperatures to land-cover and socioeconomic patterns. *Landscape Ecology, 25*(1), 17–33.

Chen, X. L., Zhao, H. M., Li, P. X., & Yin, Z. Y. (2006). Remote sensing image-based analysis of the relationship between urban heat island and land use/cover changes. *Remote Sensing of Environment, 104*(2), 133–146.

Foody, G. M., Campbell, N. A., Trodd, N. M., & Wood, T. F. (1992). Derivation and applications of probabilistic measures of class membership from the maximum-likelihood classification. *Photogrammetric Engineering and Remote Sensing, 58*(9), 1335–1341.

Fujibe, F. (2011). Urban warming in Japanese cities and its relation to climate change monitoring. *International Journal of Climatology, 31*(2), 162–173.

Gallo, K. P., McNab, A. L., Karl, T. R., Brown, J. F., Hood, J. J., & Tarpley, J. D. (1993). The use of NOAA AVHRR data for assessment of the urban heat island effect. *Journal of Applied Meteorology, 32*(5), 899–908.

Gallo, K. P., & Owen, T. W. (1999). Satellite-based adjustments for the urban heat island temperature bias. *Journal of Applied Meteorology, 38*(6), 806–813.

Huang, H., Ooka, R., & Kato, S. (2005). Urban thermal environment measurements and numerical simulation for an actual complex urban area covering a large district heating and cooling system in summer. *Atmospheric Environment, 39*(34), 6362–6375.

IPCC. (2007). *Climate change 2007: The physical science basis*. Contribution of Working Group I to the Fourth Assessment Report of the Intergovernmental Panel on Climate Change. Cambridge University Press, Cambridge, United Kingdom and New York, NY, USA.

Landsat Project Science Office. (2001). *Landsat 7 science data user's handbook* (Goddard Space Flight Center). Available online at: http://Landsathandbook.gsfc.nasa.gov/handbook/handbook_htmls/chapter11/chapter11.html. Accessed September 15, 2010.

Landsberg, H. E. (1981). *Urban climate*. New York: Academic.

Lillesand, T., Kieffer, R. W., & Chipman, J. (2007). *Remote sensing and image interpretation* (6th ed.). New York, US: Joahn Wiley and Sons.

Liu, L., & Zhang, Y. (2011). Urban heat island analysis using the landsat TM data and ASTER data: A case study in Hong Kong. *Remote Sensing, 3*(7), 1535–1552.

Lu, Y., Feng, X., Xiao, P., Shen, C., & Sun, J. (2009, May). Urban heat island in summer of Nanjing based on TM data. In *Urban remote sensing event, 2009 Joint* (pp. 1–5). IEEE.

Ma, Q., Wu, J., & He, C. (2016). A hierarchical analysis of the relationship between urban impervious surfaces and land surface temperatures: spatial scale dependence, temporal variations, and bioclimatic modulation. *Landscape Ecology, 31*(5), 1139–1153.

Mallick, J., Kant, Y., & Bharath, B. D. (2008, July). Estimation of land surface temperature over Delhi using Landsat-7 ETM. *J. Ind. Geophys. Union 12*(3), 131–140.

Mitchell, B. C. (2011). Urbanization and land surface temperature in Pinellas County, Florida.

Ogashawara, I., & Bastos, V. D. S. B. (2012). A quantitative approach for analyzing the relationship between urban heat islands and land cover. *Remote Sensing, 4*(11), 3596–3618.

Peng, S., Piao, S., Ciais, P., Friedlingstein, P., Ottle, C., Bréon, F. M., et al. (2011). Surface urban heat island across 419 global big cities. *Environmental Science and Technology, 46*(2), 696–703.

Quattrochi, D. A., & Luvall, J. C. (1999). Thermal infrared remote sensing for analysis of landscape ecological processes: Methods and applications. *Landscape Ecology, 14*(6), 577–598.

Roth, M., Oke, T. R., & Emery, W. J. (1989). Satellite-derived urban heat islands from three coastal cities and the utilization of such data in urban climatology. *International Journal of Remote Sensing, 10* (11), 1699–1720.

Rinner, C., & Hussain, M. (2011). Toronto's urban heat island—Exploring the relationship between land use and surface temperature. *Remote Sensing, 3*(6), 1251–1265.

Streutker, D. R. (2003). Satellite-measured growth of the urban heat island of Houston, Texas. *Remote Sensing of Environment, 85*(3), 282–289.

Sun, Q., Wu, Z., & Tan, J. (2012). The relationship between land surface temperature and land use/land cover in Guangzhou, China. *Environmental Earth Sciences, 65*(6), 1687–1694.

Tehrany, M. S., Pradhan, B., & Jebur, M. N. (2013). Remote sensing data reveals eco-environmental changes in urban areas of Klang Valley, Malaysia: Contribution from object based analysis. *Journal of the Indian Society of Remote Sensing, 41*(4), 981–991.

Tran, H., Uchihama, D., Ochi, S., & Yasuoka, Y. (2006). Assessment with satellite data of the urban heat island effects in Asian mega cities. *International Journal of Applied Earth Observation and Geoinformation, 8*(1), 34–48.

USGS. (2010b). New earth explorer. Available online at: http://earthexplorer.usgs.gov/

Voogt, J. A., & Oke, T. R. (2003). Thermal remote sensing of urban climates. *Remote Sensing of Environment, 86*(3), 370–384.

Weng, Q. (2001). A remote sensing? GIS evaluation of urban expansion and its impact on surface temperature in the Zhujiang Delta, China. *International Journal of Remote Sensing, 22*(10), 1999–2014.

Weng, Q. (2009). Thermal infrared remote sensing for urban climate and environmental studies: Methods, applications, and trends. *ISPRS Journal of Photogrammetry and Remote Sensing, 64*(4), 335–344.

Weng, Q., Lu, D., & Schubring, J. (2004). Estimation of land surface temperature–vegetation abundance relationship for urban heat island studies. *Remote Sensing of Environment, 89*(4), 467–483.

Xian, G., & Crane, M. (2006). An analysis of urban thermal characteristics and associated land cover in Tampa Bay and Las Vegas using Landsat satellite data. *Remote Sensing of Environment, 104*(2), 147–156.

Xiao, R., Weng, Q., Ouyang, Z., Li, W., Schienke, E. W., & Zhang, Z. (2008). Land surface temperature variation and major factors in Beijing, China. *Photogrammetric Engineering and Remote Sensing, 74*(4), 451.

Xiong, Y., Huang, S., Chen, F., Ye, H., Wang, C., & Zhu, C. (2012). The impacts of rapid urbanization on the thermal environment: A remote sensing study of Guangzhou. *South China. Remote Sensing, 4*(7), 2033–2056.

Xu, H. (2007). Extraction of urban built-up land features from Landsat imagery using a the matic oriented index combination technique. *Photogrammetric Engineering & Remote Sensing, 73*(12), 1381–1391.

Zhou, W., Qian, Y., Li, X., Li, W., & Han, L. (2014). Relationships between land cover and the surface urban heat island: seasonal variability and effects of spatial and thematic resolution of land cover data on predicting land surface temperatures. *Landscape Ecology, 29*(1), 153–167.

MIX
Papier aus verantwortungsvollen Quellen
Paper from responsible sources
FSC® C105338

If you have any concerns about our products,
you can contact us on
ProductSafety@springernature.com

In case Publisher is established outside the EU,
the EU authorized representative is:
Springer Nature Customer Service Center GmbH
Europaplatz 3, 69115 Heidelberg, Germany

Printed by Libri Plureos GmbH
in Hamburg, Germany